The Physiology of Polyamines

Volume I

Editors

Uriel Bachrach, Ph.D.
Department of Molecular Biology
Hebrew University-Hadassah Medical School
Jerusalem, Israel

Yair M. Heimer, Ph.D.
Institute for Desert Research
Ben Gurion University of the Negev
Sede Boker, Israel

CRC Press, Inc.
Boca Raton, Florida

Library of Congress Cataloging-in-Publication Data

The Physiology of polyamines.

Includes bibliographies and index.
1. Polyamines—Physiological effect. I. Bachrach, Uriel. II. Heimer, Yair M. 1940- . [DNLM: 1. Polyamines—physiology. QU 60 P578]
QP801.P638P48 1989 574.19'24 88-11875
ISBN 0-8493-6808-1 (v. 1)
ISBN 0-8493-6809-X (v. 2)

Direct all inquiries to CRC Press, Inc., 2000 Corporate Blvd., N.W., Boca Raton, Florida, 33431.

International Standard Book Number 0-8493-6808-1 (v. 1)
International Standard Book Number 0-8493-6809-X (v. 2)

Library of Congress Number 88-11875
Printed in the United States

PREFACE

More than 300 years ago Antoni van Leeuwenhoek described crystals of biological materials derived from human semen. A further 250 years were to pass before the structure of these substances, which are now known as spermine and spermidine, has been established. During the 20th century, these amines, as well as other polyamines, have been isolated from practically every microbial, plant, or animal cell analyzed. Subsequently, their metabolic pathways and the various enzymes involved in polyamine synthesis and degradation have been characterized.

Recently, the importance of polyamines has been recognized because of their close relationship to cell proliferation and carcinogenesis. This led to an explosive growth in publications on polyamines, on their biosynthetic enzymes, and data on their physiological functions. During the past decade more than ten International Polyamine Meetings took place. The proceedings of those meetings were published in various books, including the four-volume *Advances in Polyamine Research* (Raven Press, New York).

The diversity of the aspects of polyamine research makes it practically impossible for a single author to summarize the present state of art in the field. In these volumes, the various aspects are reviewed by experts, including molecular biologists, biochemists, botanists, and clinicians.

We believe that they will provide useful information to those who are interested in the study of growth regulation and differentiation in normal and neoplastic systems and will stimulate further research in the field, mainly by attracting the attention of students and young scientists.

It is our hope that this book will also help in clarifying some mysteries of nature and contribute to the welfare of mankind.

The Editors
March 1988

THE EDITORS

Uriel Bachrach is Professor of Molecular Biology at the Hebrew University-Hadassah Medical School, Jerusalem. Dr. Bachrach received his M.Sc. in Biochemistry and Microbiology and his Ph.D. from the Hebrew University. He has been appointed as a Visiting Professor by various universities in the U.S. and Europe and worked as a Visiting Scientist at research institutions in the U.S., England, France, and Italy. From 1978 to 1980 Dr. Bachrach acted as the President of the Israel Biochemical Society and was the representative of Israel in various International Biochemical Organizations. He was appointed as Consultant in Biotechnology to the National Research Council of the State of Israel and was a member of the Board of Governors of the Hebrew University. Dr. Bachrach organized several International Meetings on Polyamines, was the editor or coeditor of various books dealing with polyamine research. He published more than 170 papers in the fields of molecular biology, cell biology, microbiology, and biochemistry.

Yair M. Heimer is Associate Professor of Biochemistry at the Institute for Desert Research, Ben-Gurion University of the Negev, Sede Boker, Israel. Dr. Heimer obtained the B.Sc. degree in physiology, botany, and chemistry, the M.Sc. degree in physiology from the Hebrew University, Jerusalem, and the Ph.D. degree in Biochemistry from Michigan State University, East Lansing. From 1971 to 1984 he was on the staff of the Nuclear Research Center, Negev, Israel and thereafter joined Ben-Gurion University. Dr. Heimer published over 40 papers in biochemistry, cell biology, and physiology.

CONTRIBUTORS

Enzo Agostinelli, D.Sc.
Researcher
Department of Biochemical Sciences
and
C.N.R. Center for Molecular Biology
University of Rome "La Sapienza"
Rome, Italy

Israel David Algranati, Ph.D.
Professor of Molecular Biology
Instituto de Investigaciones Bioquimicas
University of Buenos Aires
Buenos Aires, Argentina

Bruno Barbiroli, M.D., Ph.D.
Professor and Chairman
Institute of Biological Chemistry
University of Modena
Modena, Italy

Susan Bardócz, Ph.D.
Assistant Professor
Department of Biochemistry
University Medical School of Debrecen
Debrecen, Hungary

Claudio Marcello Caldarera, Ph.D.
Professor of Biochemistry
Department of Biochemistry
University of Bologna
Bologna, Italy

Evangelos S. Canellakis, Ph.D.
Professor
Department of Pharmacology
Yale University School of Medicine
New Haven, Connecticut

Maria Cartenì-Farina, Ph.D.
Professor
Institute for Biochemistry of Macromolecules
University of Naples, First Medical School
Naples, Italy

Carlo Clo
Associate Professor
Department of Biochemistry
University of Bologna
Bologna, Italy

Philip Coffino, Ph.D.
Professor
Departments of Medicine and Microbiology
University of California
San Francisco, California

Fulvio Della-Ragione, M.D., Ph.D.
Biochemistry Researcher
Institute for Biochemistry of Macromolecules
University of Naples, First Medical School
Naples, Italy

C.-C. Fan, Ph.D.
Assistant Professor
Department of Neurology
Northwestern University Medical School
and
V.A. Lakeside Medical Center
Chicago, Illinois

Burt G. Feuerstein, Ph.D., M.D.
Assistant Professor
Departments of Laboratory Medicine and Pediatrics
University of California
San Francisco, California

Flavio Flamigni, M.D.
Biochemistry Researcher
Department of Biochemistry
University of Bologna
Bologna, Italy

Sara H. Goldemberg, Ph.D.
Instituto de Investigaciones
Bioquimicas "Fundacion Campomar"
Buenos Aires, Argentina

A. D. Goldstone, Ph.D.
Associate Professor
Departments of Neurology
and Molecular Biology and Biochemistry
Northwestern University Medical School
and
V.A. Lakeside Medical Center
Chicago, Illinois

Carlo Guarnieri
Professor of Biochemistry
Department of Biochemistry
University of Bologna
Bologna, Italy

Shin-ichi Hayashi, Ph.D.
Professor
Department of Nutrition
Jikei University School of Medicine
Tokyo, Japan

Olle Heby, Ph.D.
Associate Professor
Department of Zoophysiology
University of Lund
Lund, Sweden

Zafar Iqbal, Ph.D.
Associate Professor
Department of Neurology
Northwestern University Medical School
and
V.A. Lakeside Medical Center
Chicago, Illinois

Samson T. Jacob, Ph.D.
Professor
Department of Pharmacology
Hershey Medical Center
Pennsylvania State University
Hershey, Pennsylvania

Chaim Kahana, Ph.D.
Scientist
Department of Virology
Weizmann Institute of Science
Rehovot, Israel

Harold Koenig, M.D., Ph.D.
Professor
Department of Neurology
Northwestern University Medical School
and
V.A. Lakeside Medical Center
Chicago, Illinois

Chung Y. Lu, Ph.D.
Biochemist
Department of Neurology
Northwestern University Medical School
and
V.A. Lakeside Medical Center
Chicago, Illinois

Giordana Marcozzi, D. Sc.
Researcher
ENI Ricerche
University of Rome "La Sapienza"
Rome, Italy

Laurence J. Marton, M.D.
Professor and Chairman
Department of Laboratory Medicine
Professor
Department of Neurological Surgery
University of California School of Medicine
San Francisco, California

Lisa McConlogue, Ph.D.
Scientist
Department of Molecular Biology
Cetus Corporation
Emeryville, California

José D. Méndez, Ph.D.
Researcher
Biomedical Investigation Unit
Division of Biology of Reproduction
National Medical Centre
Mexican Institute of Social Security
Mexico City, Mexico

Gabriele Mezzetti, Ph.D.
Professor
Institute of Biological Chemistry
University of Modena
Modena, Italy

Bruno Mondovi, Ph.D.
Professor
Department of Biochemical Sciences
and
C.N.R. Centre for Molecular Biology
Institute of Applied Biochemistry
University of Rome "La Sapienza"
Rome, Italy

David M. L. Morgan, Ph.D.
Section of Vascular Biology
Clinical Research Center
Harrow, Middlesex, England

David R. Morris, Ph.D.
Professor
Department of Biochemistry
University of Washington
Seattle, Washington

Maria Stella Moruzzi, Ph.D.
Professor
Institute of Biological Chemistry
University of Modena
Modena, Italy

Anthony E. Pegg, Ph.D.
Professor
Departments of Physiology and Pharmacology
Hershey Medical Center
Pennsylvania State University
Hershey, Pennsylvania

Pierluigi Riccio
Researcher
Department of Biochemical Sciences
and
C.N.R. Centre for Molecular Biology
Institute of Applied Biochemistry
University of Rome "La Sapienza"
Rome, Italy

Carmen Rossoni
Associate Professor
Department of Biochemistry
University of Bologna
Bologna, Italy

Diane H. Russell, Ph.D.
Professor
Department of Pharmacology and Therapeutics
University of South Florida College of Medicine
Tampa, Florida

Nikolaus Seiler, Ph.D.
Professor
Department of Neurochemistry
Merrell Dow Research Institute
Strasbourg, France

Dean A. Stetler, Ph.D.
Assistant Professor
Department of Pharmacology
Hershey Medical Center
Pennsylvania State University
Hershey, Pennsylvania

J. J. Trout, Ph.D.
Research Assistant Professor
Department of Neurology
Northwestern University Medical School
and
V.A. Lakeside Medical Center
Chicago, Illinois

Michael W. White, Ph.D.
Senior Fellow
Department of Biochemistry
University of Washington
Seattle, Washington

H. Guy Williams-Ashman, Ph.D.
Goldblatt Professor
Ben May Institute
Department of Pharmacological and Physiological Sciences
University of Chicago
Chicago, Illinois

Vincenzo Zappia, M.D., Ph.D.
Professor
Institute for Biochemistry of Macromolecules
University of Naples First Medical School
Naples, Italy

SERIES TABLE OF CONTENTS

TABLE OF CONTENTS

Section A.
Physiological Function of Polyamines

Chapter 1

POLYAMINES AND STEROID SEX HORMONE ACTION

H. Guy Williams-Ashman

TABLE OF CONTENTS

I. HISTORICAL INTRODUCTION

Credit for the discovery of spermine is rightly attributed to van Leeuwenhoek[1] because his epoch-making letter written to the Royal Society of London in 1677 announcing the existence of spermatoza ends with a description of the gradual deposition of human seminal plasma of tiny crystals which were easily visualized with his wonderful magnifying glasses. As discussed in several treatises,[2-6] a quarter of a millenium elapsed before these crystals in human semen were shown to consist of the phosphate salt of spermine, and the molecular structure of this aliphatic polyamine and its chemical relationship to spermidine and putrescine were elucidated. Human seminal plasma is very rich in spermine which is derived almost exclusively from the secretion of the prostate gland which is added to semen at the time of ejaculation. Under normal circumstances, many other mammalian body fluids (including blood plasma, saliva, pancreatic juice, sweat, and fluids produced by the seminal vesicle, testis, epididymis, uterus, oviducts, and vagina) contain only traces of polyamines in comparison with the prostatic secretion in man and certain other species.[4-6] As considered later, the functions (if any) of polyamines in seminal plasma remain enigmatic, despite a plethora of hypotheses. Nevertheless, investigations on the prostate gland — whose development and secretory functions are utterly dependent on androgenic hormones — uncovered much of what is now known about the *de novo* biosynthesis of polyamines in mammalian tissues and the nature and endocrine regulation of the enzyme reactions that produce putrescine, spermidine, and spermine.

The finding of Rhodes and Williams-Ashman[7] that the ventral and dorsolateral (but not the anterior) lobes of the sexually mature rat prostate produce and secrete very large amounts of spermidine and spermine sparked the first extensive studies in the same laboratory of the purifications and properties of L-ornithine decarboxylase[8,9] (ODC), putrescine-activated *S*-adenosylmethionine decarboxylase[10,11] (AdoMetDC), and separate spermidine and spermine synthases.[11-14] Early investigations on adult rat ventral prostate also established that (1) the accumulation of spermidine and spermine, and the activities of the two polyamine biosynthetic decarboxylases, declined immensely after orchiectomy, and could be restored by administration of testosterone[15,16] and (2) 5′-methylthioadenosine (MTA), which is the stochiometric byproduct of the spermidine and spermine synthase reactions in prostate and other animal cells,[13] was degraded by the newly discovered and previously unsuspected enzyme MTA phosphorylase[18,19] which ruptures the N-ribose bond of MTA to yield adenine and 5-methylribose-1-phosphate (MTR-1-P). Later investigations by others[18-20] showed that MTR-1-P is effectively converted by a series of reactions back to L-methionine in many types of mammalian cells, and that the adenine liberated from MTA by the action of MTA phosphorylase is readily transformed into 5′-adenosine monophosphate (AMP) by the ubiquitously distributed enzyme adenine phosphoribosyltransferase. The latter research not only revealed that the adenosine triphosphate (ATP) and L-methionine utilized for the production of the decarboxylated AdoMet required for spermidine and spermine synthesis can be metabolically salvaged (so that no *net* expenditure of ATP and L-methionine is essential for polyamine biosynthesis) but also led to the explanation of the fact that the steady-state concentration of MTA in the rat ventral prostate and its secretion, and in all other nonmalignant animal tissues and extracellular fluids examined heretofore, are negligible in comparison with the levels of spermidine and spermine.[17-20]

After it quickly became evident that spermidine and spermine are manufactured *de novo* via the sequential actions of ODC, AdoMetDC, and the two polyamine synthases in very many other types of mammalian cells in addition to those present in the prostate, it was obvious that development of specific inhibitors of each of these four enzymes might have important applications. The availability of such drugs that would penetrate into cells might enable the levels of various polyamines in various tissues or glandular secretions to be altered in ways that perhaps could provide insight into the biological functions of putrescine, spermidine, and spermine. Studies[21,22] on rat ventral prostate enyzmes revealed that methylglyoxal bis(guanylhydrazone) (MGBG), a drug that had previously been explored as an antileukemic agent, was a very potent ($K_i < 0.1$ μ*M*) inhibitor of putrescine-activated AdoMetDC, but was without direct influence on ODC, or spermidine and spermine synthases, in concentrations as high as 1 m*M*. Inhibition of AdoMetDC by MGBG in the presence of putrescine was competitive with respect to the AdoMet substrate, and was freely reversible if the MGBG was removed by procedures such as dialysis or gel filtration. Unfortunately, despite the remarkable selectivity of MGBG as an inhibitor of only AdoMetDC compared with its lack of effect on the other three enzymes of *de novo* polyamine biosynthesis, MGBG has several other actions which obviate the use of this drug to depress the formation of decarboxylated AdoMet (and hence of the production of spermidine and spermine) in living cells or organisms. As reviewed by Williams-Ashman and Seidenfeld,[23] MGBG is a direct inhibitor of certain enzymes that oxidize certain aliphatic diamines and polyamines, and at high concentrations damages the structure and bioenergetic functions of mitochondria. Moreover, MGBG in vivo elicits paradoxically large increases in the levels of AdoMetDC (as measured after removal of the drug), apparently by stabilizing this very fast turning-over enzyme against intracellular degradation; this effect of MGBG tends to counteract the direct inhibition of AdoMetDC by the drug in living cells.[23] Nonetheless, the aforementioned influences of MGBG on AdoMetDC have been exploited fruitfully in research on this enzyme with regard to (1) piling up of very large amounts of AdoMetDC in tissues from animals treated with MGBG, so as to provide adequate starting material for purification of the enzyme and (2) use of columns of MGBG chemically attached to inert supporting materials such as Sepharose for affinity chomatographic procedures that enable AdoMetDC to be enriched several hundredfold in a single step.[24] α-Difluormethylornithine (DFMO), which Metcalf et al.[25] discovered to be a highly specific and irreversible "suicide" inhibitor of ODC has proved to be a much more valuable inhibitor of polyamine biosynthesis in studies on the dynamics of spermidine and spermine in relation to their functions in various cells.[26] Very recently, progress has been made in the design and synthesis of specific inhibitors of spermidine and spermine synthases.[27]

Many aspects of the foregoing investigations on the enzymology and pharmacology of *de novo* polyamine production in mammalian tissues are considered in various other chapters in this book, and will not be discussed in detail here. Rather, this essay concerns attempts to unravel the relationships of polyamines to the actions of steroid sex hormones on various genitourinary tract tissues of male and female eutherian mammals. Steroid sex hormones also exert profound influences on the differentiation and functions of (1) the mammary gland and (2) the anterior pituitary as well as the hypothalamus and certain other regions of the brain, which are not discussed here because so little is known about the roles of polyamines in the physiology of the hypothalamico-hypophyseal axis.

II. ANDROGENIC INFLUENCES ON THE METABOLISM AND SECRETION OF POLYAMINES IN MALE ACCESSORY SEX GLANDS

A. Accumulation and Secretion of Polyamines in the Prostate Gland of Certain Species: Physiological Implications

Before considering the profound effects of androgens on the levels and biosynthesis of polyamines in cells of various male accessory glands of reproduction, an apparently unique capacity of the prostate glands of a restricted number of mammalian species to secrete large quantities of spermidine and/or spermine will be discussed in relation to the possible physiological utility of polyamines in semen.

The prostate gland of sexually mature men is clearly the almost exclusive source of the exceptionally high concentrations of spermine in human seminal plasma, in which spermine can be present in concentrations of even more than 10 m*M*.[2,5,6,28-32] Putrescine and spermidine levels in human seminal plasma are respectively only about 7 and 4% of those of spermine.[30] In the adult rat, the secretions of the dorsolateral and ventral lobes of the prostate contain roughly equivalent amounts of spermidine and spermine at concentrations of several millimolars, but very little putrescine.[7,16,33-35] By contrast, the secretions of rat anterior prostate (coagulating) gland and seminal vesicles contain hardly any spermidine or spermine.[7,36] The total polyamines (i.e., those present both intracellularly as well as extracellularly in stored secretions) in adult rat ventral prostate decline after orchiectomy and can be restored by treatment with testosterone, as is only to be expected from the great androgen dependency of the epithelial cells of male accessory sex glands to undergo complete functional differentiation and perform their secretory functions.[4,37-40] Reliable estimates of the intracellular concentrations of polyamines in the secretory epithelium vs. the fibromuscular stroma cells of these organs are not available. However, studies utilizing histochemical procedures that apparently specifically detect spermidine plus spermine hint that these polyamines are present predominantly in the cytoplasm of nondividing epithelial cells and in the secretions stored in the lumens of the ventral and dorsal lobes of rat prostate.[41] In contrast to the human and rat prostate glands, the levels of putrescine, spermidine, and spermine in all lobes of the prostates of adult mice, guinea pigs, and rabbits were found to be well within the range of intracellular polyamines in a large number of other tissues.[28] Moreover, bull and dog semens are essentially devoid of polyamines.[2,4,5] Thus, extensive secretion of polyamines is not an attribute of fully functional prostate glands in all species of eutherian mammals.

Two general questions arise from the presence of large amounts of polyamines of prostatic origin in the seminal plasma of man and the rat: (1) what are the mechanisms responsible for secretion of polyamines in the latter species? and (2) do the polyamines in the seminal plasmas of certain species play a significant role in reproduction?

Elaboration of secretions by the prostate and seminal vesicles occurs by apocrine mechanisms, in which proteins and certain low molecular weight substances eventually discharged from the cytoplasm of epithelial cells are packed into membrane-bound secretory granules that finally fuse with the plasmalemma during the exocytotic process.[37,39] Whether secretory granules in human and rat ventral prostate are very rich in polyamines has not yet been examined experimentally.[41] It would be interesting to study this eventuality in the prostates of a variety of species that either secrete polyamines copiously or do not. Human and rat ventral prostate secretions contain high concentrations of both polyamines and citrate (but not other acids of the tricarboxylic acid cycle); contrariwise, the canine prostate hardly secretes either citrate or polyamines.[4,37,40] However, comparative studies on polyamines and citrate in the secretions of various male accessory sex glands do not lend credence to any general relationship between polyamine and citrate secretion.[40] The rat ventral prostate produces an androgen-induced and organ-specific glycoprotein that binds spermine (but not spermidine or putrescine) tenaciously, and is extraordinarily rich in aspartyl residues.[42,43] However, no spermine-binding protein was detectable in human prostate. These observations, together with the very low levels of the spermine-binding protein in the ventral prostate of sexually mature rats,[42] make it improbable that this protein is germane to polyamine secretion by the prostates of a restricted number of species. It seems more likely that a key factor in the secretion of polyamines is the exceptionally high activities of polyamine biosynthetic enzymes in the nongrowing prostates of adults of those species in which this takes place.[16,31]

A variety of hypotheses attributing functional significance to the polyamines in human and rat semen have been promulgated, but many of them are not based on compelling evidence. Evaluation of these notions must take into account the following features of the physiology of seminal emission and the transport of spermatozoa in the male and female genital tracts of these species under coital circumstances. Semen ejaculation in the rat and humans is a very swift process requiring prior penile erection and which is completed in a few seconds; spermatozoa from the cauda epididymis and the vas deferens are propelled into the prostatic urethra where they are almost immediately diluted by preformed prostatic fluid, and then by seminal vesicle secretion, so that the semen is squirted down the penile urethra into the vaginal barrel, where the seminal plasma is diluted by vaginal secretions. In the rat tens of millions, and in man hundreds of millions, of spermatozoa normally enter the vagina at coitus. Human semen when first emitted is a friable, watery gel which is deposited close to the entrance of the fairly long cervix; the seminal gel liquefies within 30 min due to the action of proteases in seminal plasma. Immediately after ejaculation, rat semen clots into a hard solid mass which does not spontaneously liquefy. Semen clots are formed by enzymatic coagulation of seminal vesicle secretion proteins as considered below. In man, apparently only a relatively small proportion of spermatozoa in a normal ejaculate penetrate through the cervix into the uterine lumen, whereas as many as one half or even more of the spermatozoa in rat semen are transported through the short cervix. Transport of spermatozoa through the lumens of the uterus and oviducts is facilitated by contractions of smooth muscles surrounding these organs which waft the sperms through uterine and oviductal fluids, rather than being primarily due to sperm motility (which does, however, facilitate the penetrate of spermatozoa through egg vestments and the fertilization process, and, at least in man, enhances transcervical transport of sperms). Thus, all components of seminal plasma are progressively diluted by vaginal, uterine, and oviductal secretions, so that only minute amounts of sperms and seminal plasma constituents ever arrive at the ampulla of the oviduct where fertilization occurs. Although some of the prostaglandins in human seminal plasma (which are derived from vesicular secretions[39]) may perhaps enhance sperm transport to the uterus by evoking myometrial contractions, there is no evidence that seminal polyamines exert such effects. The foregoing considerations suggest that the substantial number of published studies (see References 4 and 44 to 46 for reviews) in which effects of exogenous polyamines, at the concentrations they are present in rat or human semen, on the motility, and metabolism or capacity to undergo the acrosome reaction of isolated, washed spermatozoa suspended in artificial diluents, are of dubious significance from the standpoint of regulation of fertility.

Another feature of seminal polyamines in man is that these substances are oxidized by enzyme(s) in semen that appear to be largely derived from seminal vesicle secretion. Hölttä et al.[44] purified a "diamine oxidase" from human semen 1780-fold. At each stage of purification, the relative rates of oxidation (measured by H_2O_2 production) of putrescine, spermidine, spermine, and histamine were fairly similar. However, it remains unsettled whether a single enzyme catalyzes the oxidation of all of the substrates. The reaction products of spermine or spermidine oxidation were not characterized rigorously, but some indications were obtained that spermine was first oxidized to the dialdehyde $OHC(CH_2)_2NH(CH_2)_4NH(CH_2)_2CHO$, and spermidine was converted to the monoaldehyde $OHC(CH_2)_2NH(CH_2)_4NH_2$, as occurs during the oxidation of these polyamines by an enzyme in the blood plasma of ruminants.[45] These aldehyde oxidation products that are believed to be formed by the human seminal plasma enzyme[4,5,44,46] are well known to undergo nonenzymatic decomposition to acrolein and other products which have been postulated to be responsible for the peculiar smell emitted from human ejaculates that are allowed to stand at room temperature.[2,4,5] Parenthetically it may be mentioned that appreciable diamine oxidase activity (determined with putrescine as substrate) was reported to be associated with human spermatozoa as well as seminal plasma[47] but it was not shown conclusively that washed sperm cells oxidized spermine or spermidine. The oxidation products formed by action of the bovine plasma polyamine oxidase were shown to inhibit the fructose metabolism of spermatozoa and may depress sperm motility.[5,47] A suggestion that such oxidized polyamines might be exploited for contraceptive purposes[48] is to be deplored, because these iminoaldehydes form covalent adducts with nucleic acids[45] and might well, therefore, be mutagenic and carcinogenic.

Bacterial infections of the urinary tract are much more frequent in women than men. That prostatic fluid and semen exert antibacterial activity is well documented, and it has often been proposed that this phenomenon represents a unique defense mechanism in the male to prevent bacteria growing in the urethra and bladder, since tiny amounts of prostatic secretion normally dribble constantly into the urethra in the absence of ejaculation. Lysozyme[49] and spermine[50,51] have been proposed to be the active antibacterial agents in prostatic fluid or semen; the evidence in favor of these propositions is not at all conclusive. Investigatons by Fair and co-workers[52] identified free Zn ions as the "prostatic antibacterial factor" in man and dog.

Another extracellular function of polyamines in prostatic secretion and semen that has been postulated relates to the regulation of the postejacultory coagulation of semen, which in rodents clearly involves clotting of certain bulk proteins in seminal vesicle secretion by enzymes of prostatic origin. In the rat[53-56] and guinea pig[54,57] the clotting of vesicular secretion proteins is evoked by apparently unique forms[54,58] of transglutaminase secreted by the anterior prostate or so-called coagulating gland. The enzymatic semen coagulation process in these species involves extensive formation of ϵ-(γ-glutamyl)lysine cross-bridges between seminal vesicle secretion protein precursors of the clotted material. Experiments with isolated prostatic enzymes and vesicular secretion proteins in reconstituted systems indicated that putrescine, spermidine, and spermine are incorporated covalently into vesicular secretion proteins with the formation of both *N*-mono-ϵ-(γ-glutamyl) and *N,N-bis* (γ-glutamyl) adducts of the polyamines.[36,53,59] Moreover, spermidine and spermine in concentrations present in rat dorsolateral and ventral prostate secretions also inhibit the production of insoluble coagula of vesicular secretion proteins, which suggests that the polyamines compete with protein lysine residues in the transglutaminase-catalyzed production of ϵ-(γ-glutamyl)lysine cross-links. There is no doubt that the latter type of interchain (and conceivably also intrachain) cross-bridges are mainly responsible for the extreme insolubility of rat and guinea pig copulation plugs that remain in the vagina of these species after coitus. But it is not clear whether in the rat (whose prostatic secretions are rich in spermine and spermidine, unlike in the guinea pig) cross-links formed from *N,N-bis* (γ-glutamyl)-polymine bridges between polypeptide chains also contribute to the rigidity of the seminal clots. Williams-Ashman and Canellakis[54,59] marshalled several lines of evidence to support their hypothesis that enzymatic formation of *N*-mono-(γ-glutamyl)-polyamine adducts during emission of semen through the male urethra could attenuate a too expolsive cross-linking of vesicular secretion proteins due to formation of ϵ-(γ-glutamyl)lysine bond formation, and thereby prevent blockage of rat male urethra with very insoluble proteins when ejaculation takes place. Whether the very high concentrations of spermine in human semen and prostatic fluids might by similar mechanisms prevent obstruction of the prostatic and penile urethra during ejaculation is purely a matter of conjecture. It is evident that (1) vesicular secretion proteins are precursors of the insoluble gels evident in human semen immediately after ejaculation[4,60-62] (the subsequent liquefaction of human semen is due to proteinases of prostatic origin in seminal plasma[37,62,63]) and (2) human seminal plasma contains an active transglutaminase of prostatic origin.[63a] However, the extent to which the seminal gel in man results from transglutaminase-mediated protein cross-linking has not yet been determined. It may be added that the formation of firm rigidified semen clots resulting in copulation plugs in the vagina occur in a variety of other eutherian species (e.g., mouse and certain insectivores and nonhuman primates) but not in others such as the bull, dogs, cats, and some other carnivores.[37,53] Williams-Ashman et al.[39,53-55,64] discussed the possible relationships to fertility of the postejaculatory clotting of semen in the vagina to yield copulatory plugs.

B. Androgenic Stimulation of Polyamine Production in Rat Prostate and Seminal Vesicle

In an extensive survey of polyamine biosynthetic enzymes in various tissues of sexually mature rats, Raina et al.[65] found that the ventral lobe of the prostate was the only one of

11 nongrowing organs that were not undergoing differentiation that exhibited extremely high ODC and AdoMetDC activities, and that the levels of these enzymes were quite low in the uterus of nonpregnant adult females. ODC and AdoMetDC activities in the small intestine, in which the epithelium proliferates, were substantial, but not nearly as great as in the ventral prostate. It was observed in many laboratories that rapidly growing tumors, and proliferating normal and malignant cell lines in culture, usually had very high ODC activities, but were not always exceptionally rich in AdoMetDC.[66-69] The levels of both polyamine biosynthetic decarboxylases in rat seminal vesicle (whose full functional differentiation and growth at puberty are, like those of all lobes of the prostate, completely dependent on circulating androgens) were reported to be about one tenth of the activities of these enzymes in ventral prostate.[34,70] As noted above, rat seminal vesicle secretion is virtually devoid of polyamines. ODC activities are very low in the rat anterior prostate (coagulating) gland[9] which also does not secrete spermidine or spermine.[7,36] Within a few days after orchiectomy of adult rats, ODC activities in the ventral prostate and seminal vesicle declined to very low values, but can be restored to adult normal or even higher levels by daily administration of testosterone propionate to the castrates.[15,16,33-35,70] The levels of spermidine and spermine synthases in adult male rat accessory sex glands are strikingly greater than those of ODC and AdoMetDC, suggesting that the activities of the two polyamine biosynthetic decarboxylases are more rate-limiting to polyamine formation in vivo.[16,65,71] The activity of the two polyamine synthases decline more slowly after castration in comparison with ODC and AdoMetDC, in keeping with the much faster rates of intracellular degradation of the decarboxylases. The levels of spermidine synthase (but not spermine synthase) in both ventral prostate and seminal vesicle of orchiectomized rats increase fewfold after treatment with testosterone, but the androgen-induced alterations of spermidine synthase are more sluggish and of much lower magnitude than those of ODC and AdoMetDC.[71]

Several investigations have concerned effects of continual administration of DFMO on the androgenic regulation of putrescine, spermidine, and spermine levels and ODC activities in rat male accessory sex glands.[33-35] DFMO is a highly specific inhibitor of ODC (K_i about 40 μM) which at maximal concentrations inactivates the enzyme irreversibly in cell-free systems with a half-life of roughly 3 min; because of the very high rate of intracellular turnover of ODC, cells must be continually supplied with DFMO (which is quickly cleared from the organism) if it is desired to keep the enzyme activity at a very low level. When present in the drinking water at concentratons of 20 g/ℓ DFMO accumulates in normal rat prostate and seminal vesicle in levels of 0.1 to 0.3 μmol/g,[33] which cause more than 90% inhibition of ODC activity of soluble tissue extracts, but striking elevations in the activity of AdoMetDC (7-fold) and of the level of decarboxylated AdoMet (450-fold).[34] The latter observations accord with a marked decline in putrescine and spermidine concentrations (with much lesser lowering of spermine) as a result of DFMO treatment, since there is considerable evidence that high intracellular levels of spermidine depress the production of AdoMetDC[69,72] and, with the great lowering of intracellular putrescine and spermidine by DFMO, decarboxylated AdoMet would only be expected to accumulate because it cannot be utilized effectively as substrate for spermidine and spermine synthases.[72,73] In castrated rats, DFMO administration almost abrogated the enhancement of ODC and the levels of putrescine and spermidine, and diminished the rise in spermine, organ weight, and RNA content that is normally evoked by testosterone in ventral prostate and seminal vesicle.[34,35] Comparable studies on effects of administration of the AdoMetDC inhibitor MGBG on androgen-induced polyamine synthesis and growth by male accessory sex glands did not prove to be enlightening.[23,35]

When plasma testosterone levels in adult males are sharply decreased immediately following excision of the testes, all lobes of the prostates as well as the seminal vesicles gradually involute. This entails shriveling of epithelial cells with decline in their endoplasmic reticulum membranes and RNA content and cessation of their secretory functions, and also a loss of a major proportion of the epithelium, which is usually accompanied by some increase in the ratio of fibromuscular to epithelial cell elements, as evidenced both histologically and by a marked fall in the total DNA content of the glands. Restoration of prostate and seminal vesicle size promoted by treatment of castrated animals with androgens involves a transient phase of replication of both epithelial and stromal cells,[74-78] and also involves hypertrophy of epithelial cells with reestablishment of their secretory capacities. From the results of investigations cited in the two preceding paragraphs and other studies,[31,79] there has been a tendency to conclude that the increases in intracellular polyamines and ODC activities induced by androgens in male accessory glands correlate more with restoration of epithelial cell size and secretory functions (regardless of whether or not a particular gland or lobe thereof actually secretes large amounts of spermidine and/or spermine) than with the hyperplasia of various cellular elements that contributes to the overall restoration in organ size. Furthermore, although spermidine and spermine added to cell-free systems can profoundly modify a multitude of enzymic processes involved in eukaryotic DNA, RNA and protein synthesis and turnover,[80,81] and various messenger-independent protein kinase reactions,[82,83] the sparse evidence available does not point to any paramount intermediary roles of polyamines in androgen-induction of DNA replication, RNA transcription and processing, or translation of mRNAs cognate to specific proteins during male accessory gland growth and functional differentiation, although spermidine and spermine undoubtedly exert supportive actions on these macromolecular biosyntheses.

However, all previous experimental approaches to unraveling relationships of polyamines to prostate development need to be reexamined in the light of new insights obtained by Sugimura et al.[84-86] into the cell biology of mouse prostate growth postnatally and the castration-induced degeneration and androgen-promoted regeneration of the prostate in adult mice. Evidently, the patterns of duct development and branching that occur during the first 15 days after birth are at this early period of extrauterine life only minimally dependent on androgens, in contrast to the strict androgen dependency of both the initial morphogenesis of the prostate from the urogenital sinus of the fetus[87,88] and the full growth and functional differentiation at puberty and the maintenance of the adult prostate thereafter.[37] It was shown that the dynamics and degree of ductal branching in the early postnatal period were quite different between the dorsolateral (DLP) and ventral (VP) lobes of the mouse prostates.[84] When the mouse prostate that has fully developed under androgenic stimulation degenerates following orchiectomy of adult mice, about 35% of the ductal tips and branch points become lost in the distal regions (close to the capsule) whereas the more proximal regions (nearer to the urethra) survive in an atrophic state. Testosterone-stimulated regeneration of the VP returns the gland to its former dimensions with only moderate ductal distention, whereas in the DLP florid epithelial infoldings and ductal distension occurs.[85] The DNA synthetic activity (as reflected in autoradiographs of cells exposed to labeled thymidine) in both lobes during the prepubertal period of normal development was almost totally confined to the distal ducts; there was hardly any nuclear DNA synthesis in the normal adult prostate. But when the involuted glands in castrates started to grow in response to testosterone, DNA synthesis was confined to the distal ducts over the 1st day of androgen treatment, and by the 3rd day was evident in both proximal and distal ducts, and then subsided to sporadic focal labeling confined largely to distal ducts.[86] That there are marked morphological and biochemical differences between various lobes of the prostate in many mammalian species has long since been recognized,[4,37,38,40] but very little is known about regional differences in the physiology of epithelial cells within the same lobe of the glands. The foregoing observations suggest that the production of polyamines and their biosynthetic enzymes should be examined in epithelial cells of varying districts within specific lobes of the prostate in different species with respect to both the intracellular as well as extracellular (i.e., secreted) levels of spermidine and spermine, which should be feasible with application of more recently developed ultramicrobiochemical and histochemical analytical techniques.[41,87,88] It would be valuable to examine the polyamine metabolism of prostate tissue budding out from urethral primordia derived from the urogenital sinus during the androgen-dependent fetal morphogenesis of prostate gland.[89]

It is now well established, as overviewed by Pegg,[69] that many types of mammalian cells can convert spermine back to spermidine, and transform spermidine synthase or spermine synthase reactions, whose equilibrium constants are overwhelmingly in favor of polyamine synthesis. Rather, spermidine and spermine are first acetylated by acetyl-CoA on the amine group attached to the propylamine (N^1) moieties by a spermidine/spermine acetyltransferase (SAT); the N^1-acetyl derivatives are then oxidized by a flavoprotein intracellular polyamine oxidase (PAO) at the secondary amino sites on the N^1 acetylated polyamines to yield putrescine from N^1-acetylspermidine and spermidine from N^1-acetyl spermine with formation in both instances of 3-acetamidopropanal. The SAT-catalyzed acetylation of polyamines appears to be the rate-limiting reaction of this polyamine interconversion pathway. The SAT enzyme is readily inducible in several tissues by various toxic agents, excess polyamines, and MGBG, and, like ODC and AdoMetDC, has a remarkably short half-life in vivo;[90] in contrast the activity of PAO is greatly in excess of SAT in many resting tissues. The overall reactions promoted by SAT plus PAO probably serve to prevent spermidine and spermine from accumulating in excessive concentrations in most cells, since when intracellular levels of these polyamines become greatly enhanced the activities of ODC and AdoMetDC (which favor *de novo* synthesis of polyamines) are repressed, while the activities of SAT and the operation of the ''back conversion'' pathway are usually elevated. For obvious reasons it would be of interest to investigate in depth the regulation of PAO and SAT in organs such as human and rat ventral prostate which in sexually mature creatures secrete large quantities of spermidine and spermine. Sporadic observations indicate that in adult rat ventral prostate (1) PAO is fairly active[91] and (2) SAT is very low but is enhanced over 20-fold several hours after i.p. administration of large amounts of spermidine (100 mg/kg) though not to as great an extent as in liver.[90]

Many lines of evidence[69] indicate that the abundance of ODC and AdoMetDC molecules in eukaryotic cells is subject to regulation at both transcriptional and translational levels. ODC activities also are controlled by an apparently specific ODC-antizyme inhibitory protein,[92] and are influenced at a posttranslational level[69] by phosphorylation[93] or otherwise unknown mechanisms.[94] The recent availability of cDNA probes cognate to the mRNAs for ODC[94-97] and AdoMetDC[99] in mammalian cells has already permitted studies on the control of transcription and translation of these messages in certain tissues. Such investigations on ODC production in male accessory sex glands are in their infancy, but have been fairly extensive with respect to the androgenic regulation of ODC in mouse kidney now to be discussed.

III. EFFECTS OF ANDROGENS ON ORNITHINE DECARBOXYLASE AND POLYAMINE METABOLISM IN MOUSE KIDNEY

It has long been known that the metanephric kidney of many mammalian species is responsive to androgens in both sexes, and that this is especially apparent in laboratory mice. The cells of the kidney proximal convoluted tubules and Bowman's capsule are bigger in adult male than female mice; orchiectomy results in regression of the adult male kidney to the size of the organ in females.[100-102] Recent research on androgenic regulation of polyamine metabolism and particularly the induction of ODC in the kidneys of various strains of mice has been very instructive for the following reasons.

1. Administration of testosterone and other androgens swiftly elicits enormous elevations in ODC activity in the kidneys of males and females, from which the enzyme has

been purified in good yields to apparent homogeneity.[103,104] The properties of mouse kidney ODC have been examined extensively.[69,102,103]

2. The amino acid sequence of mouse ODC was deduced from the nucleotide sequences of corresponding cDNAs.[105,106]
3. Immunocytochemical methods[107] and procedures based on the covalent attachment of radioactivity from labeled DFMO to the enzyme[108] have permitted studies on the intracellular localization of the enzyme, which is largely present in the cytoplasm of proximal and distal convoluted tubules, with 10 to 15% also associated with the nucleolus in cell nuclei.
4. Superior methodologies for estimation of the levels and turnover of occupied nuclear and cytosolic androgen receptors in mouse kidney[109] and their interactions with different androgenic steroids and antiandrogenic drugs[110] have been well worked out, so that correlative studies on relationships between functional nuclear androgen receptors and the expression of genes for ODC and certain other androgen-induced proteins in kidney were made possible.[102,109,110] These investigations were complemented by incisive examination of effects of testosterone on the turnover of ODC in mouse kidney.[111]
5. Most importantly, the mouse renal system differs profoundly in two ways from the prostate in the peculiarities of their responses to androgens. First, circulating testosterone is not converted significantly to 5α-dihydrotestosterone (DHT) by mouse kidney. Whereas most of the testosterone that acts androgenically on the prostate in coverted therein by steroid 5α-reductase to DHT, which acts in prostatic cell nuclei via a DHT-androgen receptor complex, in mouse kidney (which is essentially devoid of steroid 5α-reductase) testosterone acts by binding to the androgen receptor as such.[101,102] However, the mouse kidney responds to exogenous DHT dramatically in addition to being influenced directly by either exogenous or endogenous testosterone. In this regard, note that (1) DHT binds more firmly to the androgen receptor than testosterone, (2) in many bioassays, DHT is a more potent androgen than testosterone, and (3) there seems to exist only a single gene for the mammalian androgen receptor that is located on the X-chromosome, despite the demonstration of multiple forms of the androgen receptor that probably arise from aggregations and limited proteolytic nicking of the receptor protein.[38] Second, the growth-promoting actions of androgens on mouse kidney reflect entirely hypertrophic responses of appropriate epithelial cells. Therefore, in contrast to the prostate where both cell proliferation and cell hypertrophy are involved in the growth of the organ, the changes in renal polyamine metabolism and ODC in the mouse are not complicated by occurrence of nuclear DNA synthesis and cell multiplication.

Some fundamental aspects of ODC induction by androgens in mouse kidney that emerged from studies cited in the foregoing paragraph are as follows. The degree of enhancement of ODC activities by large doses of testosterone is more extensive in females or castrated males than in normal adult males, and varies considerably among different strains of inbred mice. Part of the augmentation of ODC appears to reflect a decreased rate of intracellular degradation of the enzyme because the apparent half-life of ODC in vivo, as monitored by the rate of decline in activity after treatment of the animals with cycloheximide, was prolonged 4-fold when ODC levels were increased 60-fold by treating male mice of the BALB/c strain with excessive testosterone (the half-life but not the total activity of AdoMetDC was also elevated by the androgen).[111] DHT administration also enhanced renal ODC, whereas 17β-estradiol, progesterone, and cortisol were ineffective; testosterone was without influence on ODC in the kidneys of *Tfm* mutant mice which are completely insensitive to endogenous or exogenous androgens because of defects in the androgen receptor.[102] Increases in ODC mRNA levels (as measured by appropriate cDNA hybridizations) parallel those of ODC activity and immunoreactive protein, suggesting that augmented transcription (and/or processing and stabilization) of the ODC message contributed extensively to the greatly heightened ODC activity evoked by the androgen, in addition to the aforementioned prolongation of the half-life of the enzyme.[102] The nonsteroidal and antiandrogenic drug Flutamide severely attenuated ODC induction by testosterone, again indicating that the androgen effects are receptor-mediated.[110] In the latter contexts, it must be emphasized that in a number of other mammalian biological systems, ODC can be induced by suitable stimuli without demonstrable changes in tissue levels of ODC mRNA or in the rate of intracellular turnover of the enzyme (see Reference 69). There is indeed growing evidence that translation of ODC mRNA is influenced by many factors, including the intracellular levels of putrescine, spermidine, and spermine.[69] How occupied androgen receptors interacting with the chromatin of mouse kidney nuclei can differentially enhance the transcription of mRNAs for ODC and certain other proteins remains enigmatic. Although it has not yet been demonstrated with respect to the ODC gene, indications that progestin and cortisol receptors can selectively interact with specific nucleotide sequences on appropriate steroid-inducible genes that lie 5′-upstream from the transcription initiation sites and thereby promote gene transcriptions may well be germane to this problem.[112] It may be added that amplification of the number of copies of the ODC gene per genome appears to occur under certain conditions in several types of cultured mammalian cells that greatly overproduce ODC.[95,97,113,114] However, there are no indications for such ODC gene amplification in mouse kidney or rat ventral prostate when the activity of the enzyme is spectacularly elevated in response to androgens. There are several examples of increases in the levels of ODC mRNA in various mammalian cells occurring with no concomitant increase in the number of copies of the ODC gene.[69]

It is well established that induction of ODC in murine kidney is accompanied by severalfold increases in the content of putrescine, but with much smaller changes in spermidine and spermine.[109,115] On the basis of experiments of questionable design, Goldstone et al.[116] concluded that the testosterone-stimulated hypertrophy of mouse kidney is at least partially mediated by induction of ODC and increased cellular polyamine levels. Comparable studies by Berger and Porter,[117] which similarly involved scrutiny of effects of DFMO on the responses to androgen, cast grave doubt on the latter hypothesis.

IV. POLYAMINES IN RELATION TO HORMONAL CONTROL OF TESTICULAR, EPIDIDYMAL AND SPERMATOZOAL PHYSIOLOGY

In the testis of sexually mature mammals, spermatogenesis is confined to the seminiferous tubules which contain differentiating male germ cells embedded in supporting Sertoli cells and free spermatozoa, while the biosynthesis and secretion of testosterone takes place in the Leydig cells that are present outside the seminiferous tubules. Sperm production in the tubules is regulated by androgens and pituitary follicle stimulating hormone (FSH) and testosterone secretion by Leydig cells is controlled by pituitary luteinizing hormone (LH). Macindoe and Turkington[118] reported in 1973 that during the onset and development of spermatogenesis through the juvenile and pubertal ages of young rats, the activity of ODC was maximal at 3 days after birth when only Sertoli cells were present in seminiferous tubules, and ODC became reduced with the later appearance of differentiating germ cells. AdoMetDC activity increased steadily during spermatogenesis, reaching a plateau with the production of mature testicular spermatozoa. Total spermidine concentrations paralleled the activity of ODC and also increased during spermatid differentiation. ODC was not detectable in spermatid-rich fractions separated by sedimentation through a continuous serum albumin gradient; a single peak representing nearly all of the total ODC was observed in a rapidly-sedimenting cell fraction containing Sertoli cells. These observations[118] are rather at variance with some more recent studies by Tsai et al.,[119,120] who found no more than traces of ODC in Sertoli cells, testicular germ cells, and epididymal spermatozoa, whereas ODC and polyamines (especially putrescine) were present in easily measurable levels in Leydig cells. However, ODC-like molecules that reacted with what was believed to be a monospecific polyclonal antibody directed against ODC was evident by staining procedures in cultured Sertoli cells, Leydig cells, pachytene spermatocytes, round spermatids, and epididymal spermatozoa, despite the inability to demonstrate high ODC activity in any cellular elements except Leydig cells. The possibility that substances with inhibitory actions on ODC activity were present in Sertoli and other cells that failed to decarboxylate L-ornithine was not examined. Dias[121] observed transglutaminase (measured by calcium-dependent covalent incorporation of labeled putrescine into casein) in homogenates of rat Leydig cells and seminiferous tubules, and in Sertoli cells cultured in serum-free media; the enzyme activity was present in both cytosol and membrane fractions. On the basis of inhibition by the transglutaminase substrate monodansyl cadaverine of certain responses of Sertoli cells to FSH, it was proposed that transglutaminase may play a facilitatory rather than mandatory role in activation of Sertoli cell functions by FSH. Blixonas et al.[122] demonstrated a spermine-binding component in membrane preparations from bovine testes; the spermine-binding material (which could perhaps be phospholipid rather than proteinaceous in nature) did not closely resemble the soluble rat ventral prostate spermine-binding protein.[42] According to Bamberg et al.[123] unwashed spermatozoa from the corpus epididymis of the bull contained much more spermine than putrescine or spermidine, and correspondingly lower amounts of these polyamines were detected sperms from the cauda of the epididymis, but it was not proven that the polyamines were actually inside the spermatozoal cells or that other types of cells (e.g., leukocytes or sloughed-off epididymal epithelial elements) might be present as contaminants. ODC was not detected in epididymal sperm cells. Comparably, Pulkkinnnen et al.[47] reported that significant amounts of spermine were associated with unwashed human ejaculated spermatozoa; since the spermine was largely removed by washing with NaCl (0.5 *M*) or $MgCl_2$ (0.1 *M*), it is quite likely that this polyamine was not inside the sperm cells but rather represented residual spermine from the spermine-rich human seminal plasma that was bound to the outer faces of spermatozoal cell surfaces. The claim of Atmar et al.[124] that bovine epididymal spermatozoa contain a messenger-independent protein kinase that phosphorylates an endogenous 70-kdalton protein in reactions stimulated by spermidine and spermine has not received independent validation. It is quite clear from all of the aforementioned meager and in some cases contradictory data that the relationships of polyamines to testicular functions and their hormonal regulation, and (if any) to spermatozoal functions, need to be examined much more thoroughly and incisively.

Matsuzaki and co-workers[125] made the interesting observation that the hamster epididymis — a very androgen-dependent organ — contains high concentrations of N^1-acetylspermidine and lesser amounts of *sym*-homospermidine. The latter substance, which was rigorously characterized, has never been found in other mammalian tissues and was not detected in the chow diet given to the hamsters; it is, therefore, conceivable that the hamster epididymis can synthesize *sym*-homospermidine. Whether the epididymal N^1-acetylspermidine was present in spermatozoa in the lumen of the organ or in epididymal cells as such was not reported.

V. OVARIAN HORMONES AND POLYAMINE METABOLISM IN OVIDUCT AND UTERUS

Many accounts of phenomenological studies on effects of ovarian hormones on polyamine metabolism and biosynthetic enzymes in extragonadal organs of the female reproductive tract have been published over the last 15 years. Early studies in several laboratories disclosed stimulatory effects of 17β-estradiol on spermidine[126,127] and ODC[127-128] levels in the uterus of ovariectomized female rats and also in chick oviduct. Kaye et al.[129] showed that injection of 0.5 μg of 17β-estradiol into juvenile female evoked a 14- to 25-fold maximal increase in uterine ODC at 4 hr after administration of the hormone.

Many other investigations on estrogenic regulation of uterine polyamines and their biosynthetic decarboxylases have been documented more recently, but only a few of these studies will be considered here. Rorke et al.[130] examined the effects of estradiol and of two nonsteroidal antiestrogens (C1628, Parke Davis, and U23469, Upjohn) on immature female rats at more prolonged time intervals. The two aforementioned antiestrogenic drugs compete against strong estrogen agonists such as estradiol for binding to the same site on the estrogen receptor, and these antiestrogens when administered alone provoke certain weak estrogen agonist actions. Noteworth in the latter regard is that (1) such antiestrogens solo stimulate uterine epithelial cell growth without significantly affected other uterine cell types, whereas estradiol enhances growth of both epithelial and myometrial cell elements; (2) although estradiol and the antiestrogens bind to the same estrogen receptor protein, the receptor complexes with antiestrogens may interact with nuclear chromatin in different ways from estradiol-receptor complexes; and (3) the dynamics of nuclear accumulation of receptor complexes of the antiestrogens (or their metabolically activated metabolites) are different from those of estradiol-receptor complexes. It was observed[130] that following a single dose of estradiol, two peaks of ODC activity (roughly 500-fold increases) occurred at 5 and 16 hr, whereas antiestrogen solo evoked slower enhancement of ODC with only one major peak of enzyme activity at about 24 hr. Daily treatment with estradiol over 3 days elicited continued enhancement of ODC whereas continued antiestrogen administration stimulated uterine ODC only during the 1st day. Pretreatment with antiestrogen did not block the 5-hr peak of ODC elicited by estradiol but depressed the 16-hr peak. However, uterine spermidine and spermine (but not putrescine) levels per unit wet weight were elevated to about the same extent (1.5- to 2-fold) in animals treated with estadiol, or antiestrogen, or estradiol together with antiestrogen. Rorke and Katzenellenbogen[131] found under comparable conditions that when DFMO was administered continually in the drinking water for 2 or 6 days before then during estradiol treatment, increases in uterine ODC and spermidine normally caused by estradiol were completely suppressed; however, the estradiol-induced increase in uterine wet weight was the same in the absence or presence of DFMO. It was concluded that uterine ''growth'' stimulation by estradiol is not dependent on induced changes in uterine ODC or spermidine. By contrast, in experiments with ovariectomized adult female rats, Marshall and Senior[132] found that DFMO (which alone did not affect uterine blood flow, or uterine wet or dry weight) markedly attenuated the stimulatory influence of estradiol on wet and dry uterine weights at 6 hr after administration of the estrogen.

Recently, Cheng and Pollard[133] reported that at the peak of uterine DNA synthesis evoked by estradiol in ovariectomized mice, the levels of the mRNAs for ODC and the c-*ras*^{H} proto-oncogene were increased; progesterone administration nullified the estradiol-induced wave of DNA synthesis but did not have much effect of the levels of the two mRNAs. It was concluded that in uterine epithelial cells, estradiol regulates the transcription of the ODC gene and the c-*ras*^{H} proto-oncogene independently of cell proliferation.

The only unequivocal conclusion that can be drawn from all of the studies discussed in this section is that our comprehension of the relationships of polyamines and ODC to estrogen-stimulated cell proliferation and differentiation in the uterus remains very nebulous.

ADDENDUM

Two very recent reports are of interest for further research on mechanisms by which steroid sex hormones can regulate expression of genes coding for polyamine biosynthetic enzymes at both transcriptional and translational levels. First, Katz and Kahana[134] have cloned the entire active ODC gene that can be amplified in mouse myeloma 653-1 cells by exposure to DFMO, with resultant massive overproduction of the enzyme. The transcriptionally active mouse ODC gene contains 11 introns and 12 exons (only exons 3 to 12 code for the ODC protein). Exon 12 also corresponds to the 3′ noncoding regions of the 2 species of ODC mRNA which are formed by alternative utilization of 2 polyadenylation signals separated from each other by 422 nucleotides. The transcription initiation site of the mouse ODC gene was mapped. The portion of the authentic 5′ noncoding region encoded by exon 1 is very rich in G + C, and can assume potential secondary structures that might be in translational regulation of ODC mRNA, notably with regard to direct inhibitory effects of spermidine or spermine. Second, Pegg et al.[135] have shown that the mRNA for mammalian AdoMetDC codes for synthesis of a proenzyme of 37,000 in molecular weight that is then converted to the catalytically active form of AdoMetDC (with formation of the pyruvoyl prosthetic group at the N-terminus) comprised of a 32,000 molecular weight subunit, by reactions that are directly enhanced by putrescine. The extent to which the proenzyme accumulates *in vivo* in androgen-stimulated rat ventral prostate cells was highly influenced by the putrescine content of the gland. In normal control prostates, the levels of the AdoMetDC proenzyme comprised about 4% of the total mRNA translation products. In animals treated with DFMO, which lowers ventral prostate putrescine levels to negligible levels, the proenzyme accumulated to roughtly 25% of the total amounts of mRNA translation products. Conversely, in rats treated with MGBG (which greatly increases putrescine concentrations in ventral prostate), virtually all of the enzyme was present in the catalytically active form of subunit molecular weight 32,000. Thus putrescine clearly serves not only as a direct, allosteric-like activator of AdoMetDC, but also stimulates conversion of the proenzyme to the catalytically active form of AdoMetDC.

An incisive review by Danzin and Mamont[136] provides a comprehensive discussion of effects of androgens on polyamine accumulation and secretion in rat ventral prostate and seminal vesicle in relation to growth of these organs. Fozzard[137] has overviewed the contragestational effects of DFMO and other drugs that inhibit ornithine decarboxylase in several species of eutherian mammals. Early embryonic development in these mammals is highly dependent on the actions of estrogens and estrogens on the uterus, placenta, and the embryo, including the actions of these hormones on polyamine turnover. The halt in embryonic development evoked by DFMO occurs mainly at the stage of gastrulation, and seems to be largely due to effects of the drug on putrescine and spermidine production in the embryo itself.

Extensive investigations by Park and Folk[138,139] have established that spermidine is a key precursor of the hypusine moiety present in peptide linkage in the eIF-4D initiation factor of eukaryotic protein biosynthesis. In this regard, it would be of interest to study relationships of hormonal regulation of protein biosynthesis in androgen-dependent organs such as the prostate gland to the formation of hypusine in, and the functions of, the eIF-4D initiation factor.

REFERENCES

1. **Van Leeuwenhoek, A.,** Observationes, D. Anthonii Lewenhoeck, de Natis e Semine genitali. Animalculis, *Phil. Trans. Roy. Soc. London,* 12, 1040, 1677.
2. **Rosenheim, O.,** The isolation of spermine phosphate from semen and testis, *Biochem. J.,* 18, 1253, 1924.
3. **Dudley, H. W., Rosenheim, O., and Starling, W. W.,** The constitution and synthesis of spermidine, a newly discovered base isolated from animal tissues, *Biochem. J.,* 21, 97, 1927.
4. **Mann, T.,** *The Biochemistry of Semen and of the Male Reproductive Tract,* Methuen, London, 1974, 194.
5. **Tabor, H. and Tabor, C. W.,** Spermidine, spermine and related amines, *Pharmacol. Rev.,* 16, 245, 1964.

6. **Williams-Ashman, H. G.,** Nicolas Louis Vaquelin (1763—1829): reflections on his pioneer contribution to the chemistry of human seminal fluid, *Invest. Urol.*, 2, 605, 1965.
7. **Rhodes, J. B. and Williams-Ashman, H. G.,** Observations on polyamines in male accessory glands of reproduction, *Med. Exp.*, 10, 281, 1964.
8. **Pegg, A. E. and Williams-Ashman, H. G.,** Biosynthesis of putrescine in the prostate gland of the rat, *Biochem. J.*, 108, 533, 1968.
9. **Jänne, J. and Williams-Ashman, H. G.,** On the purification of and properties of L-ornithine decarboxylase from rat prostate and effects of thiol compounds on the enzyme, *J. Biol. Chem.*, 246, 1725, 1971.
10. **Pegg, A. E. and Williams-Ashman, H. G.,** Stimulation of the decarboxylation of *S*-adenosylmethionine by putrescine in mammalian tissues, *Biochem. Biophys. Res. Commun.*, 30, 76, 1968.
11. **Pegg, A. E. and Williams-Ashman, H. G.,** On the role of *S*-adenosylmethionine in the biosynthesis of spermidine in rat prostate, *J. Biol. Chem.*, 244, 682, 1969.
12. **Pegg, A. E. and Williams-Ashman, H. G.,** Enzymic synthesis of spermine in rat prostate, *Arch. Biochem. Biophys.*, 137, 156, 1970.
13. **Jänne, J., Schenone, A., and Williams-Ashman, H. G.,** Separation of two proteins required for synthesis of spermidine from *S*-adenosyl-L-methionine and putrescine in rat prostate, *Biochem. Biophys. Res. Commun.*, 42, 758, 1971.
14. **Raina, A. and Hannonen, P.,** Separation of enzyme activities catalyzing spermidine and spermine synthesis in rat brain, *FEBS Lett.*, 16, 1, 1971.
15. **Pegg, A. E. and Williams-Ashman, H. G.,** Rapid effects of testosterone on prostatic polyamine-synthesizing enzyme systems, *Biochem. J.*, 109, 32P, 1968.
16. **Pegg, A. E., Lockwood, D. H., and Williams-Ashman, H. G.,** Concentrations of putrescine and polyamines and their enzymic synthesis driving androgen-induced prostatic growth, *Biochem. J.*, 117, 17, 1970.
17. **Pegg, A. E. and Williams-Ashman, H. G.,** Phosphate-stimulated breakdown of 5′-methylthioadenosine in rat ventral prostate, *Biochem. J.*, 115, 241, 1969.
18. **Williams-Ashman, H. G., Seidenfeld, J., and Galletti, P.,** Trends in the biochemical pharmacology of 5′-deoxy-5′-methylthioadenosine, *Biochem. Pharmacol.*, 31, 277, 1982.
19. **Williams-Ashman, H. G.,** Metabolic significance of 5′-deoxy-5′-methylthioadenosine in relation to polyamine turnover in normal and malignant mammalian cells, in *Recent Progress in Polyamine Research*, Selmeci, L., Brosnan, M. E., and Seiler, N., Eds., Akademiai Kiado, Budapest, 1985, 231.
20. **Schlenk, F.,** Methylthioadenosine, *Adv. Enzymol. Relat. Areas Mol. Biol.*, 54, 196, 1983.
21. **Williams-Ashman, H. G. and Schenone, A.,** Methylglyoxal bis(guanylhydrazone) as a potent inhibitor of mammalian and yeast *S*-adenosylmethionine decarboxylases, *Biochem. Biophys. Res. Commun.*, 46, 288, 1972.
22. **Corti, A., Dave, C., Williams-Ashman, H. G., Mihich, E., and Schenone, A.,** Specific inhibition of the enzymic decarboxylation of *S*-adenosylmethionine by methylglyoxal bis(guanylhydrazone) and related substances, *Biochem. J.*, 139, 351, 1974.
23. **Williams-Ashman, H. G. and Seidenfeld, J.,** Aspects of the biochemical pharmacology of methylglyoxal bis(guanylhydrazone), *Biochem. Pharmacol.*, 35, 1217, 1986.
24. **Williams-Ashman, H. G. and Pegg, A. E.,** Aminopropyl group transfers in polyamine biosynthesis, in *Polyamines in Biology and Medicine*, Morris, D. R. and Marton, L. J., Eds., Marcel Dekker, New York, 1981, 41.
25. **Metcalf, B. W., Bey, P., Danzin, C., Jung, M. J., Casara, P., and Ververt, J. P.,** Catalytic irreversible-inhibition of mammalian ornithine decarboxylase by substrate and product analogs, *Am. Chem. Soc.*, 100, 2551, 1978.
26. **Pegg, A. E. and Williams-Ashman, H. G.,** Biosynthesis of putrescine, in *Polyamines in Biology and Medicine*, Morris, D. R. and Marton, L. J., Eds., Marcel Dekker, New York, 1981, 3.
27. **Pegg, A. E. and Williams-Ashman, H. G.,** Pharmacologic interference with enzymes of polyamine biosynthesis and of 5′-methylthioadenosine metabolism, in *Inhibition of Polyamine Biosynthesis: Biological Significance and Basis for New Therapies*, McCann, P. P., Pegg, A. E., and Sjoerdsma, A., Eds., Academic Press, New York, 1987, 33.
28. **Williams-Ashman, H. G., Jänne, J., Coppoc, G. L., Geroch, M. E., and Schenone, A.,** New aspects of polyamine biosynthesis in eukaryotic organisms, *Adv. Enzyme Regul.*, 10, 225, 1972.
29. **Fair, W. R., Clark, R. B., and Wehner, N. A.,** Correlation of seminal polyamine levels and semen analysis in the human, *Fertil. Steril.*, 23, 38, 1972.
30. **Jänne, J., Hölttä, E., Haaranen, P., and Elfving, K.,** Polyamines and polyamine-metabolizing enzyme activities in human semen, *Clin. Chim. Acta*, 48, 393, 1973.
31. **Williams-Ashman, H. G. and Lockwood, D. H.,** Role of polyamines in reproductive physiology and sex hormone action, *Ann. N.Y. Acad. Sci.*, 171, 882, 1970.
32. **Anderson, R. V. and Fair, W. R.,** Physical and chemical determinations of prostatic secretions on benign hyperplasis, prostatitis, and adenocarcinoma, *J. Urol.*, 121, 295, 1976.

33. **Danzin, C., Jung, M. J., Claverie, N., Grove, J., Sjoerdsma, A., and Koch-Weser, J.,** Effects of 2-difluoromethylornithine, an enzyme-activated irreversible inhibitor of ornithine decarboxylase, on testosterone-induced regeneration of prostate and seminal vesicle in castrated rats, *Biochem. J.*, 180, 507, 1979.
34. **Danzin, C., Claverie, N., Wagner, J., Grove, J., and Koch-Weser, J.,** Effect on prostatic growth of 2-difluoromethylornithine, an effective inhibitor of ornithine decarboxylase, *Biochem. J.*, 202, 175, 1982.
35. **Käpyaho, K., Kallio, A., and Jänne, J.,** Differential effects of 2-difluoromethylornithine and methylglyoxal bis(guanylhydrazone) on the testosterone-induced growth of ventral prostate of castrated rats, *Biochem. J.*, 219, 811, 1984.
36. **Folk, J. E., Park, M. H., Chung, S. I., Schrode, J., Lester, E. P., and Cooper, H. L.,** Polyamines as physiological substrates for transglutaminases, *J. Biol. Chem.*, 255, 3695, 1980.
37. **Price, D. and Williams-Ashman, H. G.,** The accessory reproductive glands of mammals, in *Sex and Internal Secretions*, Vol. 1, 3rd ed., Young, W. C., Ed., Williams & Wilkins, Baltimore, 1961, 366.
38. **Williams-Ashman, H. G.,** Perspectives in the male sexual physiology of eutherian mammals, in *Physiology of Reproduction*, Knobil, E. and Niall, J. D., Eds., Raven Press, New York, 1988, 727.
39. **Williams-Ashman, H. G.,** Regulatory features of seminal vesicle development and functions, *Curr. Top. Cell. Regul.*, 22, 202, 1983.
40. **Mann, T. and Lutwak-Mann, C.,** *Male Reproductive Function and Semen*, Springer-Verlag, New York, 1981.
41. **Hougaard, D. M. and Larsson, L.-I.,** Localization and possible functions of polyamines in protein and peptide secreting cells, *Med. Biol.*, 64, 89, 1986.
42. **Hiipakka, R. A., Chen, C., Schilling, K., Oberhauser, A., Saltzman, A., and Liao, S.,** Immunochemical characterization of the androgen-dependent spermine-binding protein of the rat ventral prostate, *Biochem. J.*, 218, 513, 1984.
43. **Chang, C., Hiipakka, R., and Liao, S.,** Spermine-binding protein of rat ventral prostate: phosphorylation, cDNA structure, and gene expression, *Abstr. Int. Congr. Polyamines in Life Sciences*, Japanese Society of Polyamine Research, Jikei University School of Medicine, Tokyo, 1986, 43.
44. **Hölttä, E., Pulkkinen, P., Elfving, K., and Jänne, J.,** Oxidation of polyamines by diamine oxidase from human seminal plasma, *Biochem. J.*, 145, 373, 1975.
45. **Bachrach, U.,** Oxidation of polyamines and diamines, in *Polyamines in Biology and Medicine*, Morris, D. R. and Marton, L. J., Eds., Marcel Dekker, New York, 1981, 151.
46. **Pulkkinen, P., Sinervirta, R., and Jänne, J.,** Mechanism of action of oxidized polyamines on the metabolism of human spermatozoa, *J. Reprod. Fert.*, 51, 399, 1977.
47. **Pulkkinen, P., Kanerva, S., Elfving, K., and Jänne, J.,** Association of spermine and diamine oxidase activity with human spermatozoa, *J. Reprod. Fertil.*, 43, 49, 1975.
48. **Pulkkinen, P.,** Specific inhibition of spermatozoal energy metabolism by oxidized spermine, *Contraception*, 17, 423, 1978.
49. **Taylor, P. W. and Morgan, H. R.,** Antibacterial substances in human semen and prostatic fluid, *Surg. Gynecol. Obstet.*, 94, 662, 1952.
50. **Gurevitch, J., Rozansky, R., Weber, D., Brezezinsky, A., and Eckerling, B.,** The role of spermine in the inhibition of *Staphylococcus aureus* by human semen, *J. Clin. Pathol.*, 4, 360, 1951.
51. **Razin, S. and Rozansky, R.,** The responsibility of spermine for the antibacterial action of spermine, *Arch. Biochem. Biophys.*, 81, 36, 1959.
52. **Fair, W. R. and Parrish, R. F.,** Antibacterial substances in prostatic fluid, in *The Prostatic Cell: Structure and Function: Part A*, Murphy, G. P., Sanberg, A. A., and Karr, J. P., Eds., Alan R. Liss, New York, 1981, 247.
53. **Williams-Ashman, H. G.,** Transglutaminases and the clotting of mammalian seminal fluids, *Mol. Cell. Biochem.*, 58, 51, 1984.
54. **Williams-Ashman, H. G., Beil, R. E., Wilson, J., Hawkins, M., Grayhack, J., Zunamon, A., and Weinstein, N. K.,** Transglutaminases in mammalian reproductive tissues and fluids: relation to polyamine metabolism and semen coagulation, *Adv. Enzyme Regul.*, 18, 239, 1980.
55. **Williams-Ashman, H. G., Wilson, J., Beil, R. E., and Lorand, L.,** Transglutaminase reactions associated with the rat semen clotting system: modulation by macromolecular polyamines, *Biochem. Biophys. Res. Commun.*, 79, 1192, 1977.
56. **Fawell, S. E., Pappin, D. J. C., McDonald, C. J., and Higgins, S. J.,** Androgen-regulated proteins of rat seminal vesicle secretion constitute a structurally related family present in the copulatory plug, *Mol. Cell. Endocrinol.*, 45, 205, 1986.
57. **Williams-Ashman, H. G., Notides, A. C., Pabalan, S. S., and Lorand, L.,** Transamidase reactions involved in the enzymic coagulation of semen: isolation of γ-glutamyl-ϵ-lysine dipeptide from dotted secretion protein of guinea pig seminal vesicle, *Proc. Natl. Acad. Sci. U.S.A.*, 69, 2322, 1972.
58. **Lorand, L. and Conrad, S. M.,** Transglutaminases, *Mol. Cell. Biochem.*, 58, 9, 1984.

59. **Williams-Ashman, H. G. and Canellakis, Z. N.**, Transglutaminase-mediated covalent attachment of polyamines to proteins: mechanisms and potential physiological significance, *Physiol. Chem. Phys.*, 12, 457, 1980.
60. **Lilja, H., Laurell, C.-B., and Teppsson, T.-O.**, Characterization of the predominant basic protein in human seminal plasma, one cleavage product of the major seminal vesicle protein, *Scand. J. Clin. Lab. Invest.*, 44, 439, 1984.
61. **Lilja, H. and Laurell, C. B.**, The predominant protein in human seminal coagulate, *Scand. J. Clin. Lab. Invest.*, 45, 635, 1985.
62. **Boonsaeng, V.**, Molecular structure of the human seminal cooagulum: the role of proteolysis, *Andrologia*, 18, 252, 1986.
63. **Lilja, H. and Laurell, C. B.**, Liquefaction of coagulated human semen, *Scand. J. Clin. Lab. Invest.*, 44, 447, 1984.
63a. **Esposito, C., Metafora, S., Fusco, A., Gentile, V., and Porta, R.**, Studies on human semen transglutaminase, in *Biomedical Studies of Natural Polyamines*, Caldarera, C. M., Clô, C., and Guarnieri, C., Eds., Editrice CLUEB, Bologna, Italy, 1986, 205.
64. **Gotterer, G., Ginsburg, D., Schulman, T., Banks, J., and Williams-Ashman, H. G.**, Enzymatic coagulation of semen, *Nature*, 176, 1209, 1955.
65. **Raina, A., Pajula, R.-L., and Eloranta, T.**, A rapid assay method for spermidine and spermine synthases. Distribution of polyamine-synthesizing enzymes and methionine adenosyltransferase in rat tissues, *FEBS Lett.*, 67, 252, 1976.
66. **Williams-Ashman, H. G., Coppoc, G. L., and Weber, G.**, Imbalance in ornithine metabolism in hepatomas of different growth rates as expressed in formation of putrescine, spermidine and spermine, *Cancer Res.*, 32, 1924, 1972.
67. **Russell, D. H. and Durie, B. G. M.**, *Polyamines as Biochemical Markers of Normal and Malignant Growth*, Raven Press, New York, 1978.
68. **Williams-Ashman, H. G. and Canellakis, Z. N.**, Polyamines in mammalian biology and medicine, *Persp. Biol. Med.*, 22, 421, 1979.
69. **Pegg, A. E.**, Recent advances in the biochemistry of polyamines in eukaryotes, *Biochem. J.*, 234, 249, 1986.
70. **Piik, P., Rajamaki, P., Guha, S. K., and Jänne, J.**, Regulation of L-ornithine decarboxylase and *S*-adenosylmethionine decarboxylase in rat ventral prostate and seminal vesicle, *Biochem. J.*, 168, 379, 1977.
71. **Käpyaho, K., Pösö, H., and Jänne, J.**, Role of propylamine transferases in hormone-induced stimulation of polyamine biosynthesis, *Biochem. J.*, 192, 59, 1980.
72. **Mamont, P., Danzin, C., Wagner, J., Siat, M., Joder-Ohlenhusch, A.-M., and Claverie, N.**, Accumulation of decarboxylated *S*-adenosylmethionine in mammalian cells as a consequence of inhibition of putrescine synthesis, *Eur. J. Biochem.*, 123, 499, 1982.
73. **Pegg, A. E., Pösö, H., Shuttleworth, K., and Bennett, R. A.**, Effect of inhibition of polyamine synthesis on the content of decarboxylated *S*-adenosylmethionine, *Biochem. J.*, 202, 519, 1982.
74. **Coffey, D. S., Shimazaki, J., and Williams-Ashman, H. G.**, Polymerization of deoxyribonucleotides in relation to androgen-induced prostatic growth, *Arch. Biochem. Biophys.*, 124, 184, 1968.
75. **Tuohimma, P. and Niemi, M.**, Cell renewal and mitogenic activity of testosterone in male sex accessory glands, in *Male Accessory Sex Organs*, Brandes, D., Ed., Academic Press, New York, 1974, 329.
76. **Coffey, D. S.**, The effects of androgens on DNA and RNA synthesis in sex accessory tissue, in *Male Accessory Sex Organs*, Brandes, D., Ed., Academic Press, New York, 1974, 307.
77. **Williams-Ashman, H. G., Tadolini, B., Wilson, J., and Corti, A.**, Polynucleotide polymerizations and prostate proliferation, *Vitam. Horm.*, 33, 39, 1975.
78. **Lesser, B. and Bruchovsky, N.**, The effects of testosterone, 5α-dihydrotestosterone and adenosine 3′,5′-monophosphate on cell proliferation and differentiation in rat prostate, *Biochem. J.*, 142, 429, 1974.
79. **Takyi, E. E. K., Fuller, D. J. M., Donaldson, L. J., and Thomas, G. H.**, Deoxyribonucleic acid and polyamine synthesis in rat ventral prostate, *Biochem, J.*, 162, 87, 1977.
80. **Tabor, C. W. and Tabor H.**, 1,4-Diaminobutane (putrescine), spermidine and spermine, *Annu. Rev. Biochem.*, 45, 275, 1976.
81. **Tabor, C. W. and Tabor H.**, Polyamines, *Annu. Rev. Biochem.*, 53, 749, 1984.
82. **Ahmed, K., Goueli, S. A., and Williams-Ashman, H. G.**, Characteristics of polyamine stimulation of cyclic nucleotide-independent protein phosphokinase reactions, *Biochem. J.*, 71, 1175, 1985.
83. **Ahmed, K. I., Goueli, S. A., and Williams-Ashman, H. G.**, Mechanisms and significance of polyamine stimulation of various protein kinase reactions, *Adv. Enzyme Regul.*, 25, 401, 1986.
84. **Sugimura, Y., Cunha, G. R., and Donjacour, A. A.**, Morphogenesis of ductal networks in the mouse prostate, *Biol. Reprod.*, 34, 961, 1986.
85. **Sugimura, Y., Cunha, G. R., and Donjacour, A. A.**, Morphological and histological study of castration-induced degeneration and androgen-induced regeneration in mouse prostate, *Biol. Reprod.*, 34, 573, 1986.

86. **Sugimura, Y., Cunha, G. R., Donjacour, A. A., Bigsby, R. M., and Brody, J. R.,** Whole-mount autoradiography of DNA synthetic activity during postnatal development and androgen-induced regeneration of rat prostate, *Biol. Reprod.*, 34, 985, 1986.
87. **Hougaard, D. M., Fujiwara, K., and Larsson, L.-I.,** Polyamine cytochemistry: comparisons between cytochemical, autogradiographic, immunocytochemical and chemical results in the prostate, *Histochem. J.*, 18, 321, 1986.
88. **Persson, K., Rosengren, E., and Sundler, F.,** Localization of ornithine decarboxylase by immunocytochemistry, *Biochem. Biophys. Res. Commun.*, 104, 1196, 1982.
89. **Wilson, J. D., George, F. W., and Griffin, J. E.,** The hormonal control of sexual development, *Science*, 211, 1278, 1981.
90. **Persson, L. and Pegg, A. E.,** Studies on the induction of spermidine/spermine N^1-acetytransferase using a specific antiserum, *J. Biol. Chem.*, 259, 12364, 1984.
91. **Seiler, N., Bolkenius, F. N., Knödgen, B., and Mamont, P.,** Polyamine oxidase in rat tissues, *Biochim. Biophys. Acta*, 615, 480, 1980.
92. **Caldarera, C. M., Zappia, V., and Bachrach, U., Eds.,** *Advances in Polyamine Research*, Vol. 3, Raven Press, New York, 1981.
93. **Meggio, F., Flamingni, F., Caldarera, C. M., Guaneri, C., and Pinna, L. A.,** Phosphorylation of rat heart ornithine decarboxylase by type-2 casein kinase, *Biochem. Biophys. Res. Commun.*, 122, 997, 1984.
94. **Mitchell, J. L. A., Qasha, P., and Mahan, D. W.,** Alternative charge states of mammalian ornithine decarboxylase and their polyamine-induced inactivation, in *Recent Progress in Polyamine Research*, Selmeci, L., Brosnan, M. E., and Seiler, N., Eds., Akademiai Kiado, Budapest, 1985, 55.
95. **Kahana, C. and Nathans, D.,** Isolation of a cloned cDNA encoding mammalian ornithine decarboxylase, *Proc. Natl. Acad. Sci. U.S.A.*, 81, 3645, 1984.
96. **Kontula, K. K., Torkkeli, T. K., Bardin, C. W., and Jänne, O. A.,** Androgen induction of ornithine decarboxylase mRNA in mouse kidney as studied by complementary DNA, *Proc. Natl. Acad. Sci. U.S.A.*, 81, 731, 1984.
97. **McConlogue, L., Gupta, M., Wu, L., and Coffino, P.,** Molecular cloning and expression of mouse ornithine decarboxylase gene, *Proc. Natl. Acad. Sci. U.S.A.*, 81, 540, 1984.
98. **Berger, F. G., Szymanski, P., Read, E., and Watson, G.,** Androgen-regulated ornithine decarboxylase mRNAs of mouse kidney, *J. Biol. Chem.*, 259, 7141, 1984.
99. **Mach, M., White, M. W., Neubauer, M., Degen, J. L., and Morris, D. R.,** Isolation of a cDNA clone encoding *S*-adenosylmethionine decarboxylase, *J. Biol. Chem.*, 261, 11697, 1986.
100. **Kochakian, C. D., Hill, J., and Anouma, S.,** Regulation of protein biosynthesis in mouse kidney by androgens, *Endocrinology*, 72, 354, 1963.
101. **Bardin, C. W. and Caterall, J. F.,** Testosterone: a major determinant of extragenital sexual dimorphism, *Science*, 211, 1285, 1981.
102. **Jänne, O. A., Kontula, K. K., Isomaa, V. V., and Bardin, C. W.,** Ornithine decarboxylase mRNA in mouse kidney: a low abundancy gene product regulated by androgens with rapid kinetics, *Ann. N.Y. Acad. Sci.*, 438, 72, 1984.
103. **Seely, J. E., Pösö, H., and Pegg, A. E.,** Purification of ornithine decarboxylase from kidneys of androgen-treated mice, *Biochemistry*, 21, 226, 1982.
104. **Isomaa, V. V., Pajunen, A. E. I., Bardin, C. W., and Jänne, O. A.,** Ornithine decarboxylase in mouse kidney: purification, characterization and radioimmunological determination of the enzyme protein, *J. Biol. Chem.*, 258, 6735, 1983.
105. **Gupta, M. and Coffino, P.,** Mouse ornithine decarboxylase: complete amino acid sequence deduced from cDNA, *J. Biol. Chem.*, 260, 2941, 1985.
106. **Kahana, C. and Nathans, D.,** Nucleotide sequence of murine ornithine decarboxylase mRNA, *Proc. Natl. Acad. Sci. U.S.A.*, 82, 1673, 1985.
107. **Persson, L., Rosengren, E., Sundler, F., and Uddman, R.,** Immunocytochemical demonstration of ornithine decarboxylase, *Adv. Polyamine Res.*, 4, 571, 1983.
108. **Zagon, I. C., McLaughlin, P. J., Seely, J. E., Hoeksma, G. W., and Pegg, A. E.,** Autoradiographic localization of ornithine decarboxylase in mouse kidney by use of radiolabeled α-difluoromethylornithine, *Cell Tissue Res.*, 235, 371, 1984.
109. **Pajunen, A. E. I., Isomaa, V. V., Jänne, O. A., and Bardin, C. W.,** Androgenic regulation of ornithine decarboxylase activity in mouse kidney and its relationship to changes in cytosol and nuclear androgen receptor concentrations, *J. Biol. Chem.*, 257, 8190, 1982.
110. **Kontula, K. K., Seppanen, P. J., Van Duyne, P., Bardin, C. W., and Jänne, O. A.,** Effect of a nonsteroidal antiandrogen, Flutamide, on androgen receptor dynamics and ornithine decarboxylase gene expression in mouse kidney, *Endocrinology*, 118, 226, 1985.
111. **Seely, J., Pösö, H., and Pegg, A. E.,** Effect of androgens on turnover of ornithine decarboxylase in mouse kidney, *J. Biol. Chem.*, 257, 7541, 1982.

112. **Scheidereit, C., Krauter, P., von der Ahe, D., Janisch, S., Rabenau, O., Cato, A. C. B., Suske, G., Westphal, H. M., and Beato, M.,** Mechanism of gene regulation by steroid hormones, *J. Steroid Biochem.*, 24, 19, 1986.
113. **Alhonen-Hongisto, L., Kallio, A., Sinervita, R., Seppanen, P., Kontala, K. K., Jänne, O. A., and Jänne, J.,** Difluoromethylornithine-induced amplification of ornithine decarboxylase genes in Ehrlich ascites carcinoma cells, *Biochem. Biophys. Res. Commun.*, 126, 734, 1985.
114. **Pohjanpelto, P., Hölttä, E., Jänne, O. A., Knuutila, S., and Alitalo, K.,** Amplification of ornithine decarboxylase gene in response to polyamine deprivation in Chinese hamster ovary cells, *J. Biol. Chem.*, 260, 8532, 1985.
115. **Persson, L.,** Putrescine turnover in kidneys of testosterone-treated mice, *Biochim. Biophys. Acta*, 674, 204, 1981.
116. **Goldstone, A., Koenig, H., and Lu, C.,** Testosterone-dependent sexual dimorphism of the mouse kidney is mediated by polyamines, *Biochem. Biophys. Res. Commun.*, 104, 165, 1982.
117. **Berger, F. G. and Porter, C. W.,** Putrescine does not mediate the androgen-response in mouse kidney, *Biochem. Biophys. Res. Commun.*, 138, 771, 1986.
118. **Macindoe, J. H. and Turkington, R. W.,** Hormonal regulation of spermidine formation during spermatogenesis, *Endocrinology*, 92, 595, 1973.
119. **Tsai, Y.-H. and Lin, S. N.,** Differential release of polyamine by cultured rat Sertoli cells, *J. Androl.*, 6, 348, 1985.
120. **Qian, Z.-U., Tsai, Y.-H., Steinberger, A., Lu, M., Greenfield, A. R. L., and Haddox, M. K.,** Localization of ornithine decarboxylase in rat testicular cells and epididymal spermatozoa, *Biol. Reprod.*, 33, 1189, 1985.
121. **Dias, J. A.,** Transglutaminase activity in testicular homogenates and serum-free Sertoli cell cultures, *Biol. Reprod.*, 33, 835, 1985.
122. **Blixonas, K. X., Dattatreymurty, B., and Dias, J. A.,** A specific and saturable spermine-binding component present in bovine testis membranes, *Biol. Reprod.*, 34, 841, 1986.
123. **Bamberg, E., Weiser, M., and Desser, H.,** Polyamines in bovine epididymal spermatozoa, *J. Reprod. Fertil.*, 45, 363, 1975.
124. **Atmar, V. J., Kuehn, G. D., and Casillas, I. R.,** A polyamine-dependent protein kinase from bovine epididymal spermatozoa, *J. Biol. Chem.*, 256, 8275, 1981.
125. **Matsuzaki, S., Hamana, K., Imai, K., and Matsuura, K.,** Occurrence in high concentrations of N^1-acetylspermidine and *sym*-homospermidine in the hamster epididymis, *Biochem. Biophys. Res. Commun.*, 107, 307, 1982.
126. **Moulton, B. C. and Leonard, S. L.,** Hormonal effects on spermidine levels in male and female reproductive organs of the rat, *Endocrinology*, 84, 1461, 1969.
127. **Russell, D. H. and Taylor, R. H.,** Polyamine synthesis and accumulation in ovariectomized rats after estradiol-17β administration, *Endocrinology*, 88, 1397, 1970.
128. **Cohen, S., O'Malley, B. W., and Stastny, M.,** Estrogenic induction of ornithine decarboxylase *in vivo* and *in vitro*, *Science*, 170, 336, 1970.
129. **Kaye, A. M., Icekson, I., and Lindner, H. R.,** Stimulation by estrogens of ornithine and *S*-adenosylmethionine decarboxylases in immature rat uterus, *Biochim. Biophys. Acta*, 252, 150, 1971.
130. **Rorke, E. A., Kenra, K. L., and Katzenellenbogen, B. S.,** Relationships among uterine growth, ornithine decarboxylase activity and polyamine levels: studies with estradiol and antiestrogens, *Mol. Cell. Endocrinol.*, 38, 31, 1984.
131. **Rorke, E. A. and Katzenellenbogen, B. S.,** Dissociated regulation of growth and ornithine decarboxylase activity by estrogen in the rat uterus, *Biochem. Biophys. Res. Commun.*, 122, 1186, 1984.
132. **Marshall, K. and Senior, J.,** Ornithine decarboxylase inhibition and the aterotrophic response to oestrogen in rats, *J. Reprod. Fertil.*, 76, 597, 1986.
133. **Cheng, S. V. Y. and Pollard, J. W.,** c-*ras*^{H} and ornithine decarboxylase are induced by oestradiol-17β in the mouse uterine luminal epithelium independently of the proliferative status of the cell, *FEBS Lett.*, 196, 309, 1986.
134. **Katz, A. and Kahana, C.,** Isolation and characterization of the mouse ornithine decarboxylase gene, *J. Biol. Chem.*, 263, 7604, 1988.
135. **Pegg, A. E., Wiest, L., and Pajunen, A.,** Detection of proenzyme form of *S*-adenosylmethionine decarboxylase in extracts from rat prostate, *Biochem. Biophys. Res. Commun.*, 150, 788, 1988.
136. **Danzin, C. and Mamont, P. S.,** Polyamine inhibition *in vivo* and in organ growth and repair, in *Inhibition of Polyamine Metabolism*, McCann, P. P., Tegg, A. E., and Sjoerdsma, A., Eds., Academic Press, New York, 1987, 141.
137. **Fozzard, J. R.,** The contragestational effects of ornithine decarboxylase inhibition, in *Inhibition of Polyamine Metabolism*, McCann, P. P., Pegg, A. E., and Sjoerdsma, A., Eds., Academic Press, New York, 1987, 187.

138. **Park, M. H. and Folk, J. E.,** Biosynthetic labelling of hypusine in mammalian cells, *J. Biol.Chem.*, 261, 14108, 1986.
139. **Park, M. H.,** Regulation of biosynthesis of hypusine in Chinese hamster ovary cells, *J. Biol. Chem.*, 262, 12730, 1987.

Chapter 2

POLYAMINES AND HUMAN REPRODUCTION

José D. Méndez

TABLE OF CONTENTS

I. INTRODUCTION

Mammalian cells contain significant amounts of the polyamines putrescine, spermidine, and spermine, which play different roles in various tissues. Although the physiological function of these amines is still not well understood at the molecular level, recent studies have shown that their concentrations are tightly regulated and that normal cellular growth, multiplication, and differentiation require polyamines.[1] In several systems the requirement of these aliphatic amines for growth and differentiation has been established and data supporting their role in DNA, RNA, and protein biosynthesis and/or regulation are extensive. During embryogenesis and in other rapidly growing systems cell proliferation is closely related to polyamine biosynthesis and accumulation.[2]

Most of the polyamine biosynthetic enzymes are localized in the cytosol. However, some controversy has arisen over the subcellular distribution of ornithine decarboxylase (ODC) which catalyzes the initial step in polyamine biosynthesis. Studies in which enzyme activity was assayed in extracts of different subcellular fractions suggested that ODC is present almost exclusively in the cytosolic compartment of the cells.[3-6] Recent biochemical studies indicated that ODC is present in the nucleolus and plays an important role in its activity.[7] Emanuelsson and Heby, using complex formation with labeled D-L-α-difluoromethylornithine (DFMO), an enzyme-activated irreversible inhibitor of ODC,[8] demonstrated that ODC is present not only in the cytoplasm, but also in the nucleoplasm of metabolically active polychaete cells.[9] Moreover, it has been suggested that ODC plays a multifunctional role in both polyamine synthesis and ribosomal gene expression.[7,9] Polyamines increase the efficiency of several biosynthetic reactions during the cell cycle.[10,11] The biosynthetic pathway for polyamines has been extensively studied and a great number of functions have been attributed to these substances.[12-14] In addition, numerous hormones increase the rate of polyamine biosynthesis in specific target tissues, suggesting that polyamines may in part function as mediators of hormone action.

In mammals ODC catalyzes the formation of putrescine from ornithine. This reaction is described elsewhere.[12-15] Another decarboxylase, *S*-adenosyl-L-methionine decarboxylase (SAMD), and spermine and spermidine synthases are also needed for polyamine biosynthesis. Amine oxidases are implicated in polyamine catabolism.[1,12,16]

Ornithine is available for these reactions and can be formed within the cell by the action of arginase.[17] It is possible that arginase, which plays an important role in the urea cycle, is present in extrahepatic tissues to supply the ornithine required for polyamine synthesis. For this reason it is believed that arginase may be one of the enzymes which regulate the initial phase of polyamine biosynthesis, in addition to its well-known function in the urea cycle.

In recent years direct evidence has been obtained that ODC (and/or polyamine generated through its action) play an essential role in the early phases of reproductive processes. In mice, rat, and rabbits, oral administration of DFMO results in the inhibition of ODC activity and in a subsequent block in embryonic development.[2] Similar results are observed during the i.p. or intrauterine administration of DFMO into rats.[18,19]

The role of polyamines in human reproductive systems has been reviewed by Williams-Ashman[20,21] and Shet and Moodbidri.[22]

In this review the role of polyamines in human reproductive physiology will be discussed.

II. POLYAMINE METABOLISM IN THE WOMAN

In women, the polyamine metabolism has been studied during normal menstrual cycle, in maternal blood, urine, placenta, and amniotic fluid as normal pregnancy progresses, as well as in women with spontaneous abortion and preeclampsia.

Table 1
ODC ACTIVITY AND POLYAMINE LEVELS IN TISSUES OF NORMAL HUMAN ENDOMETRIUM AND MYOMETRIUM

Tissue sample	ODC activity (picomoles of putrescine per milligram of protein)	Polyamines (picomoles per milligram of DNA)		
		Putrescine	Spermidine	Spermine
Normal endometrium (n = 27)	723 ± 286.4	40.2 ± 17.1	94.5 ± 31.2	125.0 ± 36.1
Normal myometrium (n = 12)	Nondetectable	10.8 ± 3.5	62.5 ± 19.1	113.2 ± 30.9

Note: Values are mean ± SD and n = number of samples.

From Romano, M., Cecco, L., Santacrose, M. A., Cerra, M., Rasquinelli, R., Pagnano, A. M., and Paladini, A., *Advances in Polyamines in Biomedical Sciences,* Caldarera, C. M. and Bachrach, U., Eds., CLUEB, Bologna, Italy, 1984, 145.

A. Menstrual Cycle

ODC activity and polyamine levels have been determined in normal human endometrium.[23] ODC activity is high in the normal endometrium, and is low in the myometrium (Table 1). The activity of ODC in the endometrium has been associated with high biological activity due to its continuous morphological and functional modifications. In contrast, the myometrium, having a different biological activity, showed no detectable ODC activity and lower values of polyamine concentration. The concentrations of polyamines are higher in the endometrium than in the myometrium. In both, spermine concentrations are higher than those of spermidine and spermidine is higher than putrescine.

Osterberg et al.[24] studied the fluctuations in urinary putrescine, spermidine, and spermine during the menstrual cycle of healthy women and found that the excretion of all three polyamines (''total polyamines'') was higher during menstruation.[24] In some cases urinary polyamines were elevated during the early follicular phase.[24] In addition to the increased polyamine excretion observed during menstruation all the women studied by these authors exhibited one or several midcycle peaks in polyamine excretion during the expected time of ovulation. During the luteal and follicular phases, polyamines accumulated in the urine of some women. However, the elevated levels of urinary polyamines in these phases was not consistently present in all women and does not appear to reflect events related to the menstrual cycle. Rather, these peaks may be a function of diet. Preliminary observations suggested that the composition of the food may modify the excretion of urinary polyamines.[24] It is tempting to speculate that the increase in urinary polyamines during menstruation may be related to the necrosis of endometrial cells. Indeed, polyamines were shown to accumulate in the extracellular fluid as a result of cell death.[25]

B. Polyamine Levels in Placenta

Polyamine concentrations increase in the plasma and urine of pregnant women as a function of the gestational age.[26,27] The placenta plays a central role in fetal growth and development; therefore, this tissue could be a primary target for the increased polyamine biosynthesis observed during pregnancy. Porta et al.[28] studied the polyamine content of human placenta at different stages of fetal development. Figure 1 shows that the highest concentrations of putrescine in the placenta were found soon after its formation. During placental development putrescine levels declined and were negligible prior to term. Spermidine concentrations declined during gestation and remained fairly constant at low levels from the 25th week of pregnancy. On the other hand, spermine levels increased continuously almost until term.

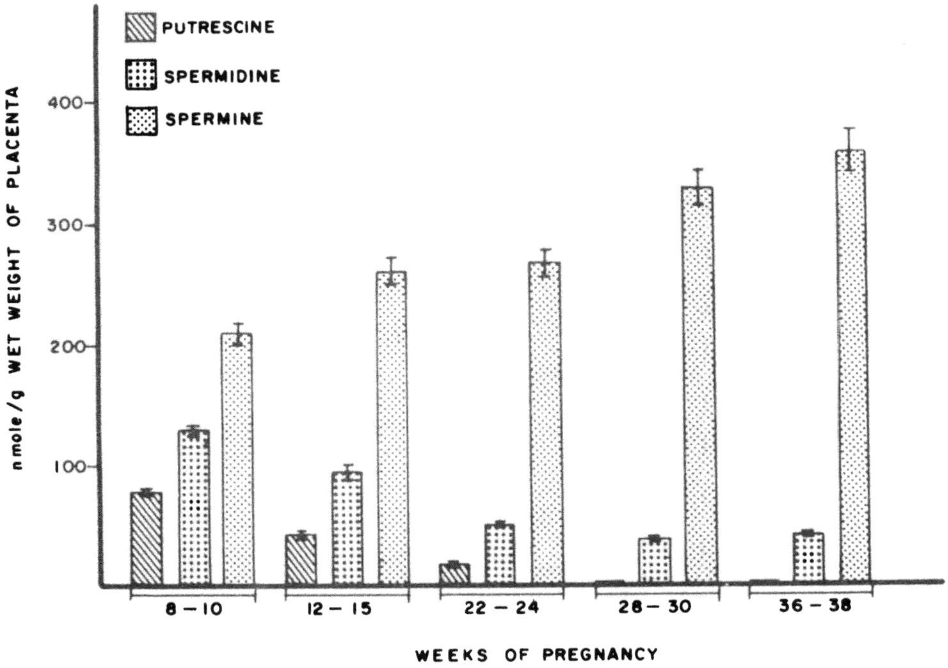

FIGURE 1. Polyamine pattern in human placenta during pregnancy. (From Porta, R., Servillo, L., Abbruzzese, A., and Della Pietra, G., *Biochem. Med.*, 19, 143, 1978.)

The increased synthesis of polyamines and their accumulation in human placenta may be related to the extensive protein synthesis required for growth and the production of protein hormones. These may be associated with early events in the development of the placenta and to its function in the later part of the pregnancy.[27]

Recent studies using physiological concentrations of polyamines indicated that these amines stimulate the phosphorylation of specific proteins in human placenta extracts.[29] The potency order of the three polyamines, spermine > spermidine > putrescine coincided with their levels in the placental tissues during gestation.[28] The spermine-induced phosphoproteins in human placenta are distinct from both cAMP and Ca-dependent phosphorylations. Spermine inhibits cAMP and Ca-dependent phosphorylation in the placenta. Polyamines may therefore act both directly by mediating phosphorylation by a polyamine-dependent reaction, or indirectly by inhibiting phosphorylation induced by other known inducers. These observations raise the possibility that polyamines or their metabolites are primary effectors for a specific cascade of events associated with pregnancy.[29]

C. Amine Oxidases and Pregnancy

Enzymes involved in the oxidation of amines have long been associated with pregnancy. Thus diamine, spermidine, and polyamine oxidase activities showed a progressive increase in gestational age to 21 weeks, or beyond, declining to very low levels in the 3 to 4 days postpartum.[30-33] The placenta is usually regarded as the source of these enzymes, although some papers suggest that may be of decidual origin.[34] Several attempts have been made to purify and characterize the placental amine oxidases and determine their substrate specificity.[35,36] It was observed that both a mono- and diamine oxidase from placental homogenates oxidized several amines, although the affinity of enzyme for substrate varied considerably. Thus, it is believed that there is a wide group of catalytic proteins capable of oxidizing amines, and another which uses polyamines or their acetylated derivatives as substrates.

Morgan[37] has proposed that this apparent lack of specificity may be a consequence of the lack of data on the conditions in which these enzymes function in vivo, leading to confusion upon the definition and classification of amine oxidases in the literature. Amine oxidases have been classified according to their prosthetic groups in two classes: monoamine oxidases or flavin adenine dinucleotide-amine oxidases (FAD-AO) and the copper-amine oxidases (Cu-AO) or diamine oxidases.[38] The term polyamine oxidases was employed to describe amine oxidases able to catalyze oxidative deamination of spermidine and spermine regardless of enzymes acting on mono- or diamines as substrates.[16, 34]

In human pregnancy, diamine oxidase which oxidizes both histamine and putrescine circulates in maternal serum. Its levels rise during the progress of pregnancy. This enzyme originally was named histaminase as it used histamine as substrate. Later it was named diamine oxidase due to its activity on putrescine.[40]

In other investigations, Gahl and Pitot[41] demonstrated that the enzyme obtained and partially purified from human pregnancy serum has affinity for putrescine and spermidine, and it acted on a wide range of substrates of which N^1-acetylspermidine was the best. This suggests that N^1-acetylspermidine constitutes an important metabolite in human pregnancy because this molecule may be recycled to maintain the polyamine pool. Moreover, amine oxidase could participate in the conversion of products that might be toxic for the organism, as it has been proposed for the polyamine oxidase from rat liver, which also uses N^1-acetylspermidine as substrate.

Gahl et al.[32] also reported the presence of spermidine oxidase in serum of pregnant women. They used labeled spermidine as substrate and separated the reaction products by ion-exchange chromatography. They determined some kinetic constants which were similar to those of diamine oxidase.

The activity of human serum spermidine oxidase can be detected during pregnancy as early as 8 weeks after the last menstrual cycle and increase with gestational age parallel to the increase in the amine oxidase activity, reaching a plateau at 20 weeks of gestation. On the other hand, spermine oxidase activity has been detected in maternal serum at the 10th week of gestation. This suggests that it may be the same enzyme, the polyamine oxidase.[33]

Polyamine oxidase activity has also been measured in amniotic fluid and in fetal membranes between 15 and 40 weeks of gestation.[42] Enzyme activity increased as pregnancy progressed, similar to that found in maternal serum, although the levels were lower (Figure 2). Fetal membranes, corion and amnion, also showed the presence of enzyme activity, with lower levels compared with those of decidua, but significantly higher than those found in placenta. This suggested that the enzyme found in amniotic fluid might be the consequence of a diffusion from the decidua through the membranes.[42]

D. Polyamine Oxidase and Abortion

Polyamine oxidase activity is highest in the intervillous circulation where the first and closest contact between fetal and maternal surfaces takes place.[33,43] This led to the hypothesis that polyamine oxidase has a protective function in the physiology of pregnancy and that amine oxidase at elevated levels protects mother and fetus from biogenic amines at high concentrations. Another alternative hypothesis arose from the suggestion that the placenta liberates an immunosuppresive factor which "switches off" potentially harmful maternal lymphocytes.[37] Some studies in vitro have shown that the action of polyamine oxidase on polyamines yields noncytotoxic compounds which are inhibitors of cell proliferation.

Human pregnancy serum also inhibits lymphocyte proliferation in vitro in the presence of added spermine,[44] the effect being proportional to the polyamine oxidase content.[45] Several observations, such as the suppresive effects of polyamines spermidine and spermine on components of the immune system in vitro and the observation that the placenta is rich in spermine and in polyamine oxidase activity in retroplacental blood, suggest that products

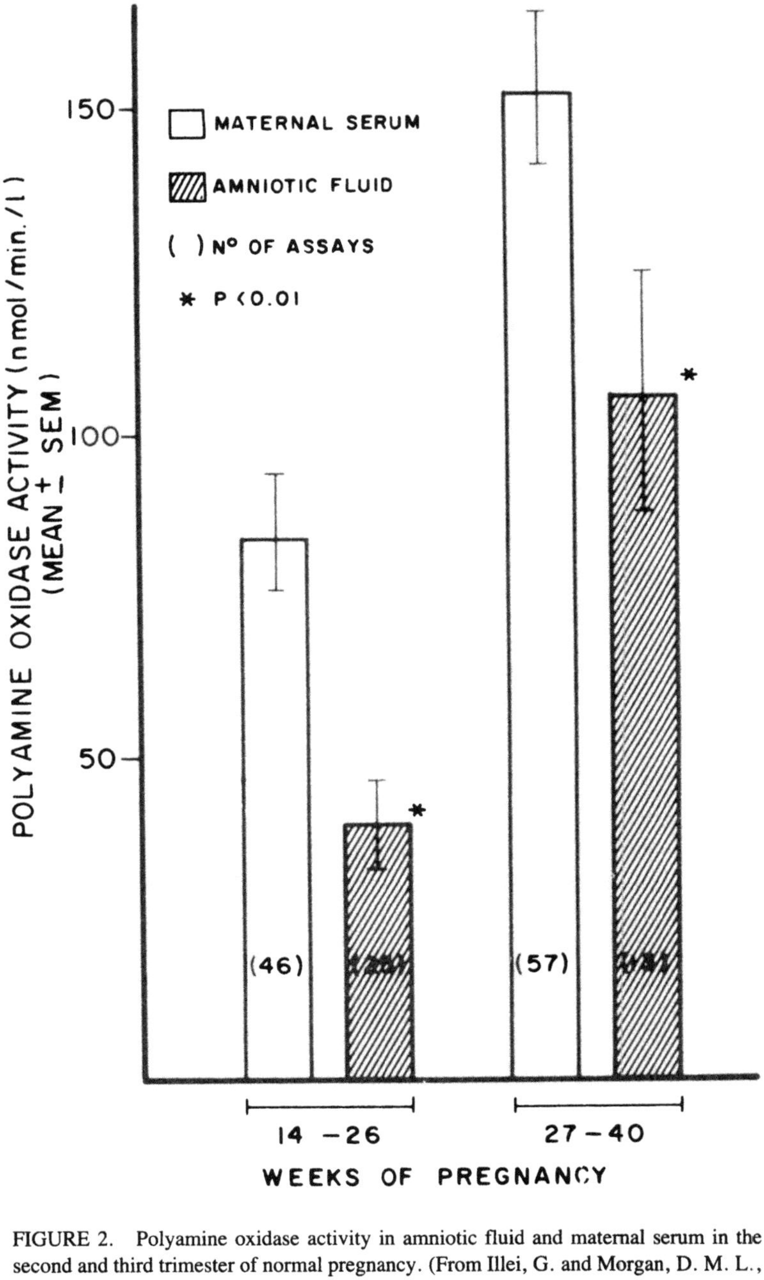

FIGURE 2. Polyamine oxidase activity in amniotic fluid and maternal serum in the second and third trimester of normal pregnancy. (From Illei, G. and Morgan, D. M. L., *Br. J. Obstet. Gynaecol.*, 87, 413, 1980.)

of the interaction of polyamines and polyamine oxidase contribute to the protection of fetus against maternal immune rejection because of the possible involvement of immunological factors in recurrent and spontaneous abortion.[46,47]

Illei and Morgan[48] studied a group of patients who aborted spontaneously between 11 and 22 weeks of pregnancy. The results of this investigation indicated that serum polyamine oxidase activity was significantly lower in patients with spontaneous abortion than the control group (Figure 3). Although these authors pointed out that the low levels of this enzyme

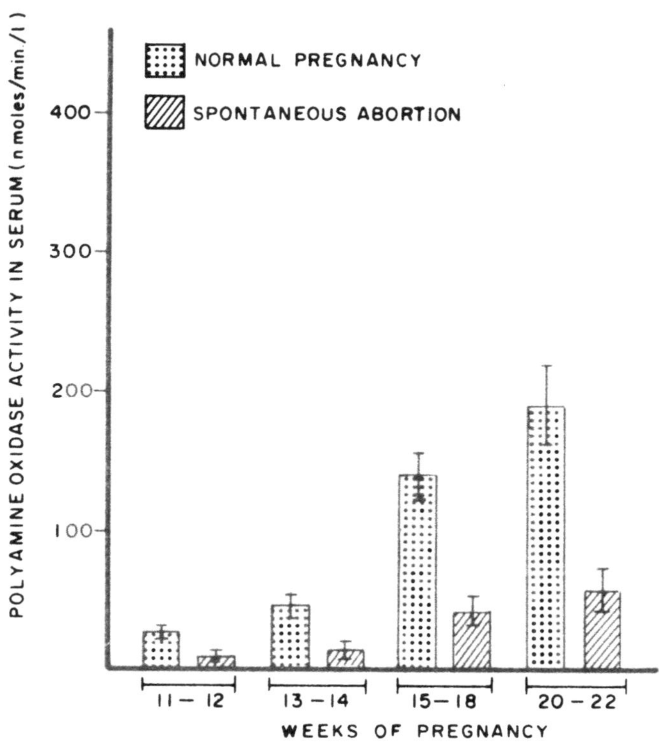

FIGURE 3. Polyamine oxidase activity in serum during normal pregnancy and spontaneous abortion. (From Illei, G. and Morgan, D. M. L., *Br. J. Obstet. Gynaecol.*, 89, 199, 1982.)

found in women that had an abortion were not necessarily linked to each other, this suggests that certain levels of polyamines-polyamine oxidase interaction may play an important role in the maintenance of normal pregnancy by protecting the fetus from maternal rejection, thus participating in the embryonic growth. Other evidence[49] also supports the notion that malfunction of a localized immunological barrier may be related to reduced amine oxidase activity and may lead to spontaneous abortion.[37]

E. Preeclampsia

Preeclampsia, also named gestosis or toxemia, usually appears in the woman after the 24th week of pregnancy. In some cases it appears in the first 24 hr postpartum.[50] Although the etiology of preeclampsia is unknown, its pathophysiology and impact produced in target organs are well defined.[51] Toxemic patients show hypertension concomitant with severe alterations in vascular structure, proteinuria, and edema. In complicated cases convulsions and coma are found. Several pathological manifestations have been associated with this syndrome, such as low platelet count, alterations in coagulation mechanisms, increased plasma viscosity, elevated levels of immunoglobulin G, and rheumatoid factor.[52] Furthermore, changes in nucleic acid and protein levels have been observed in placentae.[53]

In uterine veins of toxemic women trophoblast levels are 20-fold higher compared with those found in normal pregnant women. This may induce a challenge of the immunologic response of the mother vs. the fetus. It was also demonstrated that migration of macrophages may be inhibited by toxemic microsomal fractions of placental origin.[54]

It has been well established that polyamines may modify the structure and function of nucleic acids. Thus, alterations in polyamine metabolism may result in an abnormal synthesis

Table 2
POLYAMINE OXIDASE IN PREGNANCY SERUM OF NORMAL AND TOXEMIC WOMEN

Women	Polyamine oxidase activity (picomoles of Δ^1-pyrroline per milligram of protein per hour)
Normal (n = 20)	6.91 ± 2.53
Toxemic (n = 20)	8.48 ± 3.32

Note: Values are mean ± SD and n = number of women.

of proteins and other molecules, such as blood circulating in an immune complex way. In preeclamptic women, the permeability of cellular membranes is altered, causing a protein leak out to the urine. This process may be regulated by polyamines.[55,56]

During normal pregnancy, polyamine oxidase increases together with spermine, while the concentrations of putrescine and spermidine are reduced. It is possible that spermidine binds to enzyme to produce the complex polyamine oxidase-polyamine which exerts an inhibitory effect regulating the placental growth and limiting their extension to avoid the myometrium invasion. In toxemic women, the myometrial vessels show swelling in the intimate layer and hyperplasia in the muscular layer;[50] by this reason an abnormality in the trophoblast invasion is suspected.[37] These changes in spiral vessels insure the intervillous irrigation-provoking vasoconstriction that rises the pressure and increases the vascular permeability, in addition to causing an abnormal interaction between platelet and vessel endothelium.

To gain a better understanding of the correlation between polyamine metabolism and preeclampsia, the activity of serum polyamine oxidase was determined in preeclamptic patients. These patients, who were selected by their arterial pressure of 160/110 mg Hg or higher, maintained for a time of 6 hr associated to edema extended to abdomen or generalized, and proteinuria higher than 3.0 g/ℓ.[57] The symptoms were diagnosed only after the 35th week of reliable amenorrhea, and the cases of toxemia postpartum were excluded. The control group consisted of women with normal pregnancy. The values for polyamine oxidase activity obtained in preeclamptic patients showed no significant differences compared with the control group (Table 2). It is however recommended that the activity of the enzyme should be measured from the first trimester of pregnancy to determine whether the enzyme undergoes early significant changes.

Between weeks 39 and 41 (65%) and week 43 (5%) high incidences of toxemia cases were found, while in women with normal pregnancy it was uniformly distributed along the third trimester (Figure 4). This observation suggests that weeks 39 to 41 are the critical weeks of pregnancy in which toxemia appears.

1. Perspectives on the Inhibition of Uterine Ornithine Decarboxylase Activity

The essential roles of polyamines in normal and neoplastic growth in culture and in animal tissues led to the development of specific inhibitors of polyamine synthesis.[11,13] Until very recently the biosynthesis and hence the tissue concentrations of polyamines were not significantly inhibited by those drugs. However, with the synthesis of DFMO,[8] the situation changed dramatically. This compound irreversibly and specifically inhibits ODC (LD_{50} mice and rats: PO 5 g/kg, IP 3 g/kg). It is capable of depleting cellular polyamine concentrations in vivo.[58-60] DFMO has been used extensively to explore the biological roles of polyamines in a variety of physiological and pathological processes. It has been logical, therefore, to explore the physiological significance of the polyamines, using DFMO in systems exhibiting rapid cell growth and differentiation.

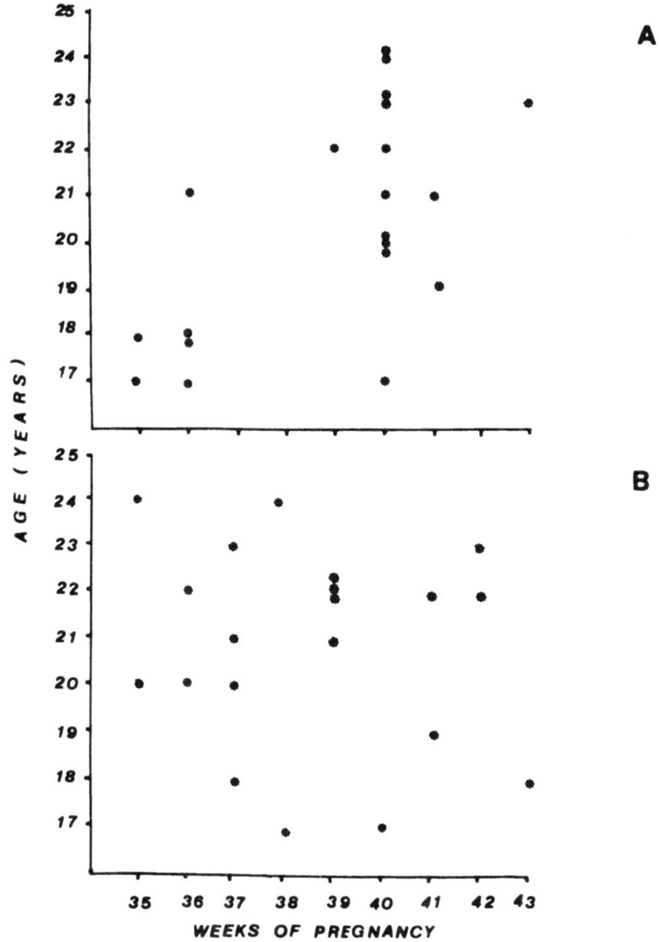

FIGURE 4. Between weeks 39 and 41 high incidences (65%) of toxemia cases were found (A). In women with normal pregnancy it was uniformly distributed along the third trimester (B).

Early mammalian embryogenesis represents one of the most actively dividing and differentiating cell systems, and numerous studies have described a substantial increase in ODC activity associated with this process.[2] In the mouse uterus, ODC activity begins to increase shortly after nidation and reaches a peak on day 8 of gestation. During this time, putrescine and spermidine concentrations also increase, but spermine levels are hardly affected. The peak of the biochemical changes corresponds to the sudden increase in embryonic growth. This is associated with early somite formation which in the mouse takes place on days 7 to 8. The effects of treatment with DFMO during days 5 to 8 of gestation were unequivocal; the increase in ODC activity and putrescine and spermidine concentrations was abolished and embryonic development failed to progress beyond day 7. The arrested embryo was subsequently resorbed or lost from the uterus.[2,61] These effects of DFMO can be completely reversed by the simultaneous administration of putrescine whose half-life is relatively long,[62] implicating inhibition of putrescine biosynthesis as the mechanism of the countergestational effect. DFMO also inhibits early embryonic development in rats when administered in pre- and postimplantation period[2,18,19] and in rabbits during postimplantation period.[2] Thus, an increase in ODC activity leading to a rapid rise in putrescine concentration appears to be absolutely essential during a critical period after implantation for continued mammalian embryonal growth.

On this basis, DFMO has also been proposed as a postcoitally effective antifertility agent in females,[18] which provides a new research area in contraception.

On the other hand, several studies have been carried out on inhibition of tumor growth by DFMO.[63,64] The therapeutic value obtained with experimental animals can be applied clinically, since altered polyamine levels may be involved in malignant tissue transformation.[65] For example, it has been reported that in endometrial carcinoma, polyamine levels are higher than in normal endometrium.[23]

III. PHYSIOLOGY OF POLYAMINES IN THE MALE REPRODUCTIVE SYSTEM

Research on physiological roles of polyamines related with the male reproductive tract has been discussed in numerous papers. Most of the information obtained comes from laboratory experiments with animals. In the reproductive systems of men the presence of polyamines has attracted the attention of many investigators, who have studied particularly the polyamine metabolism in prostate, semen, and testis. Recently, polyamines have been associated with capacitation and fertilization processes.

A. Testis

Susuki et al.[66] determined the activity of polyamine oxidase in several human organs. It was highest in the liver, followed by the testis (339 ± 80 nmol/g wet weight per 30 min). High activity in testis may be related to the process of spermatogenesis.[67]

B. Prostate

In man and in some other mammals, seminal spermine is derived from the prostate gland.[68] Analysis of polyamines in whole prostatic tissue from sexually mature animals revealed that large amounts of spermidine and spermine are present in the prostate.[69] Other observations indicate that various lobes of prostate gland of rabbit, mouse, guinea pig, and dog exhibit spermidine and spermine contents that fall within the range found in most other nongenital tissues. Ventral and dorsolateral lobes of the rat prostate are indeed depositories of large amounts of spermidine and spermine, similar to those found in human prostate.[70]

The enzymes involved in polyamine biosynthesis, ODC,[71] SAMD,[72] spermidine synthase, and spermine synthase have been purified from rat prostate,[73] and to some extent from human prostate (particularly SAMD).[74]

C. Semen

Normal human semen contains spermine in concentrations of 5 to 15 m*M*,[68,75] and smaller amounts of spermidine are present in human seminal fluid,[75] which also contains low but readily detectable quantities of putrescine and 1,3-diaminopropane. The molar ratio of spermine to spermidine in human semen is greater than 12.[21]

The fact that the concentration of spermine in human semen is much higher than in any other tissues and body fluid, led to various speculations as to their physiological role. It was found that spermine does not affect the motility of human and bovine sperm in vitro.[76] Other reports indicated that spermine has neither a beneficial nor a deleterious effect on the motility or metabolism of mammalian spermatozoa when added in vitro.[68] On the other hand, some investigators reported that spermine activates the motility of human and rabbit spermatozoa.[77,78] Evidence has been obtained by in vitro studies that spermine also enhances the activity of seminal maltase, which is involved in the degradation of glycogen, and which increases glucose utilization by sperm, at the same time reducing fructose utilization.[22]

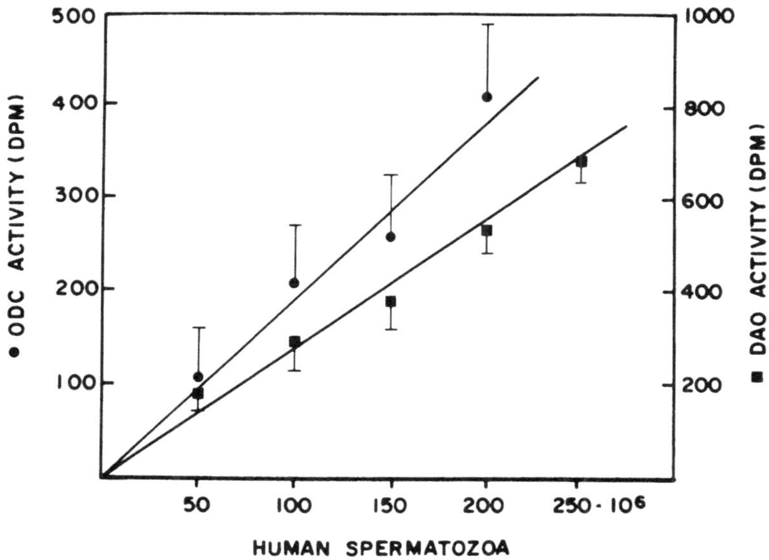

FIGURE 5. Ornithine decarboxylase (ODC) and diamine oxidase (DAO) activities at different concentrations of human spermatozoa.

Addition of physiological amounts of spermine to spermatozoal suspensions caused a significant increase in cAMP levels,[22] a nucleotide that induces capacitation in human spermatozoa.[79] The effect of spermine on cAMP levels may be explained by the fact that this polyamine inhibits phosphodiesterase activity of cAMP, besides activating adenylate cyclase activity.[80]

It has also been observed that the effect of spermine on cAMP levels could be further enhanced by prostaglandin E_2 (PgE_2), and has been demonstrated that PgE_2 produces stimulation of ODC activity.[81]

1. Correlation of Seminal Polyamine and Metabolizing Enzyme Levels with Sperm Count and Motility

Some attempts have been made to elucidate a physiological or pathological role of spermine and its metabolic enzymes by studying the correlation between polyamine concentrations in the human seminal plasma and sperm count and motility. A significant correlation between the spermine concentration of human seminal plasma and the total sperm count in the same semen specimens was noted,[82] although this correlation was not confirmed.[83]

A complex type of correlation betwen ODC, SAMD,[84] and diamine oxidase activities and sperm count has also been observed.[83] The enzyme activities were high in semen samples with sperm counts of less than 30 million per milliliter and those samples in which the sperm count exceeded 60 million per milliliter, and low in samples having sperm counts between 30 to 60 million per milliliter.

Some samples showing poor initial sperm motility have been found to contain high diamine oxidase activity and low spermine content.[85] A positive correlation between the spermine content of seminal plasma and motility has also been noted.[82] These observations led to the speculation that the spermine in semen might have a physiological role with respect to spermatozoa via its oxidized derivatives, which have a high affinity for sperm.[86] It has been observed that human seminal plasma and sperm contain both ODC and diamine oxidase activities[83,87] (Figure 5). The latter is capable of oxidizing polyamines and might produce alterations in quantitative determination of polyamines.[88] The possibility that the spermine normally present in human semen might be oxidized by enzymes to aldehyde products that are known to be toxic for a variety of eukaryotic cells,[12] including human sperm,[69] might be important for the metabolism survival and fertilizing capacity of postejaculatory sperm. Recent studies indicated that polyamines stimulate the acrosome reaction in guinea pig.[89]

2. Inhibition of Human Spermatozoa Ornithine Decarboxylase and its Effect on In Vitro Fertilization

Experiments on fertilization were carried out in Biggers medium (pH 7.4) containing 0.2% bovine serum albumin and 5% CO_2 atmosphere, incubated at 37°C. Of 40 ovocytes, 2 fractions obtained from adult hamsters superovulated with gonadotropin were incubated in presence of 1.6 million per milliliter of untreated spermatozoa (control) or with preincubated spermatozoa in presence of 100 μg DFMO. Fertilization was studied at 2, 4, and 8 hr. ODC of treated spermatozoa was inhibited by approximately 75% within 30 min.[90] Fertilization results indicated a formation of male pronucleus fourfold higher in the case of fertilized oocytes with treated spermatozoa. Motility and affinity of the spermatozoa to oocyte was not modified by this treatment. These results suggest that ODC and/or the generated products of its action might have a regulatory role on postfertilization processes.

IV. URINARY POLYAMINES IN PROSTATE CANCER

Carcinoma of the prostate is considered a cause of high incidence of mortality. Some biochemical markers have been used for their early detection and treatment.[91]

Of particular interest is the fact that the highest concentrations of polyamines found in the human body occur in the prostate.[68] This observation, plus reports of elevated urinary polyamines in some patients with far advanced malignancies of other organ systems,[92] led to investigation of the relationship between prostate carcinoma and urinary excretion of polyamines.

In early studies, Fair et al.[93] observed that patients with prostatic cancer excrete more spermidine than normal controls. The increase in spermidine excretion noted in those patients with more undifferentiated neoplasms was interesting. Of the 34 urine specimens from patients with grade II or higher cancer, 30 had elevated spermidine levels. This finding was in contrast with the observation that only one of ten urine specimens in patients with grade I histologic changes showed increased spermidine levels. These results tend to indicate that a grade I carcinoma appears to be a more slowly growing lesion and, hence, might be expected to have less biochemical activity, such as ribonucleic acid synthesis and polyamine production, than grade III or IV lesions.

In other patients, the elevated spermidine values found in three of four patients with a single prostatic nodule indicate that polyamine levels may be of value as a test for detecting prostatic malignancy at a potentially curable stage.

As a screening test for prostatic carcinoma, urine represents an obvious advantage because it is readily obtained without involving physician time.[93]

V. POLYAMINES IN PROSTATITIS TREATMENT

The high content of polyamines in the human prostatic hyperplasia was shown early by Dunzendorfer and Russell.[94] Subsequently it has been reported that DFMO decreases the normal weight gain of the ventral prostate and seminal vesicles of growing rats and blocks the testosterone-induced regrowth of ventral prostate in castrated rats.[95] The inhibition of ODC by DFMO was studied in patients with chronic nonsuppurative prostatitis.[96] Partial clinical response and remission of gross hematospermia was evident after 18 g/day of DFMO, when administered orally for 1 month. During the DFMO treatment period, a substantial decrease in prostate size was noticed. In two out of ten patients, a 16% reduction of hematocrit and hemoglobin was reported. This was reversed 2 months after stopping the therapy.[96]

Administration of DFMO in the drinking water (20 g/ℓ) to adult rats for 3 days led to marked changes in polyamine metabolism in ventral prostate.[97] A 90% inhibition of ODC activity is accompanied by an approximately 80% reduction of the concentration of putrescine and spermidine and by a 36% decrease in spermine. These observations may partly explain the mechanism of DFMO in human-treated prostatitis.

VI. SUMMARY AND CONCLUSIONS

It is evident that polyamines participate in normal and pathological processes. Their highest levels are found in those tissues with increased metabolic activity. The presence and participation of polyamines in human reproductive systems have been well established, and it is known that both physiological and biochemical processes are regulated by polyamines.

1. The endometrium which undergoes continuous morphological and functional changes contains ODC at high activities.
2. Polyamines stimulate the phosphorylation of specific proteins in the placenta.
3. Polyamines apparently have a protective function in physiology of pregnancy.
4. Polyamines participate in other situations such as preeclampsia patients.
5. The selective inhibition of polyamine biosynthesis in early pregnancy may have practical applications, also as the treatment of malignancies of the reproductive organs.

REFERENCES

1. **Pegg, A. E. and McCann, P. P.,** Polyamine metabolism and function, *Am. J. Physiol.,* 243, C 212, 1982.
2. **Fozard, J. R., Part, M. L., Prakash, N. J., Grove, J., Schechter, P. J., Sjoerdsma, A., and Koch-Weser, J.,** L-Ornithine decarboxylase: an essential role in early mammalian embryogenesis, *Science,* 208, 505, 1980.
3. **Mc Cormick, F.,** Polyamine metabolism in enucleated mouse L-cells, *J. Cell Physiol.,* 93, 285, 1977.
4. **Murphy, B. J. and Brosman, M. E.,** Subcellular localization of ornithine decarboxylase in liver of control and growth-hormone treated rats, *Biochem. J.,* 157, 33, 1976.
5. **Ono, M., Inoue, H., Suzuki, F., and Takeda, Y.,** Studies on ornithine decarboxylase from the liver of thioacetamide-treated rats. Purification and some properties, *Biochim. Biophys. Acta,* 284, 285, 1972.
6. **Pegg, A. E. and Williams-Ashman, H. G.,** Biosynthesis of putrescine in the prostate gland of the rat, *Biochem, J.,* 108, 533, 1968.
7. **Atmar, V. J. and Kuehn, G. D.,** Phosphorylation of ornithine decarboxylase by a polyamine-dependent protein kinase, *Proc. Natl. Acad. Sci. U.S.A.,* 78, 5518, 1981.
8. **Metcalf, B. N., Bey, P., Danzin, C., Jung, M. J., Casara, P., and Vevert, J. P.,** Catalytic irreversible inhibition of mammalian ornithine decarboxylase (E.C. 4.1.1.17) by substrate and product analogs, *J. Am. Chem. Soc.,* 100, 2551, 1978.
9. **Emanuelsson, H. and Heby, O.,** Nuclear ornithine decarboxylase. Electron microscope autoradiographic identification of the active enzyme using α-[5-^{3}H] difluoromethylornithine, *Eur. J. Cell Biol.,* 30, 288, 1983.
10. **Heby, O. and Andersson, G.,** Polyamines and the cell cycle, in *Polyamines in Biomedical Research,* Gaugas, J. M., Ed., John Wiley & Sons, New York, 1980, 17.
11. **Heby, O.,** Role of polyamines in the control of cell proliferation and differentiation, *Differentiation,* 14, 11, 1981.
12. **Bachrach, U.,** *Functions of Naturally Occurring Polyamines,* Academic Press, New York, 1973.
13. **Jane, J., Posso, H., and Raina, A.,** Polyamines in rapid growth and cancer, *Biochim. Biophys. Acta,* 473, 241, 1978.

14. **Hopkins, D. and Manchester, K. L.,** Polyamines: molecules in search of a role?, *South Afr. J. Sci.*, 78, 25, 1982.
15. **Tabor, C. W. and Tabor, H.,** Polyamines, *Annu. Rev. Biochem.*, 53, 749, 1984.
16. **Höltta, E.,** Oxidation of spermidine and spermine in rat liver: purification and properties of polyamine oxidase, *Biochemistry*, 16, 91, 1977.
17. **Pegg, A. E. and Williams-Ashman, H. G.,** Biosynthesis of putrescine, in *Polyamines in Biology and Medicine*, Morris, D. R. and Marton, L. J., Eds., Marcel Dekker, New York, 1981, 3.
18. **Reddy, P. R. K. and Rukmini, V.,** α-Difluoromethylornithine as a postcoitally effective antifertility agent in female rats, *Contraception*, 24(2), 215, 1981.
19. **Méndez, J. D., Díaz-Flores, M., Durán, G., and Hicks, J. J.,** Inhibition of rat embryonic development by the intrauterine administration of α-difluoromethylornithine, *Contraception*, 28(1), 93, 1983.
20. **Williams-Ashman, H. G., Pegg, A. E., and Lockwood, D. H.,** Mechanisms and regulation of polyamine and putrescine biosynthesis in male genital glands and other tissues of mammals, in *Advances in Enzyme Regulation*, Vol. 7, Weber, G., Ed., Pergamon Press, London, 1969, 291.
21. **Williams-Ashman, H. G. and Lockwood, D. H.,** Role of polyamines in reproductive physiology and sex hormone action, *Ann. N.Y. Acad. Sci.*, 171, 882, 1970.
22. **Shet, A. R. and Moodbidri, S. B.,** Significance of polyamines in reproduction, *Adv. Sex. Horm. Res.*, 3, 51, 1977.
23. **Romano, M., Cecco, L., Santacrose, M. A., Cerra, M., Rasquinelli, R., Pagnano, A. M., and Paladini, A.,** Polyamine levels and ODC activity in the normal human endometrium and myometrium, in the myoma and the endometrial carcinoma, in *Advances in Polyamines in Biomedical Sciences*, Caldarera, C. M. and Bachrach, U., Eds., CLUEB, Bologna, Italy, 1984, 145.
24. **Osterberg, S., Rose, S., and Heby, O.,** Urinary polyamine excretion during the menstrual cycle, *Clin. Chem.*, 24(5), 769, 1978.
25. **Heby, O. and Andersson, G.,** Tumor cell death: the probable cause of increased polyamine levels in physiological fluids, *Acta Pathol. Microbiol. Scand. Sect. A*, 86, 17, 1978.
26. **Russell, D. H., Giles, H. R., Christian, C. D., and Campbell, J. L.,** Polyamines in amniotic fluid, plasma, and urine during normal pregnancy, *Am. J. Obstet. Gynecol.*, 132, 649, 1978.
27. **Hiramatsu, Y., Eguchi, K., Yonezaw, M., Hayase, R., and Sekeba, K.,** Alteration of red blood cells polyamines during pregnancy and neonatal period, *Biol. Neonate*, 40, 136, 1981.
28. **Porta, R., Servillo, L., Abbruzzese, A., and Della Pietra, G.,** Automated chromatographic analysis of human placenta polyamines, *Biochem. Med.*, 19, 143, 1978.
29. **Moore, J. J., Cardaman, R. C., and Lundgren, D. W.,** Spermine-enhanced protein phosphorylation in human placenta, *Proc. Soc. Exp. Biol. Med.*, 176, 313, 1984.
30. **Swanberg, H.,** Histaminase in pregnancy, *Acta Physiol. Scand.*, 23 (Suppl. 79), 1, 1950.
31. **Southren, A. L., Kobayashi, Y., Carmody, N. C., and Weingold, A. B.,** Serial measurements of plasma diamine oxidase during normal human pregnancy by an improved method, *Am. J. Obstet. Gynecol.*, 95, 615, 1966.
32. **Gahl, W. A., Vale, A. M., and Pitot, H. C.,** Spermidine oxidase in human pregnancy serum: probable identity with diamine oxidase, *Biochem. J.*, 201, 161, 1982.
33. **Illei, G. and Morgan, D. M. L.,** Polyamine oxidase activity in human pregnancy serum, *Br. J. Obstet. Gynaecol.*, 86, 878, 1979.
34. **Illei, G. and Morgan, D. M. L.,** The distribution of polyamine oxidase activity in the fetomaternal compartments, *Br. J. Obstet. Gynaecol.*, 86, 873, 1979.
35. **Bardsley, W. G., Crabbe, M. J. C., and Scott, I. V.,** The amine oxidase of human placenta and pregnancy plasma, *Biochem. J.*, 139, 169, 1974.
36. **Amicosante, G., Oratore, A., Crito, C., and Finazzi Agro, A.,** Rapid characterization and partial purification of various animal amine oxidases, *Experientia*, 40, 1140, 1984.
37. **Morgan, D. M. L.,** Commentary: amine oxidases and pregnancy, *Br. J. Obstet. Gynaecol.*, 89, 177, 1982.
38. **Mondovi, B., Guerrieri, P., Costa, M. T., and Sabatini, S.,** Amine oxidase inhibitors and biogenic amines metabolism, in *Advances in Polyamine Research*, Vol. 3, Caldarera, C. M., Zappia, V., and Bachrach, U., Eds., Raven Press, New York, 1981, 75.
39. **Morgan, D. M. L.,** Polyamine oxidase and cellular interactions, in *Advances in Polyamine Research*, Vol. 3, Caldarera, C. M., Zappia, V., and Bachrach, U., Eds., Raven Press, New York, 1981, 65.
40. **Andersson, A. C., Henningsson, S., and Rosengren, E.,** Diamine oxidase, in *Polyamines in Biomedical Research*, Gaugas, J. M., Ed., John Wiley & Sons, Chichester, England, 1980, 273.
41. **Gahl, W. A. and Pitot, C.,** Acetylated polyamines as substrates for human pregnancy serum diamine oxidase, *Life Sci.*, 29 (21), 2177, 1981.
42. **Illei, G. and Morgan, D. M. L.,** Polyamine oxidase activity in amniotic fluid and fetal membranes, *Br. J. Obstet. Gynaecol.*, 87, 413, 1980.

43. **Klopper, A. and Hughes, G.,** Placental secresion of oestrogens and protein hormones, *Arch. Gynäkol.*, 225, 171, 1978.
44. **Gaugas, J. M. and Cursen, P.,** Polyamine interaction with pregnancy serum in suppression of lymphocyte transformation, *Lancet, i,* 18, 1978.
45. **Morgan, D. M. L. and Illei, G.,** Polyamine-polyamine oxidase interaction: part of maternal protective mechanism against fetal rejection, *Br. Med. J.*, 280, 1295, 1980.
46. **Bresniham, B., Grogor, R. R., Oliver, M., Lewkonia, R. M., Hughes, G. R. V., Lovins, R. E., and Faulk, W. P.,** Immunological mechanism for spontaneous abortion in systemic lupus erythematosus, *Lancet,* ii, 1205, 1977.
47. **Taylor, C. and Faul, W. P.,** Prevention of recurrent abortion with leucocyte transfusions, *Lancet,* ii, 68, 1981.
48. **Illei, G. and Morgan, D. M. L.,** Serum polyamine oxidase activity in spontaneous abortion, *Br. J. Obstet. Gynaecol.*, 89, 199, 1982.
49. **Labib, R. S. and Tomasi, T. B.,** Enzymatic oxidation of polyamines relationship to immunosuppresive properties, *Eur. J. Immunol.*, 11, 266, 1981.
50. **López-Llera, M.,** Fisiología patológica de la toxemia del embarazo, in "La toxemia del embarazo", *Temas de Terapéutica 1977,* Publicación particular limitada, Mexico, 1978, 1.
51. **De Voe, S. J. and O'Shaughnessy, R.,** Clinical manifestations and diagnosis of pregnancy-induced hypertension, *Clin. Obstet. Gynecol.*, 27(4), 836, 1984.
52. **Mathews, J. D. and Mason, T. W.,** Plasma viscosity and pre-eclampsia, *Lancet,* 8, 409, 1974.
53. **Samour, M. B., Ibrahim, F. K., Fattah, M. M. A., and Ramadan, M. A.,** Nucleic acids and protein changes in normal and pre-eclamptic placentae, *Acta Obstet. Gynecol. Scand.*, 58, 535, 1979.
54. **Toder, V., Eichenbrenner, I., Amit, S., Serr, D., and Nebel, L.,** Cellular hyperreactivity to placenta in toxemia of pregnancy, *Eur. J. Obstet. Gynec. Reprod. Biol.*, 9(6), 535, 1979.
55. **Koening, H., Goldstone, A., and Lu, Ch. Y.,** Polyamines regulate calcium fluxes in a rapid plasma membrane response, *Nature,* 305(5934), 530, 1983.
56. **Elgavish, A., Wallace, R. W., Pillion, D. J., and Meezan, E.,** Polyamines stimulate D-glucose transport in isolated renal brush-border membrane vesicles, *Biochim. Biophys. Acta,* 777, 1, 1984.
57. **Méndez, J. D., Espinoza, L. M., Arratia, C., and Hicks, J. J.,** unpublished data.
58. **Mamont, P. S., Bey, P., and Koch-Weser, J.,** Biochemical consequences of drug-induced polyamine deficiency in mammalian cells, in *Polyamines in Biomedical Research,* Gaugas, J. M., Ed., John Wiley & Sons, Chichester, England, 1980, 147.
59. **Fozard, J. R., Part, M. L., Prakash, N. J., and Grove, J.,** Inhibition of murine embryonic development by α-difluoromethylornithine, an irreversible inhibitor of ornithine decarboxylase, *Eur. J. Pharmacol.*, 60, 379, 1980.
60. **Slotkin, T. A., Seidler, F. J., Trepanier, P. A., Withmore, W. L., Larea, L., Barnes, G. A., Weigel, S. J., and Bartolome, J.,** Ornithine decarboxylase and polyamines in tissues of neonatal rat: effects of α-difluoromethylornithine, a specific irreversible inhibitor of ornithine decarboxylase, *J. Pharmacol. Ther.*, 222(2), 741, 1982.
61. **Fozard, J. R. and Koch-Weser, J.,** Pharmacological consequences of inhibition of polyamine biosynthesis with D,L-α-difluoromethyl ornithine, *Trends Pharmacol. Sci.*, 3(3), 107, 1982.
62. **Barkai, U. and Kraicer, P. F.,** Definition of period of induction deciduoma in the rat using ornithine decarboxylase as a marker of growth onset, *Int. J. Fertil.*, 23(2), 106, 1978.
63. **Prakash, N. J., Schechter, P. J., Grove, J., and Koch-Weser, J.,** Effect of α-difluoromethylornithine, an enzyme-activated irreversible inhibitor of ornithine decarboxylase, on L1210 leukemia in mice, *Cancer Res.*, 38, 3059, 1978.
64. **Prakash, N. J., Schechter, P. J., Mamont, P. J., Grove, J., Koch-Weser, J., and Sjoerdsma, A.,** Inhibition of EMTG tumor growth by interference with polyamine biosynthesis; effects of α-difluoromethylornithine, an irreversible inhibitor of ornithine decarboxylase, *Life Sci.*, 26, 101, 1980.
65. **Russell, D. H.,** Clinical relevance of polyamines, *Crit. Rev. Clin. Lab. Sci.*, 18(3), 261, 1983.
66. **Suzuki, O., Matsumoto, T., and Katsumata, Y.,** Determination of polyamine oxidase activities in human tissues, *Experientia,* 40, 838, 1984.
67. **McIndoe, J. H. and Turkington, R. W.,** Hormonal regulation of spermidine formation during spermatogenesis in the rat, *Endocrinology,* 92, 595, 1973.
68. **Mann, T.,** *The Biochemistry of Semen and of the Male Reproductive Tract,* John Wiley & Sons, New York, 1964, 193.
69. **Tabor, H. and Tabor, C. W.,** Spermidine, spermine, and related amines, *Pharmacol. Rev.*, 16, 245, 1964.
70. **Rhodes, J. B. and Williams-Ashman, H. G.,** Observations on polyamines in male accessory glands of reproduction, *Med. Exp.*, 10, 281, 1964.
71. **Jänne, J. and Williams-Ashman, H. G.,** On the purification of L-ornithine decarboxylase from rat prostate and effects of thiol compounds on the enzyme, *J. Biol. Chem.*, 246, 1725, 1971.

72. **Jänne, J. and Williams-Ashman, H. G.,** Dissociation of putrescine-activated decarboxylation of *S*-adenosyl-L-methionine from the enzymic synthesis of spermidine and spermine by purified prostatic enzyme preparations, *Biochem. Biophys. Res. Commun.*, 42, 222, 1971.
73. **Jänne, J., Schenone, A., and Williams-Ashman, H. G.,** Separation of two proteins required for synthesis of spermidine from *S*-adenosyl-L-methionine and putrescine in rat prostate, *Biochem. Biophys. Res. Commun.*, 42, 758, 1971.
74. **Zappia, V., Carteni-Farina, M., and Della Pietra, G.,** *S*-Adenosyl-methionine decarboxylase from human prostate, *Biochem. J.*, 129, 703, 1972.
75. **Huggins, C.,** The physiology of the prostate gland, *Physiol. Rev.*, 25, 281, 1945.
76. **Redenz, E.,** Uber die wirkung des sperminsauf die bewegung der spermatozoen, *Arch. Ges. Physiol.*, 216, 605, 1927.
77. **Harrison, G. A.,** Spermine in human tissues, *Biochem. J.*, 25, 1855, 1931.
78. **White, I. G.,** The effect of some seminal constituents and related substances on diluted mammalian spermatozoa, *Aust. J. Biol. Sci.*, 7, 379, 1954.
79. **Hicks, J. J., Pedrón, N., and Rosado, A.,** Modifications of human spermatozoa glycolysis by cyclic adenosine monophosphate (cAMP), estrogens, and follicular fluid, *Fertil. Steril.*, 23(12), 886, 1972.
80. **Shah, G. V. and Sheth, A. R.,** Inhibition of phosphodiesterase activity of human spermatozoa by spermine, *Experientia*, 34(8), 980, 1977.
81. **Madhubala, R. and Redy, P. R. K.,** Stimulation of ornithine decarboxylase activity by prostaglandins in the isolated cells of immature rat testis, *FEBS. Lett.*, 122(2), 197, 1980.
82. **Fair, W. R., Clark, R. B., and Wehner, N.,** A correlation of seminal polyamine levels and semen analysis in the human, *Fertil. Steril.*, 23(1), 38, 1972.
83. **Jänne, J., Höltta, E., Haaranen, P., and Elfving, K.,** Polyamines and polyamine-metabolising enzyme activities in human semen, *Clin. Chim. Acta*, 48, 393, 1973.
84. **Thakur, A. N., Sheth, A. R., and Rao, S. S.,** Polyamines in human semen: presence of *S*-adenosyl-L-methionine decarboxylase, *Clin. Chim. Acta*, 55, 377, 1974.
85. **Höltta, E., Pulkkinen, P., Elfving, K., and Jänne, J.,** Oxidation of polyamines by diamine oxidase from human seminal plasma, *Biochem. J.*, 145, 373, 1975.
86. **Pulkkinen, P., Kanerva, S., Elfving, K., and Jänne, J.,** Association of spermine and diamine oxidase activity with human spermatozoa, *J. Reprod. Fertil.*, 43, 49, 1975.
87. **Méndez, J. D.,** unpublished data.
88. **Leone, E., De Prisco, R., Annibalo, N., and La Montagna, R.,** Polyamines in human semen, in *Advances in Polyamine Research*, Vol. 3, Caldarera, C. M., Zappia, V., and Bachrach, U., Eds., Raven Press, New York, 1981, 473.
89. **Fleming, A. P. and Armstrong, D. T.,** Effects of polyamines upon capacitation and fertilization in the guinea pig, *J. Exp. Zool.*, 233, 93, 1985.
90. **Méndez, J. D.,** unpublished data.
91. **Pontes, J. E.,** Biological markers in prostatic cancer, *J. Urol.*, 130, 1037, 1983.
92. **Russell, D. H., Levy, C. C., Schimpff, S. C., and Hawk, I. A.,** Urinary polyamines in cancer patients, *Cancer Res.*, 31, 1555, 1971.
93. **Fair, W. R., Wehner, N., and Brorsson, U.,** Urinary polyamine levels in the diagnosis of carcinoma of the prostate, *J. Urol.*, 114, 88, 1975.
94. **Dunzendorfer, U. and Russell, D. H.,** Altered polyamine profiles in prostatic hyperplasia and in kidney tumors, *Cancer Res.*, 38, 2321, 1978.
95. **Danzin, C., Jung, M. J., Claverie, N., Grove, J., Sjoerdsma, A., and Koch-Weser, J.,** Effect of α-difluoromethylornithine, an enzyme-activated irreversible inhibitor of ornithine decarboxylase, on testosterone-induced regeneration of prostate and seminal vesicle in castrated rats, *Biochem. J.*, 180, 507, 1979.
96. **Dunzendorfer, U. and Feller, H.,** The effect of α-difluoromethylornithine and tartaric acid on semen count, seminal GT, FSH, LH, testosterone and steroid excretion in patients with chronic prostatitis, *Andrología*, 13(2), 100, 1981.
97. **Danzin, C., Claverie, N., Wagner, J., Grove, J., and Koch-Weser, J.,** Effect on prostatic growth of 2-difluoromethylornithine, an effective inhibitor of ornithine decarboxylase, *Biochem. J.*, 202, 175, 1982.

Chapter 3

POLYAMINES AND HEART PHYSIOLOGY

Claudio M. Caldarera, Flavio Flamigni, Carmen Rossoni, Carlo Guarnieri, and Carlo Clo

TABLE OF CONTENTS

I. INTRODUCTION

The aliphatic polyamines putrescine, spermidine, and spermine are present in cardiac muscle generally at lower concentrations than those of other mammalian tissues, with particular regard for putrescine. The biosynthesis of polyamines has been well established in the embryonic, neonatal, and adult heart, and it displays significant changes that might be correlated to the physiological, biochemical, and ultrastructural changes occurring at the various stages of developing heart. Furthermore, the possibility that changes in polyamine concentration may be relevant in the metabolic and electrophysiological adaptive property of the heart muscle is inferred from the very short half-life of the two decarboxylases, ornithine (ODC) and *S*-adenosylmethionine (SAM), and from their high capability to fluctuate rapidly in response to a variety of chemical and environmental stimuli. A rapid increase in ODC activity, leading to polyamine accumulation, is detectable shortly after the induction of heart hypertrophy, triggered by chemical, mechanical, or hormonal stimuli. In this regard putrescine accumulation can be included in the list of the earliest metabolic responses of the heart muscle to a hypertrophizing stimulus. It is very interesting that, in some experimentally-induced hypertrophy, prevention of putrescine accumulation partially blocks the increase in heart growth. These and other observations have led to the hypothesis that polyamines play an important role in the activation of protein and RNA synthesis, which are required for the hypertrophizing process.

However, more recently, increasing attention is focusing on other possible involvements of polyamines in the regulation of the metabolic and functional activity of the heart, both in normal and pathological conditions. The very early (seconds) and rapid onset of ODC activity observed in β-adrenergic stimulated heart suggests that polyamines might be related to the heart contractile mechanism. Furthermore, the findings that these polycations can directly affect several membrane functions and physiological parameters strongly support the idea that polyamines influence specific processes closely related to the peculiar function as a pump of the cardiac muscle.

Significant improvement in elucidating the physiological role of polyamines in the heart has been accomplished by the use of isolated beating heart cells in culture. These studies have revealed that polyamines influence the metabolism of cyclic nucleotides as well as the sensitivity of heart cell to different cardioactive agents.

II. POLYAMINE METABOLISM IN HEART TISSUE

A. Polyamine Content and Biosynthesis

The polyamine content in various organs of adult rats is reported in Table 1. The heart contains spermidine and spermine in similar amounts and a much smaller amount of putrescine. The levels of polyamines in the heart tissue are generally lower than in other organs, such as the liver and kidney. However, higher values of polyamines have been reported to occur in the heart muscle than in the skeletal muscles.[1] All the enzymes involved in the polyamine biosynthetic pathway are present in the heart tissue,[2] and the basal specific activity of ODC is fairly high in the heart compared to other tissues.[1,3] Nonetheless, as in other organs, the activity of this decarboxylase together with that of the SAM decarboxylase are lower than those of the two aminopropyltransferases, spermidine and spermine synthase, and this may account for the much lower level of putrescine and decarboxylated SAM with respect to spermidine and spermine.[2]

The key role played by ODC in polyamine biosynthesis is also indicated by the extreme sensitivity of this enzyme to a large variety of stimuli;[4] SAM decarboxylase activity may also change noticeably,[2,5,6] whereas spermine synthase activity is not modified at all and spermidine synthase activity is only slightly modified.[2]

Table 1
POLYAMINE CONTENT IN RAT TISSUES
(nmol/g TISSUE)

Tissue	Putrescine	Spermidine	Spermine	N^1-acetylspermidine
Heart	7 ± 1	278 ± 36	294 ± 34	n.d.
Liver	18 ± 2	1018 ± 92	949 ± 81	n.d.
Kidney	26 ± 4	381 ± 53	608 ± 61	n.d.
Spleen	38 ± 6	1115 ± 90	651 ± 54	2.5
Lung	40 ± 6	766 ± 60	334 ± 39	3.9
Brain	17 ± 2	371 ± 37	306 ± 19	n.d.
Aorta	13 ± 3	142 ± 21	249 ± 26	n.d.

Note: Polyamine analysis was performed on acid extracts from tissues of adult rats as previously described.[44] The results are the mean ± SD of 4 to 14 separate determinations; n.d., not detected (<0.2 nmol/g).

On the other hand, chronic addition of α-difluoromethylornithine (DFMO), a specific and widely used inhibitor of ODC, provoked an effective reduction in cardiac ODC activity and putrescine content, but did not modify noticeably the basal levels of heart spermidine and spermine in mature rats.[7,8] However, when a heart was stimulated into growth and polyamine synthesis activated, DFMO was found to be more effective in preventing the increase of polyamine concentrations,[7-9] resembling the situation described for other mammalian models.[10]

It seems likely that SAM decarboxylase cooperates with ODC in limiting polyamine synthesis and, when putrescine accumulates following ODC stimulation, SAM decarboxylase activity may be enhanced, ensuring the availability of decarboxylated SAM and favoring the conversion of putrescine into spermidine, as suggested by Pegg and Hibasami.[2]

Finally, it is interesting to note that alterations in cardiac polyamine metabolism may not be uniformly distributed in the various sections of the heart.[11]

B. Regulation of Polyamine Biosynthetic Enzymes

Marked changes in heart ODC activity (and eventually in the rate of polyamine synthesis) are evoked by a large variety of stimuli affecting the cardiac growth and function. Trophic hormones such as GH,[12] prolactin,[13] insulin,[14] and estradiol[15] increase ODC activity in the heart as well as in other tissues, whereas some catabolic conditions, such as diabetes,[14] fasting,[14,16] or hypercorticolism[16] reduce the ODC activity.

A characteristic of heart ODC activity is that it is enhanced in conditions which are correlated to an elevated functional activity of the cardiac muscle. Thus, catecholamine addition,[17,18] sympathetic stimulation by various agents,[18,19] physical exercise,[19,20] hyperthyroidism,[2,3] and aortic or pulmonary hypertension,[5,6] stimulate cardiac ODC activity. Conversely, forced immobilization,[21] hypothyroidism,[3] sympatholytic drugs,[19,21] and anoxia[22] have been reported to decrease heart basal ODC activity.

The sequence of biochemical signals that lead to changes in ODC activity is not clear, and it is likely that different stimuli utilize, at least in part, alternative mechanisms. In particular, the increase in ODC activity caused by catecholamine and sympathetic stimulation is mediated by β-adrenergic receptors,[21,23] the stimulation of which also increases the heart contractility. It is interesting to note that α- and β-receptors mediate opposite effects on the contractility of arterial smooth muscle, and in addition, in the aorta ODC activation by catecholamine is an α-mediated effect,[21,23] whereas β-agonists reduce basal ODC activity and prevent the induction by α-agonists.[24]

According to Womble et al.,[25,26] the ability of catecholamines to stimulate cardiac ODC activity (and cardiac growth) appears to be specifically coupled to β_2-receptors. This requires an increase in the concentration of cAMP and the activity of cAMP-dependent protein kinase (type I). On the other hand, Lau and Slotkin[27] have reported how cAMP analogues or PGE_1 injected into adult rats failed to modify heart ODC activity, and Boucek et al.[28] indicated that β-agonists failed to increase ODC activity in the absence of tension development in the isolated papillary muscle. It is also known that Ca^{2+}[29] and the integrity of the cytoskeleton function[30] are required for the response of ODC to catecholamines.

Not much is known about the molecular mechanisms of regulating ODC activity in heart tissue. However, some observations may give us clues to these mechanisms, such as the reported increased affinity for ornithine which follows various stimuli, including low doses of isoproterenol.[27,31] However, higher doses of the catecholamine or other stimuli may affect ODC activity via different mechanisms, which may include protein induction.[4,32] Moreover, rat heart tissue contains a significant amount of ODC-antizyme complex, the concentration of which may vary after isoproterenol stimulation.[33] Highly purified ODC from heart tissue of rats treated with isoproterenol possesses general characteristics which are similar to those of ODC isolated from other tissues[34] and may also be inhibited by basic polypeptides,[35] and phosphorylated by casein kinase-2.[35] It is also possible that heart ODC decarboxylates lysine, to which it binds with a lower affinity than to ornithine[34] and actually minute amounts of cadaverine may be shown to accumulate in the heart after ODC stimulation by isoproterenol.[37]

As in other tissues, SAM decarboxylase activity in the heart may increase following growth stimuli.[2,5,6] However, unlike the enzyme which is found in other organs, the cardiac SAM decarboxylase does not seem to be inhibited by treating rats with spermidine.[1]

C. Polyamine Catabolism and Interconversion

The biochemical processes which involve the metabolic degradation and excretion of polyamines have been investigated to a lesser extent and consequently are less well understood than the polyamine biosythesis; this is particularly true for the heart tissue. However, amine oxidases, which may act on polyamines,[38,39] and a diamine oxidase,[10] which may help to keep the putrescine level low, have been shown to occur in the heart. The activities of these amine oxidases increase during cardiac growth,[38-40] when putrescine and polyamine synthesis is stimulated.

Acetylation of polyamines may be a means to facilitate the excretion of polyamines from the cell, or it may represent the first step in their catabolism. The formation of N^1-acetylated derivatives appears to be the limiting step in the conversion of spermine and spermidine into putrescine.[10] The activities of spermidine/spermine N^1-acetyltransferase and polyamine oxidase, which are responsible for this interconversion process, are rather low in rat heart;[41,42] however, administration of large doses of spermidine leads to an increase in heart putrescine content.[43]

N^1-Acetylspermidine is usually undetectable in the heart,[44] but it rapidly accumulates after isoproterenol treatment.[45] Under these conditions it can be shown that part of spermidine is converted into putrescine; however, the formation of putrescine via this pathway is of limited importance compared to the *de novo* synthesis from ornithine via ODC, which is remarkably stimulated by isoproterenol.[45] Interestingly, N^1-acetylspermidine also accumulates in the hearts of rats which have been treated with dexamethasone, which contrarily produces a decrease in ODC activity.[33]

III. HEART FUNCTION AND POLYAMINES

A. Polyamines in Myocardial Growth

Normally, effective homeostatic mechanisms adjust the amount of cardiac muscle to the body mass which needs to be perfused at rates determined by physiological factors, such as the physical activity and the nutritional status. Chronic increases in workload provoked, for example, by endurance training, by mechanical obstruction, or by pharmacological perturbations produce an increase in the mass of the muscular walls of the appropriate chamber(s). Except for a brief period after birth, the proliferative capacity of cardiac myocytes is negligible and hypertrophy (increased cell size) represents the main adaptive process in the mature myocardium.[30] Enhancement of RNA and protein synthesis is a prominent feature in cardiac hypertrophy, whereas DNA replication is mainly confined to the postnatal development.[27,30]

As in other mammalian tissues or cells,[10,46] changes in the rate of polyamine synthesis in the heart are usually correlated to and preceeded by changes in the synthesis of nucleic acid and proteins and tissue growth.[2,26,29] We have mentioned in the previous section how hormones and nutritional states which favor protein synthesis in the heart, as well as the condition of enhanced cardiac contractility (eventually leading to hypertrophy), cause an increase in ODC activity and polyamine biosynthesis. Conversely, conditions associated with a low rate of macromolecular synthesis in the heart usually show a decrease in polyamine synthesis. For instance, Chideckel et al.[3] have found a good correlation between the functional state of the thyroid and ODC activity in the heart and in other organs of adult rats, which is in accordance with the known effect of the thyroid hormones on protein synthesis. Moreover, whereas treatment of *adult* rats with thyroid hormones induces cardiac hypertrophy, hyperthyroidism elicited an initial elevation of heart ODC activity in *neonatal* rats, followed by reduced ODC activity coupled with reduced RNA synthesis and a subsequent deficit in heart weight.[47] The administration of triidothyronine has been shown to alter the development of the ODC response to sympathetic stimulants as well as retarding cardiac growth; a similar effect was also exerted by methadone.[48]

In vitro experiments with cell-free systems have shown that polyamines might positively affect various steps involved in nucleic acid and protein biosynthesis.[46] In accordance with this, the perfusion of isolated hearts with polyamines has been reported to increase the rate of incorporation of [^{3}H] ribose, [^{14}C] phenylalanine, and [^{14}C] acetate into myocardial RNA,[39] protein,[49] and histone,[50] respectively. Furthermore, perfusion with methylglyoxal bis(guanylhydrazone) (MGBG) caused a decrease in the synthesis of polyamines and RNA and also a decrease in the acetylation of histone fractions.[51]

All together, these studies suggest that an early and marked increase in polyamine synthesis occurring in all types of experimentally induced cardiac hypertrophy[2,5,13,20,23,27,52,53] plays an important role in promoting macromolecular synthesis and cardiac growth. On the other hand, the changes in polyamine and macromolecular synthesis may be separate responses to the primary stimuli, and the rate of polyamine formation may merely reflect the level of protein synthesis.

Recently, some inhibitors of ODC were used in vivo in order to ascertain the role of polyamines in cardiac hypertrophy (Table 2). However, the results of these studies have not completely clarified the problem. In fact, continued exposure of rats to 1,3-diaminopropanol completely prevented an increase in the concentration of putrescine and polyamines in response to tyroxine, but did not prevent the increase in cardiac mass.[2] A similar lack of effect on cardiac growth was also obtained with both tyroxine and triidothyronine in the presence of DFMO, an entirely different and less toxic inhibitor of ODC.[7,54] On the other hand, treatment with DFMO prevented the increases in putrescine and spermidine concentrations and attenuated significantly the cardiac growth induced by isoproterenol.[8,54] DFMO also affected neonatal development of heart to some extent, although the deficit in weight appeared in the heart later than in the brain or kidney.[9] It should be noted that in general the ODC inhibitors were able to prevent the increase in polyamine content, but substantial amounts of spermine and spermidine remained in all cases.[2,7-9]

In conclusion these results seem to indicate that at least some cardiac growth occurs in the absence of significant increases of polyamines, with apparent differences which depend

Table 2
EFFECT OF ODC INHIBITORS ON THE CARDIAC GROWTH

Experimental model	ODC inhibitor	Effect on cardiac growth	Ref.
Triidothyronine-induced hypertrophy	DFMO	None	54
Tyrosine-induced hypertrophy	Diaminopropanol	None	2
Tyrosine-induced hypertrophy	DFMO	None	7
Isoproterenol-induced hypertrophy	DFMO	Inhibition	8,54
Isoproterenol-induced hypertrophy	Diaminopropane	Inhibition	55
Aortic constriction-induced hypertrophy	Diaminopropane	Inhibition	55
Neonatal development	DFMO	Inhibition	9

on the growth stimuli used. It is possible that various stimuli may lead to different rates of cardiac growth by different mechanisms which require different levels of polyamines.

B. Effect of Polyamines on Membrane Processes and Heart Contractility

Although several studies have been focused on the effect of polyamines on macromolecular synthesis, polyamines have also been shown to affect other biochemical processes in vitro.[46] For instance, polyamines interact with the negatively charged phospholipids (PLs) of biomembranes,[56] they may protect membranes from lipid peroxidation, or influence various membrane functions.[57,58] Spermine binds to polyphosphoinositides with a high affinity, comparable to that for single-stranded RNAs,[59] and it is possible that the Ca^{2+} binding to membranes might be influenced by these interactions.

According to Koenig et al.,[60] polyamines play a messenger function in a rapid response of the plasma membrane to various stimuli by increasing Ca^{2+} influx and mobilizing intracellular Ca^{2+} possibily via a cation-exchange reaction. In particular, in the heart, isorpoterenol induced a rapid increase in ODC activity and polyamine synthesis. These processes were reported to be necessary for stimulation of Ca^{2+} fluxes and membrane transport processes induced by isoproterenol.[61] Actually, polyamines have been shown to stimulate the uptake of amino acids in isolated hearts.[49] An early increase (within few minutes) in ODC activity has been observed following perfusion of rabbit hearts with noradrenaline.[62,63] ODC is apparently released from isolated rat hearts under conditions in which other enzymes or smaller molecules are not[64] and it may be speculated that some of the ODC molecules are in close relationship with the membrane.

Furthermore, polyamines appear to be involved in the heart damage induced by Ca^{2+} paradox.[65] Spermine appears to regulate Ca^{2+} cycling of rat liver mitochondria.[66,67] Moreover, spermine binds to heart submitochondrial particles with high affinity and activates mitochondrial ATPase activity.[68] It has also been reported how MGBG and polyamines inhibit carnitine-dependent oxidation of palmitate in homogenated rat hearts.[69]

In the previous section, we mentioned the effects of the contractility state of the heart on polyamine metabolism and, more recently, some studies have been carried out on the effects of polyamine on heart contractility. Research by Keskemeti et al.[70] indicates that polyamines may influence cardiac membrane events which are responsible for controlling action potentials and ionic currents. In particular, spermine and spermidine (at concentrations of 10^{-4})

to 10^{-5} *M*) may increase the resting potential of the heart membrane and maximum rate of rise in the action potential of some isolated preparations of atria or papillary muscles. However, these effects were prevented by the addition of indomethacin, a prostaglandin synthetase inhibitor. It may also be noted that polyamines have been shown to affect prostaglandin synthesis.[71] Spermidine and spermine, however, did not affect the slow Ca^{2+} current.[70]

Bazzani et al.[72] have reported the effects of polyamines on the force of contraction of perfused rat ventricle strips. Spermidine and spermine (at 10^{-4} to 10^{-5} *M*) caused a sharp and short-lived positive inotropic effect, immediately followed by a permanent and more marked fall of the developed tension. Finally, significant cardiovascular changes can be produced by injecting polyamines into the whole animal.[73] Spermidine was more effective than spermine and when given i.v. at doses of 0.1 to 1 mg/kg caused hypotension due to histamine release and bradycardia due to a direct cardiac action. Moreover, spermidine increased and spermine decreased the carotid-sinus baroreceptor reactivity.

In conclusion recent research suggests that polyamines may affect several membrane functions and may exert significant effects on various physiological parameters of the heart. More work is needed to extend and deepen these results and to ascertain the physiological relevance of these findings.

IV. POLYAMINES AND CULTURED HEART CELLS

A. The Heart Cell in Culture

During the past decade a great deal of attention has been focused in our laboratory to clarify the interrelationship between polyamines and other intracellular factors, such as the cyclic nucleotides (cAMP and cGMP) and Ca^{2+}. These factors are closely involved in the control of macromolecular synthesis, growth, division, differentiation, and cellular functions.[74-78] In fact, together with these factors, polyamines appear to play an important role in signal transmission and transduction for the activation of many biochemical processes and physiological functions by different effectors, including hormones and other biologically active substances.

In the heart cell, which according to Goldberg et al.[79] represents a bidirectional control system, Ca^{2+} and cAMP as well as cGMP and cAMP generally mediate opposing metabolic and functional responses.[77,80-82] In particular the two nucleotides have been recognized as possible final mediators of contractile function, being cAMP and cGMP respectively engaged as positive or negative mediators of inotropic and chronotropic stimuli. Furthermore, Ca^{2+} is involved in the control of cyclic nucleotide levels by inhibiting adenylate cyclase[83] and stimulating cAMP-dependent phosphodiesterase (cAMP-PDE)[84] and guanylate cyclase[85] activities. In the heart, polyamines mediate rapid increase of free cytosolic Ca^{2+} induced by hormonal stimuli.[61] It follows, therefore, that a better understanding of the interplay among polyamine and these factors will contribute to our knowledge of the influence of polyamines on cardiac performance. The heart cells in culture provide an appropriate model system for this purpose, since they retain the specialized characteristics of differentiation, such as myosin-synthesizing activity,[86] beating ability,[87] or sensitivity to cardioactive agents,[88,89] as well as to changes of oxygen availability.[90] In addition, the absence of blood vessels and sympathetic nerves excludes certain interferences which are otherwise present in other experimental systems. Confluent and serum-starved heart cells from chick embryos have been utilized in our studies, as they constitute a population of nondividing cells synchronized in the G phase of the cell cycle.[91]

B. Regulation of Polyamine Metabolism in Cultured Heart Cells

The ratio between the concentrations of the different polyamines in quiescent heart cell cultures (spermine > spermidine > putrescine) only partially reflects that of the whole heart.

In these cultures the molecular ratio spermine or spermidine/putrescine is much lower than in the heart tissue.[92] Moreover, cellular polyamine content strictly depends on the presence and amount of serum or PLs and on other crucial elements such as Ca^{2+} and oxygen.

A progressive increase in the basal content of polyamine is observed in cultures incubated in the presence of increasing concentrations of serum.[93] Serum readdition (10%) to serum-starved cells causes a marked onset of ODC activity with a peak between 4 to 6 hr,[92,94] followed by the accumulation of polyamines, mainly putrescine and spermidine, and by the rise of protein, RNA, and DNA synthesis.[95,96] The failure of serum to stimulate macromolecular synthesis in the presence of specific inhibitors of polyamine synthesis suggests that polyamine accumulation is a requirement and not simply a result of the action of serum on heart cells in culture. The positive influence of polyamines on the gene activity of heart cells is supported by (1) the existence of a close correlation between cellular polyamine content and the rate of histone acetylation[97] which represent an important gene derepression mechanism[98] and (2) the ability of exogenously administered polyamines (at micromolar concentrations) to substitute for serum in stimulating the incorporation of thymidine into DNA.[99]

The lipid component of serum plays an essential role in ensuring optimal growth and the long-term maintenance of cultured heart cell functions.[100] Evidence has been provided that addition of PLs, such as phosphatidylcholine, phosphatidylethanolamine, or phosphatidylserine, to serum-starved heart cells[96,101] leads to a significant increase of polyamine content by rapidly stimulating ODC and SAM-decarboxylase activities.[96] This is accompanied by the onset of DNA synthesis. The hypothesis of a cause-effect relationship between polyamine accumulation and induction of DNA synthesis by PL is supported by the inhibition of both of these events in the presence of DFMO.[96] Although no direct evidence is available concerning the rationale of polyamine accumulation by PL, it is conceivable that by virtue of their nature of polycations, the newly synthesized polyamines might serve, in part, to neutralize the negative phosphate groups of PL and/or to protect them against peroxidative reactions, as observed with other model systems.[56,57]

ODC activity and polyamine accumulation in cultured heart cells can be induced, not only by a cAMP-independent mechanism (serum or PL), but also by factors such as noradrenaline,[92] isoproterenol,[94,102] glucagon,[103] or PGE[104] whose action leads to an early and rapid increase of cAMP content. The involvement of cAMP in ODC induction is supported by the marked increase of activity observed by exposing the cells to a cAMP derivative, such as dibutyryl-cAMP.[92]

cAMP-mediated effectors apparently induce ODC in the heart muscle[62,63] and in cultured heart cells they also cause its activation within few minutes as observed PGE_1 or isoproterenol-stimulated cells (unpublished data).

Furthermore, basal ODC activity and polyamine content of heart cells closely depend on the presence of extracellular Ca^{2+}.[94] Ca^{2+}-starved cells (for 20 hr), exhibit lower ODC activity than Ca^{2+}-treated cells. ODC activity can be rapidly increased by the addition of $CaCl^2$ (1.8 m*M*) or by refeeding the cells with fresh Ca^{2+}-containing medium.

As reported for other cell types in culture,[105,106] extracellular Ca^{2+} also plays a relevant role in both cAMP-independent (serum) and cAMP-dependent mechanisms (isoproterenol) of induction of ODC in cultured heart cells.[94] However, some differences have been observed between the two. The depletion of Ca^{2+} strongly (but not completely) reduces the effect of serum, suggesting that serum can partially act independently on this ion. It has been suggested[107-109] that the sequence of events leading to the induction of ODC by serum or other mitogen stimuli, may start from the activation of membrane-bound phospholipase C, which causes the hydrolysis of phosphoinosides producing diacylglycerides (DAGs). These, in turn, activate protein kinase C (PKc), a Ca^{2+} and PL-dependent enzyme which is apparently involved in ODC induction. However, PKc activation can be ensured by DAG also in the absence of an increase of cellular Ca^{2+}.[108,110] This probably accounts for the capability of serum to stimulate ODC in Ca^{2+}-starved cells or in the presence of the Ca^{2+} chelator EGTA. A distinct feature of Ca^{2+}-starvation in isoproterenol-stimulated heart cells is the slight time-dependent increase of ODC activity, which is in striking contrast to the marked and transient onset of activity occurring in the presence of the ion.[94] The observed property of polyamines in stimulating cAMP degradation in the presence of Ca^{2+} (see below) might account for these differences.

Basal ODC activity and polyamine content of cultured heart cells are inversely related to the environmental oxygen tension (pO_2).[97,111] A progressively lower ODC activity is observed by exposing confluent cultures to increasing pO_2 (5, 20, and 80%).[111] Furthermore, isoproterenol-mediated induction of the enzyme is prevented when the cells are maintained under high pO_2 (80%). Since hyperoxia also blocks ODC induction by dibutyryl-cAMP, it is conceivable that O_2 does not affect the β-receptor system, but most likely affects the redox state of the SH groups of the enzyme.[111,112] The property of oxygen to influence both basal and stimulated ODC activity might account for its capability to regulate macromolecular synthesis and division of heart cells in culture.[97,113] Moving, growing heart cells from high to low pO_2 results in a significant increase in polyamine content, paralleled by the onset of RNA and DNA synthesis as well as of the rate of histone acetylation.[97] In view of the property of polyamines to stimulate histone acetylation,[50] together with other important gene derepression mechanisms,[114] one can speculate that in the heart cell polyamines play a relevant role in the control of gene activity which is operated by O_2.

C. Polyamines and Cyclic Nucleotides

The intracellular contents of cAMP and cGMP respond specifically to growth conditions. In quiescent heart cells, the addition of serum[92,93,115,116] or PLs,[97,117] while stimulating polyamine accumulation and DNA synthesis, causes the depletion of cAMP and the accumulation of cGMP, resulting in a net fall of the cAMP/cGMP ratio. This last event is generally considered as the appropriate stimulus for the cell to proliferate, and in the heart cell it negatively correlates with the contractile property.

The possible involvement of intracellular polyamines in the control of basal cyclic nucleotide metabolism can be inferred from the observation that specific inhibition of polyamine synthesis is associated with a higher cAMP level and with a lower cGMP level[95,116] which are consequences of changes in the activity of the specific cyclases and phosphodiesterases (PDEs).[104,116] Furthermore, under condition of inhibition of polyamine synthesis,[93,95,96,116] serum and PLs exhibit significantly lower efficiency in modulating cyclic nucleotide contents. This supports the idea that polyamine accumulation is an important step in the mechanism of action of serum and PLs. These data suggest that the antiproliferative effect of polyamine inhibitors[119] may be attributed (at least partially) to their effect on cyclic nucleotides. The cyclic nucleotides cAMP and cGMP represent, respectively, a negative and a positive signal for the induction of cell growth.[78,118]

Substantial evidence for the involvement of polyamines in the control of basal cyclic nucleotide contents is seen by the rapid decrease of cAMP and the rapid accumulation of cGMP occurring shortly after (within minutes) exposure of quiescent and serum-starved heart cells to extremely low doses of individual polyamines (0.1 μM PTC, 1 μM SPM, or 10 μM SPD),[115,120] i.e., in the range of those found in physiological fluids.[121] The effect is still maintained after a long-term incubation (2 hr), when polyamines accumulate inside the cells.[122] This is probably due to an energy-dependent process as reported for Girardi heart cell cultures[123] or isolated rat hepatocytes.[124] Higher doses of the amines are less effective in reducing cAMP, but further stimulate cGMP accumulation.[122] Extracellular Ca^{2+} is required for the regulation of both the nucleotides by polyamines, further supporting the existence of a close mutual interaction between all these factors.[94]

The capability of polyamines to modulate cyclic nucleotide concentrations has also been observed with other cell types[125] and is carried out through rapid changes of the specific cyclases and activities. While cGMP-PDE is reduced by exposure of heart cells to each amine,[115,120,122] guanylate cyclase activity significantly increases as a consequence of an increased affinity for its substrate.[120,122] In both cases the enzyme present in the soluble fraction of the cells is preferentially affected by polyamines, particularly SPM and SPD. Conversely, membrane-bound adenylate cyclase is strongly reduced by polyamines, in a dose-inverse manner, as supported also by in vitro experiments.[104,122] Kinetic analyses carried out with the particulate enzyme from chick embryo hearts indicate that each amine reduces the affinity of the enzyme for ATP, as well as its catalytic property.[104] At the same dose (1 μM), SPM exhibits the highest efficiency, followed by SPD and PTC. It is likely that this order may be related to the net positive charge of each amine and therefore to its capability to intereact with the negative charged groups of cell membrane PLs, which play key roles in determining the organization and function of the enzyme as well as its hormone-induced activation.[126] Indeed, evidence has been recently provided by us that the adenylate cyclase response of heart cells to PGE_1 or isoproterenol is inversely related to the amount of cellular polyamines.[104] Conversely, serum-mediated inhibition of the enzyme is favored in polyamine-enriched cells (unpublished data), thus supporting the view that these polycations play a role in determining the sensitivity of heart cells to different stimuli, as discussed below.

The inhibition of basal adenylate cyclase activity by micromolar concentrations of exogenous polyamines is paralleled by the rise of both soluble and, especially, particulate cAMP-PDE activity of quiescent heart cells[92,122] that account for the fall of cAMP content. The low affinity form of this enzyme, which is present in both the fractions of the cells and can be stimulated by Ca^{2+}-calmodulin complex,[127] appears to represent a more specific target for the action of each amine.[92] Again the effect of polyamines is inversely related to the dose used and is carried out through a net increase in the affinity of the enzyme for cAMP.[92]

Furthermore, in primary heart cell cultures polyamines have been shown to play a crucial role in both cAMP-dependent and cAMP-independent mechanisms of induction of cAMP-PDE. Addition of noradrenaline, isoproterenol, or dibutyryl-cAMP to the cultures stimulates ODC activity and polyamine accumulation, followed by a significant increase of cAMP-PDE activity.[92,102] The failure of all of these drugs to induce the enzyme under conditions of inhibition of polyamine synthesis suggests that polyamine accumulation is a requirement for the cAMP-mediated induction of the enzyme, which represents a regulatory mechanism by which cAMP can control its own level of accumulation. This might provide a rationale for the stimulatory effect of cAMP on ODC. It is interesting to note that a cAMP-activated protein kinase seems to be essential for both ODC[128] and cAMP-PDE[129] induction by cAMP.

Similarly, induction of PDE by serum, which operates via a cAMP-independent mechanism, is prevented by specific inhibitors of polyamine synthesis.[92] In both of these mechanisms endogenous polyamines seem to be maximally involved in the regulation of the low affinity form of PDE, in agreement with the observation in the presence of exogenous polyamines.

D. Polyamines and Heart Cell Sensitivity to Cardioactive Agents

In various cell types, cAMP and cGMP seem to mediate many metabolic and physiological responses, often to opposite direction, as in the case of contraction or relaxation of the heart cell.[130] Therefore, it follows that the metabolic and contractile activity of the heart cell closely depends on the magnitude of its sensitivity to different cyclic nucleotide-mediated cardioactive agents. This, in turn, is determined, other than by the state of integrity of cell membrane and of its receptor systems, by factors which are involved in the regulation of synthesis and degradation of one or both the nucleotides.

By virtue of the observed property of polyamines to affect the enzymatic machinery of cAMP and cGMP, polyamines might be regarded as potential regulators of heart cell sensitivity. Experimental evidence clearly supports this hypothesis. Exogenous polyamines (at micromolar concentrations) strongly reduce the cAMP response of cultured heart cells to noradrenaline,[125] glucagon,[103] or isoproterenol,[131] while they enhance the cAMP depletion by serum or insulin (unpublished data). An impaired ability of cAMP-mediated effectors to accumulate cAMP has been observed with other cell types, both normal and neoplastic, in culture.[125] The generality of the phenomenon, together with the capability of polyamines to affect basal cyclic nucleotide contents, seems to exclude specific interactions of polyamines with membrane receptors. It is more likely that the changes in cellular sensitivity may be related to the changes in the activity of cyclases and PDEs observed in the presence of polyamines and, in the case of adenylate cyclase activators, they may be also related to altered receptor-cyclase coupling, arising from an interaction between polyamines and the PL component of the enzyme.

Furthermore, evidence has been recently provided that the magnitude of cellular sensitivity is markedly influenced by changes in the intracellular amount of polyamines. In fact, accumulation of polyamines, obtained by pretreating the cells either with single amines (2 hr) or with fetal calf serum (20 hr), is associated with an increased sensitivity to agents such as insulin, serum, acetylcholine, or morphine which cause a fall in cAMP and an increase in cGMP content.[93,95,122,131,132] Conversely, polyamine-enriched cells exhibit a much lower sensitivity to cAMP-mediated effectors such as isobutylmethylxanthine, glucagon, noradrenaline, isoproterenol, or PGE_1 than do polyamine-poor cells.[93,104,122,131,132]

The possible involvement of cellular polyamines in determining heart cell sensitivity is clearly supported by the fact that inhibition of polyamine accumulation leads to an improved or an impaired responsiveness to cAMP or cGMP-mediated effectors, respectively.[93,95,103,104,131] Based on these observations, it is apparent that exogenous polyamines must enter the cells to exert their regulatory effect and that polyamines also influence the metabolic and functional responses evoked by different cyclic nucleotide-mediated effectors.

In this regard, recent experiments have indicated that in quiescent heart cell cultures, glucagon or isoproterenol-mediated induction of tyrosine aminotransferase (TAT), which requires cAMP accumulation and protein kinase activation,[133] is strongly constrained by exogenous polyamines as well as in polyamine-enriched cells.[103,131,134] Since administered polyamines cause a slight and transient increase of basal TAT activity, the impaired ability of the two stimulants to induce the enzyme is very likely related to their impaired efficiency in accumulating cAMP under the different experimental conditions. On the other hand, TAT inducibility is significantly increased during inhibition of polyamine synthesis, which also promotes a longer lasting TAT response to glucagon or isoproterenol.[103,131,135] This prompts the suggestion that the newly synthesized polyamines, resulting from stimulation of ODC by these two hormones, might play an important role in restoring basal TAT activity, possibly by increasing cAMP degradation and/or by inhibiting protein kinase activity. Conversely, serum-mediated induction of TAT is completely prevented by blocking polyamine synthesis,[131,135] thus suggesting that polyamine accumulation is a requirement for the cAMP-independent mechanism of TAT induction.

The scheme below summarizes our present hypothesis on the role of polyamines in the two mechanisms of TAT induction.

In conclusion, these findings support the concept that exogenous or endogenous polyamines might exert important regulatory functions on the metabolic activity and on the electrophysiological property of the heart cell by affecting cyclic nucleotide metabolism and cellular sensitivity to different cardioactive agents. These properties of polyamines might also be taken into consideration when explaining phenomena such as (1) the negative inotropic effect elicited by exogenous polyamines on perfused rat ventricle strips,[72] (2) the impaired

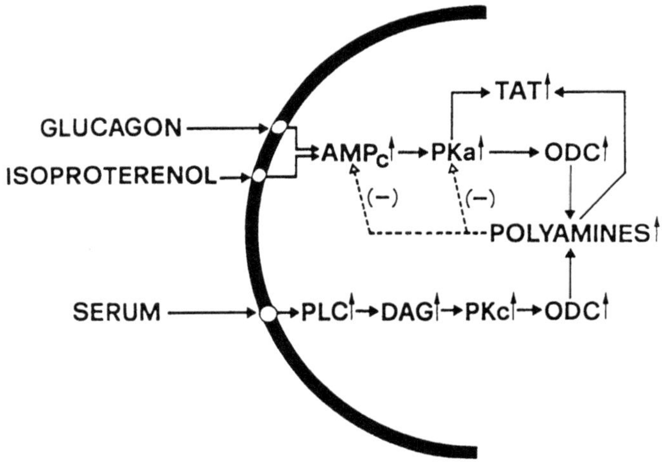

functional performance and the reduced responsiveness to catecholamines of hypertrophied heart,[136] or (3) the desensitization of the stimulant action of isoproterenol observed with cultured heart cells.[89]

Obviously, a great deal of work remains to be done before the role of polyamines in the physiology of the heart will be understood. Hopefully, isolated beating heart cells in culture will be useful in guiding these investigations.

ACKNOWLEDGMENTS

This work was supported by a grant from the Consiglio Nazionale della Ricerche, Rome, Progetto Finalizzato ''Oncologia'' (N. 85.002099.44) and by grants from Ministero Pubblica Istruzione (40%).

REFERENCES

1. **Pösö, H. and Pegg, A. E.,** Differences between tissues in response of *S*-adenosylmethionine decarboxylase to administration of polyamines, *Biochem. J.*, 200, 629, 1981.
2. **Pegg, A. E. and Hibasami, H.,** Polyamine metabolism during cardiac hypertrophy, *Am. J. Physiol.*, 239, E372, 1980.
3. **Chideckel, E. W., Rozovski, S. J., and Belur, E. R.,** Thyroid-hormone effects on ornithine decarboxylase, *Biochem. J.*, 192, 765, 1980.
4. **Flamigni, F., Rossoni, C., Stefanelli, C., and Caldarera, C. M.,** Polyamine metabolism and function in the heart, *J. Mol. Cell. Cardiol.*, 18, 3, 1986.
5. **Russell, D. H., Shiverick, K. T., Hamrell, B. B., and Alpert, N. R.,** Polyamine synthesis during initial phases of stress-induced cardiac hypertrophy, *Am. J. Physiol.*, 221, 1287, 1971.
6. **Feldman, J. M. and Russell, D. H.,** Polyamine biogenesis in left ventricle of the rat heart after aortic constriction, *Am. J. Physiol.*, 222, 1199, 1972.
7. **Pegg, A. E.,** Effect of α-difluoromethylornithine on cardiac polyamine content and hypertrophy, *J. Mol. Cell. Cardiol.*, 13, 881, 1981.
8. **Bartolome, J. V., Trepanier, P. A., Chait, E. A., and Slotkin, T. A.,** Role of polyamines in isoproterenol-induced cardiac hypertrophy: effects of α-difluoromethylornithine, an irreversible inhibitor of ornithine decarboxylase, *J. Mol. Cell. Cardiol.*, 14, 461, 1982.
9. **Slotkin, T. A., Seidler, F. J., Trepanier, P. A., Whitmore, W. L., Lerea, L., Barnes, G. A., Wiegel, S. J., and Bartolome, J.,** Ornithine decarboxylase and polyamines in tissues of the neonatal rat: effects of α-difluoromethylornithine a specific irreversible inhibitor of ornithine decarboxylase, *J. Pharmacol. Exp. Ther.*, 222, 741, 1982.

10. **Pegg, A. E.,** Recent advances in the biochemistry of polyamines in eukaryotes, *Biochem. J.*, 234, 249, 1986.
11. **Selmeci, L., Bardocz, S., and Egyed, Z.,** Isoproterenol-induced responses in the regional metabolism of the rat heart, *Biochem. Biophys. Res. Commun.*, 108, 292, 1982.
12. **Sogani, R. K., Matsushita, S., Mueller, J. F., and Raben, M. S.,** Stimulation of ornithine decarboxylase activity in rat tissues by growth hormone and by serum growth factor from rats infested with spargana of *Spirometra mansonoides, Biochim. Biophys. Acta*, 279, 377, 1972.
13. **Richards, J. F.,** Ornithine decarboxylase activity in tissues of prolactin-treated rats, *Biochem. Biophys. Res. Commun.*, 63, 292, 1975.
14. **Conover, C. A., Rozovski, S. J., Belur, E. R., Aoki, T. T., and Ruderman, N. B.,** Ornithine decarboxylase activity in insulin-deficient states, *Biochem. J.*, 192, 725, 1980.
15. **Das, R. and Kanungo, M. S.,** Activity and modulation of ornithine decarboxylase and concentrations of polyamines in various tissues of rats as a function of age, *Exp. Gerontol.*, 17, 95, 1982.
16. **Chideckel, E. W. and Belur, E. B.,** The effect of catabolic states induced by hyperthyroidism, hypercorticolism, and fasting on muscle ornithine decarboxylase activity, *Endocrinology*, 107, 918, 1980.
17. **Warnica, J. W., Antony, P., Gibson, K., and Harris, P.,** The effect of isoprenaline and propranolol on rat myocardial ornithine decarboxylase, *Cardiovasc. Res.*, 9, 793, 1975.
18. **Fuller, R. W. and Hemrick, S. K.,** Studies of noradrenergic influence on cardiac ornithine decarboxylase in rats, *J. Mol. Cell. Cardiol.*, 10, 1031, 1978.
19. **Bareis, D. L. and Slotkin, T. A.,** Responses of heart ornithine decarboxylase and adrenal catecholamines to methadone and sympathetic stimulants in developing and adult rats, *J. Pharmacol. Exp. Ther.*, 205, 164, 1978.
20. **Rettig, G., Gibson, K., and Harris, P.,** Effects of propranolol on rat myocardial 1-ornithine decarboxylase activity during exercise and hypoxia, *Res. Commun. Chem. Pathol. Pharmacol.*, 18, 67, 1977.
21. **Thadani, P. and Schanberg, S. M.,** Effect of stress and sympathetic activity on rat cardiac and aortic ornithine decarboxylase activity, *Life Sci.*, 25, 1009, 1979.
22. **Casti, A., Corti, A., Orlandini, G., Drusiani, F., and Caldarera, C. M.,** Polyamine biosynthesis in rabbit perfused heart under various degrees of oxygen tension, *Life Sci.*, 23, 1337, 1978.
23. **Johnson, M. D., Grignolo, A., Kuhn, C., and Shauberg, S. M.,** Hypertension and cardiovascular hypertrophy during chronic catecholamine infusion in rats, *Life Sci.*, 33, 169, 1983.
24. **Majesky, M. W., Yang, H. L., and Juchan, M. R.,** Interaction of alpha and beta adrenergic stimulation on aortic ornithine decarboxylase activity, *Life Sci.*, 36, 153, 1985.
25. **Womble, J. R., Larson, D. F., Copeland, J. G., and Russell, D. H.,** β_2-Receptors are coupled to the induction of myocardial protein synthesis and growth, *J. Nutr. Growth Cancer*, 1, 39, 1983.
26. **Womble, J. R. and Russell, D. H.,** Catecholamines-stimulated β_2-receptors couples to ornithine decarboxylase induction and to cellular hypertrophy and proliferation, *Adv. Polyamine Res.*, 4, 549, 1983.
27. **Lau, C. and Slotkin, T. A.,** Regulation of rat heart ornithine decarboxylase: change in affinity for ornithine evoked by neuronal, hormonal, and ontogenetic stimuli, *Mol. Pharmacol.*, 16, 504, 1979.
28. **Boucek, R. J., Jr., Olson, R. D., and Boerth, R. C.,** Relationship between the in situ activity of ornithine decarboxylase and contractile function of the rabbit papillary muscle, *Basic Res. Cardiol.*, 78, 631, 1983.
29. **Guarnieri, C., Flamigni, F., Muscari, C., and Caldarera, C. M.,** Involvement of calcium ions in the activation of ornithine decarboxylase by isoprenaline evaluated in situ in the perfused rat heart, *Biochem. J.*, 212, 241, 1983.
30. **Limas, C. J.,** Biochemical aspects of cardiac hypertrophy, *Fed. Proc. Fed. Am. Soc. Exp. Biol.*, 42, 2716, 1983.
31. **Millan, N. M., Rowe, P. M., and Siegel, F. L.,** The effects of injection of water or hypertonic saline on ornithine decarboxylase activity of neonatal rat heart and brain, *Life Sci.*, 26, 1807, 1980.
32. **Flamigni, F., Guarnieri, C., and Caldarera, C. M.,** Heart ornithine decarboxylase from control and isoproterenol-treated rats: kinetic properties, multiple forms and subcellular distribution, *Gen. Pharmacol.*, 17, 31, 1986.
33. **Flamigni, F., Stefanelli, C., Guarnieri, C., and Caldarera, C. M.,** Modulation of ornithine decarboxylase activity and ornithine decarboxylase-antizyme complex in rat heart by hormone and putrescine treatment, *Biochim. Biophys. Acta*, 882, 377, 1986.
34. **Flamigni, F., Guarnieri, C., and Caldarera, C. M.,** Characterization of highly purified rat heart ornithine decarboxylase, *Biochim. Biophys. Acta*, 802, 245, 1984.
35. **Flamigni, F., Guarnieri, C., Marmiroli, S., and Caldarera, C. M.,** Inhibition of rat heart ornithine decarboxylase by basic polypeptides, *Biochem. J.*, 229, 807, 1985.
36. **Meggio, F., Flamigni, F., Caldarera, C. M., Guarnieri, C., and Pinna, L. A.,** Phosphorylation of rat heart ornithine decarboxylase by type-2 casein kinase, *Biochem. Biophys. Res. Commun.*, 122, 997, 1984.
37. **Flamigni, F., Meggio, F., Guarnieri, C., Stefanelli, C., Pinna, L. A., and Caldarera, C. M.,** Ornithine decarboxylase regulation and polyamine metabolism in the heart tissue following catecholamine stimulation, in *Int. Conf. Polyamines in Life Science*, Lake Yamanaka, Japan, 1986, 75.

38. **Caldarera, C. M., Casti, A., Rossoni, C., and Visioli, O.,** Polyamines and noradrenaline following myocardial hypertrophy, *J. Mol. Cell. Cardiol.*, 3, 121, 1971.
39. **Caldarera, C. M., Orlandini, G., Casti, A., and Moruzzi, G.,** Polyamine and nucleic acid metabolism in myocardial hypertrophy of the overloaded heart, *J. Mol. Cell. Cardiol.*, 6, 95, 1974.
40. **Perin, A., Sessa, A., and Desiderio, M. A.,** Polyamine levels and diamine oxidase activity in hypertrophic heart of spontaneously hypertensive rats and of rats treated with isoproterenol, *Biochim. Biophys. Acta*, 755, 344, 1983.
41. **Grillo, M. A., Dianzani, U., and Pezzali, D. C.,** Acetylation of polyamines in chicken brain and retina, *Int. J. Biochem.*, 16, 1349, 1984.
42. **Seiler, N., Bolkenius, F. N., Knödgen, B., and Mamont, P.,** Polyamine oxidase in rat tissues, *Biochim. Biophys. Acta*, 615, 480, 1980.
43. **Matsui, I., Pöso, H., and Pegg, A. E.,** Conversion of exogenous spermidine into putrescine after administration to rats, *Biochim. Biophys. Acta*, 719, 199, 1982.
44. **Stefanelli, C., Carati, D., and Rossoni, C.,** Separation of N^1 and N^8-acetylspermidine isomers by reversed-phase column liquid chromatography after derivatization with dansyl chloride, *J. Chromatogr.*, 375, 49, 1986.
45. **Stefanelli, C., Carati, D., Rossoni, C., Flamigni, F., and Caldarera, C. M.,** Accumulation of N^1-acetylspermidine in heart and spleen of isoprenaline-treated rats, *Biochem. J.*, 237, 931, 1986.
46. **Tabor, C. W. and Tabor, H.,** Polyamines, *Annu. Rev. Biochem.*, 53, 749, 1984.
47. **Lau, C. and Slotkin, T. A.,** Regulation of ornithine decarboxylase activity in the developing heart of euthyroid or hyperthyroid rats, *Mol. Pharmacol.*, 18, 247, 1980.
48. **Slotkin, A. S.,** Ornithine decarboxylase as a tool in developmental neurobiology, *Life Sci.*, 24, 1623, 1979.
49. **Gibson, K. and Harris, P.,** The in vitro and in vivo effects of polyamines on cardiac protein biosynthesis, *Cardiovasc. Res.*, 8, 668, 1974.
50. **Casti, A., Guarnieri, C., Dall'Asta, R., and Clo, C.,** Effect of spermine on acetylation of histones in rabbit heart, *J. Mol. Cell. Cardiol.*, 9, 63, 1977.
51. **Caldarera, C. M., Casti, A., Guarnieri, C., and Moruzzi, G.,** Regulation of ribonucleic acid synthesis by polyamines, *Biochem. J.*, 152, 91, 1975.
52. **Krelhaus, W., Gibson, K., and Harris, P.,** The effects of hypertrophy, hypobaric conditions and diet on myocardial ornithine decarboxylase activity, *J. Mol. Cell. Cardiol.*, 7, 63, 1975.
53. **Matsushita, S., Sogani, R. K., and Raben, M. S.,** Ornithine decarboxylase in cardiac hypertrophy in the heart, *Circ. Res.*, 31, 699, 1972.
54. **Bartolome, J., Huguenard, J., and Slotkin, T. A.,** Role of ornithine decarboxylase in cardiac growth and hypertrophy, *Science*, 210, 793, 1980.
55. **Caldarera, C. M., Davalli, P., Flamigni, F., Clo, C., Toni, R., and Guarnieri, C.,** Inhibition of rat heart hypertrophy by 1-3 diaminopropane, in *Advances in Studies on Heart Metabolism*, Caldarera, C. M. and Harris, P., Eds., CLUEB, Bologna, Italy, 1982, 519.
56. **Tadolini, B., Cabrini, L., Landi, L., Varani, E., and Pasquali, P.,** Polyamine binding to phospholipid vesicles and inhibition of lipid peroxidation, *Biochem. Biophys. Res. Commun.*, 122, 550, 1984.
57. **Kitada, M., Igarashi, K., Horose, S., and Kitagawa, H.,** Inhibition by polyamines of lipid peroxide in rat liver microsomes, *Biochem. Biophys. Res. Commun.*, 87, 388, 1979.
58. **Elgavish, A., Wallace, R. W., Pillion, D. J., and Meezan, E.,** Polyamines stimulate D-glucose transport in isolated renal brush-border membrane vesicles, *Biochim. Biophys. Acta*, 777, 1, 1984.
59. **Tadolini, B. and Varani, E.,** Interaction of spermine with polyphosphoinositides containing liposomes and myo-inositol 1,4,5 triphosphate, *Biochem. Biophys. Res. Commun.*, 135, 58, 1986.
60. **Koenig, H., Goldstone, A., and Lu, C. Y.,** Polyamines regulate calcium fluxes in a rapid plasma membrane response, *Nature*, 305, 530, 1983.
61. **Fan, C. C. and Koenig, H.,** β-Adrenergic stimulation of calcium fluxes and membrane transport processes in rat heart is mediated by polyamines, *Fed. Proc. Fed. Am. Soc. Exp. Biol.*, 43, 735, 1983.
62. **Casti, A., Corti, A., Reali, N., Mezzetti, G., Orlandini, G., and Caldarera, C. M.,** Modifications of major aspects of myocardial ribonucleic acid metabolism as a response to noradrenaline, *Biochem. J.*, 168, 333, 1977.
63. **Muscari, C., Stefanelli, C., Speroni, E., Carati, D., and Rossoni, C.,** Rapid activation of ornithine decarboxylase in rabbit heart and thoracic aorta perfused with noradrenaline and inhibiting effect of the calcium antagonist diltiazem, in *Biomedical Studies of Natural Polyamines*, Caldarera, C. M., Clo, C., and Guarnieri, C., Eds., CLUEB, Bologna, Italy, 1986, 65.
64. **Astancolle, S., Baciottini, F., Davalli, P., Piccinini, G., Casti, A., and Corti, A.,** Ornithine decarboxylase in perfused rat heart, *J. Mol. Cell. Cardiol.*, 18, 223, 1986.
65. **Koenig, H.,** Polyamine mediate calcium-induced heart cell damage in the calcium paradox, *Int. Conf. Polyamines in Life Sciences*, Japan, 1986, 97.

66. **Toninello, A., Di Lisa, F., Siliprandi, D., and Siliprandi, N.,** Protective and restorative effects of spermine in the membrane potential of rat liver mitochondria, in *Advances in Polyamines in Biomedical Science,* Caldarera, C. M. and Bachrach, U., Eds., CLUEB, Bologna, Italy, 1984, 31.
67. **Nicchitta, C. V. and Williamson, J. R.,** Spermine. A regulator of mitochondrial calcium cycling, *J. Biol. Chem.,* 259, 1278, 1984.
68. **Solaini, G. and Tadolini, B.,** Spermine binding to submitochondrial particles and activation of adenosine triphosphate, *Biochem. J.,* 218, 495, 1984.
69. **Nikula, P., Alhonen-Hongisto, L., Seppänen, P., and Jänne, J.,** Inhibition of long-chain fatty acid oxidation by methylglyoxal bis (guanylhydrazone), *Biochem. Biophys. Res. Commun.,* 120, 9, 1984.
70. **Keskemeti, V., Marko, R., Selmeci, L., and Kelemen, K.,** Effects of polyamines on cardiac transmembrane potentials and transmembrane currents, in *Recent Progress in Polyamine Research,* Selmeci, L., Brosnan, M. E., and Seiler, N., Eds., Akademiai Kiado, Budapest, 1985, 203.
71. **Igarashi, K., Homma, R., Tokuno, H., Kitada, M., Kitagawa, H., and Hirose, S.,** Effect of polyamines on prostaglandin synthesis in various cell-free systems, *Biochem. Biophys. Res. Commun.,* 103, 659, 1981.
72. **Bazzani, C., Genedani, S., and Bertolini, A.,** Effect of polyamines on perfused rat heart contractility, *Pharm. Res. Commun.,* 17, 873, 1985.
73. **Marmo, E., Berrino, L., Cazzola, M., Filippelli, A., Cafaggi, G., Persico, N., Spadaro, R., and Nistico, G.,** Cardiovascular and respiratory effect of spermidine and spermine: an experimental study, *Biomed. Biochim. Acta,* 43, 509, 1984.
74. **Caldarera, C. M., Pignatti, C., and Clo, C.,** Polyamines in cultured heart cells, in *The Heart Cell in Culture,* Vol. 2, Pinson, A., Ed., CRC Press, Boca Raton, Fla., 1987, 95.
75. **Heby, O.,** Role of polyamines in the control of cell proliferation and differentiation, *Differentiation,* 14, 1, 1981.
76. **Pegg, A. E. and McCann, P. P.,** Polyamine metabolism and function, *Ann. J. Physiol.,* 243, C212, 1982.
77. **Berridge, M. J.,** The interaction of cyclic nucleotides and calcium in the control of cellular activity, *Adv. Cyclic Nucleotide Res.,* 6, 1, 1975.
78. **Chlapowski, F. J., Kelly, L. A., and Butcher, R. W.,** Cyclic nucleotides in cultured cells, *Adv. Cyclic Nucleotide Res.,* 6, 245, 1975.
79. **Goldberg, N. D., Haddox, M. K., Dunham, E., Lopez, C., and Hadden, J. W.,** The Yin Yang hypothesis of biological control. Opposing influence of cyclic GMP and cyclic AMP in the regulation of cell proliferation and other biological processes, in *Control of Proliferation in Animal Cells,* Clarkson, B. and Baserga, R., Cold Spring Harbor, Cold Spring Harbor, N.Y., 1974, 609.
80. **Krause, E. G. and Wollenberger, A.,** Cyclic nucleotides and heart, in *Cyclic 3′:5′-Nucleotides: Mechanism of Action,* Cramer, H. and Schultz, J., Eds., John Wiley & Sons, London, 1977, 229.
81. **Vapaatalo, H., Metsä-Keletä, T., Parantainen, J., Palo-Oja, T., Kangasaho, M., and Laustiola, K.,** The role of cyclic nucleotides and prostaglandins in heart function, *Acta Biol. Med. Germ.,* 37, 785, 1978.
82. **Harary, I., Renaud, J. F., Sato, E., and Wallace, G. A.,** Calcium ions regulate cyclic AMP and beating in cultured cells, *Nature,* 261, 60, 1976.
83. **Robinson, G. A., Butcher, R. W., and Sutherland, E. W.,** The catecholamines, in *Cyclic AMP,* Robinson, G. A., Butcher, R. W., and Sutherland, E. W., Eds., Academic Press, New York, 1971, 146.
84. **Teo, T. S. and Wang, J. H.,** Mechanism of activation of a cyclic adenosine 3′:5′ monophosphate phosphodiesterase from bovine heart by calcium ions, *J. Biol. Chem.,* 248, 5950, 1973.
85. **Wallach, D. and Pastan, I.,** Stimulation of membranes guanylate cyclase by concentrations of calcium that are in the physiological range, *Biochem. Biophys. Res. Commun.,* 72, 860, 1976.
86. **Desmond, W., Jr. and Harary, I.,** In vitro studies of beating heart cells in culture. XV. Myosin turnover and the effect of serum, *Arch. Biochem. Biophys.,* 151, 285, 1972.
87. **McLean, M. J. and Sperelakis, N.,** Retention of fully differentiated electrophysiological properties of chick embryonic heart cell in culture, *Dev. Biol.,* 50, 134, 1976.
88. **Moura, A. M. and Simkius, H.,** Cyclic AMP level in cultured myocardial cells under the influence of chronotropic and inotropic agents, *J. Mol. Cell. Cardiol.,* 7, 71, 1975.
89. **Sperelakis, N. and Pappano, A. J.,** Physiology and pharmacology of developing heart cells, *Pharmac. Ther.,* 22, 1, 1983.
90. **De Luca, M. A., Ingwall, J. S., and Bittl, J. A.,** Biochemical responses of myocardial cells in culture to oxygen and glucose deprivation, *Biochem. Biophys. Res. Commun.,* 59, 749, 1974.
91. **Frelin, C. and Padieu, P.,** Pleiotypic response of heart cells in culture to serum stimulation, *Biochemie,* 58, 953, 1976.
92. **Clo, C., Tantini, B., Coccolini, M. N., and Caldarera, C. M.,** Involvement of polyamines in cyclic AMP metabolism in heart cell cultures, *Adv. Polyamine Res.,* 3, 337, 1981.
93. **Clo, C., Tantini, B., Pignatti, C., Marmiroli, S., and Caldarera, C. M.,** Relevance of cellular polyamines in the response of heart cells to cyclic nucleotide-mediated effectors, in *Peptide Hormones, Biomembranes and Cell Growth,* Bolis, L., Verna, R., and Frati, L., Eds., Plenum Press, New York, 1984, 207.

94. **Pignatti, C., Tantini, C., Manfroni, S., Clo, C., and Caldarera, C. M.,** Regulation of ornithine decarboxylase activity in myocardial cell cultures: role of Ca^{2+}, in *Biomedical Studies of Natural Polyamines,* Caldarera, C. M., Clo, C., and Guarnieri, C., Eds., CLUEB, Bologna, Italy, 1986, 53.
95. **Clo, C., Coccolini, M. N., Tantini, B., and Caldarera, C. M.,** Increased sensitivity of heart cell cultures to norepinephrine after exposure to polyamine synthesis inhibitors, *Life Sci.,* 27, 67, 1980.
96. **Pignatti, C., Tantini, B., Rossoni, C., Turchetto, E., and Clo, C.,** Involvement of polyamines in phospholipid-induced changes of cyclic nucleotide contents and macromolecular synthesis in cultured heart cells, *Adv. Polyamine Res.,* 4, 331, 1983.
97. **Clo, C., Orlandini, G. C., Guarnieri, C., and Caldarera, C. M.,** Role of oxygen on growth rate and gene activity in cultured chick-embryo heart cells, *Biochem. J.,* 154, 253, 1976.
98. **Allfrey, V. G., Frankkner, R., and Mirsky, A. E.,** Acetylation and methylation of histones and their possible role in the regulation of RNA synthesis, *Proc. Natl. Acad. Sci. U.S.A.,* 51, 786, 1964.
99. **Clo, C., Orlandini, G. C., Casti, A., and Guarnieri, C.,** Polyamines as growth stimulating factors in eukariotic cells, *Ital. J. Biochem.,* 25, 94, 1976.
100. **Renaud, J. F., Scana, A. M., Kazazoglou, T., Lombet, A., Romez, G., and Lazdunski, N.,** Normal serum and lipoprotein deficient serum give different expression of excitability, corresponding to different stages of differentiation, in chick cardiac cells in culture, *Proc. Natl. Acad. Sci. U.S.A.,* 79, 7768, 1982.
101. **Caldarera, C. M., Pignatti, C., Tantini, B., Marmiroli, S., and Clo, C.,** Phospholipid-induced increase of polyamine synthesis in heart cell cultures, *J. Mol. Cell. Cardiol.,* 15 (suppl. 3), 2, 1983.
102. **Clo, C., Tantini, B., Pignatti, C., and Rossoni, C.,** A possible mediating role of polyamines in cyclic AMP-phosphodiesterase stimulation in confluent heart cell cultures, in *Advances in Studies on Heart Metabolism,* Caldarera, C. M. and Harris, P., Eds., CLUEB, Bologna, Italy, 1982, 139.
103. **Clo, C., Manfroni, S., Pignatti, C., Tantini, B., and Caldarera, C. M.,** Involvement of polyamines in glucagon-mediated induction of tyrosine aminotransferase in chick embryo heart cell cultures, in *Biomedical Studies of Natural Polyamines,* Caldarera, C. M., Clo, C., and Guarnieri, C., Eds., CLUEB, Bologna, Italy, 1986, 127.
104. **Clo, C., Pignatti, C., Manfroni, S., Marmiroli, S., and Tantini, B.,** Modulation of membrane-bound adenylate cyclase by polyamines, in *Biomedical Studies of Natural Polyamines,* Caldarera, C. M. and Guarnieri, C., Eds., CLUEB, Bologna, Italy, 1986, 117.
105. **Canellakis, Z., Theoharides, T. C., Boudy, P. K., and Thiarhos, E. T.,** The role of calcium in the induction of ornithine decarboxylase in rat HTC cells, *Life Sci.,* 29, 707, 1981.
106. **Gibbs, J. B., Hsu, C. Y., Terasaki, W. L., and Brooker, G.,** Calcium and microtubule dependence for increased ornithine decarboxylase activity stimulated by β-adrenergic agonists, dibutyryl cyclic AMP, or serum in a rat astrocytoma cell line, *Proc. Natl. Acad. Sci. U.S.A.,* 77, 995, 1980.
107. **Takai, Y., Nikkawa, U., Kaibuchi, K., and Nischizuka, Y.,** Membrane phospholipid metabolism and signal transduction for protein phosphorylation, *Adv. Cyclic Nucleotide Protein Phosphorylation Res.,* 18, 119, 1984.
108. **Kuramoto, A., Otani, S., Matsui, I., and Morisawa, S.,** Induction of ornithine decarboxylase by treatment of guinea pig lymphocytes with phospholipase C, *FEBS Lett.,* 151, 233, 1983.
109. **Otani, S., Matsui, I., Kuramoto, A., and Morisawa, S.,** Induction of ornithine decarboxylase in guinea pig lymphocytes by the divalent cation ionophore A 23187, *Eur. J. Biochem.,* 147, 27, 1985.
110. **Takai, Y., Kischimoto, A., Kikkawa, U., Mori, T., and Nischzuka, Y.,** Unsaturated diacylglycerol as a possible messenger for the activation of calcium-activated, phospholipid dependent protein kinase system, *Biochem. Biophys. Res. Commun.,* 91, 1218, 1979.
111. **Caldarera, C. M., Flamigni, F., Muscari, C., Clo, C., and Guarnieri, C.,** Evidence for sulphydrylic involvement in the inactivation of rat heart ornithine decarboxylase by oxygen radicals and hyperoxia, *Adv. Polyamine Res.,* 4, 525, 1983.
112. **Guarnieri, C., Flamigni, F., Davalli, P., Clo, C., and Caldarera, C. M.,** Inhibition of rat heart ornithine decarboxylase by superoxide radicals, *Ital. J. Biochem.,* 31, 63, 1982.
113. **Hollemberger, M.,** Effect of oxygen on growth of cultured myocardial cells, *Circ. Res.,* 28, 148, 1971.
114. **Das, R. and Kanungo, M. S.,** Effects of polyamines on in vitro phosphorylation and acetylation of histones of the cerebral cortex of rats of various age, *Biochem. Biophys. Res. Commun.,* 90, 708, 1979.
115. **Clo, C., Tantini, B., Pignatti, C., and Caldarera, C. M.,** Increased cGMP content in confluent and serum-restricted heart cell cultures exposed to polyamines, *J. Mol. Cell. Cardiol.,* 15, 139, 1983.
116. **Clo, C., Pignatti, C., Tantini, B., Marmiroli, S., and Caldarera, C. M.,** Cyclic nucleotide metabolism in quiescent chick embryo heart cell cultures treated with α-difluoromethylornithine, in *Advances in Polyamines in Biomedical Science,* Caldarera, C. M. and Bachrach, U., Eds., CLUEB, Bologna, Italy, 1984, 11.
117. **Clo, C., Tantini, B., Turchetto, E., Manfroni, S., and Pignatti, C.,** Modulation of cyclic nucleotide metabolism by exogenous phospholipids in heart cell cultures, *J. Mol. Cell. Cardiol.,* 15 (suppl. 3), 3, 1983.

118. **Pastan, I. H., Johnson, G. S., and Anderson, W. B.,** Role of cyclic nucleotides in growth control, *Annu. Rev. Biochem.*, 44, 491, 1975.
119. **Heby, O. and Jänne, J.,** Polyamine antimetabolites: biochemistry, specificity and biological effects of inhibitors of polyamine synthesis, in *Polyamines in Biology and Medicine,* Morris, D. R. and Marton, L. J., Eds., Marcel Dekker, New York, 1981, 243.
120. **Clo, C., Tantini, B., Pignatti, C., Guarnieri, C., and Caldarera, C. M.,** Regulation of cyclic nucleotide metabolism by polyamines in heart cell cultures, *Adv. Polyamine Res.*, 4, 667, 1983.
121. **Jänne, P., Pösö, H., and Raina, A.,** Polyamines in rapid growth and cancer, *Biochem. Biophys. Acta,* 473, 241, 1978.
122. **Caldarera, C. M., Tantini, B., Marmiroli, S., Pignatti, C., and Clo, C.,** Polyamine control of cellular cyclic nucleotide response, in *Recent Progress in Polyamine Research,* Selmeci, L., Brosnan, M. E., and Seiler, N., Eds., Akademiai Kiado, Budapest, 1985, 109.
123. **Alfheim, I. and Hougslo, J. K.,** Studies on the in vitro uptake of polyamines by Girardi heart cells, *Acta Pharmacol. Toxicol.*, 46, 171, 1980.
124. **Auberger, P., Samson, N., and Le Cam, A.,** Inhibition of hormonal induction of tyrosine aminotransferase by polyamines in freshly isolated rat hepatocytes, *Biochem. J.*, 216, 679, 1983.
125. **Clo, C., Caldarera, C. M., Tantini, B., Benalal, D., and Bachrach, U.,** Polyamines and cellular adenosine 3′:5′-cyclic monophosphate, *Biochem. J.*, 182, 641, 1979.
126. **Lehotay, D. C., Lo, H. K., and Levey, G. S.,** Activation of adenylate cyclase: requirement for phospholipids, in *Cyclic 3′:5′ Nucleotides: Mechanism of Action,* Cramer, H. and Schultz, J., Eds., John Wiley & Sons, London, 1977, 1.
127. **Strada, S. J. and Thompson, W. J.,** Multiple forms of cyclic nucleotide phosphodiesterases: anomalies or biological regulators?, *Adv. Cyclic Nucleotide Res.*, 9, 265, 1978.
128. **Byus, C. V. and Russell, D. H.,** Ornithine decarboxylase activity. Control of cyclic nucleotides, *Science,* 187, 650, 1979.
129. **Bourne, H. R., Coffino, P., Melman, K. L., Tomkins, G. M., and Weinstein, Y.,** Genetic analysis of cyclic AMP in a mammalian cell, *Adv. Cyclic Nucleotide Res.*, 5, 771, 1975.
130. **George, W. J., Busuttil, R. W., Paddock, R. J., White, L. A., and Ignarro, L. J.,** Opposing regulatory influence of cyclic guanosine monophosphate and cyclic adenosine monophosphate in the control of cardiac muscle contraction, in *Recent Advances in Studies on Cardiac Structure and Metabolism,* Roy, P. E. and Harris, P., Eds., University Park Press, Baltimore, 1975, 243.
131. **Clo, C., Manfroni, S., Pignatti, C., Tantini, B., Marmiroli, S., and Caldarera, C. M.,** Relevance of polyamine in the regulation of tyrosine aminotransferase in cultured heart cells, *Int. Conf. Polyamines in Life Science,* Japan, 1986, 207.
132. **Clo, C., Pignatti, C., Tantini, B., Marmiroli, S., and Caldarera, C. M.,** Cyclic nucleotide response of chick embryo heart cell cultures after preincubation with polyamines, *Ital. J. Biochem.*, 34, 203A, 1985.
133. **Liu, A. Y. C.,** Role of cyclic AMP-dependent protein kinase in the induction of tyrosine aminotransferase, *J. Biol. Chem.*, 225, 4421, 1980.
134. **Clo, C., Manfroni, S., Tantini, B., Pignatti, C., and Caldarera, C. M.,** Relevance of polyamines in the regulation of tyrosine aminotransferase (TAT) activity of cultured heart cells, *J. Mol. Cell. Cardiol.*, 18 (suppl. 2), A23, 1986.
135. **Clo, C., Pignatti, C., Tantini, B., Manfroni, S., Marmiroli, S., and Caldarera, C. M.,** Regolazione della sensibilita' ormonale di cellule cardiache in coltura ad opera di poliamine, *Cardiologia,* 32, 301, 1987.
136. **Bristow, M. R., Ginsburg, R., Minobe, W., Cubicciotti, R. S., Sageman, W. S., Lurie, K., Billingham, M. E., Harrison, D. C., and Stinson, E. B.,** Decreased catecholamine sensitivity and β-adrenergic-receptor density in failing human hearts, *New Engl. J. Med.*, 307, 205, 1982.

Chapter 4

POLYAMINES, HORMONE RECEPTORS, AND CALCIUM FLUXES

Harold Koenig, A. D. Goldstone, Chung Y. Lu, Zafar Iqbal, C.-C. Fan, and J. J. Trout,

TABLE OF CONTENTS

I. INTRODUCTION

Many hormones and neurotransmitters exert some of their biological actions by elevating the cytosolic Ca^{2+} concentration and generating Ca^{2+} signals in target cells.[1,2] Ca^{2+}-dependent responses are typically rapid and transient and involve binding of Ca^{2+} to calmodulin (or other Ca^{2+}-binding proteins) and subsequent interaction of the Ca^{2+}-calmodulin complex with specific or multifunctional calmodulin-dependent protein kinases or other enzymes and proteins.[3] Ca^{2+} mediates a wide variety of short-term processes such as endocytosis, exocytosis (secretion), membrane transport, contraction, locomotion, and the phosphorylation of diverse enzyme, cytoskeletal, and membrane proteins.[1,2] Ca^{2+} is also important for the promotion of long-term cellular events, such as growth, replication, and differentiation, which become apparent only after latencies of hours to days.[4]

A. Agonist-Mediated Regulation of Cytosolic Ca^{2+}

Figure 1 delineates the principal mechanisms for the regulation of cytosolic Ca^{2+} in mammalian cells. Hormones and other agonists increase cytosolic Ca^{2+} by interacting with surface receptors and augmenting the entry of extracellular Ca^{2+} and/or mobilizing intracellular Ca^{2+}.[1,2] Agonist-regulated Ca^{2+} entry occurs via voltage-dependent or -independent Ca^{2+} channels, through the Na^+ channel, or by the Na^+/Ca^{2+} exchanger.[1,2] Intracellular Ca^{2+} is mobilized mainly from the endoplasmic (sarcoplasmic) reticulum,[5] but some Ca^{2+} may also originate from the mitochondria and the plasmalemma.[6] To avert a pathological accumulation of Ca^{2+} and preserve calcium homeostasis, cells possess a plasmalemmal Ca^{2+}-ATPase pump of high affinity and low transport capacity which maintains cytosolic Ca^{2+} at a resting level of 0.05 to 0.15 μM.[7] Certain hormones inhibit this Ca^{2+} pump, and apparently prolong the elevation in cytosolic Ca^{2+} through this mechanism.[8] The Na^+/Ca^{2+} exchanger with low affinity and high capacity also assists in the extrusion of Ca^{2+}.[9]
In addition, cytomembranes and soluble components buffer or bind Ca^{2+}. Of particular importance in this regard is the endoplasmic (sarcoplasmic) reticulum, which removes Ca^{2+} via a high affinity Ca^{2+}-ATPase pump distinct from that in the plasmalemma, and serves as a rapid and sensitive modulator of Ca^{2+} in the physiological range (0.1 to 1 μM).[10] Mitochondria take up Ca^{2+} through a putative uniporter of low affinity and high transport capacity that is driven by the proton motive force. For this and other reasons, mitochondria are thought to act as a large-scale, long-term regulator primarily to prevent excessive fluctuations of cytosolic Ca^{2+}.[11]

B. Second Messengers

It has long been thought that agonist-regulated effects on Ca^{2+} fluxes are mediated by one or more intracellular messengers generated in or near the plasma membrane. Much recent evidence has implicated inositol 1,4,5-trisphosphate, produced by receptor-mediated hydrolysis of phosphatidylinositol 4,5-bisphosphate by a phosphodiesterase (phospholipase C), as a second messenger involved in the release of Ca^{2+} from the endoplasmic reticulum.[12] This pathway generates yet another putative messenger, diacylglycerol, which accumulates in the plasma membrane where it activates protein kinase *c*, a Ca^{2+}- and phospholipid-dependent enzyme which phosphorylates membrane and other proteins.[13] In many cell types the Ca^{2+} pathway of agonist action appears to interact synergistically with the diacylglycerol pathway to produce physiological responses.[13] Receptor-mediated activation of adenylate cyclase produces cyclic AMP, the most ancient of the second messengers, which regulates a number of multipurpose cyclic AMP-dependent protein kinases, and also acts synergistically with the Ca^{2+}-calmodulin messenger system in some cell systems to evoke physiological responses.[14] In the cardiac myocyte β-adrenergic receptor activation is thought to

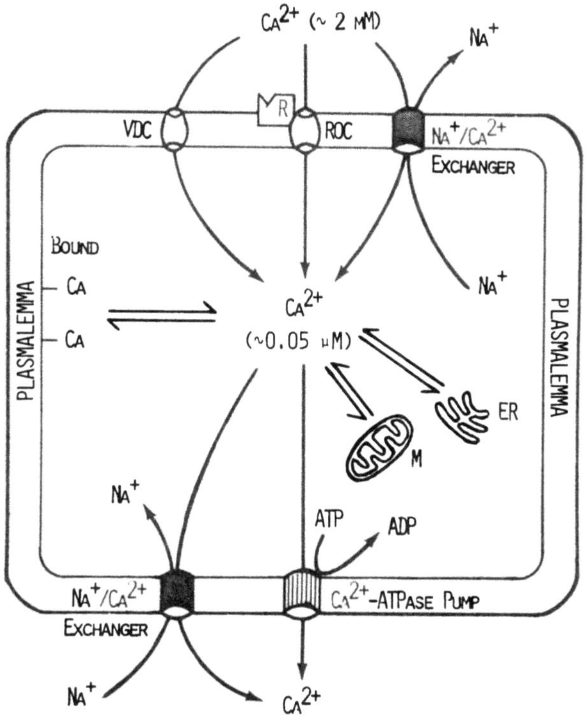

FIGURE 1. Scheme of cellular mechanisms for regulation of cytosolic Ca^{2+}: M, mitochondria; ER, endoplasmic reticulum; VDC, voltage-dependent channel; ROC, receptor-operated channel; and R, receptor.

increase Ca^{2+} influx through a cyclic AMP-dependent phosphorylation of a Ca^{2+} channel protein(s) or protein(s) closely associated with the channel.[15] Despite the rapid advance in knowledge in this area, many important questions remain unanswered. These include the molecular mechanisms underlying agonist-mediated Ca^{2+} influx or inhibition of the Ca^{2+} pump, the precise mechanisms by which inositol, 1,4,5-trisphosphate mobilizes intracellular Ca^{2+}, the occurrence of other messengers, and the coupling of receptors to messenger formation.

II. ROLE OF POLYAMINES AS SECOND MESSENGERS

A. Introduction

Ornithine decarboxylase (ODC), the first and rate-limiting enzyme in the polyamine biosynthetic pathway, features a striking inducibility by a wide variety of stimuli, including all known classes of hormones, and an exceptionally short half-life (10 to 20 min).[16,17] Classically the increase in ODC commences 1 to 2 hr after the administration of a hormone or other growth stimulus, peaks at 4 to 8 hr, and leads to sequential increase in putrescine, spermidine, and ribosomal RNA synthesis and accumulation.[17] These hormonally induced increases in ODC activity generally reflect *de novo* synthesis of enzyme protein by transcriptional and/or translational regulatory mechanisms.[16,17] There is mounting evidence that ODC is also subject to very rapid regulation of its catalytic activity by an as yet uncharacterized posttranslational process. In 1983, Koenig et al. reported that the steroid hormone testosterone[18] and the β-adrenergic agonist isoproterenol[19] induce a rapid (<30 sec), transient increase in ODC activity followed by an early (<2 min) increase in putrescine, spermidine, and spermine in mouse kidney cortex. These stimuli also evoke an early increase in Ca^{2+} influx, mobilization of intracellular Ca^{2+}, and stimulation of Ca^{2+}-dependent membrane transport (endocytosis, amino acid transport, hexose transport).[20-22] Rapid polyamine synthesis has been shown to be mandatory for the mediation of these effects, suggesting that polyamines may serve as second messengers in short-term hormonal responses.[18,19] Subsequent studies have confirmed and extended these findings. These results support a new model for signal transduction and stimulus-response coupling in which polyamines are messengers that regulate Ca^{2+} signals, and mediate diverse Ca^{2+}- and polyamine-sensitive responses to hormones, neurotransmitters, and other stimuli.[18,19,23,24] Some of these studies are here briefly surveyed.

B. Role of Polyamines in Androgenic Hormone Action

1. Kidney Cortex

Testosterone induces a number of phenotypic effects in mouse kidney that require receptor occupancy, are characterized by enhanced RNA and protein synthesis and increases in β-glucuronidase, ODC, and several other specific proteins, and lead to a large growth response within a few days.[25-27] Studies in the authors' laboratory showed that the androgenic response in mouse kidney involves enhanced activity of the lysosomal-vacuolar system in proximal tubule cells which is manifested morphologically in enhanced autophagy, an accumulation of hypertrophied, membrane-filled lysosomes (myeloid bodies), and exocytosis of these lysosomes into the tubule lumen and biochemically in increased tissue levels of numerous lysosomal enzymes, a dramatic lysosomal enzymuria and proteinuria,[28,29] and decreased enzyme latency and membrane stability of kidney lysosomes.[30,31] Testosterone also induces alterations in proximal tubule mitochondria that are expressed as changes in size, fine structure,[29] and equilibrium density (unpublished data), a concomitant increase in the inner mitochondrial enzyme cytochrome *c* oxidase,[29] and enhanced cortical tissue respiration.[29a] These morphological and biochemical features of androgenic hormone action are readily observed in the kidney cortex of untreated male mice, and are abolished within a week following orchiectomy, indicating that the sexual dimorphism of this organ is mediated by endogenous testosterone.[29] Male mice exhibit higher levels of kidney ODC and polyamines and excrete larger amounts of polyamines into the urine than female mice, and this sex difference is also abolished by orchiectomy and restored by testosterone administration.[32] α-Difluoromethylornithine (DFMO), a specific, enzyme-activated, irreversible inhibitor of ODC,[16,17,33] suppressed the testosterone-induced increase in kidney ODC and polyamines and urinary excretion of polyamines, inhibited the increment in kidney cytochrome *c* oxidase, lysosomal enzymes, and lysosomal enzymuria, and attenuated the renal hypertrophy. Exogenous putrescine raised kidney polyamine levels and abrogated DFMO inhibition.[32] Thus, testosterone-induced stimulation of ODC activity and polyamine synthesis appears to be essential in the mediation of the long-term effects of the hormone on proximal tubule mitochondria, lysosomes, and growth.

That testostersone also evokes a rapid response in mouse kidney proximal tubules was first suggested by the observation that the hormone induces an early (< 5 min) decrease in the equilibrium density, enzyme latency, and membrane stability of a population of kidney lysosome apparently resulting from enhanced endosome lysosome fusion.[30,31] A subsequent study in mouse kidney cortex slices disclosed that nanomolar concentrations of testosterone induce a rapid (<1 min) stimulation of endocytosis, amino acid transport, and hexose transport, monitored by the temperature-sensitive uptake of horseradish peroxidase (HRP), α-[^{14}C] aminoisobutyrate (AIB), and 2-deoxy-[^{3}H] glucose (DG) respectively.[20] This membrane transport response showed testosterone concentration dependence and steroid specificity, and was absent in the tfm/Y mouse, which lacks androgen receptors, indicating that it is androgen receptor mediated.[20] Androgenic stimulation of membrane transport functions is dependent on extracellular Ca^{2+} and apparently involves Ca^{2+} channel activity as it is verapamil sensitive.[22] Testosterone rapidly stimulates the influx and efflux of $^{45}Ca^{2+}$ and mobilizes mitochondrial $^{45}Ca^{2+}$ into the cytosol, presumably leading to a rise in free cytosolic Ca^{2+} concentration.[22]

Testosterone also evokes a rapid (<30 sec), transient stimulation of ODC activity and an accumulation of all three polyamines in mouse kidney in vivo and in mouse kidney cortex slices in vitro (Figure 2).[18] The ODC inhibitor DFMO was used to assess the physiological role of these changes in polyamines. DFMO suppressed the testosterone-induced increase in tissue ODC and polyamines and abolished the stimulated increment in Ca^{2+} fluxes and Ca^{2+}-dependent membrane transport (Figure 3). Conversely exogenous putrescine (0.5 m*M*), the product of ODC activity, negated DFMO inhibition and restored the increase in polyamines, Ca^{2+} fluxes, and membrane transport. These results indicate that rapid polyamine synthesis is obligatory for the testosterone-mediated increase in Ca^{2+} movements and Ca^{2+}-dependent membrane transport, and support the hypothesis that polyamines serve as intracellular messengers to control Ca^{2+} fluxes and Ca^{2+}-dependent cell responses.[18]

2. Heart and Heart Myocytes

The hearts of rats and mice also display a testosterone-dependent sexual dimorphism similar to that observed in the kidney,[34] and this long-term effect apparently is mediated by polyamines.[35] We therefore studied the short-term effects of testosterone in female rat heart slices.[35a] Testosterone (10 n*M*) induced an abrupt (<30 sec) rise in ODC activity which peaked at 60 sec and then declined, but remained elevated for 30 min. Putrescine, spermidine, and spermine increased by 30 sec, peaked at 2 min, and then declined but they were also still elevated at 30 min. Concomitantly, testosterone stimulated $^{45}Ca^{2+}$ influx and efflux at 30 sec, and Ca^{2+}-dependent endocytosis, hexose transport, and amino acid transport by 60 sec. All of these acute effects of testosterone were blocked by DFMO (5 m*M*), and putrescine (0.5 m*M*) nullified the DFMO effect, thereby demonstrating that rapid ODC-mediated polyamine synthesis is an essential prerequisite for the mediation of acute heart responses to androgenic stimulation.

Similar experiments were conducted in acutely isolated rat ventricular myocytes, where testosterone (10 n*M*) evoked an even earlier (<5 sec) rise in ODC activity which peaked (twofold) at 30 to 60 sec and then declined.[35a,36] Putrescine, spermidine, and spermine increased almost as rapidly as ODC and then also declined. In pilot experiments using the fluorescent Ca^{2+} indicator dye quin 2, testosterone (10 n*M*) induced an early (<8 sec) rise in free cytosolic Ca^{2+} concentration in isolated cardiomyocytes, and this increase was prevented by removal of extracellular Ca^{2+} or by DFMO, suggesting Ca^{2+} and polyamine dependence.[38] These observations led to the prediction that testosterone might influence cardiac contractility and rate by modulating Ca^{2+} fluxes and cytosolic Ca^{2+} levels. This prediction was confirmed in experiments employing the isolated rat heart perfused retrograde through the aorta in the Langendorff mode.[37,38] Testosterone (10 n*M*) was observed to stimulate cardiac contraction and increase heart rate after a lag period of about 15 sec. An acute (<5 to 10 sec), transient increase (1.5- to 2-fold) in ODC and polyamine levels (Figure 4) preceded the contractile response. DFMO (5 m*M*) blocked the testosterone-evoked increase in ODC and polyamines and the stimulation of contractility and rate, whereas putrescine (0.5 m*M*) reversed the DFMO effect. These findings demonstrate that testosterone exerts an acute positive inotropic and chronotropic effect on rat heart that evidently is mediated by polyamines. In light of the central importance of Ca^{2+} in excitation-contraction coupling,[39] it appears likely that the effects on contractility involve a polyamine-dependent stimulation of Ca^{2+} fluxes and an elevation of free cytosolic Ca^{2+} in ventricular myocytes and cells of the conducting system.

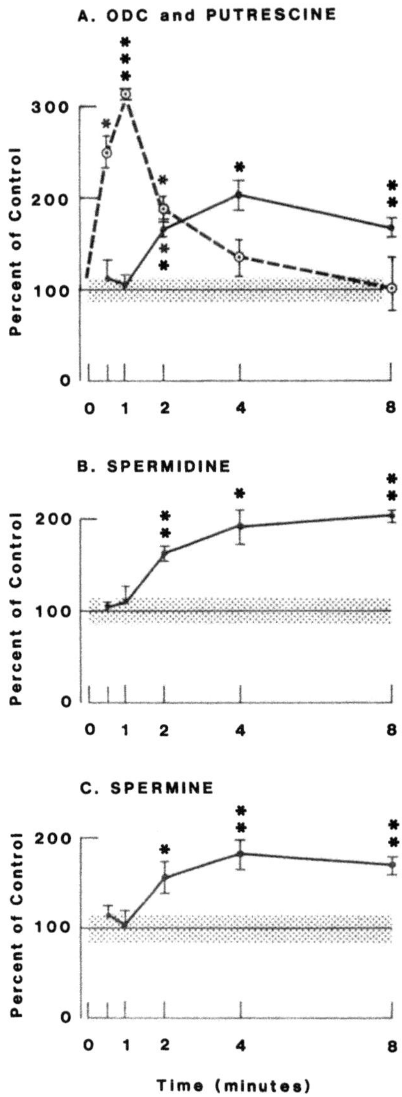

FIGURE 2. Testosterone induces an acute, transient rise in ODC activity and an increase in polyamine levels in mouse kidney cortex in vitro. ODC (⊙ — — — ⊙). Tissue slices from female mouse kidney cortex are incubated in a physiological salt medium with 10 n*M* testosterone for given times. Data are means ± SEM. *, **, ***: $p < 0.05$, 0.01, 0.001 (vs. incubation matched controls). (Reprinted by permission from Koenig, H., Goldstone, A., and Lu, C. Y., *Nature*, 305, 530. Copyright ©1983 Macmillan Journals Limited.)

C. Role of Polyamines in β-Adrenergic Responses

1. Kidney Cortex

The β-adrenergic agonist 1-isoproterenol induces an acute Ca^{2+}- and β-adrenoceptor-dependent stimulation of endocytosis, hexose transport, and amino acid transport in mouse kidney cortex involving proximal tubule cells.[21] This membrane transport response includes

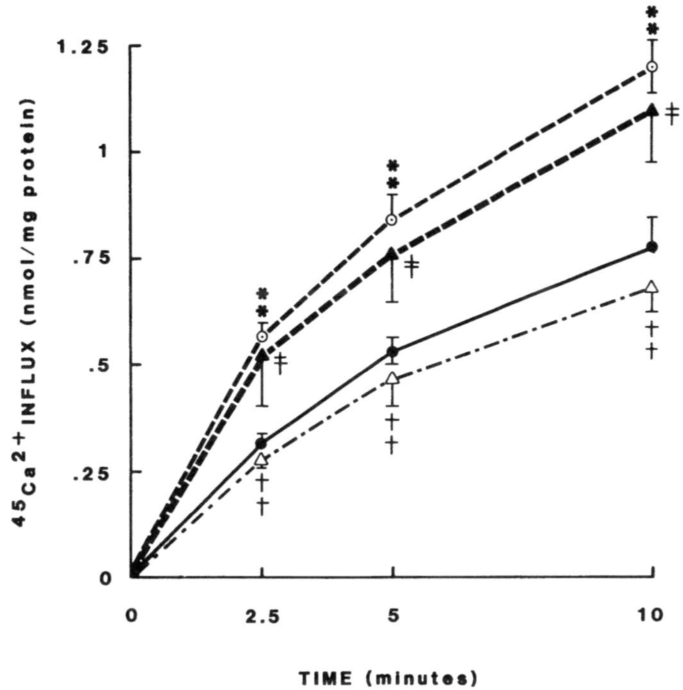

A

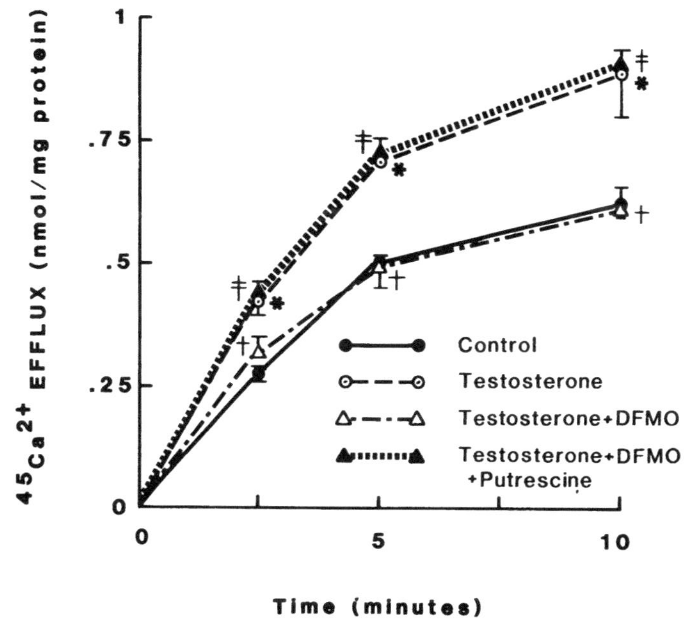

B

FIGURE 3. DFMO abolishes the testosterone-induced increase in ^{45}Ca influx and efflux in kidney cortex, and putrescine reverses DFMO inhibition. (A) Cortex slices from mouse kidney were preincubated in medium with 5 m*M* DFMO, DFMO + 500 μ*M* putrescine, or no agent for 10 min at 37°C. At zero time slices were transferred to fresh incubation medium containing ^{45}Ca with 10 n*M* testosterone and additives as shown and incubated for the given times. Data are means ± SEM. (B) Cortex slices were preincubated for 30 min at 37°C in medium with ^{45}Ca. Additives were present in the last 10 min of preincubation and during the incubation with 10 n*M* testosterone which was initiated at zero time. Data are means ± SEM. *, **, ***: $p < 0.05$, 0.01, 0.001 (vs. control). †, ††, †††: $p < 0.05$, 0.01, 0.001 (vs. testosterone). ‡, ‡‡, ‡‡‡: $p < 0.05$, 0.01, 0.001 (vs. testosterone + DFMO). (Reprinted by permission from Koenig, H., Goldstone, A., and Lu, C. Y., *Nature*, 305, 530. Copyright© 1983 Macmillan Journals Limited.)

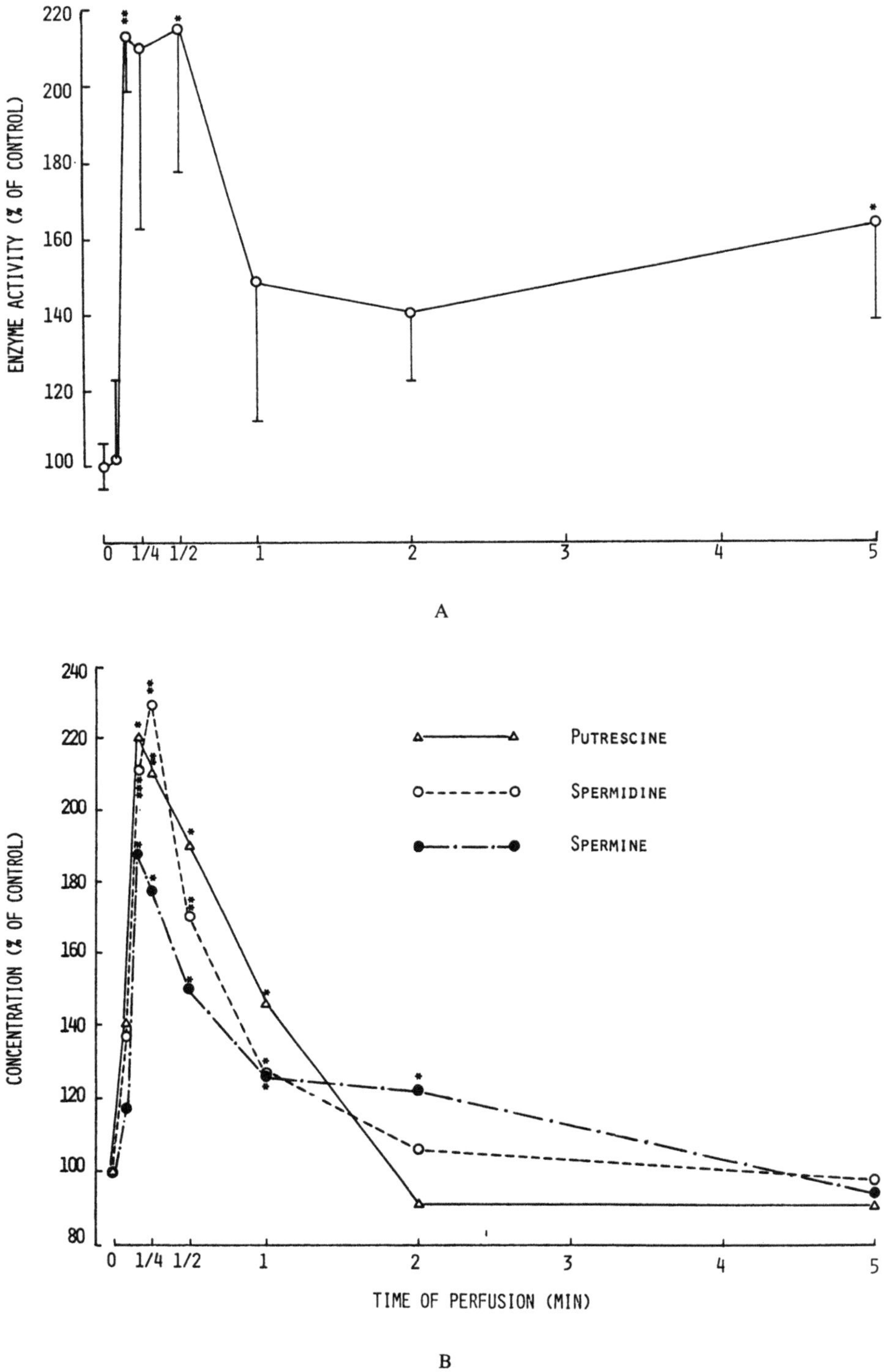

FIGURE 4. Testostersone induces an acute, transient increase in ODC activity and polyamine levels in the isolated perfused rat heart. Hearts excised from anesthetized, heparinized female rats were perfused with a modified Krebs-Henseleit buffer in the Langendorff mode. At the given times after infusion of 10 n*M* testosterone hearts were freeze-clamped and analyzed for (A) ODC and (B) polyamines. Data are means ± SEM. *, **: $p < 0.05$, 0.01 (vs. control perfusions).

an early (<30 sec) increase in Ca^{2+} influx and efflux, and a redistribution of mitochondrial calcium into the cytosol. This response requires a verapamil-sensitive transmembrane transport of extracellular Ca^{2+}, and is mimicked by the calcium ionophore A23187, suggesting that cytosolic Ca^{2+} is involved in stimulus-permeability coupling. Isoproterenol also induces an acute (<30 sec) transient rise in ODC activity which peaks (threefold) at 90 sec, followed by early increases in putrescine, spermidine, and spermine levels (Figure 5).[19] The β-adrenergic antagonist propranolol and DFMO block isoproterenol stimulation of ODC activity, polyamine synthesis Ca^{2+} fluxes (Figure 6), calcium redistribution, and membrane transport processes. Putrescine reverses the effects of DFMO. Therefore, β-adrenoceptor-mediated polyamine synthesis is essential for coupling these several responses to β-adrenergic stimulation.[19]

2. Heart and Heart Myocytes

β-Adrenergic modulation of voltage-sensitive Ca^{2+} channels in cardiac myocytes mediates a slow inward current carried by Ca^{2+} after a 5- to 10-sec delay that profoundly influences cardiac function and is of major importance in the neuroregulatory control of heart action.[40] Ca^{2+} channel modulation and enhanced Ca^{2+} influx are thought to be mediated by a cyclic AMP-dependent protein phosphorylation which increases the number of functional Ca^{2+} channels per cell and alters the kinetics of channel opening and closing.[15] However, the molecular mechanisms underlying the control of Ca^{2+} channels remain only partially understood. In studies involving rat heart slices, isoproterenol was found to induce an early concentration-, Ca^{2+}-, and β-adrenoceptor-dependent stimulation of $^{45}Ca^{2+}$ influx (<30 to 60 sec) and efflux, endocytosis, hexose transport, and amino acid transport.[41,42] Isoproterenol (0.1 μM) also induced a rapid (<30 sec) elevation in ODC and polyamine levels. DFMO (5 mM) suppressed isoproterenol stimulation of ODC activity, polyamine accumulation, Ca^{2+} fluxes, and membrane transport, and putrescine (0.5 mM) negated DFMO suppression. These data are consistent with the hypothesis that polyamines are important in β-adrenoceptor-mediated regulation of cytosolic Ca^{2+} and Ca^{2+}-dependent cell responses by enhancing Ca^{2+} influx via Ca^{2+} channels and/or Na^+-Ca^{2+} exchange, and by mobilizing internal Ca^{2+}.[19,41,42]

According to this model, the positive inotropic effect of isoproterenol and other β-adrenergic agonists should result from a polyamine-directed increase in cytosolic Ca^{2+} in cardiac myocytes. This prediction was examined in the isolated rat heart perfused in the Langendorff mode. Infusion of isoproterenol (1 nM) induced an increase in heart contractility and rate after a lag period of 15 to 20 sec, whereas ODC and polyamine levels in clamp-frozen heart increased within 5 to 10 sec and peaked (1.5- to 2-fold) at 30 sec (Figure 7). DFMO (5 mM) blocked isoproterenol stimulation of heart ODC activity, polyamine synthesis, contractility, and rate, and putrescine (0.5 mM) negated the DFMO effect, thus confirming that rapid polyamine synthesis is mandatory for the positive inotropic and chronotropic effects of isoproterenol.[38,43]

Acutely isolated rat ventricular myocytes prepared by enzymatic digestion were used to investigate further the cellular and molecular mechanisms of β-adrenergic stimulation.[36,36a,44] In isolated cardiac myocytes isoproterenol (10 nM) induced an immediate (<5 sec) transient increase in ODC activity which peaked (twofold) at 30 sec and then decreased. The concentrations of all three polyamines also increased transiently. Isoproterenol also induced a rapid (<60 sec) stimulation of Ca^{2+}-dependent endocytosis, hexose transport, and amino acid transport in myocytes. In studies with the fluorescent Ca^{2+} indicator dye fura 2, isoproterenol (10 nM) induced an acute (<8 sec) fall in free cytosolic Ca^{2+}, and this fall was prevented in Ca^{2+}-free medium or in the presence of DFMO (10 mM). DFMO also abolished isoproterenol stimulation of ODC activity, polyamine synthesis, Ca^{2+} fluxes, and membrane transport, and putrescine (0.5 mM) reversed the DFMO effect. These data show

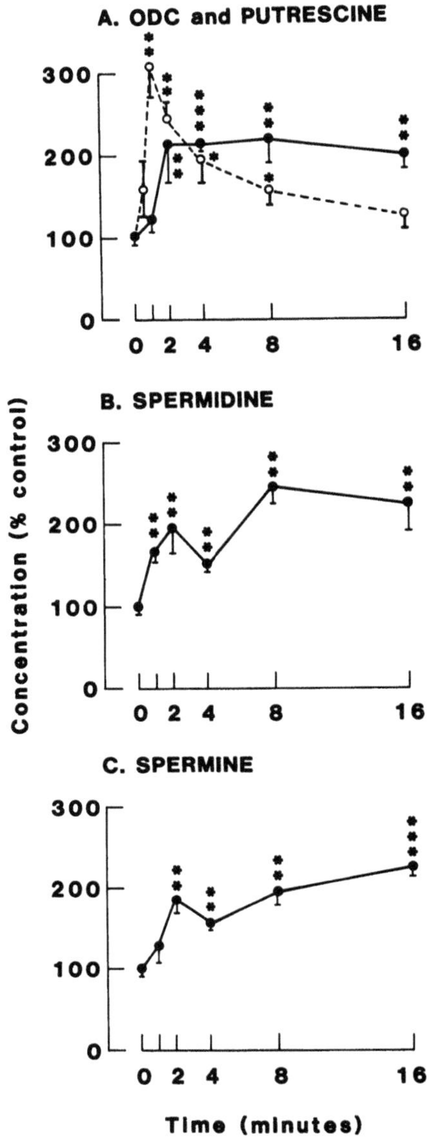

FIGURE 5. Isoproterenol induces an acute, transient increase in ODC activity and an early accumulation of polyamines in mouse kidney cortex in vitro. ODC (○ ---- ○). Tissue slices were incubated with 1 μM isoproterenol for given times. Results are means ± SEM. *, **, ***: p <0.05, 0.01, 0.001 (vs. zero-time control). (From Koenig, H., Goldstone, A., and Lu, C. Y., *Proc. Natl. Acad. Sci. U.S.A.*, 80, 7210, 1983).

that nanomolar concentrations of isoproterenol evoke a rapid, Ca^{2+}- and receptor-dependent stimulation of ODC activity and polyamine synthesis in ventricular myocytes. This rapid polyamine synthesis appears to be obligatory for the mediation of the various physiological effects of β-adrenergic stimulation.

A

B

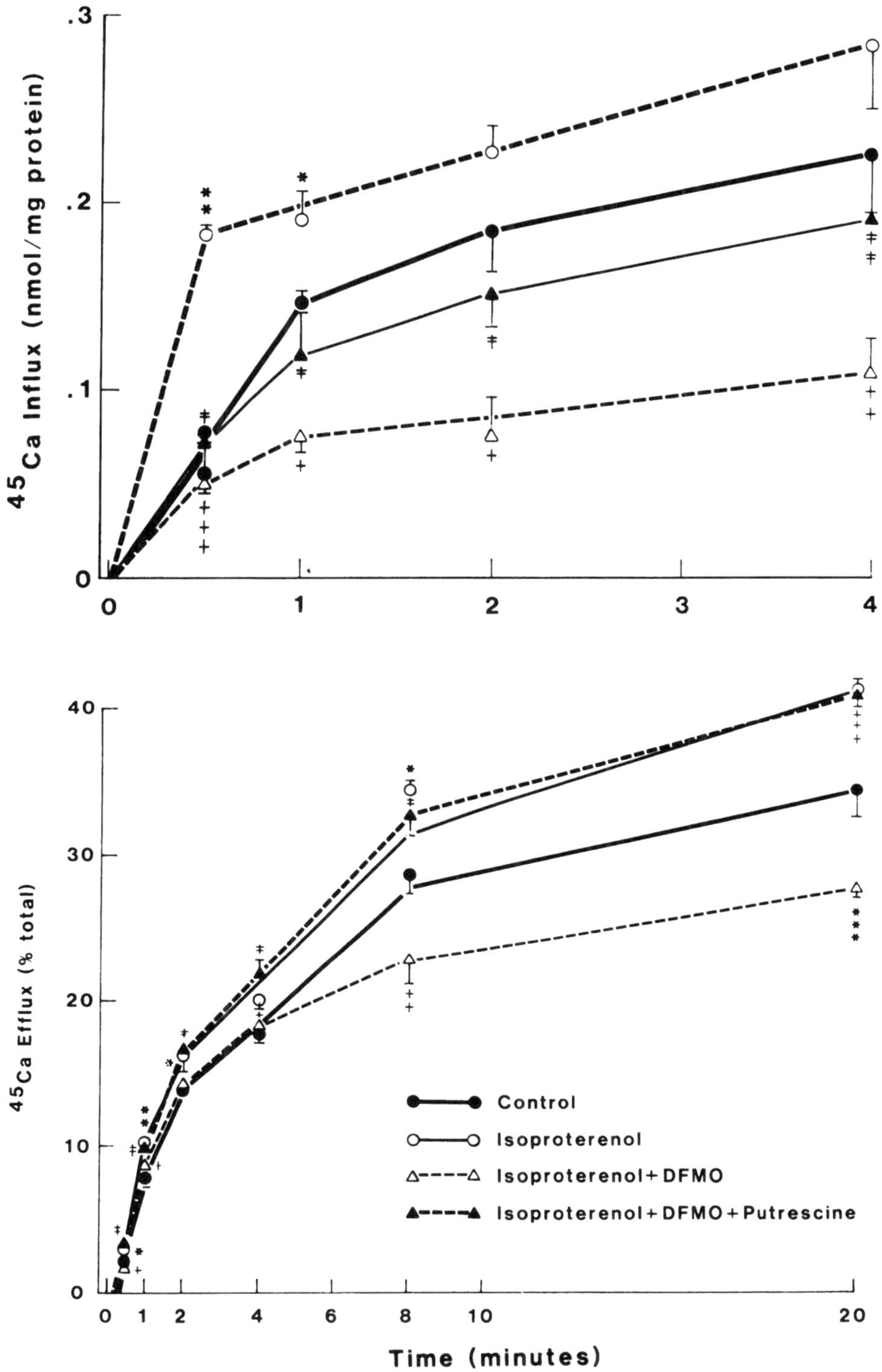

FIGURE 6. DFMO abolishes the isoproterenol-induced increase in ^{45}Ca influx and efflux in kidney cortex, and putrescine reverses DFMO inhibition. The experimental protocol was similar to that given in Figure 3. At zero time 1 μM isoproterenol was added. (A) ^{45}Ca influx, and (B) ^{45}Ca efflux. The data are means ± SEM. The key to the interpretation of significance is given in Figure 3. (From Koenig, H., Goldstone, A., and Lu, C. Y., *Proc. Natl. Acad. Sci. U.S.A.*, 80, 7210, 1983).

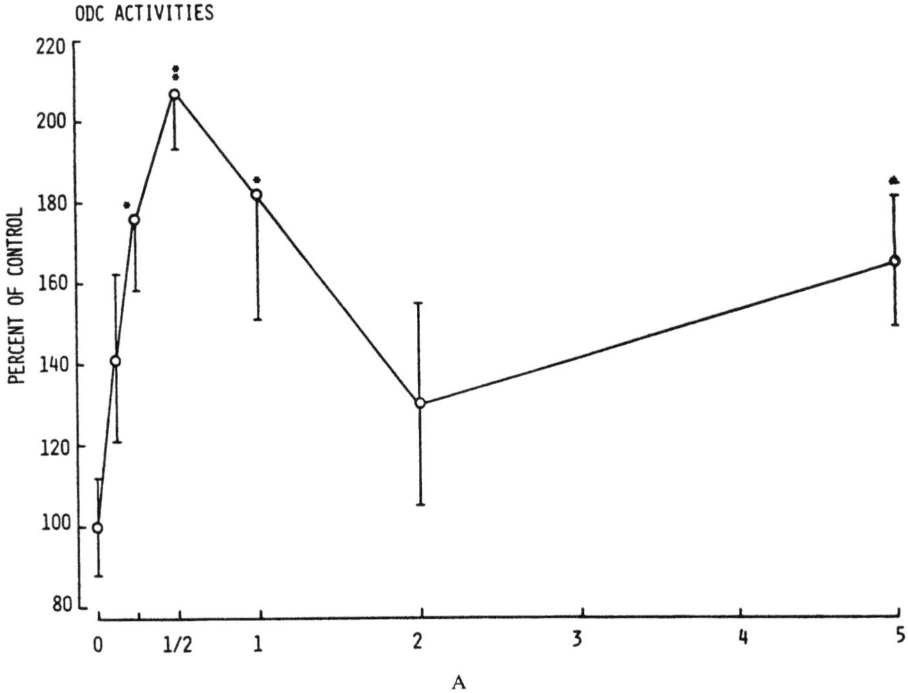

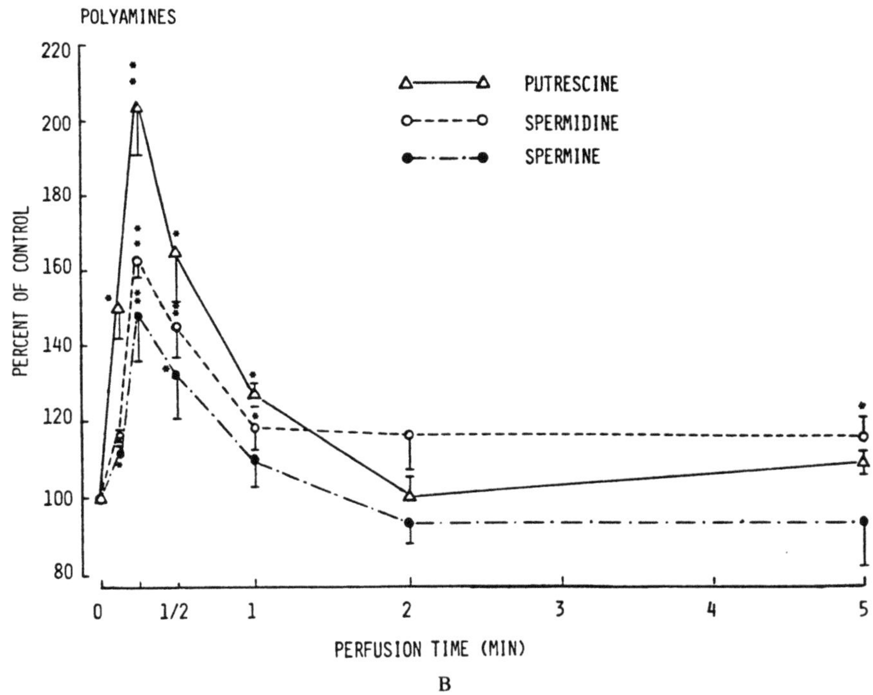

FIGURE 7. Isoproterenol evokes an acute, transient elevation in ODC activity and polyamine levels in the isolated perfused rat heart. Excised female rat hearts were perfused with a modified Krebs-Henseleit buffer in the Langendorff mode. Hearts were freeze-clamped at the given times after infusion of 1 n*M* isoproterenol for biochemical analysis of (A) ODC and (B) polyamines. Data are means ± SEM. *, **: $p < 0.05$, 0.01 (vs. control perfusions).

D. The Role of Polyamines in Insulin Responses

The binding of insulin to its receptor in target cells induces changes in several membrane transport processes and in the activities of numerous enzymes regulating the metabolism of carbohydrates, lipids, and amino acids, as well as long term effects on RNA, DNA, and protein hypothesis.[45] Many of the acute effects of insulin upon metabolism involve pathways regulated by protein kinases and protein phosphatases controlling reversible enzyme phosphorylations. Despite extensive efforts, the molecular mechanisms underlying the effects of the hormone are still largely unknown. Recent studies have indicated that insulin, on interacting with its receptor in the plasma membrane, causes the generation and release of several related acid- and heat-stable molecules of low molecular weight (500 to 2000 daltons) that appear to act as mediators for many intracellular effects of insulin.[46-48] However, the chemical nature of these mediators and their mode of formation remain unresolved. Recent studies in the authors' laboratory strongly suggest that these mediators are in fact polyamines.

1. Kidney Cortex, Liver, and Arterial Myocytes

Mouse kidney cortex, an insulin-sensitive tissue, was employed to investigate the putative mediator role of polyamines in insulin action in some detail.[49-51] Insulin rapidly and transiently increased the rates of endocytosis, amino acid transport, and glucose transport, as monitored by the temperature-sensitive uptake of HRP, ^{14}C-AIB, and ^{3}H-DG, respectively, in kidney cortex slices. Stimulation of these transport processes varied with insulin concentration and was maximal at 200 microunits per milliliter (1.3 n*M*), suggesting receptor dependence. Insulin stimulation of membrane transport was Ca^{2+}-dependent, as it was attenuated or abolished by Ca^{2+} deletion, the Ca^{2+} chelator EGTA, the calcium antagonist 1 m*M* La^{3+}, and the calcium channel blocker verapamil. Insulin stimulation of pyruvate dehydrogenase activity also was rapid and transient and followed by inhibition at 10 min. Other workers have described dose- and time-dependent biphasic insulin responses and inhibitory, as well as stimulatory, mediators.[46-48] Insulin evoked an early (<30 sec) increase in $^{45}Ca^{2+}$ influx and efflux which peaked by 1 min and then declined to near basal levels by 4 min. Insulin also effected a redistribution of cellular ^{45}Ca, manifested by a reduction in mitochondrial ^{45}Ca and an increase in cytosolic ^{45}Ca. These data suggest that insulin may transiently increase the cytosolic Ca^{2+} concentration.

Insulin (1.3 n*M*) induced an early, transient accumulation of putrescine, spermidine, and spermine in kidney cortex slices (Figure 8). All three polyamines increased significantly by 30 sec, remained elevated for about 2 min, and then decreased to basal levels by 5 min. An abrupt (<15 sec) increase in ODC activity preceded the rise in polyamines, peaked (2.3-fold) at 1 min, and then rapidly declined to near basal levels. The insulin-induced increase in ODC activity and polyamine levels displayed concentration dependence with maximum elevations occurring at 1.3 n*M* insulin. This polyamine accumulation was blocked by the suicide ODC inhibitor DFMO, indicating that it was due to ODC-mediated polyamine synthesis, and not a consequence of decreased polyamine catabolism. The insulin-induced increase in ODC activity and polyamine levels preceded the stimulation of $^{45}Ca^{2+}$ fluxes, membrane transport processes, and pyruvate dehydrogenase activation. It is also noteworthy that the dose-response curves for ODC activity and polyamine accumulation by insulin correlated well with the dose-response curve for membrane transport and Ca^{2+} influx. DFMO (5 m*M*) blocked insulin stimulation of ODC activity and polyamine synthesis and concurrently abolished the stimulation of $^{45}Ca^{2+}$ fluxes (Figure 9).

We also studied the role of polyamines in insulin stimulation of pyruvate dehydrogenase activity in mouse liver in vivo. Intraperitoneal administration of insulin (0.33 nmol/kg) to mice induced an early (<1 min), coordinate increase in liver polyamine levels and pyruvate dehydrogenase activity which peaked at 6 min and then declined. These effects were blocked by DFMO (200 mg/kg), and putrescine (100 mg/kg) negated DFMO inhibition (Figure 10).

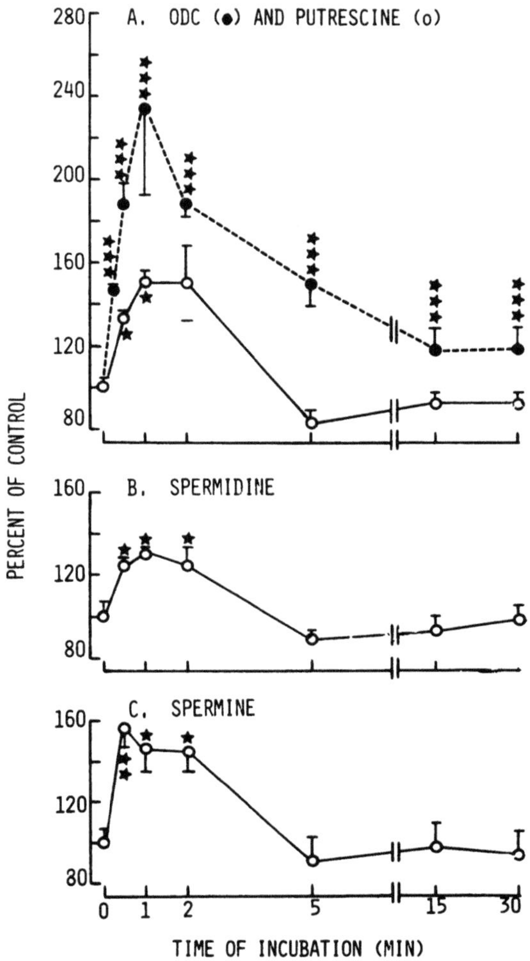

FIGURE 8. Insulin induces an acute, transient rise in ODC activity and polyamine concentrations in mouse kidney cortex in vitro. Cortex slices were incubated in a physiological salt solution with 1.3 n*M* insulin for given times. Data are means ± SEM. *, **, ***: $p < 0.05$, 0.01, 0.001 (vs. incubated controls).

Insulin also evoked a DFMO-sensitive stimulation of pyruvate dehydrogenase activity in cultured bovine aorta myocytes. This was associated with a DFMO-sensitive increase in mitochondrial function as assessed with rhodamine 123, a permeant fluorescent cationic dye which selectively stains mitochondria and monitors mitochondrial transmembrane potential in vivo.[52] These data, taken together, strongly support the hypothesis that newly synthesized polyamines mediate the rapid effects of insulin on cells, and are not merely epiphenomena of insulin action.

A series of experiments was undertaken to examine the hypothesis that the agonists in insulin "mediator" preparations, which have an average molecular weight of 500 to 2000 by molecular sieve filtration, are in fact polyamines. Insulin mediator preparations were made according to Macauley and Jarett.[53] In brief, livers were homogenized in an acidic acetate-buffered (pH 3.8) medium and the supernatants were boiled to remove most proteins. Supernatants were treated with activated charcoal to remove nucleotides, and subjected to

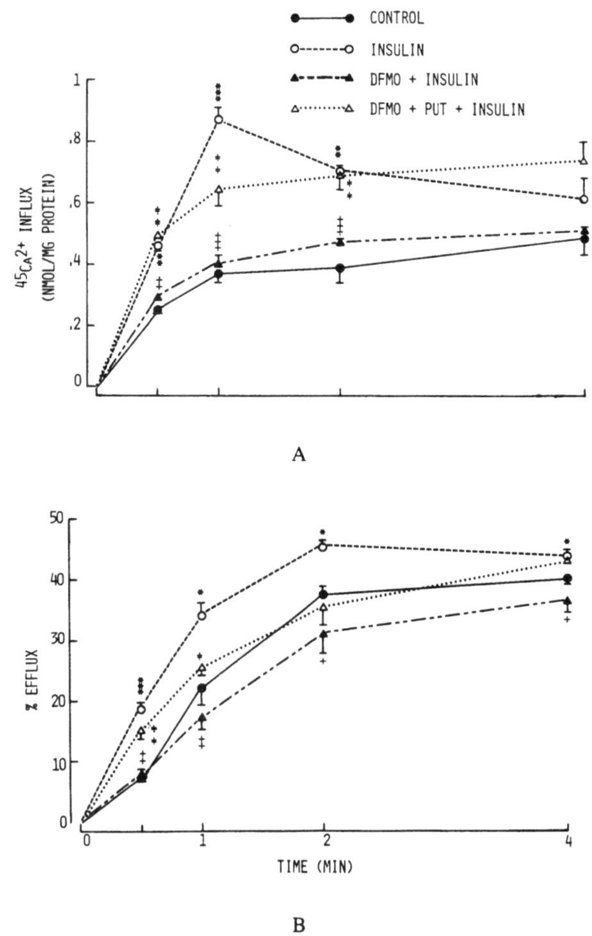

FIGURE 9. DFMO abolishes insulin stimulation of ^{45}Ca influx and efflux in kidney cortex, and putrescine reverses the DFMO effect. The experimental protocol was similar to that shown in Figure 3. At zero time 1.3 n*M* insulin was added. (A) ^{45}Ca influx and (B) ^{45}Ca efflux. The results are means ± SEM. See Figure 3 for the key to the interpretation of significance.

sequential ultrafiltration with Amicon membrane filters to remove molecules larger than 2000 (YM2 membrane) and smaller than 500 (YC05 membrane). The resultant 500 to 2000 M_r fraction was lyophilized, washed, and taken up in 1 m*M* formic acid. The various fractions were assayed for polyamines by thin layer chromatography of the dansylated derivatives. Selected fractions were assayed for pyruvate dehydrogenase stimulating activity in an isolated rat liver mitochondrial system.[53] The bulk of the polyamines, 62 to 84% of the putrescine, spermidine, and spermine present in the original liver homogenates, was recovered in the final mediator preparations. The polyamine content of these mediator preparations was substantially increased by insulin administrtion, and DFMO inhibited the insulin-induced increment of polyamines. The pyruvate dehydrogenase stimulating effect of mediator preparations was similarly enhanced by insulin administration and DFMO suppressed insulin stimulation. A puzzling aspect of these observations relates to the occurrence of the bulk of the tissue polyamines with molecular weights in the range of 88 to 202 in the 500 to 2000 molecular weight fraction. Polyamines might be retained by the Amicon YC05 membrane

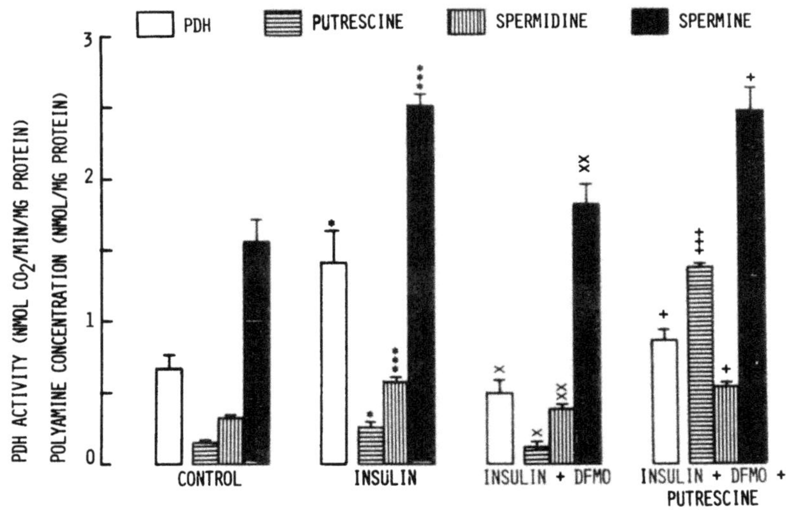

FIGURE 10. DFMO blocks the insulin-induced increase in polyamine and pyruvate dehydrogenase (PDH) levels in mouse liver, and putrescine negates the effects of DFMO. Mice received DFMO (200 mg/kg), DFMO + putrescine (100 mg/kg), or 0.9% NaCl intraperitoneally 10 min before receiving 0.33 n*M*/kg insulin. Livers were excised 6 min after insulin and analyzed for PDH and polyamines. Data are means ± SEM. See Figure 3 for the key to the interpretation of significance.

filter (500 M_r) either because they form heteroaggregates with constituents present in the tissue extracts or because of their elongated molecular shape which causes them to behave like larger molecules during molecular sieve filtration. To discriminate between these two possibilities, we examined the behavior of polyamine standards during this ultrafiltration sequence. In a typical experiment, we found that 79.6, 79.4, and 69.5% of the spermine, spermidine, and putrescine was retained by the YC05 filter, the remainder occurring in the filtrate, indicating that the polyamines behave anomalously during molecular sieve ultrafiltration. These experiments demonstrate that classical insulin mediator preparations contain a major portion of the acid-extractable tissue polyamines, and provide additional suport for the hypothesis that polyamines serve as second messengers in mediating insulin actions in target cells.

2. Heart

Preliminary experiments have been performed to assess the hypothesis that insulin may exert a direct influence on cardiac contractility via a polyamine-dependent process.[54] This hypothesis grew out of studies on the positive inotropic effects of testosterone and β-adrenoceptor stimulation, and was considered to be significant in light of published reports on the positive inotropism of insulin,[55] and the association of an insulin-sensitive cardiomyopathy with experimental diabetes.[56,57] The Langendorff-perfused rat heart was used for these studies. Insulin induced a concentration-dependent increase in cardiac contractility in normal rat heart in the concentration range of 6.5 to 650 p*M*. Insulin stimulated heart contractility after a lag period of about 20 sec, and this stimulation was rapidly suppressed by infusion of DFMO together with insulin. Insulin (0.65 n*M*) also induced an acute (<7.5 sec), transient, two- to threefold increase in the levels of ODC and polyamines in nondiabetic rat heart which peaked at 15 to 30 sec. DFMO blocked the insulin-evoked changes in ODC and polyamine levels, confirming that the accumulation of heart polyamines is mediated by ODC activation. These experiments support the hypothesis that insulin modulates rat cardiac contractility by a receptor-mediated stimulation of ODC activity and polyamine synthesis, probably by enhancing Ca^{2+} fluxes involved in excitation-contraction coupling.

E. Role of Polyamines in Thyroid Hormone Action

In classical target organs, e.g., heart, liver, and kidney, thyroid hormones exert a potent stimulating action on various metabolic processes, including nucleic acid, protein, and polyamine synthesis, oxygen consumption, and the activities and rates of synthesis of a number of enzymes involved in mitochondrial respiration. There is much evidence suggesting that thyroid hormone action is initiated by binding of triiodothyronine (T_3) to nuclear receptors.[58] However, T_3 and thyroxine (T_4) receptors have also been detected in the plasma membrane,[59,60] and T_3 receptors in the inner mitochondrial membrane,[61,62] and these have been involved in the mediation of some of the acute effects of these hormones. Recent studies in our laboratory have provided evidence suggesting that rapid effects of T_3 on membrane transport, mitochondrial activity, synaptic function, and contractility involve T_3 receptor-regulated stimulation of ODC activity and polyamine synthesis and polyamine-dependent Ca^{2+} fluxes.

1. Kidney Cortex

The active thyroid hormone T_3 evokes an early, Ca^{2+}-dependent stimulation of Ca^{2+} fluxes, endocytosis, and transport of hexose and amino acid in kidney cortex slices from hypothyroid, but not euthyroid, mouse and rat.[63,64] This rapid membrane transport response was maximum at 0.45 to 45 n*M* T_3, was blocked in a calcium-free medium, and was minimal or about absent with T_4 or the T_3 analogue reverse T_3 (rT_3), suggesting T_3 receptor and Ca^{2+} dependence. Furthermore, the basal rates of $^{45}Ca^{2+}$ influx and efflux and Ca^{2+}-dependent membrane transport processes were substantially greater in kidney cortex from euthyroid than hypothyroid rodents, indicating that thyroid hormones exert a long-term modulating action on these cellular functions. T_3 (4.5 n*M*) evoked a rapid, transient increase in ODC activity and putrescine, spermidine, and spermine levels in hypothyroid kidney cortex. DFMO abolished T_3 stimulation of polyamine synthesis, Ca^{2+} fluxes, and Ca^{2+}-dependent membrane transport (endocytosis, hexose transport, and amino acid transport), and putrescine nullified the effects of DFMO, thus corroborating that rapid polyamine synthesis is linked to T_3 stimulation of membrane transport processes.

2. Heart

Thyroid hormones exert a profound long-term influence on heart function, but their mechanisms of action are unresolved. We found that T_3 evoked a rapid (<30 sec), Ca^{2+}- and receptor-dependent stimulation of Ca^{2+} fluxes, endocytosis, amino acid transport, and hexose transport in heart slices from hypothyroid mice, but not from euthyroid mice.[65] This transport was concentration dependent and was maximum at 0.45 to 45 n*M* T_3. The basal transport of HRP, AIB, and DG by hypothyroid mouse heart was substantially less than in euthyroid mouse heart, and addition of 4.5 n*M* T_3 restored these transport processes to euthyroid levels within 5 min, whereas rT_3 (4.5 n*M*) and T_4 (450 n*M*) were without effect. T_3 (4.5 n*M*) induced an acute, transient stimulation of ODC activity and concurrently increased the levels of putrescine, spermidine, and spermine in hypothyroid mouse heart slices (Figure 11). That this rapid polyamine synthesis is essential for T_3 stimulation of Ca^{2+} fluxes and Ca^{2+}-dependent membrane transport processes is shown by the finding that DFMO (5 m*M*) blocked these effects of T_3, and putrescine (0.5 m*M*) negated the action of DFMO (Figure 12).

The impact of T_3 on cardiac contractility was assessed in the Langendorff-perfused heart from hypothryoid rats.[54] T_3 (1 n*M*) induced an acute, transient increase in heart ODC and polyamine levels commencing at ~8 sec which peaked at 30 sec. Heart contractility and rate were increased 20 to 30 sec after infusion of 1 n*M* T_3. DFMO blocked the T_3-induced increase in polyamines, contractility, and rate, suggesting that polyamines are involved in mediating the acute positive inotropic and chronotropic effects of T_3.[54]

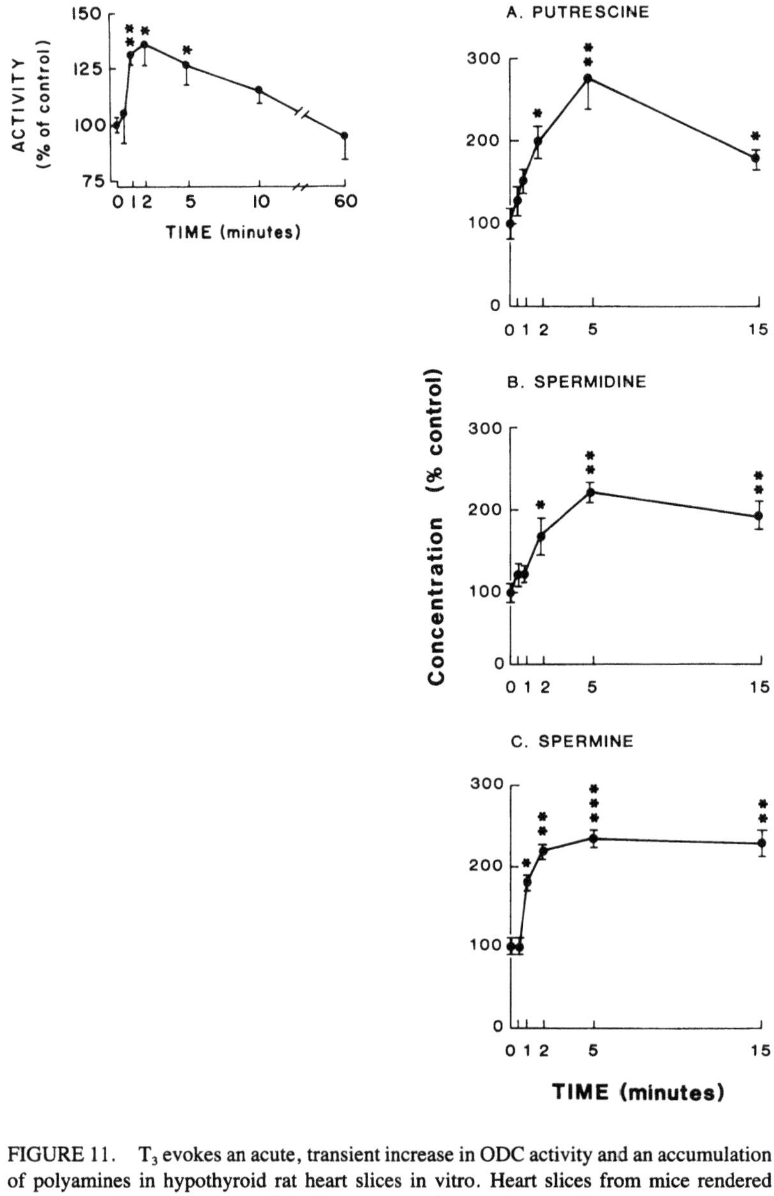

FIGURE 11. T_3 evokes an acute, transient increase in ODC activity and an accumulation of polyamines in hypothyroid rat heart slices in vitro. Heart slices from mice rendered hypothyroid by propylthiouracil feeding were incubated with 4.5 n*M* T_3 for various times. ODC values are shown. Data are means ± SEM. *, **, ***: $p < 0.05$, 0.01, 0.001 (vs. control incubations).

3. Cerebral Cortex and Neocortical Synaptosomes

Thyroid hormones are known to influence brain function in mature animals but the precise nature of this influence remains to be defined. Since dysthyroid states in man lead to profound neuropsychiatric disturbances, it was of interest to investigate the effects of thyroid hormones in mouse neocortical slices.[66,67] The rates of $^{45}Ca^{2+}$ influx, endocytosis, and transport of amino acid and hexose were 35 to 80% less in hypothyroid, than in euthryoid neocortex. T_3 (0.045 to 45 n*M*) evoked an immediate (<60 sec), concentration-dependent stimulation of $^{45}Ca^{2+}$ influx and membrane transport in hypothyroid, but not euthyroid cortex. This hormonal response was Ca^{2+}-dependent and was not elicited by either T_4 (450 n*M*) or rT_3

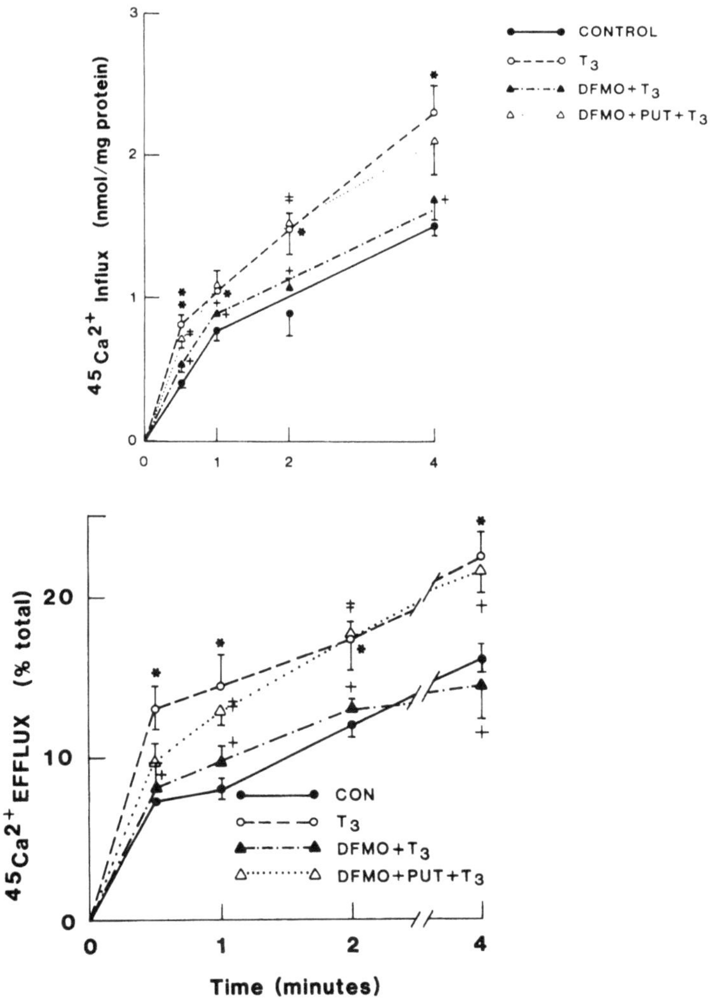

FIGURE 12. DFMO inhibits T_3 stimulation of ^{45}Ca influx and efflux in hypothyroid mouse heart and putrescine nullifies DFMO inhibition. The experimental protocol was similar to that shown in Figure 3. At zero time 4.5 n*M* T_3 was added. Data are means ± SEM. The key to the interpretation of significance is given in Figure 3.

(4.5 n*M*). It was found by ultrastructural cytochemistry of HRP uptake that T_3 stimulation of endocytosis was largely confined to nerve terminals. T_3 induced a rapid (<30 sec), transient increase in ODC activity and an early (<60 to 120 sec), 1.5- to 3-fold increase in putrescine, spermidine and spermine levels in neocortex (Figure 13). DFMO (5 m*M*) abolished the T_3-induced accumulation of polyamines and the attendant increase in $^{45}Ca^{2+}$ influx, endocytosis, and transport of hexose and amino acid in neocortical slices, and putrescine (0.5 m*M*) partially or completly negated DFMO inhibition. These results indicate that T_3 induces an acute stimulation of ODC activity and polyamine synthesis in hypothyroid neocortex, and suggest that polyamines mediate T_3 stimulation of Ca^{2+} fluxes and membrane transport in neocortex.

The observation that T_3 stimulation of HRP uptake was confined to nerve endings suggested that thyroid hormones directly modulate endocytosis and other synaptic functions. This

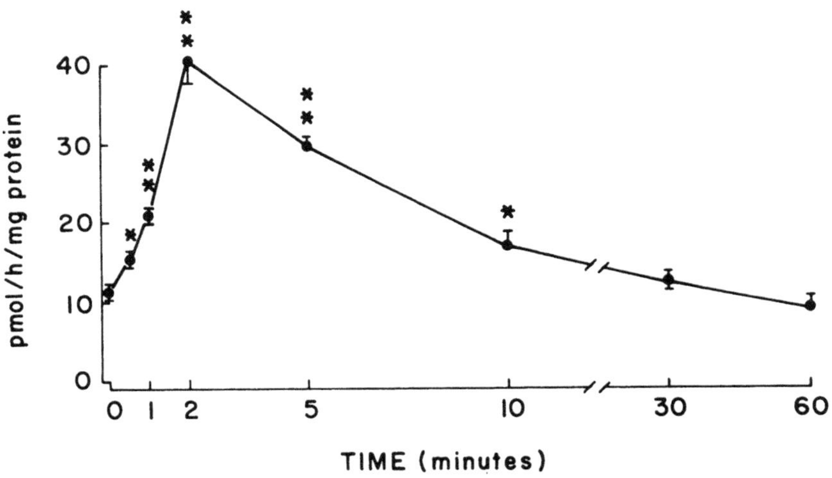

A

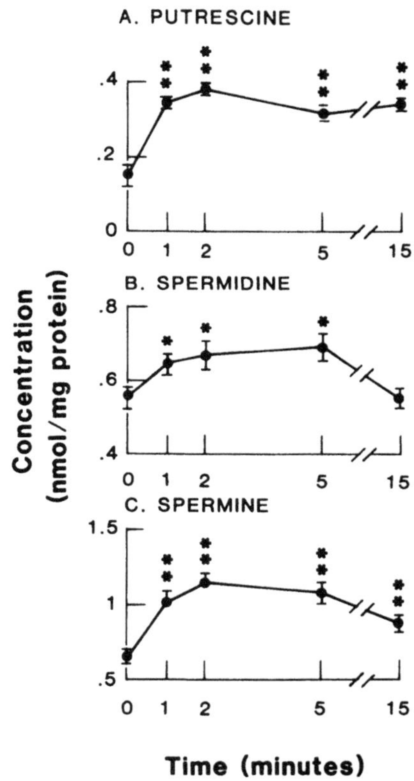

B

FIGURE 13. T_3 induces an acute, transient elevation of (A) ODC activity and (B) polyamine levels in hypothyroid mouse neocortex in vitro. At zero time 4.5 n*M* T_3 was added. Data are means ± SEM. *, **: $p < 0.05$, 0.01 (vs. control incubations).

inference was examined in synaptosomes isolated from hypothyroid rat neocortex by Ficoll-sucrose density gradient centrifugation.[68] T_3 (4.5 n*M*) induced an acute (<15 sec) increase in ODC activity which peaked (fivefold) at 30 to 60 sec and returned to basal by 2 to 5 min, and resulted in a substantial increase in putrescine, spermidine, and spermine within 1 min.[69,70] T_3 evoked an early (<15 sec) and marked increase in $^{45}Ca^{2+}$ influx and efflux, and a release of [^{3}H]norepinephrine and [^{3}H]GABA from preloaded synaptosomes. T_3 also induced an early (<15 to 30 sec) increase in endocytosis and amino acid and hexose transport in synaptosomes. DFMO (5 m*M*) inhibited these effects of T_3. Pilot studies of [^{125}I]T_3 binding revealed the presence of specific, high affinity T_3 receptor binding activity in synaptosomes from hypothyroid, but not from euthyroid, neocortex.[70] Specific T_3 binding sites have previously been demonstrated in neocortical synaptosomes[71] where they have been localized to the synaptic (plasma) membrane.[72] These findings are consistent with the hypothesis that thyroid hormones modulate neural function via a T_3 receptor-mediated regulation of ODC activity and polyamine synthesis in neocortical nerve terminals.

F. Concluding Remarks

Two major findings have emerged from these investigations: (1) diverse hormones (testosterone, the β-adrenergic agonist isoproterenol, T_3, insulin) induce a very rapid and transient increase (commencing within 5 to 15 sec and returning to near basal levels within 60 to 120 sec in isolated cells and perfused hearts) in ODC activity and the levels of putrescine, spermidine, and spermine and (2) newly synthesized polyamines appear to play an essential role in coupling enhanced Ca^{2+} fluxes and Ca^{2+}-dependent cell responses to hormonal stimulation.

The acute, transient increase in ODC activity appears to be a receptor- and Ca^{2+}-dependent change in the catalytic activity of a latent ODC which is associated, in part, with the plasma membrane. Ca^{2+} influx via Ca^{2+} channels and synthesis of prostaglandins and cyclic AMP appear to be required for androgenic and β-adrenergic stimulation of ODC activity and membrane transport functions in mouse kidney cortex, as these responses are blocked by verapamil,[21,22] dexamethasone, aspirin and 2′,5′-dideoxyadenosine, inhibitors of Ca^{2+} channels, arachidonate release, cyclooxygenase, and adenylate cyclase, respectively (unpublished data). In addition, the prostaglandins PGA_2 and PGE_2, and the cyclic AMP analogue dibutyryl-cyclic AMP have been shown to mimic the effects of testosterone and isoproterenol in stimulating ODC activity[73] and membrane transport processes in kidney cortex.[74] We have therefore suggested that this rapid, transient stimulation of ODC activity may represent a receptor-mediated, Ca^{2+}-, prostaglandin-, and cyclic AMP-dependent activation of a latent ODC via a posttranslational process, possibly a phosphorylation-dephosphorylation sequence involving ODC, an ODC regulatory protein, or an ODC binding site.[18,19] Cytoskeletal participation also appears to be important as androgenic and β-adrenergic stimulation of ODC activity and membrane transport is blocked by the microfilament-disrupting agent cytochalasin *b* and the microtubule-disrupting agent colchicine (unpublished data). The hypothetical model for receptor-mediated regulation of ODC activity shown in Figure 14 incorporates these observations.

The studies summarized in this chapter support the hypothesis that newly synthesized polyamines serve as messengers to generate Ca^{2+} signals by enhancing Ca^{2+} influx via Ca^{2+} channels and Na^+/Ca^{2+} exchange, and by mobilizing Ca^{2+} sequestered in intracellular storage sites. Possible molecular mechanisms by which polyamines could regulate transmembrane movements include one or more of the following modifications of ion channel or transporter molecules:

1. Phosphorylations and dephosphorylations by polyamine-dependent protein kinases and phosphatases

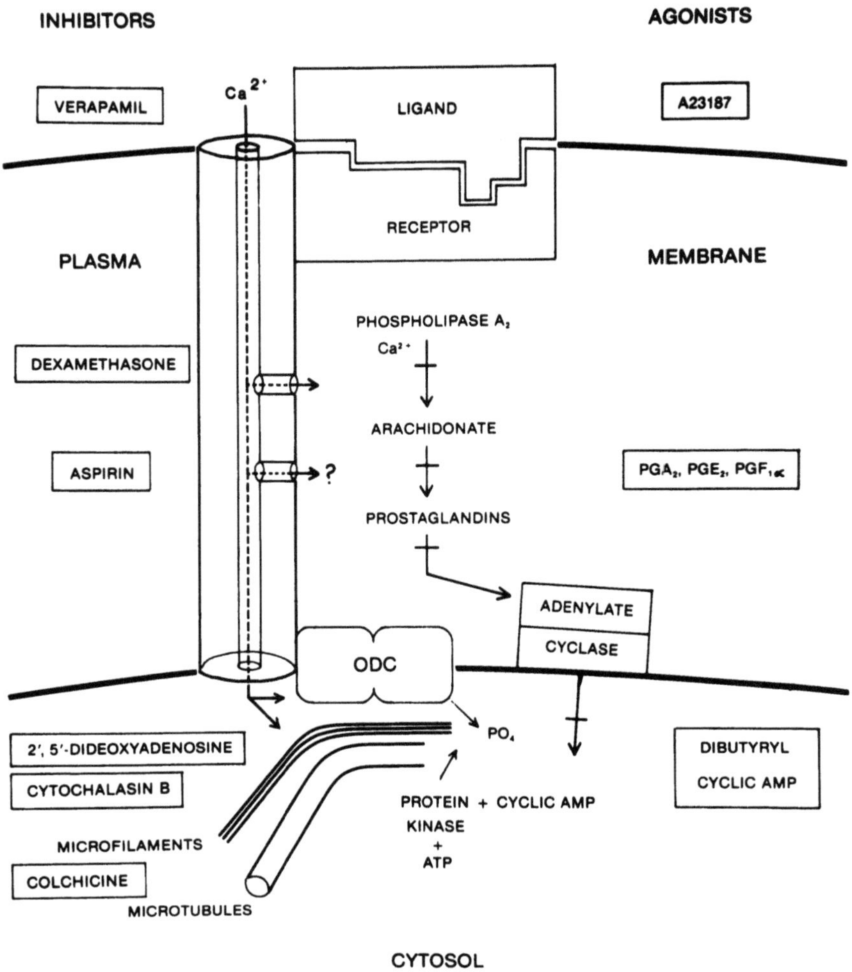

FIGURE 14. Model for signal transduction. This model is based on studies of the effects of inhibitors which have been found to suppress postulated steps in the membrane cascade evoked by androgenic and β-adrenergic stimulation of mouse kidney cortex leading to an early increase in ODC activity and the effectiveness of postulated intermediates in either negating the action of a specific inhibitor or directly stimulating ODC activity. This model is based also on the reasonable assumption that a rapid, short-lived stimulation of ODC activity via a Ca^{2+}-, prostaglandin-, and cAMP-dependent process is likely to involve a reversible phosphorylation-dephosphorylation sequence of ODC or an ODC regulatory protein. The possible participation of guanine nucleotide-binding protein in this transduction sequence is omitted from this scheme. Validation of this mechanism must await future work.

2. Polyamines binding to anionic sites and acting as counterions to decrease surface charge and potential
3. Polyamines displacing bound calcium by a cation-exchange reaction

A number of important issues remain to be resolved: (1) the precise nature of the coupling of receptors to ODC, (2) the cellular and molecular mechanism of the activation-inactivation cycle of ODC, (3) the mechanisms by which polyamines modulate transmembrane Ca^{2+} fluxes, and (4) the interrelationships between polyamines and other putative cell messengers, notably inositol trisphosphate, diacylglycerol, and cyclic AMP.

ACKNOWLEDGMENTS

The studies cited in this paper were supported by the Veterans Administration Research Service and NIH grants HL 26835 and NS 14700. We thank Mrs. T. Howell for the secretarial work and Mr. P. Joshi for technical assistance.

REFERENCES

1. **Rasmussen, H. and Barrett, P. Q.,** Calcium messenger system: an integrated view, *Physiol. Rev.,* 64, 938, 1984.
2. **Exton, J. H.,** Role of calcium and phosphoinositides in the actions of certain hormones and neurotransmitters, *J. Clin. Invest.,* 75, 1753, 1985.
3. **Manalan, A. S. and Klee, C.,** Calmodulin, *Adv. Cyclic Nucleotide Protein Phosphorylation Res.,* 18, 227, 1984.
4. **Whitfield, J. F., Boynton, A. L., Rixon, R. H., and Youdale, T.,** The control of cell proliferation by calcium, Ca^{2+}-calmodulin and cyclic AMP, *Control Anim. Cell Proliferation,* 1, 331, 1985.
5. **Berridge, M. J. and Irvine, R. F.,** Inositol triphosphate, a novel messenger in cellular signal transduction, *Nature,* 312, 315, 1984.
6. **Williamson, J. R., Cooper, R. H., Joseph, S. K., and Thomas, A. P.,** Inositol trisphosphate and diacylglycerol as intracellular second messengers in liver, *Am. J. Physiol.,* 248, C203, 1985.
7. **Chan, K.-M. and Junger, K. D.,** Calcium transport and phosphorylated intermediate of (Ca^{2+} + Mg^{2+})-ATPase in plasma membranes of rat liver, *J. Biol. Chem.,* 258, 4404, 1983.
8. **Lotersztajn, S., Epand, R., Mallat, A., Pavoine, C., and Pecker, F.,** The liver plasma membrane Ca^{2+} pump: hormone sensitivity, *Biochimie,* 67, 1169, 1985.
9. **Langer, G. A.,** Sodium-calcium exchange in the heart, *Annu. Rev. Physiol.,* 44, 435, 1982.
10. **Burgess, G. M., McKinney, J. S., Fabiato, A., Leslie, B. A., and Putney, J. W., Jr.,** Calcium pools in saponin-permeabilized guinea pig hepatocytes, *J. Biol. Chem.,* 258, 15336, 1983.
11. **Nicholls, D. and Åkerman, K.,** Mitochondrial calcium transport, *Biochim. Biophys. Acta,* 683, 57, 1982.
12. **Sekar, M. C. and Hokin, L. E.,** The role of phosphoinositides in signal transduction, *J. Membr. Biol.,* 89, 193, 1986.
13. **Nishizuka, Y.,** The role of protein kinase *c* in cell surface signal transduction and tumor promotion, *Nature,* 308, 693, 1984.
14. **Boynton, A. L. and Whitfield, J. F.,** The role of cyclic AMP in cell proliferation: a critical assessment of the evidence, *Adv. Cyclic Nucleotide Res.,* 15, 193, 1983.
15. **Tsien, R. W., Bean, B. P., Hess, P., Lansman, J. B., Nilius, B., and Nowycky, M. C.,** Mechanisms of calcium channel modulation by β-adrenergic agents and dihydropyridine calcium agonists, *J. Mol. Cell. Cardiol.,* 18, 691, 1986.
16. **Tabor, C. W. and Tabor, H.,** Polyamines, *Annu. Rev. Biochem.,* 53, 749, 1984.
17. **Russell, D. H.,** Ornithine decarboxylase: a key regulatory enzyme in normal and neoplastic growth, *Drug Metab. Rev.,* 16, 1, 1985.
18. **Koenig, H., Goldstone, A., and Lu, C. Y.,** Polyamines regulate calcium fluxes in a rapid plasma membrane response, *Nature,* 305, 530, 1983.
19. **Koenig, H., Goldstone, A., and Lu, C. Y.,** β-Adrenergic stimulation of Ca^{2+} fluxes, endocytosis, hexose transport and amino acid transport in mouse kidney cortex is mediated by polyamine synthesis, *Proc. Natl. Acad. Sci. U.S.A.,* 80, 7210, 1983.
20. **Koenig, H., Goldstone, A., and Lu, C. Y.,** Testosterone induces a rapid stimulation of endocytosis, amino acid and hexose transport in mouse kidney cortex, *Biochem. Biophys. Res. Commun.,* 106, 346, 1982.
21. **Goldstone, A. D., Koenig, H., Lu, C. Y., and Trout, J. J.,** β-Adrenergic stimulation evokes a rapid Ca^{2+}-dependent stimulation of endocytosis, hexose and amino acid transport associated with increased Ca^{2+} fluxes in mouse kidney cortex, *Biochem. Biophys. Res. Commun.,* 114, 913, 1983.
22. **Goldstone, A., Koenig, H., and Lu, C. Y.,** Androgenic stimulation of endocytosis, amino acid and hexose transport in mouse kidney cortex involves increased calcium fluxes, *Biochim. Biophys. Acta,* 762, 366, 1983.
23. **Iqbal, Z. and Koenig, H.,** Polyamines appear to be second messengers in mediating Ca^{2+} fluxes and neurotransmitter release in potassium-depolarized synaptosomes, *Biochem. Biophys. Res. Commun.,* 133, 563, 1985.

24. **Koenig, H., Goldstone, A. D., Iqbal, Z., Fan, C.-C., Lu, C. Y., and Trout, J. J.,** Rapid polyamine synthesis plays a key role in the mediation of receptor- and voltage-regulated calcium fluxes, in *Recent Progress in Polyamine Research,* Selmeci, L., Brosnan, M. E., and Seiler, N., Eds., Akademiai Kiado, Budapest, 1985, 191.
25. **Kochakian, C. D.,** Regulation of kidney growth by androgens, *Adv. Steroid Biochem. Pharmacol.,* 6, 1, 1977.
26. **Bardin, C. W., Bullock, L. P., Mills, N. C., Lin, Y.-C., and Jacob, S. T.,** The role of receptors in the anabolic action of androgens, in *Receptors and Hormone Action,* O'Malley, B. W. and Birnbaumer, L., Eds., Academic Press, New York, 1978, 83.
27. **Grahn, B., Henningsson, S. S. G., Kahlson, G., and Rosengren, E.,** Alterations in the activities of ornithine and histidine decarboxylases provoked by testosterone in mice, *Br. J. Pharmacol.,* 48, 113, 1973.
28. **Koenig, H., Goldstone, A., and Hughes, C.,** Lysosomal enzymuria in the testosterone-treated mouse. A manifestation of cell defecation, *Lab. Invest.,* 39, 329, 1978.
29. **Koenig, H., Goldstone, A., Blume, G., and Lu, C. Y.,** Testosterone-mediated sexual dimorphism of mitochondria and lysosomes in mouse kidney proximal tubules, *Science,* 209, 1023, 1980.
29a. **Blume, G., Goldstone, A., and Koenig, H.,** unpublished data.
30. **Koenig, H., Goldstone, A., Lu, C. Y., Blume, G., and Hughes, C.,** Glucocorticoids, androgens and lysosomes. Experimental studies in mouse brain, muscle and kidney, in *Neurochemistry and Clinical Neurology,* Battistin, L., Hashim, G., and Lajtha, A., Eds., Alan R. Liss, New York, 1980, 275.
31. **Goldstone, A., Koenig, H., Blume, G., and Lu, C. Y.,** Glucocorticoids and aspirin inhibit and cyprotereone acetate enhances the response of mouse kidney to androgens, *Biochim. Biophys. Acta,* 677, 133, 1981.
32. **Goldstone, A., Koenig, H., and Lu, C. Y.,** Testosterone-dependent sexual dimorphism of mouse kidney is mediated by polyamines, *Biochem. Biophys. Res. Commun.,* 104, 165, 1982.
33. **Metcalf, B. W., Bey, P., Danzin, C., Jung, M. J., Casara, P., and Vevert, J. P.,** Catalytic irreversible inhibition of mammalian ornithine decarboxylase (EC 4.1.1.17) by substrate and product analogues, *J. Am. Chem. Soc.,* 100, 2551, 1978.
34. **Koenig, H., Goldstone, A., and Lu, C. Y.,** Testosterone-mediated sexual dimorphism of the rodent heart. Ventricular lysosomes, mitochondria and cell growth are modulated by androgens, *Circ. Res.,* 50, 782, 1982.
35. **Goldstone, A., Koenig, H., and Lu, C. Y.,** Polyamines mediate the androgenic response in heart and skeletal muscle, *Trans. Am. Soc. Neurochem.,* 12, 89, 1981.
35a. **Koenig, H., Fan, C.-C., Goldstone, A. D., Lu, C. Y., and Trout, J. J.,** Polyamines mediate androgenic stimulation of calcium fluxes and membrane transport in rat heart myocytes, *Circ. Res.,* in press.
36. **Fan, C.-C., Koenig, H., Goldstone, A. D., and Lu, C. Y.,** Polyamines regulate Ca^{2+} fluxes and cytosolic Ca^{2+} concentration in cardiac myocytes subjected to androgenic and β-adrenergic stimulation, *J. Neurochem.,* 44(Suppl), S65, 1985.
36a. **Koenig, H., Goldstone, A. D., and Lu, C. Y.,** Polyamines are intracellular messengers in the β-adrenergic regulation of Ca^{2+} fluxes, $[Ca^{2+}]_i$ and membrane transport in rat heart myocytes, *Biochem. Biophys. Res. Commun.,* 153, 1179, 1988.
37. **Lu, C. Y., Koenig, H., Goldstone, A. D., and Fan, C.-C.,** Positive inotropic effect of testosterone in rodent heart: mediation by rapid polyamine synthesis, *Fed. Proc. Fed. Am. Soc. Exp. Biol.,* 44, 830, 1985.
38. **Koenig, H., Fan, C.-C., Lu, C. Y., and Goldstone, A. D.,** Polyamines as signals mediating enhanced heart contractility after androgenic and β-adrenergic stimulation, *J. Neurochem.,* 44(Suppl.), S64, 1985.
39. **Chapman, R. A.,** Control of cardiac contractility at the cellular level, *Am. J. Physiol.,* 245, H535, 1983.
40. **Reuter, H. A.,** Properties of two inward membrane currents in the heart, *Annu. Rev. Physiol.,* 41, 413, 1979.
41. **Fan, C.-C. and Koenig, H.,** β-Adrenergic stimulation of calcium fluxes and membrane transport processes in rat heart is mediated by polyamines, *Fed. Proc. Fed. Am. Soc. Exp. Biol.,* 43, 735, 1984.
42. **Fan, C.-C. and Koenig, H.,** The role of polyamines in β-adrenergic stimulation of calcium influx and membrane transport in rat heart, *J. Mol. Cell. Cardiol.,* in press.
43. **Koenig, H., Fan, C.-C., Lu, C. Y., and Goldstone, A.,** β-Adrenergic stimulation of heart contractility is mediated by polyamines, *Trans. Am. Soc. Neurochem.,* 16, 236, 1985.
44. **Fan, C.-C., Koenig, H., Goldstone, A. D., and Lu, C. Y.,** Polyamines regulate cytosolic Ca^{2+} concentration and Ca^{2+}-dependent membrane transport in β-adrenoceptor stimulated cardiomyocytes, *Fed. Proc. Fed. Am. Soc. Exp. Biol.,* 44, 1593, 1985.
45. **Czech, M. P.,** Molecular basis of insulin action, *Annu. Rev. Biochem.,* 46, 359, 1977.
46. **Larner, J.,** Insulin mediator — fact or fancy?, *J. Cyclic Nucleotide Res.,* 8, 289, 1982.
47. **Jarett, L., Kiechle, F. I., and Parker, J. C.,** Chemical mediator or mediators of insulin action: response to insulin and mode of action, *Fed. Proc. Fed. Am. Soc. Exp. Biol.,* 41, 2736, 1982.
48. **Seals, J. R.,** Intracellular mediators of insulin action, in *Molecular Bases of Insulin Action,* Czech, M. P., Ed., Plenum Press, New York, 1985, 199.

49. **Goldstone, A., Koenig, H., and Lu, C. Y.,** Insulin stimulation of Ca^{2+} fluxes and membrane transport processes in mouse kidney cortex is mediated by polyamine synthesis, *Fed. Proc. Fed. Am. Soc. Exp. Biol.*, 44, 1593, 1985.
50. **Goldstone, A. D., Koenig, H., and Lu, C. Y.,** Polyamines as mediators of insulin's action on pyruvate dehydrogenase, *Fed. Proc. Fed. Am. Soc. Exp. Biol.*, 45, 1011, 1986.
51. **Goldstone, A., Koenig, H., and Lu, C. Y.,** Polyamines as mediators of insulin action on Ca^{2+} fluxes, membrane transport and pyruvate dehydrogenase, Abstr. Int. Conf. Polyamines in Life Sciences (Tokyo), 1986, 213.
52. **Johnson, L. V., Walsh, M. L., Bockus, B. J., and Chen, L. B.,** Monitoring of relative mitochondrial membrane potential in living cells by fluorescence microscopy, *J. Cell Biol.*, 88, 526, 1981.
53. **Macaulay, S. L. and Jarett, L.,** Insulin mediator causes dephosphorylation of the α-subunit of pyruvate dehydrogenase by stimulating phosphatase activity, *Arch. Biochem. Biophys.*, 237, 142, 1985.
54. **Lu, C. Y., Koenig, H., Goldstone, A. D., and Fan, C.-C.,** Polyamines mediate acute stimulation of heart contractility by insulin and triiodothyronine, *Fed. Proc. Fed. Am. Soc. Exp. Biol.*, 45, 1011, 1986.
55. **Lee, J. C. and Downing, S. E.,** Effects of insulin on cardiac muscle contraction and responsiveness to epinephrine, *Am. J. Physiol.*, 230, 1360, 1976.
56. **Miller, T. B., Jr.,** Cardiac performance of isolated perfused hearts from alloxan diabetes rats, *Am. J. Physiol.*, 236, H808, 1979.
57. **Fein, F. S., Miller-Green, B., and Sonnenblick, E. H.,** Altered myocardial mechanics in diabetic rabbits, *Am. J. Physiol.*, 248, H729, 1985.
58. **Oppenheimer, J. H.,** Thyroid hormone action at the cellular level, *Science*, 203, 971, 1979.
59. **Maxfield, F. R., Willingham, M. C., Pastan, I., Dragsten, P., and Cheng, S.-Y.,** Binding and mobility of the cell surface receptors for 3,3′,5-triiodo-L-thyronine, *Science*, 211, 63, 1980.
60. **Cheng, S.-Y.,** Structural similarities in the plasma membrane 3,3′,5-triido-L-thyronine receptors from human, rat and mouse cultured cells. Analysis by affinity labeling, *Endocrinology*, 113, 1155, 1983.
61. **Sterling, K. and Milch, P. O.,** Thyroid hormone binding by a component of mitochondrial membrane, *Proc. Natl. Acad. Sci. U.S.A.*, 72, 3225, 1975.
62. **Goglia, F., Torresani, J., Bulgi, P., Barletta, A., and Liverini, G.,** *In vitro* binding of triiodothyronine in rat liver mitochondria, *Pfluegers Arch.*, 390, 120, 1981.
63. **Lu, C. Y., Koenig, H., Goldstone, A. D., and Trout, J. J.,** Triiodothyronine (T_3) induces a rapid Ca^{2+}-dependent stimulation of membrane transport processes and Ca^{2+} fluxes in mouse kidney cortex, *Fed. Proc. Fed. Am. Soc. Exp. Biol.*, 43, 735, 1984.
64. **Goldstone, A., Koenig, H., and Lu, C. Y.,** Triiodothyronine (T_3) stimulation of Ca^{2+} influx and membrane transport processes in mouse kidney cortex is mediated by polyamines, *Fed. Proc. Fed. Am. Soc. Exp. Biol.*, 43, 735, 1984.
65. **Koenig, H., Fan, C.-C., and Iqbal, Z.,** Polyamines mediate T_3 stimulation of Ca^{2+} influx and membrane transport, *Trans. Am. Soc. Neurochem.*, 15, 220, 1984.
66. **Iqbal, Z., Koenig, H., and Trout, J. J.,** Triiodothyronine (T_3) regulates membrane transport processes in mature neurons and nerve endings, *Neurology*, 34(Suppl. 1), 195, 1984.
67. **Iqbal, Z., Koenig, H., and Trout, J. J.,** Triiodothyronine (T_3) stimulates calcium influx and membrane transport processes in nerve cells and terminals of hypothyroid mouse cortex, *Fed. Proc. Fed. Am. Soc. Exp. Biol.*, 43, 735, 1984.
68. **Booth, R. F. G. and Clark, J. B.,** A rapid method for the preparation of relatively pure metabolically competent synaptosomes from rat brain, *Biochem. J.*, 176, 365, 1978.
69. **Iqbal, Z., Barsano, C. P., and Koenig, H.,** Triiodothyronine modulates Ca^{2+}-dependent synaptic functions in neocortex, *Trans. Am. Soc. Neurochem.*, 16, 224, 1985.
70. **Iqbal, Z. and Koenig, H.,** Triiodothyronine (T_3) regulates Ca^{2+} fluxes, hexose transport and neurotransmitter release in nerve endings, *Neurology*, 35(Suppl. 1), 257, 1985.
71. **Mashio, Y., Inada, M., Tanaka, K., Ishii, H., Naito, K., Nishikawa, M., Takahashi, K., and Imura, H.,** High affinity 3,5,3′--triiodothyronine binding to synaptosomes in rat cerebral cortex, *Endocrinology*, 110, 1257, 1982.
72. **Mashio, Y., Inada, M., Tanaka, K., Ishii, H., Haito, K., Nishikawa, M., Takahashi, K., and Imura, H.,** *Endocrinologica*, 104, 134, 1983.
73. **Koenig, H., Goldstone, A. D., and Lu, C. Y.,** Surface ligands induce a rapid activation of ornithine decarboxylase, *Trans. Am. Soc. Neurochem.*, 14, 198, 1983.
74. **Lu, C. Y., Goldstone, A. D., and Koenig, H.,** Various surface agonists induce a rapid Ca^{2+}-dependent stimulation of endocytosis, hexose and amino acid transport in kidney cortex, *Fed. Proc. Fed. Am. Soc. Exp. Biol.*, 42, 294, 1983.

Chapter 5

POLYAMINES AND CELL DIFFERENTIATION

Olle Heby

TABLE OF CONTENTS

I. INTRODUCTION

The polyamines putrescine, spermidine, and spermine play an important role in cell growth and division. This has been clearly demonstrated in experiments using polyamine auxotrophs and inhibitors of the enzymes in the polyamine biosynthetic pathway, and the topic has been the subject of many reviews.[1-5] Polyamine-deficient cells grow very slowly or not at all, but if exogenous polyamines are provided the growth-inhibited state is rapidly reversed.

Experiments with polyamine synthesis inhibitors have also shown that polyamines play an important role in differentiative processes.[1,6] In the majority of experiments, inhibitors of ornithine decarboxylase (ODC) or *S*-adenosylmethionine decarboxylase (SAMDC) have been used. The present paper attempts to review the results obtained and to summarize the effects of altered polyamine levels on the expression of various differentiated phenotypes.

The first available polyamine synthesis inhibitor, methylglyoxal-bis(guanylhydrazone) (MGBG), a cancer chemotherapeutic agent, was discovered because of its structural resemblance to spermidine.[7] MGBG was found to reduce the cellular spermidine and spermine content by inhibiting the activity of SAMDC.[8] The fact that its antiproliferative action was prevented by concurrent administration of spermidine suggested high specificity.[7] Nevertheless, MGBG has been found to exert many effects that seem to be unrelated to polyamine depletion.[9,10]

More specific inhibitors, both substrate and product analogues, have been developed for ODC. The most potent and specific inhibitor available is α-difluoromethylornithine (DFMO), an enzyme-activated irreversible inhibitor of the enzyme.[11] The high specificity of DFMO is due to its mechanism of action. Other substrate analogues, such as α-methylornithine (MO), α-hydrazinoornithine (HO), or α-hydrazino-δ-aminovaleric acid (HAVA) have been used in some studies, but these inhibitors are reversible and competitive, and usually less potent.[5,9]

To ascertain that biological effects resulting from treatment with polyamine synthesis inhibitors are indeed due to polyamine deficiency, prevention and reversal experiments with exogenous polyamines are usually performed. Regarding developmental effects it is important to emphasize, however, that if there is no reversal of DFMO-mediated effects by polyamine provision it does not necessarily mean that the effects are unrelated to polyamine deficiency. Since embryos and differentiating cells gradually become restricted in their developmental potential, reversal experiments may be inconceivable. Therefore, when studying early developmental processes it is particularly important to use polyamine synthesis inhibitors with a high degree of specificity.

Prevention experiments, in which putrescine is added at the same time as DFMO, to compensate for the DFMO-mediated inhibition of putrescine synthesis, provide important information as to the specificity of the inhibitor. Putrescine does not interfere with the uptake of DFMO, because it has a separate transport system. However, the possibility of interference with inhibitor uptake should be considered when spermidine is used to compensate for MGBG-mediated inhibition of spermidine synthesis, because MGBG uses the same transport carrier.[10]

II. OOGENESIS AND OVARIAN DEVELOPMENT

A. Amphibia

Certain hormones (progesterone or human chorionic gonadotropin) stimulate oocytes of *Xenopus laevis* to resume meiosis from their prophase arrest. This maturation process involves the dissolution of the nuclear membrane (germinal vesicle breakdown) and progression through the first meiotic division, preparing the egg for fertilization.

Hormone-induced meiotic maturation is associated with increases in ODC activity and putrescine, but not spermidine and spermine, concentration.[12,13] Prevention of this increase in putrescine synthesis by DFMO treatment causes inhibition of both maturation and ovulation. This effect is not observed if DFMO is added at the same time as the hormone,[12] only if DFMO is added in advance.[13] The fact that putrescine reverses the effects of DFMO only if provided shortly after hormone addition also suggests a role for this polyamine in a very early phase of oocyte maturation.

B. Mammals

In the rat ovary there is an increase in ODC activity and putrescine concentration late in proestrus, which is controlled by luteinizing hormone. When this increase is prevented by DFMO treatment, ovulation takes place normally as does the course of the subsequent pregnancy, resulting from mating during DFMO treatment.[14] These data provide no support for a role of the preovulatory rise in putrescine synthesis in the final stages of follicular maturation and ovulation, or in the formation and function of the corpora lutea derived from these follicles. However, there is an effect of DFMO treatment during the subsequent cycle; following mating at the next estrus there is an increase in the number of implantation sites and viable fetuses.[14] This presumably reflects an increase in the number of ovulated eggs, for which there is no explanation as yet.

III. EMBRYONIC DEVELOPMENT

A. Polychaetes

In developing embryos of the polychaete *Ophryotrocha labronica* there is a precipitous increase in ODC activity and putrescine content during the cleavage period, but a decrease in spermidine and spermine content.[15,16] Inhibition of the increase in putrescine synthesis by MO or DFMO treatment results in developmental arrest at gastrulation.[15,16] Gastrular arrest is prevented by the addition of putrescine, suggesting a specific role for this polyamine in gastrulation.

The eggs of *Chaetopterus variopedatus*, another polychaete worm, may develop almost normally without any cytoplasmic cleavage upon artificial activation by KCl treatment. MO exerts no effect on such "differentiation without cleavage".[17] During normal development the segmentation of the fertilized eggs is not affected by MO treatment, but at the time of hatching the trochophore larvae have no intestine and no apical tuft.

B. Sea Urchins

In sea urchin (*Hemicentrotus pulcherrimus* and *Anthocidaris crassispina*) eggs the ODC activity and polyamine concentrations increase following fertilization. When these increases are prevented by treatment with HO, the first and second cleavages are delayed markedly, and many eggs divide unequally.[18] The third cleavage is almost completely blocked. Exogenous ornithine and polyamines prevent the HO-mediated inhibition of egg cleavage.

Treatment of sea urchin (*Paracentrotus* and *Arbacia*) embryos with MO instead of HO results in developmental arrest at a later stage, the late blastula stage.[17] As in the experiments with HO, polyamines reduce the inhibitory effect when added at the same time as the inhibitor. Whether the discrepancies between the developmental effects are due to species and/or inhibitor differences remains to be determined.

C. Chick

Embryonic development begins while the hen egg is still in the oviduct, but it is interrupted at laying. The blastoderm is then in late cleavage. Upon incubation there are marked increases in ODC activity and polyamine concentrations.[19] The first major peak in polyamine synthesis occurs during the primitive streak stage (gastrulation) and the second during early neurulation and somitogenesis. When the increase in polyamine synthesis is inhibited by DFMO treatment, the embryos fail to develop beyond the primitive streak stage.[19] Mesoderm outgrowth and differentiation are drastically impaired.[20] These findings suggest that stimulation of polyamine synthesis is an obligatory step in the differentiation of epiblast cells into mesoderm cells, i.e., the fundamental developmental change occurring at gastrulation.

In primary culture, mesoderm from definitive-streak stage chick embryos retains its in ovo outgrowth behavior and differentiation pattern. If DFMO is added to the culture medium, however, cell attachment and outgrowth are delayed, and the rate of cell proliferation is reduced.[21] Furthermore, the differentiation into beating heart muscle tissue, erythroid cells, chondrocytes, and adipose cells is markedly delayed. Simultaneous addition of putrescine prevents or reduces the adverse effects of DFMO. The data suggest that polyamine depletion decreases the rate of mesoderm cell proliferation, and as a consequence delays the expression of differentiated cellular phenotypes. This is consistent with the contention that the fates of most mesodermal cells are already determined by the end of gastrulation and that most embryonic tissues must reach a certain mass or cell number before they can express their developmental capabilities.

D. Mammals

In the mouse uterus, the ODC activity is low until implantation of the blastocyst into the endometrium.[22,23] It increases sharply in the deciduomata (embryo plus decidua) to reach a peak during the primitive streak stage.[23] There is a concomitant increase in putrescine and spermidine concentration. Inclusion of DFMO in the drinking water from the time of implantation abolishes the increases in uterine ODC activity and polyamine concentrations and causes resorption or expulsion of the embryos.[22,23] Decidualization seems to take place normally after implantation, but subsequent embryonic development is greatly reduced. Embryonic development fails to progress beyond the primitive streak stage of normal gestation. Subsequently, signs of rejection and resorption become evident, and on the expected date of parturition there are no signs of pregnancy. These effects of DFMO on the developing embryo are prevented by the concomitant administration of putrescine, implicating inhibition of putrescine synthesis as the mechanism of the contragestational effect. DFMO exerts similar contragestational effects, which are counteracted by provision of putrescine, also in the rat and the rabbit.[22,24]

Cleavage of the fertilized ovum and blastocyst formation seem to take place normally in mice subjected to DFMO treatment.[22,23] This is in agreement with studies on preimplantation mouse embryos developing in culture in the presence of MO or DFMO.[25,26] MGBG treatment on the other hand slows DNA synthesis and mitotic rate and blocks development at an early stage of cleavage (the 8- to 16-cell stage).[26,27] Embryonic development resumes after transfer to normal medium, but cavitation, i.e., the formation of the blastocoel, is delayed.[27] Moreover, the in vitro shedding of the zona pellucida is decreased.

When diapausing mouse blastocysts are cultured in the presence of MO and/or MGBG, their attachment to the culture dish and the trophoblastic outgrowth are retarded or even blocked.[28] These adverse effects are prevented when polyamines are supplied in the medium along with the inhibitors, suggesting that polyamine synthesis is required for activation, growth, and development of diapausing blastocysts.

IV. EMBRYONAL CARCINOMA CELL DIFFERENTIATION

Biochemical investigations of early embryonic mammalian development are hampered by both the scarcity and heterogeneity of biological material. It is possible to avoid these difficulties by using embryonal carcinoma (EC) cells, which may serve as a model system to study mammalian embryogenesis and neoplasia. EC cells are the pluripotent stem cells of teratocarcinomas, germ cell tumors which contain a mixture of differentiated and undifferentiated cell types. Although EC cells are tumorigenic they can participate in normal embryonic development. In culture, EC cells can be stimulated to differentiate by various inducers, including polyamine synthesis inhibitors.[6,29-40]

Most EC cell lines, including a human cell line (NT2/D1),[40] are induced to differentiate by DFMO treatment. One cell line (Nulli-SCC1), however, fails to differentiate in response to DFMO, yet the levels of both ODC and putrescine are markedly reduced.[33,35,39] The fact that DFMO inhibits their growth suggests that induction of differentiation is not simply a result of a reduced proliferation rate.

The differentiative effects of DFMO are reversed if exogenous polyamines are added within the first few days of treatment.[29,30,32,34] Nevertheless, the differentiated phenotype is irreversible, even after removal of the inducer. A minor fraction of the cells, which decreases during the course of treatment, does not respond to DFMO.[38,40] Upon removal of the inducer, these cells start proliferating at a similar rate as untreated cells. Surprisingly, they exhibit no resistance to DFMO, but differentiate and cease to proliferate in response to DFMO just like the parental cells.[38,40] Because proliferation in the absence of polyamine synthesis is the only efficient way of reducing the cellular polyamine content, it appears that some teratocarcinoma cells may be in a quiescent state (in the GO phase), and thus refractory to polyamine depletion.

One of the most potent inducers of EC cell differentiation is retinoic acid (RA). Since retinoids can inhibit ODC activity and reduce polyamine levels[41,42] attempts have been made to determine whether RA induces EC cell differentiation by inhibiting the ODC activity. However, treatment of EC cells with RA does not result in the rapid reduction of ODC activity and polyamine levels seen in DFMO-treated cells.[32,34,35] Accordingly, putrescine, which reverses the effect of DFMO, fails to reverse RA-induced differentiation. This suggests that RA does not induce differentiation by directly interfering with polyamine synthesis.

RA-mediated inhibition of ODC activity may be due to a block in cell cycle progression prior to ODC induction, which occurs in late-G1 to early-S.[1] This is consistent with the finding that RA-treated F9 cells become arrested in the G1 phase[34] and that retinoids arrest Chinese hamster ovary cells in mid-G1.[42] DFMO treatment also blocks the traverse of the cell cycle in the G1 phase.[34] Diff(RA)$^{-}$1 cells, which are refractory to RA-induced differentiation, show no change in ODC activity and polyamine levels during RA treatment.

These studies on EC cells indicate a role for polyamines in their differentation, but not necessarily a direct one. Since DFMO-treated cells have no putrescine and spermidine to act as aminopropyl group acceptors in the spermidine and spermine synthase reactions, the concentration of the aminopropyl group donor, decarboxylated *S*-adenosylmethionine (deSAM), increases dramatically.[37] When F9 cells are treated with MGBG to prevent this increase, the antiproliferative and differentiative effects of DFMO are counteracted.[37] Therefore it may be suggested that the elevated level of deSAM may compete with *S*-adenosylmethionine in methyltransferase reactions, thus reducing the degree of DNA methylation. This hypothesis is consistent with the finding that differentiation of F9 cells is associated with demethylation of the DNA.

V. MAMMARY GLAND DIFFERENTIATION

In the mammary gland, polyamine synthesis increases during pregnancy when the epithelial cells undergo rapid proliferation, and reaches its maximum level during the lactation period. During this period, terminal differentiation of the mammary gland epithelium culminates in the synthesis of milk proteins. Similar changes are induced in mammary gland explants from midpregnant mice by provision of insulin, glucocorticoid, and prolactin. In such explants, the hormone-induced synthesis of milk proteins (casein and α-lactalbumin), is prevented by treatment with MGBG or HAVA.[43-45] The inhibition of milk protein synthesis is overcome by the addition of spermidine. Moreover, spermidine is able to substitute for the glucocorticoid needed to induce mammary gland differentiation.

In the rabbit system, the induction of milk protein synthesis does not have an absolute requirement for glucocorticoid and spermidine,[46] but it is enhanced by glucocorticoids.[47] The enhancement is dependent on spermidine, but spermidine cannot itself replace the glucocorticoid. In the rat system, there is an absolute requirement for both, but spermidine cannot substitute for the glucocorticoid.[47]

Casein synthesis is stimulated in a prolactin-like manner upon addition of a combination of spermidine and a prostaglandin to mouse mammary gland explants.[48,49] MGBG does not prevent the effect of spermidine plus a prostaglandin, but it does prevent the stimulation of casein synthesis by prolactin. These findings suggest that spermidine has a specific function in casein synthesis.

Spermidine appears to have multiple, fundamental functions in the regulation of proliferation and differentiation of the mammary gland.[43-49] Although the precise molecular mechanisms involved are still largely unknown, one finds that spermidine enhances the efficiency of the cell-free translation of mRNAs for milk proteins, and stimulate the activity of a protein kinase, which is involved in the phosphorylation of casein.[45]

VI. HEMATOPOIESIS

In adult vertebrates, the continuous formation of new blood cells is accomplished in the bone marrow by the hematopoietic stem cells. The stem cells are capable of extensive proliferation. They are engaged in both self-renewal and production of more differentiated progeny. Recent experiments indicate that polyamines may modulate the proliferation and/or differentation of hematopoietic cells.

DFMO administration to mice produces bone marrow cells that have the capacity of forming an increased number of colonies (1) in the spleens of lethally irradiated mice (pluripotent stem cells),[50,51] (2) in diffusion chambers implanted i.p. in mice (granulocytic and erythroid cells),[52] and (3) in semisolid culture media (committed stem cells of granulocytes, monocytes, and erythrocytes).[51,52] This stimulation is abrogated by administration of putrescine to the mice, suggesting that polyamines play an important role in the early stages of hematopoiesis in vivo.

Small numbers of thymocytes in coculture with hematopoietic progenitor cells from bone marrow produce a reduction in erythroid committed stem cell formation. Large numbers of thymocytes instead cause an enhancement. In this coculture system, DFMO exerts an enhancing effect.[50] This effect is due to an increase in helper thymocyte activity. Thus, DFMO seems to cause acute inhibition of suppressor thymocyte proliferation, reflected by increased helper activity. The enhancing effect of DFMO is prevented by putrescine administration.

When congenitally aplastic mice (anemic) are given bone marrow transplants, there is a more rapid engraftment of marrow from DFMO-treated donors than from untreated controls.[53] This again suggests that DFMO modifies the proliferation of the regulatory cells of erythropoiesis.

Upon extended DFMO treatment, cells other than the suppressor thymocytes may be inhibited in their proliferation. These cells may include helper thymocytes and erythroid progenitor cells. As a matter of fact, in both humans[54] and rats[55] the administration of DFMO for a longer time period (2 to 5 weeks) results in decreased peripheral blood cell elements, especially in platelets. The effects in rats are prevented by the addition of putrescine, and in humans they are promptly reversed when DFMO treatment is discontinued.

When DFMO is administered to bone marrow cultures, there is a dose-dependent inhibition of erythroid committed stem cell formation. This inhibition is not prevented by exogenous putrescine. However, addition of aminoguanidine, which blocks diamine oxidase present in the culture medium, increases the yield of committed stem cells.[51]

In primary cultures of chick embryo mesoderm, erythroid differentiation is delayed by DFMO treatment, and this adverse effect is prevented by putrescine addition.[21]

The colony stimulating factor-induced proliferation and differentiation of human granulocyte-macrophage progenitor cells is inhibited by the administration of DFMO.[56] In the presence of increasing DFMO concentrations, the arrest of myeloid differentiation occurs at progressively earlier stages of maturation. This arrest is prevented by provision of exogenous putrescine, indicating that it is indeed due to polyamine deficiency.

VII. PROMYELOCYTIC LEUKEMIA CELL DIFFERENTIATION

The human promyelocytic leukemia cell line, HL-60, may be used for the study of myeloid differentiation. These cells are tumorigenic in nude mice and show a limited capacity to differentiate spontaneously. However, a variety of substances (dimethyl sulfoxide, *N,N*-dimethyl formamide, and retinoids) cause the HL-60 cells to mature into granulocytes. These granulocytes appear to have undergone terminal differentiation, because they do not proliferate, are nontumorigenic in nude mice, and have increased phagocytic capacity. Exposure of HL-60 cells to phorbol esters induces maturation along an alternate pathway toward monocytes. These cells also appear to be terminally differentiated.

Induction of HL-60 cell differentiation increases the ODC activity and the putrescine and spermidine levels.[57,58] Exposure of the HL-60 cells to DFMO, together with either RA or a phorbol ester, results in growth inhibition, but there is no effect on the induced myeloid differentiation.[29,57,58] Thus, the increase in polyamine synthesis seems to be important for proliferation but not for differentiation of the HL-60 cells.

When the cellular polyamine content is reduced to a greater extent by prolonged DFMO treatment, HL-60 cell differentiation (induced by dimethyl sulfoxide, hexamethylene bisacetamide, butyric acid, RA, or leukocyte-conditioned medium) is inhibited.[59] Exogenous spermidine abrogates this inhibition of induced differentiation. Therefore, the polyamines seem to play an important role also in HL-60 cell differentiation. However, phorbol ester-induced HL-60 cell differentiation is not inhibited by prolonged DFMO treatment.[60] Thus, cell differentiation can proceed by pathways apparently not involving polyamines.

VIII. ERYTHROLEUKEMIA CELL DIFFERENTIATION

Friend erythroleukemia cells are induced to differentiate by a variety of compounds. These inducers may be separated into two groups: compounds that enhance polyamine synthesis (e.g., dimethyl sulfoxide and hexamethylene bisacetamide) and compounds that have no effect on polyamine synthesis.[61] Differentiation induced by these compounds is inhibited by MO, HO, MGBG, and DFMO.[61,62] This inhibition is reversed by addition of putrescine, spermidine, or spermine. These studies emphasize that, even in cell systems in which polyamines appear to be necessary for cell differentiation, differentiation may nevertheless proceed or be induced in the absence of a stimulation of polyamine synthesis. These studies suggest an important role for the polyamines in the differentiation of Friend erythroleukemia cells.

Under certain conditions DFMO stimulates hexamethylene bisacetamide-induced differentiation in murine erythroleukemia cells.[63]

IX. INTESTINAL MUCOSAL DIFFERENTIATION

The intestinal mucosa undergoes constant renewal and cell differentiation. In the young rat, there is rapid proliferation in the crypt compartment and rapid differentiation of the mucosal cells into differentiated villus cells. Administration of DFMO results in decreased cell proliferation, and the progeny shows delayed cell maturation, as measured by the appearance of mature, absorptive enzymes.[64] This inhibition of intestinal mucosal differentiation is also observed when DFMO is administered to adult rats stressed by exposure to a cytotoxic agent.

X. ADIPOCYTE DIFFERENTIATION

When exposed to an inducer (e.g., 20 to 30% serum, insulin, 1-methyl-3-isobutylxanthine, dexamethazone, or prostaglandins), mouse 3T3-L1 fibroblasts, maintained in a density-inhibited state, show rapid and irreversible differentiation into adipocytes,[65,66] the cells specialized for fat storage in adipose tissue. Adipocyte differentiation is inhibited if DFMO is added at the same time as the inducer, but not if it is added after exposure to the inducer. Thus, the antidifferentiative effect of DFMO is temporally restricted, and it does not appear to be mediated via an effect on cell proliferation. That the effect is reversible upon putrescine or spermidine addition indicates a role specifically for polyamines in this differentiation process.

XI. CHONDROCYTE DIFFERENTIATION

Parathyroid hormone stimulates differentiation of rabbit costal chondrocytes in confluent cultures, as revealed by morphological change and increased glycosaminoglycan synthesis, which are characteristic of the cartilage phenotype. A precipitous increase in ODC activity is elicited by the hormone. Inhibition of this increase with HAVA blocks differentiation.[67] In view of this fact, and because polyamines alone can stimulate chondrocyte differentiation,[68] it appears that polyamines play a regulatory role in the expression of the differentiated phenotype.

XII. MYOBLAST DIFFERENTIATION

Mouse L6 myoblasts may be induced to differentiate into myotubes by a variety of compounds (e.g., insulin or somatomedin). Differentiation is revealed by the formation of multinucleate skeletal muscle cells and by increased activity of creatine kinase, a biochemical marker for muscle cell differentiation. Growth factor-induced differentation is accompanied by a marked increase in ODC activity. Inhibition of this increase by DFMO treatment blocks the differentiation of myoblasts into myotubes.[69,70] Since DFMO is added as the myoblast cultures approach confluency, and since there is no effect on DNA synthesis, the antidifferentiative effect is not secondary to inhibition of cell proliferation. That the effect is reversed by exogenous polyamines indicates a role for polyamines also in this differentiation process.

XIII. NEURONAL DIFFERENTIATION

Primary cultures of neuroblasts, prepared from the cerebral cortex of embryonic chick brain, develop spontaneously to neurons with many cellular processes and multiple synaptic contacts. DFMO treatment does not affect neuronal differentiation at doses which deplete the cellular putrescine and spermidine content.[71] However, treatment with other enzyme-activated irreversible ODC inhibitors [6-heptyne-2,5-diamine or (E)-α-(fluoromethyl)dehydroputrescine] blocks neuronal outgrowth and differentiation. The explanation for these discrepancies may be that the latter inhibitors are product analogues. Due to their structural analogy to putrescine they do not only inhibit the ODC activity, but they may also bind to sites which are normally occupied by putrescine, thus interfering with its function.[71] This view is also supported by the finding that putrescine prevents and reverses the antidifferentiative effects of the product analogues.

Administration of DFMO to neonatal rats causes retardation of synaptic development of catecholaminergic systems in the brain.[72] DFMO treatment exerts even more profound effects on the development of peripheral sympathetic neurons. These findings indicate that the polyamines play an important role in neuronal outgrowth and synaptic maturation.

XIV. NEUROBLASTOMA CELL DIFFERENTIATION

Mouse neuroblastoma cells possess some of the biochemical properties of neurons, but they generally remain in a relatively immature state of differentiation. However, they can be induced to differentiate with a variety of compounds (e.g., dibutyryl cyclic AMP and 1-methyl-3-isobutylxanthine), exhibiting biochemical changes related to the synthesis of neurotransmitter and, eventually, a morphology characteristic of neurons. Studies with (R,S)-α-fluoromethylornithine, an enzyme-activated irreversible inhibitor of ODC, demonstrate that reduced levels of ODC activity and polyamines induce, rather than inhibit, the differentiation of neuroblastoma cells.[73,74] This induction is prevented by exogenous polyamines.

XV. MELANOMA CELL DIFFERENTIATION

Mouse melanoma cell differentiation may be induced by melanocyte stimulating hormone (MSH), and involves dramatic increases in tyrosinase activity and melanin content, as well as changes in growth characteristics and cell morphology. Treatment with DFMO and MGBG induces extensive differentiation of Cloudman S91[75,76] and Bl6[77] mouse melanoma cells, and these effects are prevented by exogenous polyamines. Interestingly, DFMO potentiates the stimulatory effect of MSH, and MSH decreases the intracellular putrescine level to a degree comparable to that observed with DFMO.[75] However, the fact that treatment of MSH-stimulated S91 cells with exogenous putrescine fails to prevent melanocyte differentiation suggests that the mechanism of action of polyamine synthesis inhibitors is different from that of MSH.

XVI. SUMMARY

The collective data obtained in experiments with polyamine synthesis inhibitors, using a large spectrum of biological systems from invertebrate embryos to human cell cultures, indicate that the polyamines play an important role in cell differentiation. In some systems it is difficult to rule out the possibility that the differentiative response is simply a reflection of a reduced growth rate, but in other systems it is clear that the response is due to a specific effect on a differentiative event. Realizing that each experimental system is unique, one may not be surprised to find that in some cell types the polyamines act as positive effectors of cell differentiation, in others as negative effectors. This dichotomy is probably also a reflection of the complexity of polyamine functions. Present work is aimed at defining these functions at a molecular level.

ACKNOWLEDGMENTS

The author's laboratory received financial support from the Swedish Natural Science Research Council, and the Berta Kamprad, the Gunnar, Arvid and Elisabeth Nilsson, and the John and Augusta Persson Research Foundations.

REFERENCES

1. **Heby, O.,** Role of polyamines in the control of cell proliferation and differentiation, *Differentiation*, 19, 1, 1981.
2. **Pegg, A. E. and McCann, P. P.,** Polyamine metabolism and function, *Am. J. Physiol.*, 243, C212, 1982.
3. **Tabor, C. W. and Tabor, H.,** Polyamines, *Annu. Rev. Biochem.*, 53, 749, 1984.
4. **Williamson, J. D. and Tyms, A. S.,** Polyamine inhibitors as antimicrobial agents?, *Med. Microbiol.*, 4, 239, 1984.
5. **Heby, O.,** Ornithine decarboxylase as target of chemotherapy, *Adv. Enz. Regu.*, 24, 103, 1986.
6. **Heby, O., Luk, G. D., and Schindler, J.,** Polyamine synthesis inhibitors act as both inducers and suppressors of cell differentiation, in *Inhibition of Polyamine Metabolism: Biological Significance and Basis for New Therapies*, McCann, P. P., Pegg, A. E., and Sjoerdsma, A., Eds., Academic Press, New York, 1987, 165.
7. **Mihich, E.,** Current studies with methylglyoxal-bis(guanylhydrazone), *Cancer Res.*, 23, 1375, 1963.
8. **Williams-Ashman, H. G. and Schenone, A.,** Methyl glyoxal bis(guanylhydrazone) as a potent inhibitor of mammalian and yeast *S*-adenosylmethionine decarboxylases, *Biochem. Biophys. Res. Commun.*, 46, 288, 1972.
9. **Heby, O. and Jänne, J.,** Polyamine antimetabolites: biochemistry, specificity, and biological effects of inhibitors of polyamine synthesis, in *Polyamines in Biology and Medicine*, Morris, D. R. and Marton, L. J., Eds., Marcel Dekker, New York, 1981, 243.
10. **Porter, C. W., Dave, C., and Mihich, E.,** Inhibition of *S*-adenosylmethionine decarboxylase as an approach in cancer therapeutics, in *Polyamines in Biology and Medicine*, Morris, D. R. and Marton, L. J., Eds., Marcel Dekker, New York, 1981, 407.
11. **Metcalf, B. W., Bey, P., Danzin, C., Jung, M. J., Casara, P., and Vevert, J. P.,** Catalytic irreversible inhibition of mammalian ornithine decarboxylase (EC 4.1.1.17) by substrate and product analogues, *J. Am. Chem. Soc.*, 100, 2551, 1978.
12. **Younglai, E. U., Godeau, F., Mester, J., and Baulieu, E. E.,** Increased ornithine decarboxylase activity during meiotic maturation in *Xenopus laevis* oocytes, *Biochem. Biophys. Res. Commun.*, 96, 1274, 1980.
13. **Sunkara, P. S., Wright, D. A., and Nishioka, K.,** An essential role for putrescine biosynthesis during meiotic maturation of amphibian oocytes, *Dev. Biol.*, 87, 351, 1981.
14. **Fozard, J. R., Prakash, N. J., and Grove, J.,** Ovarian function in the rat following irreversible inhibition of L-ornithine decarboxylase, *Life Sci.*, 27, 2277, 1980.
15. **Emanuelsson, H. and Heby, O.,** Inhibition of putrescine synthesis blocks development of the polychete *Ophryotrocha labronica* at gastrulation, *Proc. Natl. Acad. Sci. U.S.A.*, 75, 1039, 1978.
16. **Heby, O. and Emanuelsson, H.,** Role of the polyamines in germ cell differentiation and in early embryonic development, *Med. Biol.*, 59, 417, 1981.
17. **Brachet, J., Mamont, P., Boloukhère, M., Baltus, É., and Hanocq-Quertier, J.,** Effets d'un inhibiteur de la synthèse des polyamines sur la morphogénèse, chez l'Oursin, le Chétoptère et l'Algue *Acetabularia*, *C. R. Acad. Sci. Ser. D*, 287, 1289, 1978.
18. **Kusunoki, S. and Yasumasu, I.,** Inhibitory effect of α-hydrazinoornithine on egg cleavage in sea urchin eggs, *Dev. Biol.*, 67, 336, 1978.
19. **Löwkvist, B., Heby, O., and Emanuelsson, H.,** Essential role of the polyamines in early chick embryo development, *J. Embryol. Exp. Morphol.*, 60, 83, 1980.
20. **Löwkvist, B., Emanuelsson, H., and Heby, O.,** Effects of polyamine limitation on nucleolar development and morphology in early chick embryos, *Cell Differentiation*, 12, 19, 1983.
21. **Löwkvist, B., Oredsson, S. M., Holm, I., Emanuelsson, H., and Heby, O.,** Inhibition of polyamine synthesis reduces the growth rate and delays the expression of differentiated phenotypes in primary cultures of embryonic mesoderm from chick, *Cell Tissue Res.*, 249, 151, 1987.
22. **Fozard, J. R., Part, M.-L., Prakash, N. J., Grove, J., Schechter, P. J., Sjoerdsma, A., and Koch-Weser, J.,** L-Ornithine decarboxylase: an essential role in early mammalian embryogenesis, *Science*, 208, 505, 1980.
23. **Fozard, J. R., Part, M.-L., Prakash, N. J., and Grove, J.,** Inhibition of murine embryonic development by α-difluoromethylornithine, an irreversible inhibitor of ornithine decarboxylase, *Eur. J. Pharmacol.*, 65, 379, 1980.
24. **Méndez, J. D., Diaz-Flores, M., Durán, G., and Hicks, J. J.,** Inhibition of rat embryonic development by the intrauterine administration of α-difluoromethylornithine, *Contraception*, 28, 93, 1983.
25. **Alexandre, H.,** Effets de deux inhibiteurs de la biosynthèse des polyamines sur la différenciation primaire de l'oeuf de Souris, *C. R. Acad. Sci. Ser. D*, 286, 1215, 1978.
26. **Zwierzchowski, L., Czlonkowska, M., and Guzkiewicz, A.,** Effect of polyamine limitation on DNA synthesis and development of mouse preimplantation embryos *in vitro*, *J. Reprod. Fertil.*, 76, 115, 1986.

27. **Alexandre, H.,** The utilization of an inhibitor of spermidine and spermine synthesis as a tool for the study of the determination of cavitation in the preimplantation mouse embryo, *J. Embryol. Exp. Morphol.*, 53, 145, 1979.
28. **Van Winkle, L. J. and Campione, A. L.,** Effect of inhibitors of polyamine synthesis on activation of diapausing mouse blastocysts *in vitro, J. Reprod. Fertil.*, 68, 437, 1983.
29. **Heby, O., Oredsson, S. M., Olsson, I., and Marton, L. J.,** A role for the polyamines in mouse embryonal carcinoma (F9 and PCC3) cell differentiation but not in human promyelocytic leukemia (HL-60) cell differentiation, *Adv. Polyamine Res.*, 4, 727, 1983.
30. **Schindler, J., Kelly, M., and McCann, P. P.,** Inhibition of ornithine decarboxylase induces embryonal carcinoma cell differentiation, *Biochem. Biophys. Res. Commun.*, 114, 410, 1983.
31. **Heby, O., Oredsson, S. M., and Kanje, M.,** Polyamine biosynthetic enzymes as targets in cancer chemotherapy, *Adv. Enz. Regul.*, 22, 243, 1984.
32. **Jetten, A. M. and Shirley, J. E.,** Inhibition of ornithine decarboxylase by retinoic acid and difluoromethylornithine in relation to their effects on differentiation and proliferation, *Exp. Cell Res.*, 156, 221, 1985.
33. **Schindler, J., Kelly, M., and McCann, P. P.,** The response of several murine embryonal carcinoma cell lines to stimulation of differentiation by α-difluoromethylornithine, *J. Cell. Physiol.*, 122, 1, 1985.
34. **Oredsson, S. M., Billgren, M., and Heby, O.,** Induction of F9 embryonal carcinoma cell differentiation by inhibition of polyamine synthesis, *Eur. J. Cell Biol.*, 38, 335, 1985.
35. **Kelly, M., McCann, P. P., and Schindler, J.,** Alterations in polyamine metabolism during embryonal carcinoma cell differentiation, *Dev. Biol.*, 111, 510, 1985.
36. **Heby, O. and Oredsson, S. M.,** Induction of differentiation in teratocarcinoma stem cell cultures by α-difluoromethylornithine, in *Recent Progress in Polyamine Research,* Selmeci, L., Brosnan, M. E., and Seiler, N., Eds., Akadémiai Kiadó, Budapest, 1985, 363.
37. **Holm, I., Wallon, M., Thorsson, L., and Heby, O.,** Methylglyoxal-bis(guanylhydrazone) (MGBG), an inhibitor of *S*-adenosylmethionine decarboxylase, interferes with α-difluoromethylornithine (DFMO)-induced growth arrest and differentiation of embryonal carcinoma cells, *J. Cell Biol.*, 101, 246a, 1985.
38. **Heby, O., Wallon, M., Holm, I., and Thorsson, L.,** Residual proliferative potential among teratocarcinoma cells induced to differentiate into parietal endoderm cells by polyamine synthesis inhibition, *Proc. Am. Assoc. Cancer Res.*, 27, 46, 1986.
39. **Schindler, J.,** Retinoids, polyamines and differentiation,in *Retinoids and Cell Differentiation,* Sherman, M. I., Ed., CRC Press, Boca Raton, Fla., 1986, chap. 7.
40. **Heby, O., Wallon, M., Holm, I., Thorsson, L., Oredsson, S., and Stjernborg, L.,** Polyamines in teratocarcinoma stem cell differentiation, presented at Int. Conf. Polyamines in Life Sciences, Lake Yamanaka, Japan, 1986.
41. **Verma, A. K., and Boutwell, R. K.,** Vitamin A acid (retinoic acid), a potent inhibitor of 12-*O*-tetradecanoyl-phorbol-13-acetate-induced ornithine decarboxylase activity in mouse epidermis, *Cancer Res.*, 37, 2196, 1977.
42. **Haddox, M. K. and Russell, D. H.,** Cell cycle-specific locus of vitamin A inhibition of growth, *Cancer Res.*, 39, 2476, 1979.
43. **Oka, T. and Perry, J. W.,** Spermidine as a possible mediator of glucocorticoid effect on milk protein synthesis in mouse mammary epithelium *in vitro, J. Biol. Chem.*, 249, 7647, 1974.
44. **Oka, T., Perry, J. W., Takemoto, T., Sakai, T., Terada, N., and Inoue, H.,** The multiple regulatory roles of polyamines in the hormonal induction of mammary gland development, *Adv. Polyamine Res.*, 3, 309, 1981.
45. **Oka, T., Perry, J. W., and Terada, N.,** The regulatory function of spermidine in hormonal control of the development of mouse mammary gland in culture, *Fed. Proc. Fed. Am. Soc. Exp. Biol.*, 41, 3073, 1982.
46. **Houdebine, L. M., Devinoy, E., and Delouis, C.,** Role of spermidine in casein gene expression in the rabbit, *Biochimie,* 60, 735, 1978.
47. **Bolander, F. F., Jr. and Topper, Y. J.,** Relationships between spermidine, glucocorticoid and milk proteins in different mammalian species, *Biochem. Biophys. Res. Commun.*, 90, 1131, 1979.
48. **Rillema, J. A.,** Activation of casein synthesis by prostaglandins plus spermidine in mammary gland explants of mice, *Biochem. Biophys. Res. Commun.*, 70, 45, 1976.
49. **Rillema, J. A., Linebaugh, B. E., and Mulder, J. A.,** Regulation of casein synthesis by polyamines in mammary gland explants of mice, *Endocrinology,* 100, 529, 1977.
50. **Sharkis, S. J., Luk, G. D., Collector, M. I., McCann, P. P., Baylin, S. B., and Sensenbrenner, L. L.,** Regulation of hematopoiesis. II. The role of polyamine inhibition on helper or suppressor influences of the thymus, *Blood,* 61, 604, 1983.
51. **Niskanen, E., Kallio, A., McCann, P. P., and Baker, D. G.,** The role of polyamine biosynthesis in hematopoietic precursor cell proliferation in mice, *Blood,* 61, 740, 1983.

52. **Niskanen, E. O., Kallio, A., McCann, P. P., Pou, G., Lyda, S., and Thornhill, A.,** Divergent effects of ornithine decarboxylase inhibition on murine erythropoietic precursor cell proliferation and differentiation, *Cancer Res.*, 43, 1536, 1983.
53. **Sharkis, S. J.,** The role of the thymus in erythropoiesis, *Blood Cells,* 10, 223, 1985.
54. **Abeloff, M. D., Slavik, M., Luk, G. D., Griffin, C. A., Hermann, J., Blanc, O., Sjoerdsma, A., and Baylin, S. B.,** Phase I trial and pharmacokinetic studies of α-difluoromethylornithine — an inhibitor of polyamine biosynthesis, *J. Clin. Oncol.*, 2, 124, 1984.
55. **Luk, G. D., Sharkis, S. J., Abeloff, M. D., McCann, P. P., Sjoerdsma, A., and Baylin, S. B.,** Polyamine biosynthesis is required for the maintenance of peripheral blood cell elements in the rat, *Proc. Natl. Acad. Sci. U.S.A.*, 80, 5090, 1983.
56. **Verma, D. S. and Sunkara, P. S.,** An essential role for polyamine biosynthesis during human granulopoietic differentiation, *Cancer Res.*, 42, 3046, 1982.
57. **Huberman, E., Weeks, C., Herrmann, A., Callaham, M., and Slaga, T.,** Alterations in polyamine levels induced by phorbol diesters and other agents that promote differentiation in human promyelocytic leukemia cells, *Proc. Natl. Acad. Sci. U.S.A.*, 78, 1062, 1981.
58. **Luk, G. D., Civin, C. I., Weissman, R. M., and Baylin, S. B.,** Ornithine decarboxylase: essential in proliferation but not differentiation of human promyelocytic leukemia cells, *Science,* 216, 75, 1982.
59. **Sugiura, M., Schafman, T., Mitchell, T., Griffin, J., and Kufe, D.,** Involvement of spermidine in proliferation and differentiation of human promyelocytic leukemia cells, *Blood,* 63, 1153, 1984.
60. **Kufe, D. W., Griffin, J., Mitchell, T., and Shafman, T.,** Polyamine requirements for induction of HL-60 promyelocyte differentiation by leukocyte-conditioned medium and phorbol ester, *Cancer Res.*, 44, 4281, 1984.
61. **Gazitt, Y. and Friend, C.,** Polyamine biosynthesis enzymes in the induction and inhibition of differentiation in Friend erythroleukemia cells, *Cancer Res.*, 40, 1727, 1980.
62. **Sugiura, M., Shafman, T., and Kufe, D.,** Effects of polyamine depletion on proliferation and differentiation of murine erythroleukemia cells, *Cancer Res.*, 44, 1440, 1984.
63. **Meilhoc, E., Moutin, M.-J., Hugues Romani, B., and Osborne, H. B.,** Relationship between ornithine decarboxylase activity and hexamethylene bisacetamide-induced differentiation of murine erythroleukemia cells, *Exp. Cell Res.*, 162, 142, 1986.
64. **Luk, G. D., Marton, L. J., and Baylin, S. B.,** Ornithine decarboxylase is important in intestinal mucosal maturation and recovery from injury in rats, *Science,* 210, 195, 1980.
65. **Bethell, D. R. and Pegg, A. E.,** Polyamines are needed for the differentiation of 3T3-L1 fibroblasts into adipose cells, *Biochem. Biophys. Res. Commun.*, 102, 272, 1981.
66. **Erwin, B. G., Bethell, D. R., and Pegg, A. E.,** Role of polyamines in differentiation of 3T3-L1 fibroblasts into adipocytes, *Am. J. Physiol.*, 246, C293, 1984.
67. **Takano, T., Takigawa, M., and Suzuki, F.,** Role of polyamines in expression of the differentiated phenotype of chondrocytes: effect of DL-α-hydrazino-δ-aminovaleric acid (DL-HAVA), an inhibitor of ornithine decarboxylase, on chondrocytes treated with parathyroid hormone, *J. Biochem.*, 93, 591, 1983.
68. **Takano, T., Takigawa, M., and Suzuki, F.,** Role of polyamines in expression of the differentiated phenotype of chondrocytes in culture, *Med. Biol.*, 59, 423, 1981.
69. **Erwin, B. G., Ewton, D. Z., Florini, J. R., and Pegg, A. E.,** Polyamine depletion inhibits the differentiation of L6 myoblast cells, *Biochem. Biophys. Res. Commun.*, 114, 944, 1983.
70. **Ewton, D. Z., Erwin, B. G., Pegg, A. E., and Florini, J. R.,** The role of polyamines in somatomedin-stimulated differentiation of L6 myoblasts, *J. Cell. Physiol.*, 120, 263, 1984.
71. **Seiler, N., Sarhan, S., and Roth-Schechter, B. F.,** Polyamines and the development of isolated neurons in cell culture, *Neurochem. Res.*, 9, 871, 1984.
72. **Slotkin, T. A., Grignolo, A., Whitmore, W. L., Lerea, L., Trepanier, T. A., Barnes, G. A., Weigel, S. J., Seidler, F. J., and Bartolome, J.,** Impaired development of central and peripheral catecholamine neurotransmitter systems in preweanling rats treated with α-difluoromethylornithine, a specific irreversible inhibitor of ornithine decarboxylase, *J. Pharmacol. Exp. Ther.*, 222, 746, 1982.
73. **Chen, K. Y. and Liu, A. Y.-C.,** The role of polyamines in the differentiation of mouse neuroblastoma cells, *Adv. Polyamine Res.*, 4, 743, 1983.
74. **Chen, K. Y., Nau, D., and Liu, A. Y.-C.,** Effects of inhibitors of ornithine decarboxylase on the differentiation of mouse neuroblastoma cells, *Cancer Res.*, 43, 2812, 1983.
75. **Käpyaho, K. and Jänne, J.,** Stimulation of melanotic expression in murine melanoma cells exposed to polyamine antimetabolites, *Biochem. Biophys. Res. Commun.*, 113, 18, 1983.
76. **Käpyaho, K., Sinervirta, R., and Jänne, J.,** Effects of inhibitors of polyamine biosynthesis on the growth and melanogenesis of murine melanoma cells, *Cancer Res.*, 45, 1444, 1985.
77. **Sunkara, P. S., Chang, C. C., Prakash, N. J., and Lachmann, P. J.,** Effect of inhibition of polyamine biosynthesis by DL-α-difluoromethylornithine on the growth and melanogenesis of B16 melanoma *in vitro* and *in vivo, Cancer Res.*, 45, 4067, 1985.

Chapter 6

POLYAMINES IN TISSUE REGENERATION

Susan Bardócz

TABLE OF CONTENTS

I. INTRODUCTION

Regeneration is a general phenomenon in invertebrates, but it is very limited in mammals, though they are also able to regenerate to some extent. In mammals, after wounding the skin, vascular changes occur within seconds, and the injured skin is regenerated in a few days. The cells of the hemopoietic system are also able to restore the original cell number by proliferation. The regeneration of muscle, cartilage, connective tissue, and bones after their wounding are also well-known examples,[1] though the details of the process are not yet entirely elucidated.

Regeneration should involve a number of stages:

1. Detection of tissue depletion (on loss)
2. Making the decision whether the remaining cells should proliferate
3. Induction of proliferation of specific cells
4. Termination of proliferation after reaching a certain density
5. Differentiation of certain cells to permit the fulfillment of specific functions[1]

Regeneration of an organ by tissue restructuring is an even more complicated process because of cellular heterogeneity. Humoral factors and hormones influence their physiological function and metabolic operation.

In spite of their complexity, some important organs such as the liver, kidney, or lung are able to regenerate, that is, restore their size and function, usually by compensatory growth.[1]

During regeneration, several metabolic changes appear. The new cells need energy for proliferation, different precursors to synthesize nucleic acids, proteins, membranes, etc. on a well-regulated way. Polyamines are obligatory components of the cells. They fulfill different roles; as polycations they stabilize the negatively charged structures and stimulate the synthesis of nucleic acids and proteins. They probably have a regulatory role in proliferation as well as in differentiation. Ornithine decarboxylase (ODC) is the first, usually rate-limiting enzyme of polyamine biosynthesis. Increased ODC activity and elevated polyamine content were found during different types of tissue regeneration appearing in the regenerating organ within hours.[2] Polyamines must have important function in regeneration, most probably in initiating the process.

Jänne et al.[2] reviewed polyamine research on the field of tissue regeneration in 1978. In this chapter some of the new results that have been published since then in connection with regeneration are summarized and some new explanations concerning compensatory hyperplasia during liver regeneration and compensatory renal growth are given. Compensatory lung growth as well as axonal transport and regeneration have not been reviewed before. The question of heart hypertrophy is discussed in Chapter 3.

II. POLYAMINES IN AXONAL TRANSPORT AND REGENERATION

The nerve cell body never regenerates, but the axons are able to do so. They can restore the motor function and regenerate their axons in rats after axotomy. An elevation of ODC activity was observed under such conditions[3] and also in the sympathetic neurons of rats during recovery from axonal injury.[4]

A twofold increase of ODC activity in rat brain is a common phenomenon after metabolic, mechanical, thermal, or chemical injury.[5]

In rats the presence of polyamines is necessary for the recovery of severed motorneurons.[6] The polyamines may have a role in axonal transport as well as in regeneration.

In the optic nerve system of goldfish, spermidine and spermine are extensively transported along the axons in contrast to putrescine. After optic nerve crush, a rapid transport system is turned on preferentially for putrescine. Spermidine and spermine transport also intensifies during nerve regeneration.[7]

A similar axonal transport system of polyamines exists in normal and regenerating axons of rats.[8] It was reported that the synthesis of polyamines is also necessary for the survival of sympathetic neurons after axonal injury.[9]

III. POLYAMINES IN REGENERATING LIVER

The regenerating liver is a model for the study of compensatory liver growth. After removing two thirds of the rodent liver, its original mass can be restored within 14 to 28 days. This capability of the liver is referred to as hepatic regeneration, the remnant part as regenerating liver. The process is in fact a compensatory hyperplasia. First, the cells in the residual lobes divide rapidly, enlarging their size, and the mass of the liver is restored.[10]

The removal of a part from the liver, i.e., partial hepatectomy, is not the only way to induce hepatic regeneration. Loss of functional liver mass can be indirectly induced by the necrotizing action of viruses or hepatotoxins.[11] This kind of injury, however, is followed by different types of regeneration or restoration as hepatectomy and referred to as hepatic regeneration following necrosis.[11]

At least three kinds of liver regeneration can be distinguished: hepatectomy of intact liver, that of damaged liver, and the repair of liver damage following chemically induced liver lesion.[11]

When tissues are injured, several factors may be released from the injured organ into the circulation, causing alterations in the activities of serum enzymes and in polyamine content.[12] The polyamine metabolism was widely studied in various types of liver regeneration. The changes of ODC activity and polyamine as well as nucleic acid synthesis seem to be different depending on the method used to influence hepatic mass.[2]

A. Regenerating After Partial Hepatectomy

Ornithine, which is a precursor of polyamine synthesis, is also used in the cells for the *de novo* synthesis of urea (arginine) and pyrimidines. The ornithine pool is shared between these pathways. The equilibrium between them is determined by the activities of ornithine-utilizing enzymes,[13,14] depending on whether the proliferation is normal or neoplastic. Fausto and co-workers[15] observed that the ornithine level is increased in the regenerating liver due to the increased flow of metabolites through urea cycle after hepatectomy.

1. Ornithine Decarboxylase

The activity of ODC is increased by a factor of 100 to 500 after partial hepatectomy. The first peak in the activity usually appears 4 hr after operation and is followed by several other maxima depending on various factors, such as age, food supply, etc. (for a review see Reference 2). A vitamin B_6-deficient diet can also delay the first 4-hr peak to a 12-hr one.[16]

The properties of ODC also appear to be altered after hepatectomy. Zuretti et al.[17] have found that the ODC protein was more stable in the regenerating liver than in the controls when it was incubated with microsomes. This may be attributed to different factors: lower inactivating capacity and lower susceptibility to inactivation brought about by structural or conformational changes on the enzyme molecules. Moreover, a different concentration of protective factors may be present in the tissue following hepatectomy.

The ODC activity can be influenced by various inhibitors. 1,3-Diaminopropane decreases ODC activity and the accumulation of putrescine after operation, but it does not alter the *S*-adenosyl-L-methione decarboxylase (SAM DC) activity, or the spermidine and spermine concentration.[18] Administration of D,L-δ-hydrazino-α-aminovaleric acid, a competitive inhibitor of ODC, resulted in a decrease of the enzyme activity and it inhibited the increase of putrescine concentration and DNA synthesis.[19] The activity of thymidine kinase, thymidilate synthesis, and the level of the dTTP pool also decreased after partial hepatectomy. The inhibitory effect of the drug can be reversed by addition of putrescine, but not with 1,3-diaminopropane or cadaverine administered prior to operation.[19]

Another inhibitor of ODC, α-difluoromethylornithine (DFMO), can also prevent ODC induction, reduces putrescine content by inhibiting ODC, and lowers the rate of DNA synthesis.[20]

The inhibitory effect of these compounds can be overcome by administration of putrescine prior to surgery.[19,20]

The administration of putrescine induces the synthesis of an ODC antizyme as well.[21] After partial hepatectomy, the first increase in ODC activity appearing 2.5 to 4 hr after operation is regulated by induction of the antizyme, whereas the following ones are regulated by different mechanisms.[22]

To make the picture even more complicated, the experimental results show that hormonal treatment also affects ODC activity in the liver (for a review see Reference 23). Corti et al.[24] suggest that catecholamines and glucocorticoids, released during surgical stress of partial hepatectomy, form the neuroendocrine link between the operational damage and the response of liver. Catecholamines are necessary for the induction of ODC, since the latter can be eliminated by inhibiting catecholamine synthesis. On the other hand, other hormones do not seem to be required for ODC induction.[2]

2. The Role of Polyamines

The dramatic enhancement of ODC activity triggers all changes in polyamine metabolism.[2] Putrescine accumulates[2,19,22,24] as a result of the increase in ODC activity. The concentration of spermidine increases (though the SAM DC activity is not elevated until the 2nd to 3rd day) accompanied with a simultaneous decrease in spermine content.[2]

The changes of polyamine concentration remained unexplained until Matsui and Pegg presented that the spermidine/spermine acetylase activity began to rise 4 hr after operation and reached a maximum about 8 hr after partial hepatectomy.[25] Whether the activity of polyamine oxidase[26] is high enough to convert the acetylated spermidine to putrescine and the acetylspermine to spermidine, causing a decrease in spermine and an increase in putrescine and spermidine pools by polyamine interconversion,[27] has not yet been examined.

The activity of diamine oxidase, a putrescine-metabolizing enzyme with very low activity in normal liver, is increasing rapidly in the regenerating liver and reaches a peak between 16 and 48 hr after operation.[28] This pathway can help to eliminate the superfluous putrescine, caused by the induction of ODC. It might even prevent antizyme induction as well.

Based on the observations of Haddox and Russell,[29] the concentration of putrescine, spermidine, and spermine conjugates are elevated in the cell nucleus after hepatectomy.

3. Polyamines and Nucleic Acid Synthesis

After partial hepatectomy DNA and RNA synthesis changes parallel with the concentration of spermidine and spermine with increasing intensity from 24 hr postoperation.[2]

According to Inoue et al.,[30] putrescine has at least two different roles in DNA synthesis after partial hepatectomy. Reversible disorder of polyamine and DNA synthesis were also observed in regenerating rat liver following hepatectomy.[31] When polyamine synthesis was inhibited by simultaneous use of ODC and SAM DC inhibitors prior to operation, the DNA synthesis was completely suspended; however, when only one of the enzymes was inhibited, the synthesis of DNA could not be prevented.[32]

Two phases are recognized in the regeneration of liver during the first 24 hr: a hypertrophy which lasts 12 to 16 hr and a subsequent hyperplasia, characterized by an enhancement in DNA synthesis at about the 24th hr. These events are followed by mitosis, 6 to 8 hr later.[10]

The first increase in ODC activity may play a role in the first phase, while the other(s) may be involved in initiating DNA synthesis.

B. Partial Hepatectomy of Damaged Liver

There are few data available on the behavior of polyamine biosynthesis and metabolism during the regeneration of necrotic liver. Matsui et al.[33] studied the effect of a carcinogen, urethane, administered immediately after surgery on the process of regeneration following partial hepatectomy. The response of ODC activity was delayed,[33] and SAM DC activity was even lower than in the controls.[34] Spermidine content changed in a biphasic manner similar to the DNA synthesis.[34]

When rats were treated with carcinogen before hepatectomy, the cAMP-dependent protein kinase was induced prior to the prolonged induction of ODC activity. Increased spermidine/spermine ratio followed the changes of ODC activity in a similar manner; subsequently the activity of γ-glutamyl transpeptidase was increased.[35]

The papers of Pösö and co-workers[36,37] give a clue for the behavior of the damaged (cirrhotic) liver, suggesting that it behaves disparately as compared to a normal one following hepatectomy.[36] Continuous treatment with ethanol inhibits ODC induction and DNA synthesis during the first days of regeneration. The differences, however, disappear later.[36] Due to the continuous ethanol treatment during regeneration, the protein content of the remnant liver became elevated as a result of decreased protein degradation. Among the proteins a decreased degradation rate of ODC and tyrosine aminotransferase[37] as well as of other macromolecules was observed.

C. Regeneration Following Hepatic Necrosis

The liver lesion caused by hepatotoxins induces liver regeneration. This kind of injury, however, is followed by an altered type of regeneration as compared to that appearing after partial hepatectomy.[38]

A low dose of thioacetamide stimulates both DNA synthesis and ODC induction without any cellular damage.[11] A high dose of the same drug can induce liver damage, as does carbon tetrachloride, resulting in a remarkable activation of enzymes of polyamine synthesis 12 to 24 hr after the operation.[2] After carbon tetrachloride treatment the intensity of polyamine acetylation is also increasing[39] without significant alterations in polyamine oxidase activity. The activation of spermidine/spermine acetylase was also observed after thioacetamide treatment,[25] similarly to that after partial hepatectomy. However, while the maximum rate of acetylation appears around 6 to 8 hr after hepatectomy, the maximum of spermidine/spermine acetylase activity is delayed to 20 to 30 hr after the treatment with toxic agents.[25] Generally, after administration of hepatotoxins the changes in polyamine metabolism are more dramatic than those found in response to tissue loss.[2]

Though there is no clear understanding of the factors which trigger and control hepatic regeneration, it is generally accepted that the influence of hormones and humoral factors is important in this respect. An ancient method, acupuncture, may bring new understanding of the phenomenon. Acupuncture of the liver produces local injury only in some cells.[38] It also increases ODC activity after the treatment as does hepatectomy, but the synthesis of DNA is delayed and it appears only 40 hr after surgery instead of 24 hr observed after partial hepatectomy. Intact and injured lobes have shown similar responses. Hepatic lesion leads to the production of substances which are transported from the injured part to the intact lobes by humoral route and may trigger regeneration.[38]

IV. POLYAMINES IN KIDNEY HYPERTROPHY

Renal growth can be induced by unilateral nephrectomy[40] in response to unilateral ureteral occlusion[41] as well as by administration of hormones[2] or nephrotoxic agents.[42] The mechanism of growth after toxic injury and during compensatory renal growth seems to be different.

A. Compensatory Renal Growth

Removal of one kidney results in enlargement of the contralateral one and is referred to as compensatory renal growth (CRG). The solitary kidney is increasing in mass during the process, about 75% of the increase is brought about by hypertrophy, i.e., the enlargement of tubule cells, and 25% is attributed to the increase of cell number, i.e., proliferation, with a simultaneous increase in DNA and RNA content.[10]

This process is associated with the enhancement in renal ODC activity. In the remaining kidney unilateral nephrectomy results in biphasic stimulation of ODC activity which is followed by a moderate accumulation of polyamines.[40] The stimulation of ODC activity might be prevented by administration with actinomycin D.[40]

Unilateral ureteral obstruction causes similar changes in renal polyamine concentration as unilateral nephrectomy.[41] The temporary unilateral occlusion of renal artery can also induce CRG in the contralateral kidney characterized by changes in the polyamine and nucleic acid content.[43]

All procedures mentioned here cause similar alterations in polyamine metabolism. These are the ODC induction, putrescine accumulation,[2] and a prevalence of spermidine concentration over spermine, a condition found otherwise in rapidly proliferating tissues.[43] The synthesis of nucleic acids is also intensified.[2]

Various treatments, such as growth hormone, serum growth factor,[44] or testosterone[45,46] can induce renal growth in rats with a simultaneous increase in ODC activity and alteration in polyamine concentration. The increase in testosterone-induced ODC activity can be prevented by actinomycin D, 1,3-diaminopropane, or by cadaverine treatment, while cycloheximide or putrescine are ineffective.[46] Inhibition of renal ODC activity by the addition of 1,3-diaminopropane suppresses the functioning of ODC and prevents putrescine as well as spermidine accumulation. In addition it even lowers spermine content and exerts nephrotoxic action by causing tissue damage.[46] Laitinen et al.[47] have shown that after testosterone treatment multiple species of ODC could be separated from mouse kidney, but the appearance of the multiple enzyme was not due to molecular changes in enzyme protein. Nandrolone, an anabolic steroid also induces kidney growth. After nandrolone treatment the activity of ODC and lysine decarboxylase were augmented and both enzymes were present in more than one isoelectric form. An increase in putrescine and spermidine concentrations was also observed.[48]

Putrescine accumulation can be induced by administration of aminoguanidine, an inhibitor of copper containing diamine oxidase (DAO). The drug leaves the spermidine and spermine levels unaltered.[49] Actinomycin D or cycloheximide prevents the increase of DAO activity.[50] There is practically no data available on the activity of other polyamine metabolizing enzymes and on the concentrations of metabolites during compensatory renal growth.

B. Effect of Nephrotoxic Agents on Renal Regeneration

Nephrotoxic agents causing tissue damage, such as folic acid,[42] suramin,[51] methyl-mercury hydroxyde,[52] or trenbolone acetate[53] may modify polyamine concentrations partly by increasing ODC activity and partly by changing the rate of nucleic acid synthesis.[54] Again, no data are available on the activity of other polyamine synthesizing or metabolizing enzymes following toxic injuries.

Biphasic stimulation of ODC was observed in kidney during both compensatory renal growth and following treatment with nephrotoxins; however, its time course was different. Moderate accumulation of putrescine and spermidine and slight changes in spermine concentration were also observed. Renal ODC is carefully regulated; its activity may be altered even by changing the blood flow rate.[55] On the other hand the alterations in enzyme activity are not always coupled with renal growth and increased DNA synthesis.[56]

Selmeci et al.[43] suggest that alterations of polyamine concentration and nucleic acid content during the late phase of compensatory renal growth are caused by a similar mechanism as in normal kidney growth, but the pattern of alterations in the same parameters occurring in damaged (postischemic) kidney was found quite different.

V. POLYAMINES IN REGENERATING LUNG

The lung is also able to regenerate, i.e., to restore its size and function, after various pulmonary damages. Similarly to the complexity of liver regeneration, which in rodents might even be strain dependent by its mechanism,[57] one must distinguish compensatory lung growth from the regenerating process elicited by toxic or mechanical injuries.

A. Compensatory Lung Growth

After removing the left upper pulmonary lobe of mice, within 2 weeks the remaining four lobes reach the original mass of the whole lung of control mice,[58] the contribution of the individual lobes being different. During the compensatory lung growth, there is a period of rapid proliferation of pneumocytes type I. These cells differentiate into pneumocytes type II and become enlarged. The increase of ODC activity is characteristic of the growth process. The activity of ODC exhibits a biphasic course showing a fivefold increase at 4 hr after surgery, then a threefold elevation at 48 hr after operation. The participation of the individual lobes is different (Figure 1). There is a correlation between the ODC activity and growth rate. The elevation of ODC activity can be prevented by cycloheximide, or actinomycin D treatments. This indicates that the increase in ODC activity was caused by *de novo* protein synthesis, regulated partly at transcriptional level, as it was found for liver.[58]

DFMO reduced the activity of ODC activity to 20 to 60% of the value measured in untreated animals, depending on the individual lobe.[59] The concentration of putrescine tripled 4 hr after lobectomy, whereas spermidine and spermine concentrations did not change significantly (unpublished).

Smith et al.[60,61] observed an accumulation of cadaverine, putrescine, spermidine, and spermine initiated by an intensive polyamine uptake of mouse lung slices in vitro. The lung may have a specific membrane carrier protein for di- and polyamines. The diamine transport system might be the same as that for paraquat.[61] The authors suggest that the transport system is turned off in the regenerating lung. It appears that after removal of a lobe this transport system may remain active, particularly after DFMO treatment, and supply the lung with polyamines.

To elucidate the nature of various factors involved in regeneration after lobectomy, left lower lobe fragments were transplanted under the renal capsule of sister mice. After a week when the transplanted fragments became vascularized and the ODC activity was similar to that in untreated mice, the animals were lobectomized, the ODC activity determined in both the transplanted and the own left lower pulmonary lobes of host mice. The ODC activity appeared elevated in both types of lobes (Figure 2) indicating that both local and humoral factors should play a role in regeneration.

B. Repair of Damaged Lung

Rats exposed to pure oxygen atmosphere develop pulmonary damage. Alterations of pulmonary ODC activity and polyamine concentrations during the repair process were correlated with the ultrastructural changes of the lung.[62] ODC activity increased by a factor of 25 on the 2nd day and returned to the normal value after a week. DFMO treatment lowered the ODC activity below control and reduced the polyamine content and the number of type I epithelial cells as well.[62]

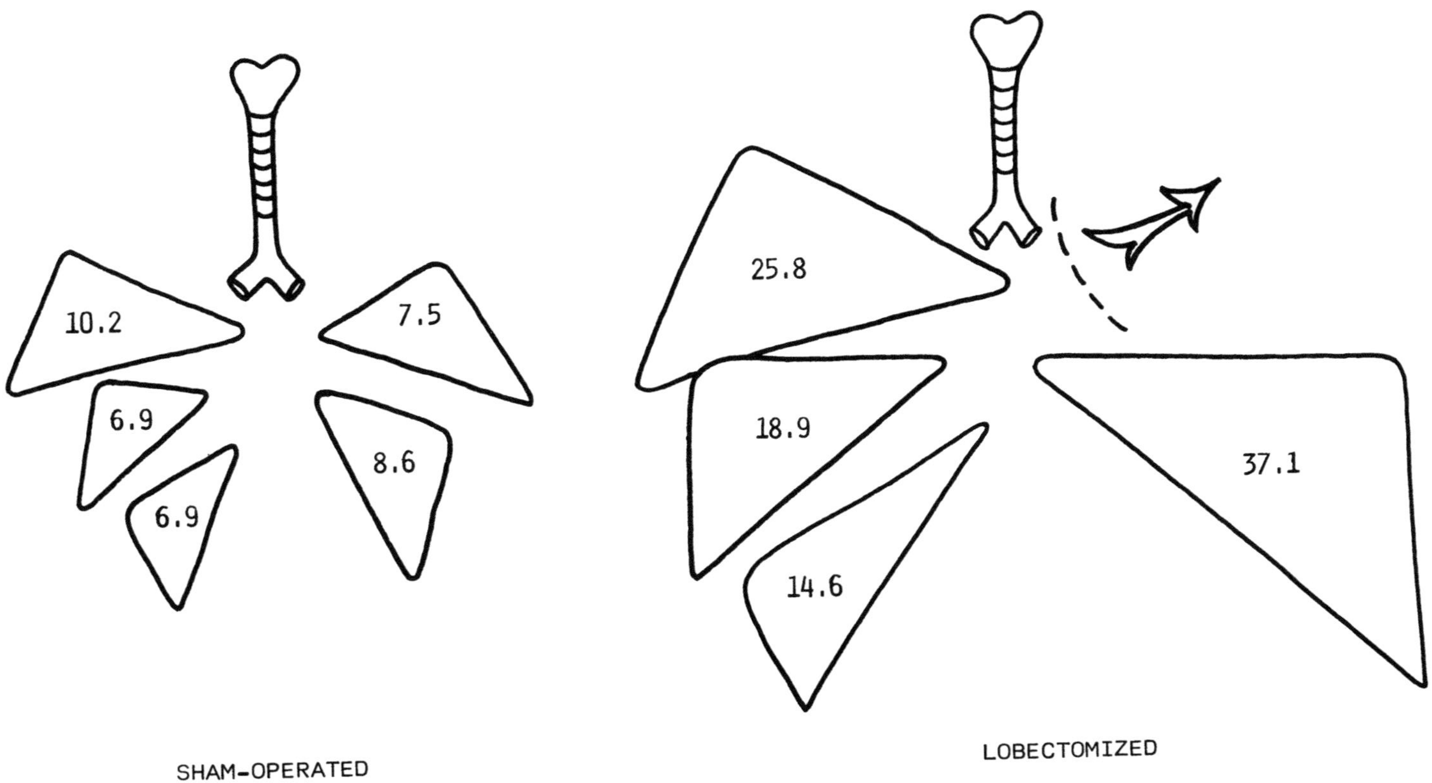

FIGURE 1. ODC activity of the individual pulmonary lobes in BALB/c female mice 4 hr after surgery. The values are in $^{14}CO_2$ pmol/hr/mg protein ± SE for four animals per group.

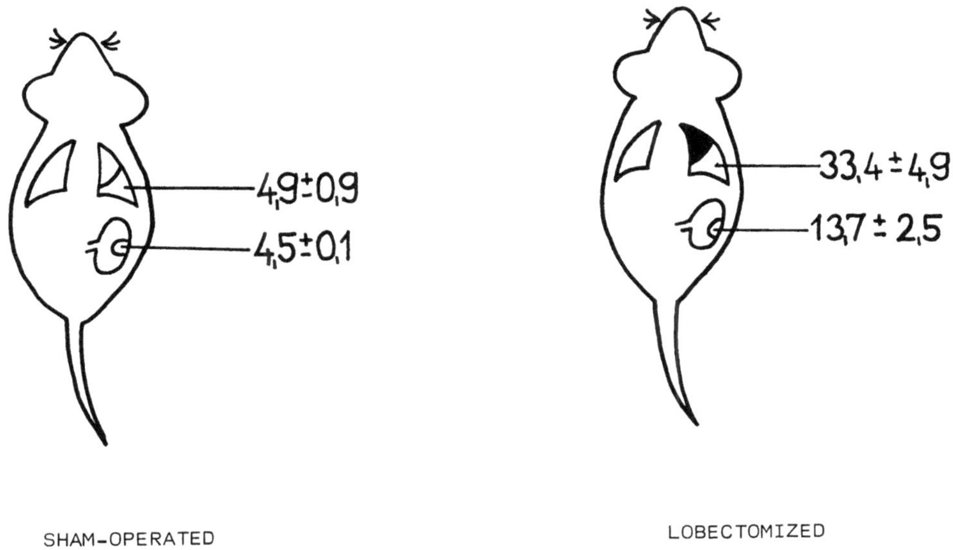

FIGURE 2. ODC activity of the left lower pulmonary lobes in BALB/c female mice 4 hr after surgery. Values are in $^{14}CO_2$ pmol/hr/mg proteins ± SE for four animals per group.

A toxic drug, monocrotaline, can also produce pulmonary damage by inducing hypertension associated with elevated ODC[63] and SAM DC activities[64] and enhanced polyamine content.[64] DFMO treatment under these conditions is able to prevent the increase in ODC activity and also the development of the monocrotaline-induced pulmonary hypertension associated with cellular proliferation and hypertrophy.[64]

VI. CONCLUSIONS

When various tissues of mammals are injured, their structure and function are restored usually by hyperplasia rather than by proliferation or hypertrophy. Age, nutrient input, and other conditions may have an influence on the process. Hormones and other humoral factors can alter the behavior of the pattern of regeneration. Increased ODC activity and the accumulation of polyamines, especially that of putrescine, are characteristic of the various types of regeneration, independently whether it is elicited by tissue loss or by toxic agents. Under such conditions the induction of ODC activity generally exhibits a biphasic pattern during the 1st day, and each of the two peaks of ODC activity are regulated by different mechanisms. ODC activity is regulated by various diamines present in the tissue. The concentration of spermidine and spermine is less affected than that of putrescine during the first hours of regeneration. Later, the concentration of spermidine is increasing and while that of spermine is decreasing their normal ratio is restored. It seems to be an attractive idea that in every kind of regeneration not only is the ODC induced, but the spermidine/spermine acetylase as well, helping to increase the putrescine and spermidine concentration, causing an early decrease in spermine pool, well before SAM DC is activated.

ODC inhibitors can prevent the elevation of ODC activity, but have negligible effect on the regenerating process itself. Combined treatment with ODC and SAM DC inhibitors is more effective in this regard.

To elicit compensatory growth, local and humoral factors of undefined nature must be present. Signals promoting proliferation and differentiation are required for response to tissue damage. Hormones, growth factors, must be activated and nutritional requirements must be fulfilled for support growth through DNA, RNA, and protein synthesis. These signals are likely to be specific for a given tissue or cell type. At least for the liver it was demonstrated that there is a liver-specific factor, since in cross-circulation experiments after partial hepatectomy DNA synthesis was initiated in the liver only and the response is absent in other organs.

Polyamines are involved some way in the regeneration process and we can only hope that a clear molecular interpretation of their exact role will emerge soon.

REFERENCES

1. **Orn, M. G. and Stocken, L. A.,** *Cell and Tissue Regeneration,* John Wiley & Sons, New York, 1984.
2. **Jänne, J., Pösö, H., and Raina, A.,** Polyamines in rapid growth and cancer, *Biochim. Biophys. Acta,* 473, 241, 1978.
3. **Tetzlaff, W. and Kreutzberg, G. W.,** Ornithine decarboxylase in motoneurons during regeneration, *Exp. Neurol.,* 89, 679, 1985.
4. **Gilad, G. M. and Gilad, V. H.,** Early rapid and transient increase in ornithine decarboxylase activity within sympathetic neurons after axonal injury, *Exp. Neurol.,* 81, 158, 1983.
5. **Dienel, G. A. and Cruz, N. F.,** Induction of brain ornithine decarboxylase during recovery from metabolic, mechanical, thermal or chemical injury, *J. Neurochem.,* 42, 1053, 1984.
6. **Sebille, A. and Bondoux-Jahan, M.,** Motor function recovery after axotomy: enhancement by cyclophosphamide and spermine in rat, *Exp. Neurol.,* 70, 507, 1980.
7. **Ingoglia, N. A., Sturman, J. A., and Eisner, R. A.,** Axonal transport of putrescine, spermidine and spermine in normal and regenerating goldfish optic nerves, *Brain Res.,* 130, 433, 1977.
8. **Lindquist, T. D., Sturman, J. A., Gould, R. M., and Ingoglia, N. A.,** Axonal transport of polyamines in intact and regenerating axons of rat sciatic nerve, *J. Neurochem.,* 44, 1913, 1985.
9. **Gilad, G. M. and Gilad, V. H.,** Polyamine biosynthesis is required for survival of sympathetic neurons after axonal injury, *Brain Res.,* 273, 191, 1983.
10. **Buchler, N. L. R. and Malt, R. A.,** *Regeneration of Liver and Kidney,* Little, Brown, Boston, 1971.
11. **Watanabe, A., Shiota, T., Hayashi, S., Sakata, T., and Nagashima, H.,** Pleiomorphism of hepatic regeneration, *Res. Exp. Med.,* 184, 49, 1984.
12. **Monson, R. C., Rodgers, B. M., Garnica, A. D., and Talbert, J. L.,** Serum polyamine and octopamine changes following partial hepatectomy and portacaval shunt, *J. Pediatr. Surg.,* 12, 1079, 1977.
13. **Weber, G.,** Imbalance in ornithine metabolism in hepatomas of different growth rates as expressed in behavior of L-ornithine carbamoil-transferase activity, *Cancer Res.,* 32, 1933, 1972.
14. **Williams-Ashman, H. G., Coppoc, G. L., and Weber, G.,** Imbalance of ornithine in hepatomas of different growth rate as expressed in formation of putrescine, spermidine and spermine, *Cancer Res.,* 32, 1924, 1972.
15. **Fausto, N., Brandt, J. T., and Kesner, L.,** Possible interactions between urea cycle and synthesis of pyrimidines and polyamines in regenerating liver, *Cancer Res.,* 35, 397, 1975.
16. **Meisler, N. T. and Thanassi, J. W.,** Vitamin B_6 metabolism and its relation to ornithine decarboxylase activity in regenerating rat liver, *J. Nutrit.,* 112, 314, 1982.
17. **Zuretti, M. F., Gravela, E., Papino, F., and Sartorio, L.,** Studies on mechanisms of ornithine decarboxylase activity regulation in regenerating liver, *Cell. Biochem. Function,* 3, 139, 1985.
18. **Pösö, H., Kallio, A., Scalabrino, G., and Jänne, J.,** Specific inhibition of the synthesis of putrescine and spermidine by 1,3-diaminopropane in rat liver in vivo, *Biochim. Biophys. Acta,* 497, 288, 1977.
19. **Kato, Y., Nishihara, H., Inoue, H., Gohda, E., and Takeda, Y.,** Polyamine metabolism and DNA synthesis in regenerating rat liver, *J. Biochem.,* 84, 1617, 1978.
20. **Pösö, H. and Pegg, A. E.,** Effect of α-difluoromethylornithine on polyamine and DNA synthesis in regenerating rat liver. Reversal of inhibition of DNA synthesis by putrescine, *Biochim. Biophys. Acta,* 696, 179, 1981.
21. **Heller, J. S. and Canellakis, E. S.,** Cellular control of ornithine decarboxylase activity by its antizyme, *J. Cell. Physiol.,* 107, 209, 1981.
22. **Scalabrino, G., Ferioli, M. E., and Modena, D.,** Decreased activity of ornithine decarboxylase antizyme during hyperplastic and neoplastic growth of rat liver, in *Advances in Polyamine Research,* Vol. 4., Bachrach, U., Kaye, A., and Chayen, R., Eds., Raven Press, New York, 1983, 79.
23. **Bachrach, U.,** Physiological aspects of ornithine decarboxylase, *J. Cell Biochem. Funct.,* 2, 6, 1984.

24. **Corti, A., Astancolle, S., and Davalli, P.,** Response of hepatic ornithine decarboxylase and polyamine concentration to surgical stress in the rat: evidence of catecholamines on glucocorticoid action, *Biochem. Biophys. Res. Commun.*, 129, 885, 1985.
25. **Matsui, I. and Pegg, A. E.,** Effect of thioacetamide, growth hormone or partial hepatectomy on spermidine acetylase, *Biochim. Biophys. Acta,* 633, 87, 1980.
26. **Höltta, E.,** Oxidation of spermidine and spermine in rat liver: purification and properties of polyamine oxidase, *Biochemistry,* 16, 91, 1977.
27. **Seiler, N.,** On the role of GABA in vertebrate polyamine metabolism, *Physiol. Chem. Phys.*, 12, 411, 1980.
28. **Sessa, A., Desiderio, M. A., Baizini, M., and Perin, A.,** Diamine oxidase activity in regenerating rat liver and 4-dimethylaminoazobenzene-induced and Yoshida AH 130 hepatomas, *Cancer Res.*, 41, 1929, 1981.
29. **Haddox, M. K. and Russell, D. H.,** Increased nuclear conjugated polyamines and transglutaminase during liver regeneration, *Proc. Natl. Acad. Sci. U.S.A.*, 78, 1712, 1981.
30. **Inoue, H., Konishi, Y., Takeda, Y., and Cihak, A.,** The role of putrescine in modulation of DNA synthesis in regenerating liver of rats pretreated with 5-aza-cytidine, *J. Biochem.*, 89, 861, 1981.
31. **Mikhailovskii, V. O., Tsiakalo, L. V., and Gulyi, M. F.,** Reversible disorder of polyamine and DNA synthesis in regenerating rat liver, *Biul. Eksp. Biol. Med.*, 90, 165, 1980.
32. **Wiegand, L. and Pegg, A. E.,** Effect of inhibitors of *S*-adenosylmethionine decarboxylase and ornithine decarboxylase on DNA synthesis in rat liver after partial hepatectomy, *Biochim. Biophys. Acta,* 517, 169, 1978.
33. **Matsui, I., Otani, S., and Morisawa, S.,** Effect of urethane on the induction of ornithine decarboxylase in regenerating rat liver, *Biochim. Biophys. Acta,* 544, 372, 1978.
34. **Matsui, I., Otani, S., and Morisawa, S.,** Effect of urethane on polyamine and DNA synthesis in regenerating rat liver, *Chem. Biol. Interact.*, 33, 35, 1980.
35. **Olson, J. W. and Russell, D. H.,** Prolonged ornithine decarboxylase induction in regenerating, carcinogen treated liver, *Cancer Res.*, 40, 4373, 1980.
36. **Pösö, H.,** Inhibition of protein degradation in regenerating rat liver by ethanol treatment, *Pharmacol. Biochem. Behav.*, 13, 83, 1980.
37. **Pösö, H. and Pösö, A. R.,** Stabilization of tyrosine aminotransferase and ornithine decarboxylase in regenerating rat liver by ethanol treatment, *FEBS Lett.*, 113, 211, 1980.
38. **Karima-Tari, F., Mahmud, I., Fukui, N., and Miura, Y.,** Induction of ornithine decarboxylase and DNA synthesis after liver injury, *Cell Mol. Biol.*, 26, 647, 1980.
39. **Pösö, H. and Pegg, A. E.,** Effect of carbon tetrachloride on polyamine metabolism in rodent liver, *Arch. Biochem. Biophys.*, 217, 730, 1982.
40. **Brandt, J. T., Pierce, D. A., and Fausto, N.,** Ornithine decarboxylase activity and polyamine synthesis during kidney hypertrophy, *Biochim. Biophys. Acta,* 279, 184, 1972.
41. **Selmeci, L., Pósch, E., Mosonyi, J., and Hársing, L.,** Effect of unilateral ureteral obstruction on renal polyamine levels in rat, *Experientia,* 37, 24, 1981.
42. **Raina, A. and Jänne, J.,** Polyamines and the accumulation of RNA in mammalian systems, *Fed. Proc. Fed. Am. Soc. Exp. Biol.*, 29, 1568, 1970.
43. **Selmeci, L., Losonczy, G., Pósch, E., and Hársing, L.,** Alterations in renal polyamine and nucleic acid contents during normal and compensatory kidney growth in rats, in *Recent Progress in Polyamine Research,* Selmeci, L., Brosnan, M. N., and Seiler, N., Eds., Akadémiai Kiadó, Budapest, 1985, 391.
44. **Sogani, R., Matsushita, S., Mueller, J., and Ruben, M.,** Stimulation of ornithine decarboxylase activity in rat tissues by growth hormone and by serum growth factor from rats infected with Spargana of Spirometra mansonoides, *Biochim. Biophys. Acta,* 279, 377, 1972.
45. **Cameron, R., Henningsson, S., Persson, L., and Rosengren, E.,** Effect of 1,3-diaminopropane on testosterone induced hypertrophy and polyamine synthesis in mouse kidney, *Acta Physiol. Scand.*, 106, 299, 1979.
46. **Pass, K. A., Bintz, J. E., and Postulka, J. J.,** Effect of testosterone on renal, ornithine decarboxylase and kidney function, in *Enzyme,* Vol. 27, Colombo, J. P. and Greengard, O., Eds., S. Karger, Basel, 1982, 108.
47. **Laitinen, S. I., Laitinen, P. H., Huhtinen, R.-L., Pulkka, A. E., and Pajunen, E. L.,** Multiple species of ornithine decarboxylase in mouse kidney. Effect of testosterone, *Biochem. Int.*, 10, 559, 1985.
48. **Anderson, A.-C., Hammar, L., Henningsson, S., and Löwendahl, G. O.,** Changes in cadaverine and putrescine metabolism in the mouse kidney induced to growth by an anabolic steroid, *Acta Physiol. Scand.*, 114, 225, 1982.
49. **Desiderio, M. A., Sessa, A., and Perin, A.,** Diamine oxidase activity inhibition and putrescine levels in renal hypertrophy, in *Advances in Polyamines in Biochemical Science,* Caldarera, C. M. and Bachrach, U., Eds., CLUEB, Bologna, Italy, 1984, 141.

50. **Desiderio, M. A., Sessa, A., and Perin, A.,** Induction of diamine oxidase activity in rat kidney during compensatory hypertrophy, *Biochim. Biophys. Acta,* 714, 243, 1982.
51. **Selmeci, L., Pósch, E., Simon, G., and Hársing, L.,** Effect of suramin (Bayer 205) on renal ornithine decarboxylase activity and polyamine concentrations in rats, *Biochem. Biophys. Res. Commun.,* 87, 323, 1979.
52. **Bartolome, J., Chait, E. A., Trepanier, P., Whitmore, W. L., Wiegel, S., and Slotkin, T. A.,** Organ specificity on neonatal methyl mercury hydroxide poisoning in the rat: effects on ornithine decarboxylase activity in developing tissue, *Toxicol. Lett.,* 13, 267, 1982.
53. **Fagan, C. and Ward, L. C.,** Renal ornithine decarboxylase activity of trenbolone acetate treated mice, *Nutr. Rep. Int.,* 29, 205, 1984.
54. **Malt, R. A.,** Humoral factors in regulation of compensatory renal hypertrophy, *Kidney Int.,* 23, 611, 1983.
55. **Uvelius, B. and Rosengren, E.,** Renal polyamine metabolism in rats with renovascular hypertension, *Acta Physiol. Scand.,* 124, 11, 1985.
56. **Rosenberg, M. R., Strom, S. C., Pachman, S., Slotkin, T. A., and Mihalopoulos, G.,** Induction of rat kidney ornithine decarboxylase by nicotinamide without a concomitant increase in DNA synthesis, *Carcinogenesis,* 7, 175, 1986.
57. **Manen, C. A., Costa, M., and Russell, D. H.,** Further evidence of cyclic AMP-mediated hypertrophy as a prerequisite of drug specific enzyme induction, *Biochem. Pharmacol.,* 27, 219, 1978.
58. **Tatár-Kiss, S., Bardócz, S., and Kertai, P.,** Changes in L-ornithine decarboxylase activity in regenerating lung lobes, *FEBS Lett.,* 175, 131, 1984.
59. **Bardócz, S., Tatár-Kiss, S., and Kertai, P.,** Effect of α-difluoromethylornithine on the ornithine decarboxylase activity and compensatory lung growth in mice, *Acta Biochim. Biophys. Hung.,* 21, 59, 1986.
60. **Smith, L. L. and Wyatt, I.,** Accumulation of diamines and polyamines into rat lung slices, *Biochem. Pharmacol.,* 31, 3029, 1982.
61. **Smith, L. L., Pratt, I., Elliott, C., and Wyatt, I.,** The accumulation of putrescine and paraquat into lung slices taken from BHT treated mice, *Toxicology,* 27, 1, 1983.
62. **Thet, L. A., Parra, S. C., and Shelburne, J. D.,** Repair of oxygen-induced lung injury in adult rats, *Am. Rev. Resp. Dis.,* 126, 174, 1984.
63. **Olson, J. W., Altiere, R. J., and Gillespie, M. N.,** Prolonged activation of rat lung ornithine decarboxylase induced pulmonary hypertension, *Biochem. Pharmacol.,* 33, 3633, 1984.
64. **Olson, J. W., Hacker, A. D., Altier, R. J., and Gillespie, M. N.,** Polyamines and the development of monocrotaline-induced pulmonary hypertension, *Am. J. Physiol.,* 247, H682, 1984.

Section B.
Polyamines and the Structure of Nucleic Acids

Chapter 7

SPECIFICITY AND BINDING IN POLYAMINE/NUCLEIC ACID INTERACTIONS

Burt G. Feuerstein and Laurence J. Marton

TABLE OF CONTENTS

I. INTRODUCTION

Interactions of the polyamines with nucleic acids have been studied since the early 1960s, when it was found that there were significant concentrations of polyamines in prokaroytes (reviewed in Reference 1). Because polyamines were known to be important for cellular proliferation and growth, a wide range of interactions of polyamines with both DNA and RNA have been studied (reviewed in References 2 to 6). These early investigations established that polyamines are associated with nucleic acids, promote the integrity of many of their biologic functions, and shield them from various stresses.

Because they influence many biochemical processes, and because some of their effects are mimicked by other intracellular cations, whether polyamines interact specifically with nucleic acids has been a central question. Obviously, polyamines can interact specifically, nonspecifically, or in a combination of both modes. In the discussion that follows, we will review selected studies of the interactions of polyamines with *t*RNA and DNA with emphasis on the questions of the specificity and binding of polyamines.

II. POLYAMINE/*t*RNA INTERACTIONS

Studies of the polyamine/*t*RNA interaction show the evolution of investigation concerning specificity in polyamine function. It is well-known that polyamines are important for the formation of a well-ordered crystals of *t*RNA (reviewed in Reference 7), and that at physiologic concentrations polyamines effectively catalyze *t*RNA aminoacylation.[8] When *t*RNA was isolated from *Escherichia coli* under various growth conditions,[9] a stoichiometric relationship was found between polyamines and *t*RNA that did not depend simply on intracellular polyamine concentrations. Physical and chemical studies of *t*RNA in the presence of polyamines (reviewed in Reference 10) strongly suggested that polyamines stabilized *t*RNA conformation through specific site binding. Analysis of X-ray diffraction data for a *t*RNA^{phe} crystal[11] confirmed that both spermine and Mg^{++} have specific binding sites (Figure 1). In this model, one spermine appears to function as a counterion to large concentrations of negative charge where two polynucleotide arms interact. Another spermine binds to four different phosphates on both sides of the major groove and interacts with the bases in the groove. Its placement appears both to narrow the major groove by 3Å and to contribute to a 25° bend in the helical axis. Note that this description is similar to that of the B-DNA major groove model discussed below.

The impression that polyamines bind specifically to *t*RNA and thereby affect its conformation has been tested by physical chemical methods in solution with specific *t*RNA; results have led to disputes about both the specificity and the cooperativity of polyamine binding. McMahon and Erdmann,[12] citing data from equilibrium dialysis experiments, question whether a change in *t*RNA conformation occurs under near physiologic salt conditions.[12] Their results can be explained without invoking either cooperativity or specific site binding. Nöthig-Laslo et al.,[13,14] however, have detected polyamine-induced changes in the affinity of *t*RNA^{tyr} for divalent cations at both near-physiologic and low salt concentrations. Results of control studies using denatured *t*RNA and single- and double-stranded synthetic polynucleotides showed that these effects depend on the structure and base sequence of the specific *t*RNA. In addition, Nilsson et al.[15] used a *t*RNA^{phe} with a fluorescent label attached to the anticodon loop and found evidence that spermine stabilizes a particular conformation of the loop. Finally, Tropp and Redfield[16] measured the solvent exchange rates of protons in specific *t*RNAs and found that spermidine stabilizes specific secondary and tertiary conformations. In control studies they showed that Mg^{++} does not always substitute for polyamines in these stabilization processes.

Biochemical studies of polyamine/*t*RNA interactions also have been conducted. Two groups found that spermine and spermidine help both to prevent enzymatic misacylations

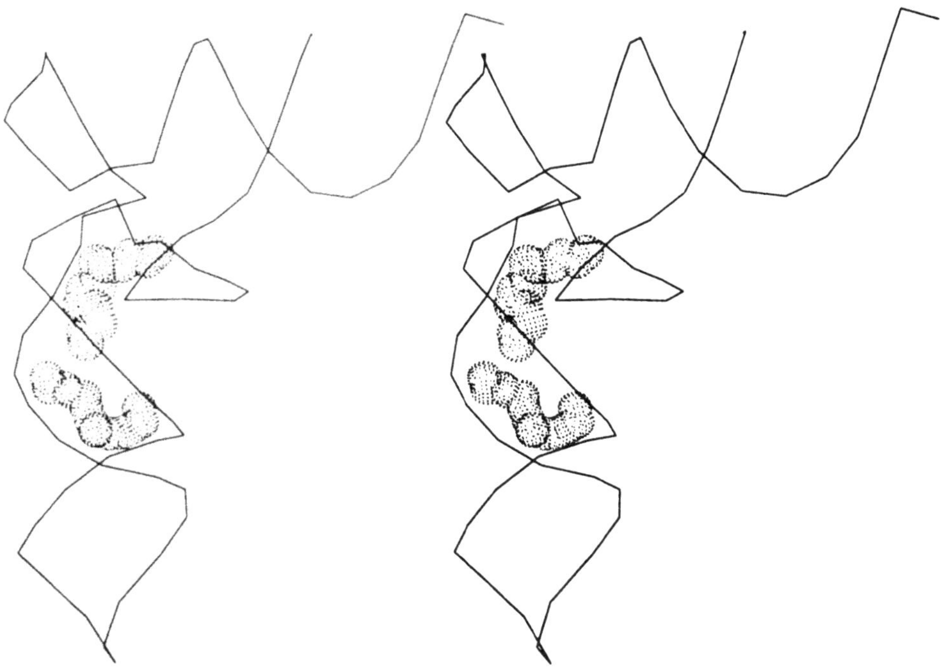

FIGURE 1. Spermine/*t*RNA^{phe} interaction. This stereodiagram represents *t*RNA by chaining the phosphorus atoms of the backbone, and represents spermine by van der Waal radii of the atoms. The upper spermine lies at the point at which two arms of *t*RNA interact, while the lower spermine interacts with both phosphates and bases, and narrows the distance between phosphate backbones.

and to improve the rate of *t*RNA aminoacylation.[8,17] In a different *t*RNA system, another group[18,19] found that spermidine more accurately promoted enzymatic splicing by a specific endonuclease of precursor *t*RNA to mature *t*RNA than did Mg^{++}. Lee and Knapp[20] investigated the secondary and tertiary structures of precursor *t*RNA substrates for this endonuclease. They speculated that polyamines and Mg^{++} in concert stabilize a consensus precursor secondary structure for *t*RNAs that enables this single endonuclease to splice efficiently and accurately in forming mature *t*RNAs. The hypothesis of a common class of functional structures for precursor *t*RNAs stabilized by polyamines stresses the importance of polynucleotide substrate conformation in enzyme recognition and emphasizes the probable importance of polyamines in this process. Thus, although evidence is incomplete, structural, physical chemical, and biochemical experimental results appear to favor involvement of polyamines in the biology and physiology of *t*RNA conformation.

III. POLYAMINE/DNA INTERACTIONS

Studies of *t*RNA illustrate how both physical chemical and biochemical methods may be used to describe specific or nonspecific polyamine activity in various systems and to provide evidence for polyamine function in stabilization of specific conformations. For our discussion of polyamine/DNA interactions, we will review models for the interaction that have been proposed, and then review recently published experimental evidence that either supports or contradicts these models.

A. Polyamine/DNA Models

Two decades ago Tsuboi[21] and Liquori et al.[22] proposed specific binding models to help explain stabilization of the DNA double-stranded helix in the presence of polyamines. In

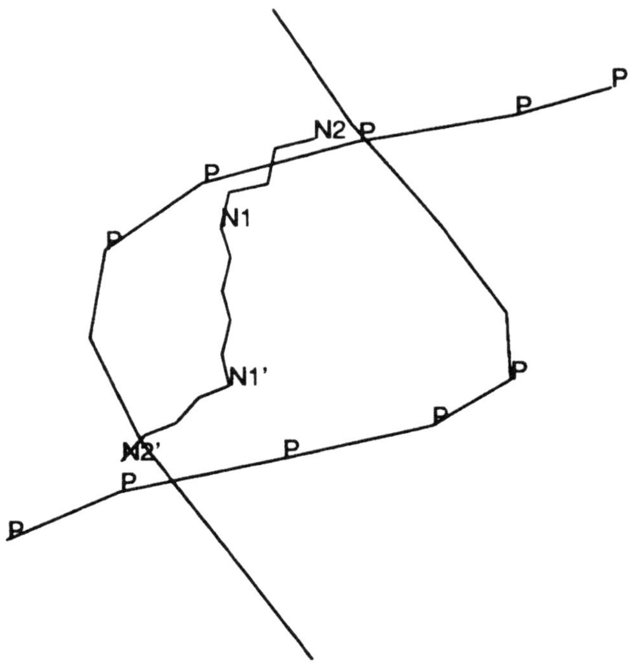

FIGURE 2. Minor groove model of spermine/DNA interaction. B DNA is represented by chaining phosphorus atoms in the backbone. Spermine tetramethylene bridges the minor groove, and primary and secondary amines interact with phosphates on the backbone. Coordinates were taken from Reference 22.

both models, the tetramethylene portion of the natural polyamines was placed to form a bridge across the minor groove of B DNA, with the positively charged amino groups interacting electrostatically with the negatively charged phosphates of DNA (Figure 2). Because no interaction between bases and polyamines had been postulated, these models did not consider the possibility of sequence specificity for the interaction. Since these models were proposed, it has been found that polyamines can cause major alterations in DNA conformation including the induction of transitions from B to both A and Z forms of DNA (see below). A theoretical model of the interaction of spermine with A DNA[23,24] proposed that spermine preferred the A over the B conformation because the aminopropyl end groups of spermine fit the shorter intrachain phosphate distances of the A conformation better than those of the B conformation. Energy calculations were carried out to confirm these observations, but again did not consider possible interactions of spermine with bases.

X-ray diffraction studies of spermine-containing crystalline hexamers of Z DNA[25,26] show that two spermine molecules interact in different patterns with both bases and phosphates of DNA and thereby stabilize portions of the double-helical structure (Figure 3). In a model of B DNA proposed by Drew and Dickerson[27] that is based on X-ray diffraction data in the crystalline dodecamer d(CGCGAATTCGCG), a single spermine bridges the major groove and interacts with both phosphates and bases of the oligomer. This type of placement is reasonable, because although a priori it might be expected that the best electrostatic interactions would occur between positions near phosphates, Pattabiraman et al.[28] have shown that the theoretical electrostatic potential of B DNA favors an initial placement of counterions in the major groove.

Using a major groove site for spermine and a position that favors an interaction between

FIGURE 3. Stereodiagram of the spermine/Z DNA interaction. Spermine, marked by its nitrogens, interacts with Z DNA in two ways. On the left, one spermine interacts along the shallow major groove, and on the right a second spermine bridges to another molecule of DNA.

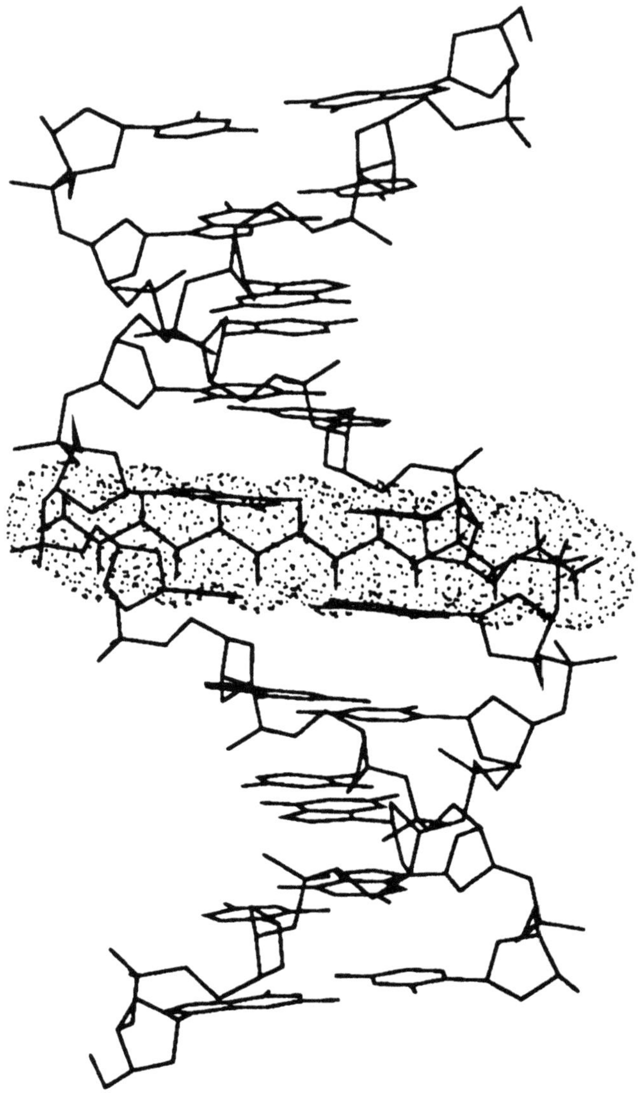

A

FIGURE 4. Major groove model of the spermine/B DNA interaction viewed looking at the major groove *en face*. Spermine, shown as van der Waal radii of its atoms, bridges across the major groove from one strand of the backbone to the other. (A) Spermine DNA complex before energy minimization and (B) spermine-DNA complex after energy minimization.

the N_7 position of guanine and the two amino groups of spermine in B DNA, we[29] used an energy minimization technique to model the spermine-DNA interaction (Figure 4). This method produced a structure bent 25° along the helical axis, which can be seen in Figure 5 by imagining the average helical axis at the bottom of the molelcule and comparing it with the axis at the top. The bend is produced by folding the major groove over spermine, which allows multiple interactions to occur between the oligomer and spermine, including the formation of hydrogen bonds between the amine groups of spermine and DNA phosphate

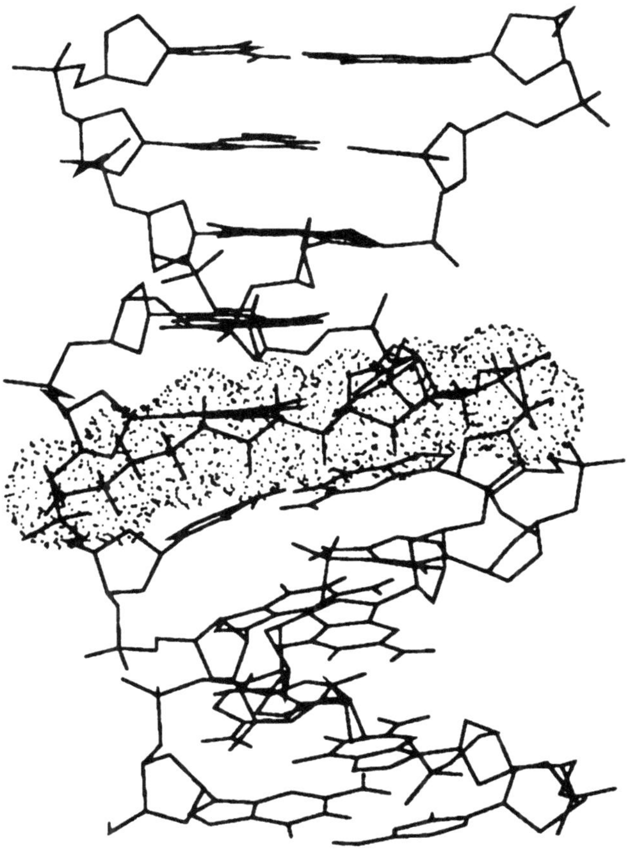

FIGURE 4B.

oxygens. In the process, the minor groove has been widened considerably and sugar puckering and intrastrand phosphate distances have been affected. The structure produced by this model appears to be analogous to the bend in the *t*RNA^{phe} crystal described above. In general, the DNA structure produced by energy minimization of spermine binding sites appeared to be more A-like than B-like. We also compared Tsuboi's and Liquori's minor groove model to our major groove model using the same energy minimization techniques. The stabilization energies achieved by spermine binding in each model suggest that the major groove model is the more favorable. In addition, structural studies (Figure 6) show that in the Liquori model a large cavity is created in the minor groove under the spermine methylene bridge. Because the methylene bridge is hydrophobic, and only water will fill the cavity, this structure appears to be less tenable than the major groove model.

Models of nonspecific interactions have also been proposed for polyamine/DNA interactions. Bloomfield and Wilson[30] have applied Manning's[31] theory of polyelectrolytes to the DNA/polyamine system by assuming that DNA is a linear distribution of negative charge and that polyamines are concentrations of point charge. In this model there is no specific binding, only a volume of space about DNA where counterions preferentially localize; in this volume, the site of counterion condensation, only "delocalized binding" occurs.

B. Physical Chemical Studies

The validity of the models has been tested by studying the effects of polyamines on the behavior of DNA. One much-investigated phenomenon is DNA condensation, which may

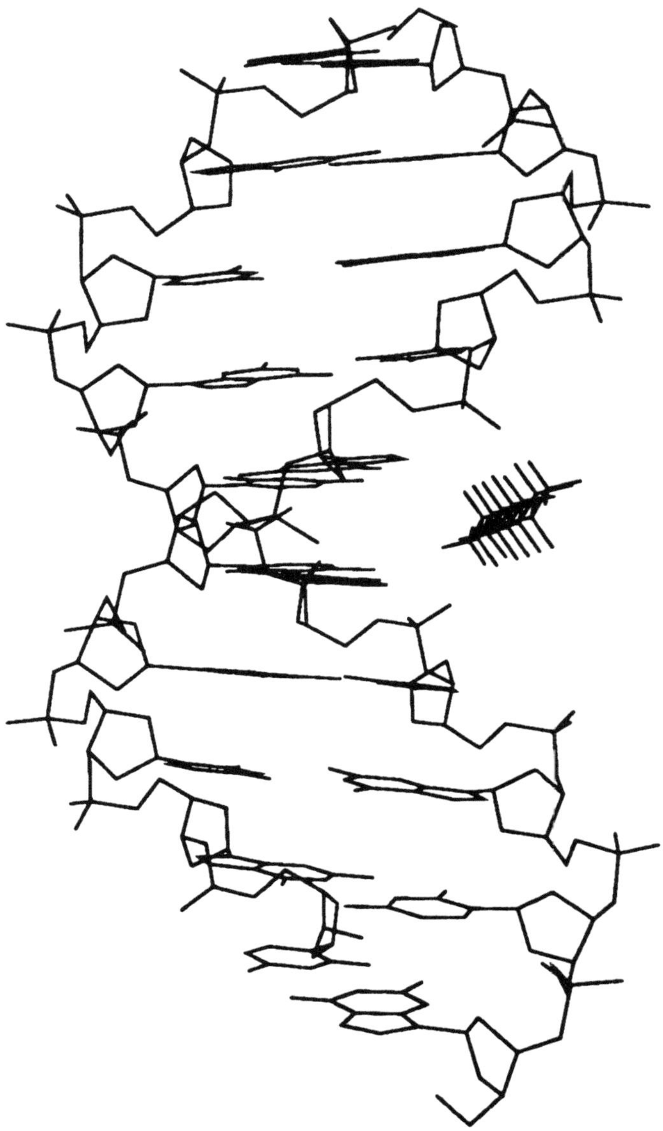

A

FIGURE 5. Major groove model of the spermine/B DNA interaction viewed down the minor (left side of DNA) and major (right side of DNA) grooves. (A) Spermine in the major groove of B DNA before energy minimization. Note the distance across the major groove. (B) DNA energy minimized complex in the presence of spermine (for clarity spermine is not shown). Note the bend in the axis of the DNA by comparing the top and bottom halves of the DNA. Also note the short distance across the major groove.

serve as a model for DNA in the nucleosomal, viral, and chromosomal condensed states. Kaiser et al.[32] first showed that in the presence of spermine, DNA is converted to a condensed form that is protected from shear. Using streaming linear dichroism and electron microscopy, Gosule and Schellman[33,34] described a compact form of DNA reversibly induced by polyamines. The structures formed after condensation were mainly DNA toruses, although other

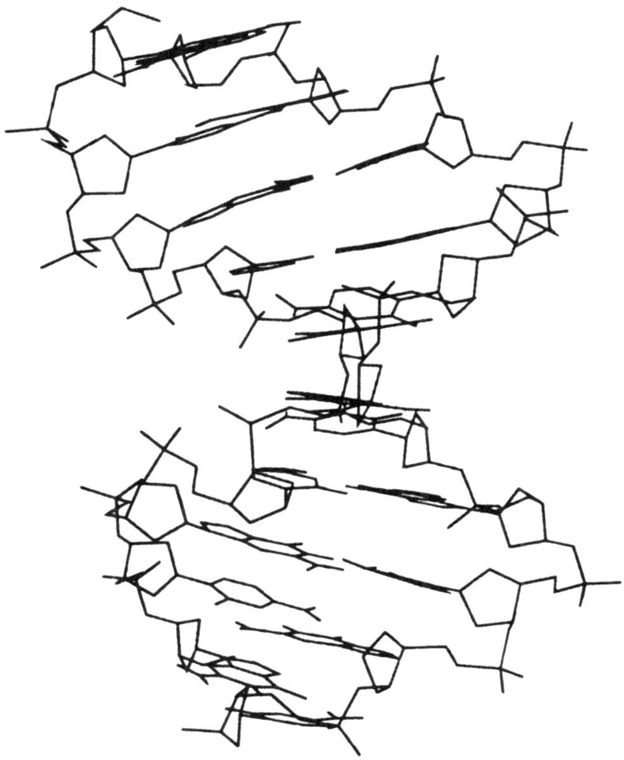

FIGURE 5B.

shapes were described that depended on the conditions of fixation and isolation.[35] The structure of these toruses has been studied using freeze fracture techniques and enzyme digestion.[36-38] Using the counterion condensation model, Bloomfield and Wilson[30] postulated that a collapse threshold would be reached when a calculated 89 to 90% of the DNA phosphate charge had been neutralized by cations. According to their calculations, putrescine and Mg^{++}, which have no condensing activity in aqueous buffer, cannot reach the theoretical 89 to 90% threshold. The counterion condensation model of delocalized binding in this system is strengthened by Porschke's[39] study of the condensation of viral DNA caused by spermine and spermidine. He showed that the binding kinetics of polyamines were too fast to allow for specific site binding; site exclusion should have slowed binding at high ligand concentration.

Results of other studies, however, have shown the presence of specific ion effects not consistent with the counterion condensation model. Several studies have shown that cations such as $Co(NH_3)_6$ and a series of spermidine analogues with the same charge as spermidine have different condensing activities.[40] Moreover, conditions have been found[41] in which theory predicts that greater than 90% neutralization of phosphate charge by divalent cations occurs, yet condensation does not take place. Baase et al.[42] used flow linear dichromism and viscometry to study the DNA/spermine complex. They found that counterion condensation could be used to describe a situation before condensation during which polyamines neutralized the negative charge on the DNA backbone. However, collapse of the DNA was poorly described by the counterion condensation formalism. Specificity of polyamine binding has been found in other studies of condensed DNA. Thomas and Bloomfield[40] noted differences in hydrodynamic radius on light scattering measurements of T_4 phage DNA that

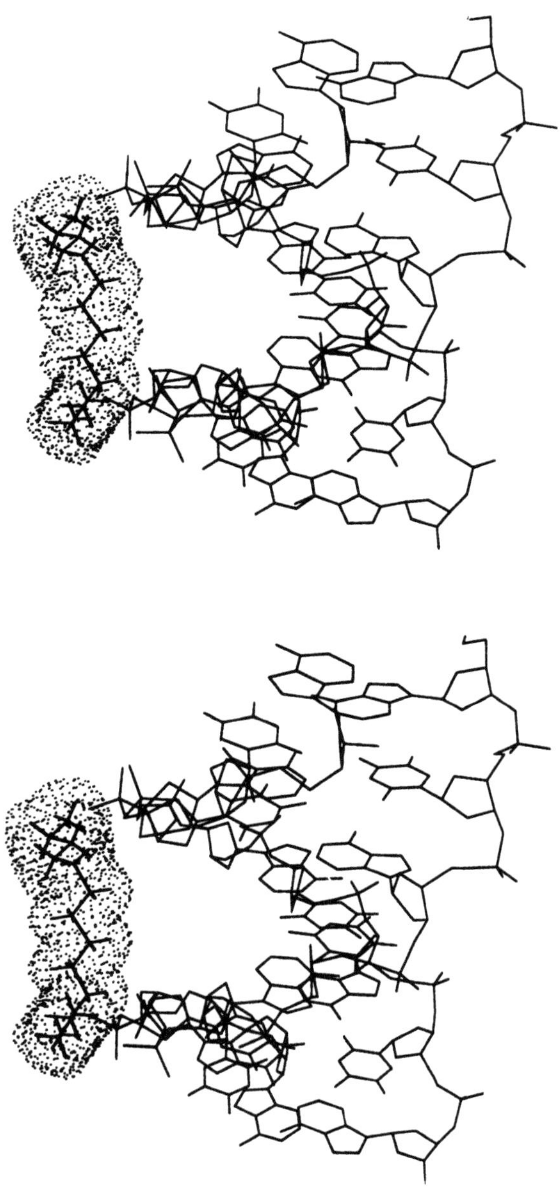

FIGURE 6. Stereodiagram of the spermine/B DNA minor groove model. Spermine bridges the minor groove, but leaves a cavity beneath the methylene bridge of spermine.

depended on whether the condensing cation was $Co(NH_3)_6$ or spermidine. Using X-ray diffraction, Schellman and Pathasarathy[43] showed that the Bragg spacing and the calculated interhelical spacing of collapsed DNA vary as a function of methylene bridge length in a series of spermidine and putrescine analogues. These results confirm earlier observations by Suwalsky et al.[44] that indicated a role for polyamines as bridges between DNA helices. Thus, both the condensing ability of the polyamines and the arrangement of DNA in the condensed state provide evidence for some specificity in the polyamine/DNA interaction.

In vitro studies of the polyamine/DNA interaction in the uncondensed state have been conducted. Early studies that used equilibrium dialysis to determine binding constants in GC- or AT-rich DNA[45] provided no evidence for sequence specificity. Later studies by Braunlin et al.[46] found that results obtained with equilibrium dialysis could be interpreted either through a specific binding model or through a counterion condensation model. Using a gel filtration system, however, Igarashi et al.[47] found that spermine bound preferentially to GC-rich DNAs and poly (dG)·poly (dC) compared with AT-rich DNAs and poly (dA)·poly (dT). Specificity of binding that is a function of base sequences and composition is more compatible with a site-specific model. Burton et al.[48] used ^{23}Na nuclear magnetic resonance to monitor displacement of Na^+ from phosphate by a variety of polyamines and polyamine analogues and found that displacement varied with polyamines and polyamine analogues of the same charge but different structure. Unfortunately, they did not monitor their solutions for condensation, but did note that precipitation interfered with some of their measurements. Even with these difficulties, their measurements of the displacement activities of spermidine and a spermidine analogue were impressively different under what appeared to be noncondensing conditions. Marquet et al.[49] used electric dichroism and birefringence measurements on calf thymus DNA to investigate the problem and found that changes in the stiffness and orientation of DNA in electric fields occurred and that a permanent dipole was formed as a function of spermine concentration. Formation of the permanent dipole moment might be explained by asymmetrical binding of spermine to DNA in a manner that is similar to Drew and Dickerson's[27] model in the crystalline dodecamer described above. However, in a nuclear magnetic resonance study of the interaction of spermine with the same Drew and Dickerson dodocamer, Wemmer et al.[50] found that the mobility of spermine is independent of the mobility of DNA. Coupling of mobilities would be expected to occur for a site-specific interaction. Recently, the same investigators[50a] found that conditions favoring condensation do change the mobility of spermine, which implies that site-specific binding is possible. The different results obtained in these studies underscore the importance of experimental conditions in the interpretation of results of in vitro studies, especially in the face of the mounting evidence of the importance of nucleic acid conformation in polyamine binding.

We[50b] have measured hydrogen/deuterium exchange kinetics of uncondensed polydeoxynucleotides in the presence of spermine. With increasing spermine concentration, the exchange rate of the deoxynucleotide imino protons with solvent protons increases in poly (dA·dT) but decreases in poly(dG·dC). This base-dependent difference in exchange rate was not found for the amino protons and was not found with either imino or amino protons using Mg^{++} or catalysts such as imidazole or Tris. This is chemical evidence for sequence specificity of polyamine binding. The results obtained with Mg^{++} in place of spermine clearly show that the interaction depends on the type of cation used.

Condensation of DNA is not the only change in conformation that is associated with the polyamines, however. Alterations in dimensions of the DNA grooves, sugar puckering, intrastrand phosphate distances, and (in the Z conformation) helical sense make the A, B, and Z families of DNA important considerations in studies of the regulation of DNA function and polyamine binding.[51] As noted above, several groups have reported that polyamines can cause transitions from one form of DNA into another. In 1978, it was reported that the A conformation was stabilized by polyamines in both fibers and solutions.[23,52] Soon thereafter, it was found that spermidine and spermine induced the left-handed Z conformation both at near physiologic salt in poly (dG-me^{5}dC)[53] and under less physiologic conditions in poly (dG-dC).[54] At very low ionic strength, it was found that poly (dG-me^{5}dC) requires only one spermine per 40 to 50 nucleotides for the transition to occur;[55] the salt dependence of the transition is very compatible with specific binding of polyamines.

The importance of cation structure in the promotion of the B to Z transition was observed by Thomas et al.,[56] who found that the Z-forming ability of N_1- and N_8-acetyl spermidine in poly (dG-me^{5}dC) varies inversely with their ability to stabilize T_4 DNA to thermal melting. These data are compatible with the presence of different binding sites for polyamines in these DNAs. While differences in nucelotide sequence, conformation, or both may have major effects on ligand binding to DNA, they were not considered in this study. Nonetheless, the results of this study emphasize the importance of polyamine structure in binding, because isomers with the same charge and only slightly different structure bind to DNA with different affinities. In another study, Thomas and Bloomfield[57] found that the thermal stability of calf thymus DNA varies as a function of the length of spermidine analogues at low salt concentration, but at physiologic salt concentrations the DNA/polyamine complexes had different thermal stability that did not depend in a simple way on length or charge of polyamines. These results also emphasize the importance of experimental conditions for analyzing polyamine/DNA interactions, and show that polyamine charge is not isolated from structure in terms of their influence on the physical properties of DNA.

C. Biochemical Studies

Polyamines have a stimulatory effect on several in vitro models of DNA synthesis[58-62] and the absence of polyamines reduces the rate of DNA synthesis.[63] Using a series of spermidine homologues in polyamine-depleted *E. coli*, Geiger and Morris[64] found that rates of both DNA replication fork movement and growth depended on homologue type at equal intracellular concentrations, which suggests some specificity in spermidine/DNA interactions. They found no evidence for a polyamine effect on the initiation of DNA synthesis. Results of a recent study of DNA polymerase alpha[65] suggest that spermine stimulates the enzyme by increasing its specificity of binding to single-stranded regions and by destabilizing the enzyme/substrate complex when elongation is blocked; it did not, however, differentiate between an effect on DNA substrate and an effect on the polymerase itself. La Duca et al.[66] more specifically addressed the effect of substrate in the DNA polymerase reaction using *E. coli* DNA polymerase III holoenzyme. They showed that spermidine stimulated enzyme activity as expected, but found that spermidine could strengthen pause sites for DNA synthesis in areas of DNA secondary structure predicted by computer models. Other pause sites not associated with the computer-predicted structures faded in the presence of spermidine. Thus, spermidine appears to stabilize specific secondary or tertiary structure in DNA that is related to enzyme activity. These results reinforce the findings of Geiger and Morris,[64] because spermidine appears to affect structures that control replication rate.

Pingoud et al.[67-68] found that spermidine and spermine mediate the endonuclease reaction in a series of restriction endonucleases. At a concentration of polyamine that does not appear to cause condensation, they found stimulation of endonuclease activity and a large increase in enzyme specificity. At concentrations at which DNA condensation takes place, however, enzyme activity decreases. Although these authors did not present evidence for site specificity of polyamine binding in this system, the increase in both the specificity and the rate of reaction in a series of different enzymes and DNA substrates strengthens the evidence that polyamines stabilize specific structures in DNA in a manner similar to the effects of polyamines in both the DNA polymerases and *t*RNA systems discussed above.

One other interesting case of the polyamine/DNA interaction has been documented in an *E. coli* gyrase system.[69] These workers found that catenation (interlocking) of DNA rings occurs under condensing conditions, while decatenation takes place when DNA is not condensed. They applied Manning's[31] counterion condensation model to their data and to other published data, and concluded that the process of catenation required interaction of DNA segments and therefore was facilitated by condensing levels of polyamines. It was apparent in the case of the gyrase that the enzymatic reaction proceeded whether or not DNA was condensed. The catenation or decatenation of circular DNA segments simply mirrored the state of condensation of DNA.

The polyamine/nucleic acid interaction also has been studied in chromatin and in the cell. In chromatin, the activity of certain nucleases,[70-72] phosphorylation of certain chromatin associated proteins,[73,74] as well as chromatin assembly[75] have been shown to be modulated by polyamines. Hung et al.[76] showed that depletion of polyamines in 9L rat brain tumor cells by treatment with α-difluoromethylornithine altered DNA conformation as measured by viscoelastometry. In addition, the hypothesis that polyamines are important for the physiologic control of the dynamics of DNA conformation has been the basis for studies of altered cytotoxicity by DNA-directed antitumor agents after pharmacologic polyamine depletion.[77] Our studies in this area represent initial probes into this complicated system. However, even results of early studies appear to underscore the importance of polyamines in the regulation of nucleic acid secondary structure.

IV. CONCLUSIONS

Mechanisms for the effects of polyamines in chromatin and cells have not been studied on the molecular level. Clearly, in order to understand the biology and physiology of polyamine/nucleic acid interactions, the problem of understanding polyamine/nucleic acid interactions in vivo must now be addressed. Although synthetic polynucleotides and polydeoxynucleotides can be used to model simple interactions of polyamines with nucleic acids, the importance of nucleic acid secondary and tertiary structure in the polyamine/DNA interaction limit their use. Meaningful secondary or tertiary structure that is important in a specific gene or *t*RNA molecule might not be present in either a small or a monotonous fragment of synthetic polynucleotide. Specific cloned sequences may therefore produce more biologically relevant results. Studies are underway in both solution and crystal phases, but these are only models for events in vivo. The state of DNA in the nucleus during replication or transcription certainly varies, and specific proteins may play prominent roles in determining which polyamine/DNA interactions occur. In addition, specific interactions may occur in the presence of various ions, including polyamines, and nonspecific interactions may also be of importance in polyamine binding and nucleic acid function.

Investigators have begun to use physical chemical measurements such as light scattering to monitor DNA conformation in biochemical assay, which allows them to differentiate between biochemical interactions that depend on DNA condensation from those that occur at lower polyamine concentrations. Defined translational, transcriptional, and DNA synthetic systems should be studied for changes in conformation with available techniques that allow a correlation of biochemical and biophysical variables. Extension of conformational techniques to chromatin and to the cell will allow study of polyamine/nucleic acid interactions with the complex mixture of intracellular proteins, nucleic acids, and lipids.

The above discussion emphasizes the current great interest in theoretical, biochemical, and biophysical studies on polyamine/nucleic acid interactions. Much of the data points to singular, specific interactions between polyamines and nucleic acids and hints at the importance of polyamines for the regulation of nucleic acid structure. We anticipate that productive interactions between theoretician, physical chemist, biochemist, and cell biologist will continue as these exciting studies are pursued.

ACKNOWLEDGMENTS

Supported in part by NIH Program Project Grant CA-13525 and National Cooperative Drug Discovery Group Grant CA-37606. We thank Hirak Basu, Richard Shafer, and Nagarajan Pattabiramin for helpful discussion. We also acknowledge Robert Langridge for the use of the UCSF Computer Graphics Laboratory (Gr RR108) and the Midas Program for preparation of many of the figures. Gary Quigley generously contributed crystal structure data on complexes between spermine and *t*RNA and Z-DNA. We thank Cindy Huff for typing the manuscript and Neil Buckley for editing it.

REFERENCES

1. **Tabor, H., Tabor, C. W., and Rosenthal, S. M.,** The biochemistry of the polyamines: spermine and spermidine, *Annu. Rev. Biochem.*, 30, 579, 1961.
2. **Tabor, H. and Tabor, C. W.,** Biosynthesis and metabolism of 1,4-diaminobutane, spermidine, spermine, and related amines, *Adv. Enzymol.*, 36, 203, 1972.
3. **Tabor, C. W. and Tabor, H.,** 1,4-Diaminobutane (putrescine), spermidine, and spermine, *Annu. Rev. Biochem.*, 45, 285, 1976.
4. **Tabor, C. W. and Tabor, H.,** Polyamines, *Annu. Rev. Biochem.*, 53, 749, 1984.
5. **Cohen, S. S.,** *Introduction to the Polyamines,* Prentice-Hall, Englewood Cliffs, N.J., 1970.
6. **Bachrach, U.,** *Function of the Naturally Occurring Polyamines,* Academic Press, New York, 1973.
7. **Dock, A.-C., Lorber, B., Moras, D., Pixa, G., Thierry, J.-C., and Giege, R.,** Crystallization of transfer ribonucleic acids, *Biochemie,* 66, 179, 1984.
8. **Loftfield, R. B., Eigner, E. A., and Pastuszyn, A.,** The role of spermine in preventing misacylation by phenylalanyl-*t*-RNA synthetase, *J. Biol. Chem.*, 256, 6729, 1981.
9. **Cohen, S. S., Morgan, S., and Streibel, E.,** The polyamine content of the RNA of *E. coli, Proc. Natl. Acad. Sci. U.S.A.*, 64, 669, 1969.
10. **Sakai, T. T. and Cohen, S. S.,** Effects of polyamines on the structure and reactivity of *t*-RNA, *Prog. Nucleic Acid Res. Mol. Biol.*, 17, 15, 1975.
11. **Quigley, G. J., Teeter, M. M., and Rich, A.,** Structural analysis of spermine and magnesium ion binding to yeast phenylalanine transfer RNA, *Proc. Natl. Acad. Sci. U.S.A.*, 75, 64, 1978.
12. **McMahon, M. E. and Erdmann, V. A.,** Binding of spermine to transfer ribonucleic acid, *Biochemistry,* 21, 5280, 1982.
13. **Nöthig-Laslo, V., Weygand, Durasevic, I., Zivkovic, T., and Kuncan, Z.,** Binding of spermine to *t*-RNA^{tyr} stabilizes the conformation of the anticodon loop and creates strong binding sites for divalent cations, *Eur. J. Biochem.*, 117, 263, 1981.
14. **Nöthig-Laslo, V., Weygand-Durasevic, I., and Kuncan, Z.,** Structural changes of yeast *t*-RNA^{tyr} caused by the binding of divalent ions in the presence of spermine, *J. Biomol. Struc. Dyn.*, 2, 941, 1985.
15. **Nilsson, L., Rigler, R., and Wintermeyer, W.,** The influence of spermine on the structural dynamics of yeast *t*-RNA^{phe}, *Biochem. Biophys. Acta,* 740, 460, 1983.
16. **Tropp, J. S. and Redfield, A. G.,** Proton exchange rates in *t*-RNA as a function of spermidine and magnesium, *Nucleic Acids Res.*, 11, 2121, 1983.
17. **Jelenc, P. C. and Kurland, C. G.,** Nucleoside triphosphate regeneration decreases the frequency of translation errors, *Proc. Natl. Acad. Sci. U.S.A.*, 76, 3174, 1979.
18. **Peebles, C. L., Gegenheimer, P., and Abelson, J.,** Precise excision of intervening sequences from precursor *t*-RNAs by a membrane-associated yeast endonuclease, *Cell,* 32, 525, 1983.
19. **Greer, C. L., Peebles, C. L., Gegenheimer, P., and Abelson, J.,** Mechanism of action of a yeast RNA ligase in *t*-RNA splicing, *Cell,* 32, 537, 1983.
20. **Lee, M. C. and Knapp, G.,** Transfer RNA splicing in *Saccharomyces cerevisiae, J. Biol. Chem.*, 260, 3208, 1985.
21. **Tsuboi, M.,** On the melting temperature of nucleic acid in solution, *Bull. Chem. Soc. Jpn.*, 37, 1514, 1964.
22. **Liquori, A. M., Constantino, L., Crescenzi, V., Elia, V., Giglio, E., Puliti, R., DeSantis, S. M., and Vitigliano, V.,** Complexes between DNA and polyamines: a molecular model, *J. Mol. Biol.*, 24, 113, 1967.

23. **Minyat, E. E., Ivanov, V. I., Kritzyn, A. M., Minchenkova, L. E., and Schyolkina, A. K.,** Spermine and spermidine-induced B to A transition of DNA in solution, *J. Mol. Biol.*, 128, 397, 1978.
24. **Zhurkin, V. B., Lysov, Y. P., and Ivanov, V. I.,** Interaction of spermine with different forms of DNA. A conformational study, *Biopolymers,* 19, 1415, 1980.
25. **Wang, G. J., Quigley, G. J., Kolpak, F. J., Crawford, J. L., vanBoom, J. H., van der Marel, G., and Rich, A.,** Molecular structure of a left-handed helical DNA fragment at atomic resolution, *Nature,* 282, 680, 1979.
26. **Quigley, G. J.,** The interaction of ions, water and drugs with DNA, in *Molecular Structure and Biological Activity,* Griffin, J. F. and Duax, W. L., Eds., Elsevier, New York, 1982, 317.
27. **Drew, H. R. and Dickerson, R. E.,** Structure of a B-DNA dodecamer III. Geometry of hydration, *J. Mol. Biol.,* 151, 535, 1981.
28. **Pattabiraman, N., Langridge, R., and Kollman, P.,** An iterative approach to placing counterions around DNA, *J. Biomol. Struc. Dyn.,* 1, 1515, 1984.
29. **Feuerstein, B. G., Pattabiramin, N., and Marton, L. J.,** Spermine DNA interactions: a theoretical study, *Proc. Natl. Acad. Sci. U.S.A.,* 83, 5948, 1986.
30. **Bloomfield, V. A. and Wilson, R. W.,** Interaction of polyamines with polynucleotides,in *Polyamines in Biology and Medicine,* Morris, D. A. and Marton, L. J., Eds., Marcel Dekker, New York, 1981, 183.
31. **Manning, G. S.,** The molecular theory of polyelectrolyte solutions with applications to the electrostatic properties of polynucleotides, *Q. Rev. Biophys.,* 11, 179, 1978.
32. **Kaiser, D., Tabor, H., and Tabor, C. W.,** Spermine protection of coliphage λ DNA against breakage by hydrodynamic shear, *J. Mol. Biol.,* 6, 131, 1963.
33. **Gosule, L. C. and Schellman, J. A.,** Compact form of DNA induced by spermidine, *J. Mol. Biol.,* 259, 333, 1976.
34. **Gosule, L. C. and Schellman, J. A.,** DNA condensation with polyamines, *J. Mol. Biol.,* 121, 322, 1978.
35. **Chattoraj, D. K., Gosule, L. C., and Schellman, J. A.,** DNA condensation with polyamines. II. Electron microscopic studies, *J. Mol. Biol.,* 121, 327, 1978.
36. **Marx, K. A. and Reynolds, T. C.,** Spermidine-condensed ØX 174 DNA cleavage by micrococcal nuclease: torus cleavage model and evidence for unidirectional circumferential DNA wrapping, *Proc. Natl. Acad. Sci. U.S.A.,* 79, 6484, 1982.
37. **Marx, K. A. and Ruben, G. C.,** Evidence for hydrated spermidine-calf thymus DNA toruses organized by circumferential DNA wrapping, *Nucleic Acids Res.,* 11, 1139, 1983.
38. **Marx, K. A. and Ruben, G. C.,** Studies of DNA organization in hydrated spermidine-condensed DNA toruses and spermidine-DNA fibers, *J. Biomol. Struc. Dyn.,* 1, 1109, 1984.
39. **Porschke, D.,** Dynamics of DNA condensation, *Biochemistry,* 23, 4821, 1984.
40. **Thomas, T. J. and Bloomfield, V. A.,** Collapse of DNA caused by trivalent cations: pH and ionic specificity effects, *Biopolymers,* 22, 1097, 1983.
41. **Widom, J. and Baldwin, R. L.,** Monomolecular condensation of λ DNA induced by cobalt hexamine, *Biopolymers,* 22, 1595, 1983.
42. **Baase, W. A., Staskus, P. S., and Allison, S. A.,** Precollapse of T7 DNA by spermidine at low ionic strength: a linear dichroism and intrinsic viscosity study, *Biopolymers,* 23, 2835, 1984.
43. **Schellman, J. A. and Pathasarathy, N.,** X-ray diffraction studies on cation-collapsed DNA, *J. Mol. Biol.,* 175, 313, 1984.
44. **Suwalsky, M., Traub, M., Shmueli, U., and Subirana, J. A.,** An X-ray study of the interaction of spermine with DNA, *J. Mol. Biol.,* 42, 363, 1969.
45. **Shapiro, J. T., Stannard, B. S., and Felsenfeld, G.,** The binding of small cations to deoxyribonucleic acid: nucleotide specificity, *Biochemistry,* 8, 3233, 1969.
46. **Braunlin, W. H., Strick, T. J., and Record, M. T.,** Equilibrium dialysis studies of polyamine binding to DNA, *Biopolymers,* 21, 1301, 1982.
47. **Igarashi, K., Sakamoto, I., Goto, N., Kashiwagi, K., Honma, R., and Hirose, S.,** Interaction between polyamines and nucleic acids or phospholipids, *Arch. Biochem. Biophys.,* 219, 438, 1982.
48. **Burton, D. R., Forsen, S., and Reimarsson, P.,** The interaction of polyamines with DNA: a ^{23}Na NMR Study, *Nucleic Acids Res.,* 9, 1219, 1981.
49. **Marquet, R., Houssier, C., and Fredericq, E.,** An electro-optical study of the mechanisms of DNA condensation induced by spermine, *Biochem. Biophys. Acta,* 825, 365, 1985.
50. **Wemmer, D. E., Srivenugopal, K. S., Reid, B. R., and Morris, D. R.,** Nuclear magnetic resonance studies of polyamine binding to a defined DNA sequence, *J. Mol. Biol.,* 185, 457, 1985.
50a. **Wemmer, D. E., Srivenugopal, K. S., and Morris, D. R.,** unpublished data.
50b. **Basu, H. S., Shafer, R. H., and Marton, L. J.,** A stopped flow H-D exchange kinetic study of spermine-polynucleotide interactions, *Nucleic Acids Res.,* 15, 5873, 1987.
51. **Nordheim, A., Pardue, M. L., Lafer, E. M., Mooler, A., Stoller, B. D., and Rich, A.,** Antibodies to left-handed Z-DNA bind to interband regions of drosophila polytene chromosomes, *Nature,* 294, 427, 1981.

52. **Huse, Y., Mitsui, Y., Iitaka, Y., and Miyakki, K.,** Preliminary x-ray studies of the interaction of salmon sperm DNA with spermine, *J. Mol. Biol.*, 122, 43, 1978.
53. **Behe, M. and Felsenfeld, G.,** Effects of methylation on a synthetic polynucleotide: the B-Z transition in poly (dG-me^{5}dC), *Proc. Natl. Acad. Sci. U.S.A.*, 78, 1619, 1981.
54. **Ivanov, V. I. and Minyat, E. E.,** The transitions between left and right handed forms of poly (dG-dC), *Nucleic Acids Res.*, 9, 4783, 1981.
55. **Chen, H. H., Behe, M. J., and Rau, D. C.,** Critical amount of oligovalent ion binding required for the B-Z transition of poly (dG-me^{5}dC), *Nucleic Acids Res.*, 12, 2381, 1984.
56. **Thomas, T. J., Bloomfield, V. A., and Canallakis, Z. N.,** Differential effects of the B to Z transition of poly(dG-me^{5}dC)poly(dG-me^{5}dC) produced by N1- and N8-acetyl spermidine, *Biopolymers*, 24, 724, 1985.
57. **Thomas, T. J. and Bloomfield, V. A.,** Ionic and structural effects on the thermal helix-coil transition of DNA complexed with natural and synthetic polyamines, *Biopolymers*, 23, 1295, 1984.
58. **Schekman, R. W., Wickner, O., Westergaard, D., Brutlag, K., Geider, K., Bertsch, L., and Kornberg, A.,** Initiation of DNA synthesis: synthesis of ØX 174 replicative form requires RNA synthesis resistant to rifampin, *Proc. Natl. Acad. Sci. U.S.A.*, 69, 1691, 1972.
59. **Schekman, K., Weiner, A., and Kornberg, A.,** Multienzyme systems of DNA replication, *Science*, 186, 987, 1974.
60. **Geider, K. and Kornberg, A.,** Conversion of the M13 viral single strand to the double-stranded replicative forms of purified proteins, *J. Biol. Chem.*, 249, 3999, 1974.
61. **Weiner, J. H., Bertsch, L., and Kornberg, A.,** The deoxyribonucleic acid unwinding protein, *J. Biol. Chem.*, 250, 1972, 1975.
62. **Christensen, C. and Baldwin, R.,** Catalysis of DNA reassociation by the *Escherichia coli* DNA binding protein, *J. Mol. Biol.*, 115, 441, 1977.
63. **Geiger, L. E. and Morris, D. R.,** Polyamine deficiency reduces the rate of DNA replication fork movement in *Escherichia coli*, *Nature*, 272, 730, 1978.
64. **Geiger, L. E. and Morris, D. A.,** Stimulation of deoxyribonucleic acid replication fork movement by spermidine analogues in polyamine-deficient *Escherichia coli*, *J. Bacteriol.*, 141, 1193, 1980.
65. **Mikhailov, V. S. and Androsova, I. M.,** Effect of spermine on interaction of DNA polymerase alpha from loach (misgurnus fossilis) eggs with DNA, *Biochem. Biophys. Acta*, 783, 6, 1984.
66. **LaDuca, R. J., Fay, P. J., Chuang, C., McHenry, C. S., and Bambara, R. A.,** Site-specific pausing of deoxyribonucleic acid synthesis catalyzed by four forms of *Escherichia coli* DNA polymerase III, *Biochemistry*, 22, 5177, 1983.
67. **Pingoud, A., Urbanke, C., Alves, J., Ehbrecht, H. J., Zabeau, M., and Gualerzi, C.,** Effect of polyamines and basic proteins on cleavage of DNA by restriction endonucleases, *Biochemistry*, 23, 5697, 1984.
68. **Pingoud, A.,** Spermidine increases the accuracy of type II restriction endonucleases, *Eur. J. Biochem.*, 147, 105, 1985.
69. **Krasnow, M. A. and Cozzarelli, N. R.,** Catenation of DNA rings by topoisomerases, *J. Biol. Chem.*, 257, 2687, 1982.
70. **Billett, M. A. and Hall, T. J.,** Cations and the accessibility of chromatin to nucleases, *Nucleic Acids Res.*, 7, 2929, 1980.
71. **Rowlatt, C. and Smith, G. J.,** Ultrastructural studies on chromatin digestion by micrococcal nuclease in the presence of polyamines, *J. Cell Sci.*, 48, 171, 1981.
72. **Vanderbilt, J. N., Bloom, K. S., and Anderson, J. N.,** Endogenous nuclease, *J. Biol. Chem.*, 257, 13009, 1982.
73. **Ahmed, K., Davis, A. T., and Goueli, S. A.,** Differential effects of polyamines on the phosphorylation of chromatin-associated proteins, *Biochem. J.*, 209, 1971, 1983.
74. **Kinohara, N., Usui, H., Yoshikawa, K., and Takeda, M.,** Effects of polyamine hydrochlorides on dephosphorylation of phosphorylated H2B histone integrated into nucleosome core particles, *J. Biochem.*, 95, 597, 1984.
75. **Bogdanova, E. S.,** H1 histone, polylysine and spermine facilitate nucleosome assembly in vitro, *FEBS Lett.*, 175, 321, 1984.
76. **Hung, D. T., Marton, L. J., Deen, D. F., and Shafer, R. H.,** Depletion of intracellular polyamine levels may alter DNA conformation in 9L rat brain tumor cells, *Science*, 221, 268, 1983.
77. **Marton, L. J.,** in *Polyamines in Clinical Disorders*, Tierney, D. F., Moderator, *West. J. Med.*, 142, 63, 1985.

Chapter 8

POLYAMINES AND INITIATION OF PROTEIN SYNTHESIS IN EUKARYOTES

Bruno Barbiroli, Gabriele Mezzetti, and Maria Stella Moruzzi

TABLE OF CONTENTS

I. INTRODUCTION

Protein synthesis is a complex multi-step process which in principle can be controlled by the availability and/or activity of any one of the enzymes and components involved in the overall process. Therefore, many factors which affect cell metabolism such as pH, hormones, and specific activators or inhibitors are candidates or determinants in the control of protein synthesis.[1-3] Indeed, a number of reports have implicated ubiquitous aliphatic polyamines among the modulators of the process.[4-6] This experimental evidence, extensively reviewed, has been obtained using several different approaches including cell-free systems, cell cultures added with inhibitors of polyamine biosynthesis, and bacterial polyamine-deficient mutants.[7-14]

From a general point of view the control of protein synthesis can be expressed as a modification of the rate of any of the single phases of the process, i.e., initiation, elongation, or termination. Recently it has been pointed out that in most cases initiation is the step primarily affected by most factors responsible for the control of protein synthesis, which in turn means that initiation can be regarded as the rate-limiting step of the whole process.[1-3] In this light we will restrict our attention to the effects of polyamines on the initiation of polypeptide chain formation and on their interactions with the components involved in this early phase.

Many of the effects of these polycharged cations can be ascribed to their ability to interact by ionic forces with nucleic acids, acidic proteins, and other acidic cellular components.[7,8,15] This raises the question concerning the specificity of the action of polyamines since other cations such as Mg^{2+} can effectively substitute them in many reactions of protein synthesis when studied in vitro. Nevertheless, it is not possible to exclude a specific interaction between polyamines and particular components taking part in the complex process of protein synthesis, since in experiments involving different molecular events it has been shown that spermine is able to bind specifically and with high affinity hormone-sensitive cytoplasmic proteins.[15-18]

Another general problem which must be taken into consideration before discussing the effects of polyamines on initiation of protein synthesis is that considerable uncertainty exists about the real in vivo concentration of polyamines in the subcellular organelles and in particular in the protein synthesis compartment because of the possible redistribution of these polycations during cell disruption. Therefore, we will devote some attention to recent developments indicating the need for more diversified experimental efforts in the exploration of molecular mechanism by which polyamines affect initiation of protein synthesis in higher organisms.

II. INTERACTION OF POLYAMINES WITH COMPONENTS INVOLVED IN THE INITIATION STEP

Before considering the specific reactions of the initiation phase of protein synthesis it seems noteworthy to discuss the relationships between polyamines, particularly spermine and spermidine, and various components involved in the process itself such as aminoacyl-tRNA and ribosomes.

It has been shown that polyamines at low concentrations can interact in vitro with various species of RNA such as tRNA and small RNAs.[7-9,19-22] In this light many papers have dealt with the binding of these polycations to tRNA in the perspective of a possible involvement of polyamines in the synthesis and/or degradation of aminoacyl-tRNA. However, the possible redistribution of polyamines during homogenate preparation raises serious experimental problems concerning the quantitative estimation of the polycations bound to tRNA. In fact, several reports are in disagreement on the stoichiometry of the binding itself. Only recently

McMahon and Erdmann[23] have proposed that there are two classes of binding sites for spermidine on tRNA molecules at low ionic strength but only one class of binding sites at high ionic strength. In terms of physiological meaning of these interactions, other authors have underlined the possibility that these polycations contribute in stabilizing the secondary and tertiary conformations of tRNA which are active in protein biosynthesis.[24]

It has been reported that polyamines participate as physiological regulators of protein biosynthesis by activating the aminoacylation of tRNA performed by aminoacyl-tRNA synthetase.[21,25] Although there is still disagreement in the literature whether or not this enzyme is active in the absence of Mg^{2+}, there is evidence that polyamines at millimolar concentrations can substitute for the Mg^{2+} requirement.[25]

Another reaction affected by the presence of polyamines is the in vitro hydrolysis of aminoacyl-tRNA.[26] It has been shown that cytosol of rat ventral prostate, a polyamine-rich tissue, contains several enzymes capable of deacylating the initiator tRNA (Met-$tRNA_f$) and other aminoacyl-tRNA with various degrees of specificity.[26] Different polyamines at concentrations compatible with cell physiology are able to inhibit selectively the activity of the above prostatic enzymes.[26]

A recent paper by Peebles et al.[27] has pointed out that spermidine is also involved directly in tRNA splicing by increasing the accuracy of excision of intervening sequences from precursor tRNA. All these data suggest that polyamines are involved not only in the aminoacyl-tRNA synthesis, but also in the overall control of aminoacyl-tRNA turnover in the cell, thus controlling the substrate availability for protein biosynthesis.

It is generally established that quite high amounts of polyamines are associated with ribosomes extracted from various eukaryotic cells.[8,28] Unwashed rabbit reticulocyte ribosomes contain approximately 0.08 μmol/mg of spermine and spermidine in a ratio of 1:3.[29] On this and other evidence it has been proposed that ribosomes in vivo also contain polyamines and that these polycations are essential for maintaining ribosomal structure and function.[28] Attempts to free ribosomes of polyamines resulted in a dramatic loss of translation capacity.[28] In this light it has been proposed that ribosome-bound polyamine can facilitate the association of ribosomal subunits in protein synthesis.[30] This effect is also shared by Mg^{2+} at millimolar concentration and polyamines cannot replace all the ribosome Mg^{2+} ions.[31]

III. THE PARTIAL REACTIONS OF PEPTIDE CHAIN INITIATION: EFFECT OF POLYAMINES

The initiation of protein synthesis is generally intended as the sequence of events which yields to an 80S·Met-$tRNA_f$·mRNA complex (Figure 1). This phase, which is thought to be the main site of regulation of the overall protein synthesis process, is a complex sequence of reactions and requires eukaryotic initiation factors (eIF), of which there are at least nine and also a number of cofactors which are necessary for the activity of the initiation factors.[1,32]

It has been established that in eukaryotic cells eIF-2 forms a ternary complex with GTP and Met-$tRNA_f$ as the first step in peptide chain initiation. The next step involves the transfer of Met-$tRNA_f$ to the 40S ribosomal subunit to form the preinitiation complex in the presence of several initiation factors. After the binding of the ternary complex to the 40S subunit, mRNA is joined to the 40S complex. This reaction involves four more initiation factors and hydrolysis of ATP. Initiation factor eIF-5 promotes joining of the 60S subunit to form finally an intact 80S initiation complex. The initiation factors are then released to recycle and GTP is hydrolyzed to GDP which is also released from the ribosome. At this point 80S-initiation complex can take part in the elongation phase of protein synthesis.

In reticulocytes and possibly in all other eukaryotic cells polypeptide chain initiation appears to be controlled largely by the state of phosphorylation of the initiation factor eIF-

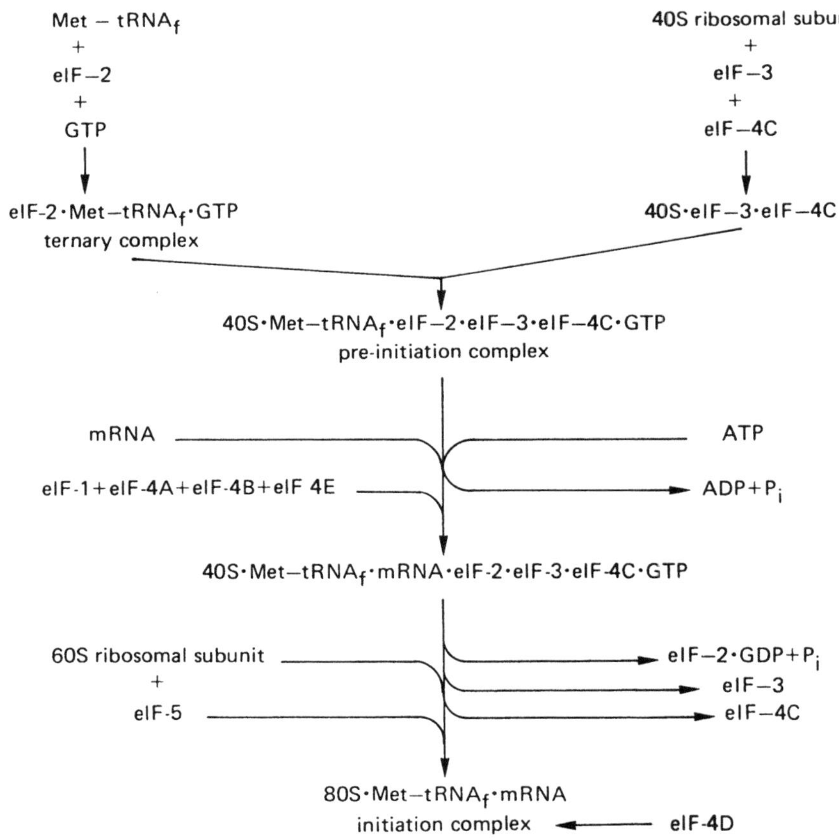

FIGURE 1. Proposed pathway of initiation of protein synthesis in eukaryotes.

2 which leads to a control of the reaction of ternary complex formation and its subsequent binding to 40S ribosomal subunit.[33-35] Assay of this binding activity is therefore generally considered extremely useful in examining the regulation of initiation of protein synthesis in various experimental systems.[36-39]

It has been reported from several laboratories including ours that polyamines are among the factors which stimulate protein biosynthesis by increasing the rate of peptide chain initiation reactions.[28,40-43] In general, polyamines appear to have at least two kinds of effects on these reactions. One appears to be a rather nonspecific effect for which divalent cations can effectively substitute. In addition these amines appear to have a more specific effect, not mimicked by other cations, on one or more steps of initiation.[28,40]

An important finding has been reported by Kuroda et al.[44] concerning the isolation and characterization of a cyclic nucleotide-independent protein kinase from bovine adrenocortical PK 380. Polyamines such as spermine and spermidine specifically and selectively inhibit the PK 380-catalyzed phosphorylation of eIF-2. Since this phosphorylation inhibits concomitantly the eIF-2 activity and the initiation of protein synthesis, the possibility exists that polyamines enhance the formation of initiation complex by inhibiting the phosphorylation of eIF-2 by PK 380.

Several lines of evidence suggest that at least one mechanism by which polyamines possibly affect the initiation phase of protein synthesis involves ribosomes and/or ribosomal particles (see Section II). On this line it has been reported that spermine and spermidine are active in decreasing the requirement for Mg^{2+} in the formation of initiation complex.[45] In addition it has been proposed by Kramer et al.[28] that one polyamine binding site on 40S ribosomal subunit may be specifically involved in the formation of preinitiation complex. The same authors also showed that in *Artemia salina* spermine has a marked stimulatory effect on in vitro binding of [^{125}I]-globin-mRNA to 40S ribosomal subunit.[28]

The 80S-initiation complex contains Met-tRNA$_f$ and mRNA and presumably no initiation factors. Such complexes can react with puromycin to form methionyl-puromycin. Therefore, the Met-tRNA$_f$ is bound in the ribosomal *P* site and the 80S-initiation complex is capable of entering the elongation phase of protein synthesis. At this step the ubiquitous initiation factor eIF-4D strongly stimulates the synthesis of methionyl-puromycin and appears to act on preformed 80S-initiation complex.[32,46,47] Recently, it has been reported that eIF-4D contains the amino acid hypusine, which is formed from lysine by a reaction which originates from spermidine.[48] There is one molecule of hypusine per molecule of eIF-4D and this composition is similar for factors derived from a variety of sources.[48] Although the role of hypusine in the eIF-4D functioning is not known this conservation suggests an important function that it has been related to the rate of cellular growth.[49]

It is widely accepted that the regulation of protein synthesis in the target tissue is undoubtedly the principal action of steroid hormones.[50] In the last years a number of studies concerning the effect of polyamines on peptide chain initiation have been performed in tissues or cells targets of steroid hormones, where the translational process is always physiologically controlled by the agonist.[37,39,43] Liang and Liao[37,51] have reported an effect of androgens within 10 min of administration on GTP-dependent Met-tRNA$_f$ binding to eIF-2 in rat prostate. This effect was apparently due to a modulation of eIF-2 activity in forming ternary complex exerted by an endogenous cytosolic fraction containing polyamines.[37] As a consequence of this finding a specific role of polyamines in the hormonal regulation of the translation process in mammalian cells has been postulated by those authors.[52]

More recently, in our laboratory we have studied the effect of polyamines on ribosomal binding of [^{35}*S*]-Met-tRNA$_f$ via ternary complex formation in two opposite hormonal conditions. We prepared cytosols, as sources of initiation factors, from intestine either of normal and vitamin D-deprived (rachitic) chicks and we studied the eIF-2 directed binding of [^{35}*S*]-Met-tRNA$_f$ to ribosomal particles in this target tissue, both in the absence or presence of physiological concentration of spermine. As shown in Figure 2, the binding of [^{35}*S*]-Met-tRNA$_f$ to 40S ribosomal subunit, as measured by sucrose density gradient centrifugation, was absolutely dependent on the presence of the respective cytosol protein factors and GTP. The ability of initiation factors to support the initiation of protein synthesis by ribosomal particles was higher in normal animal (panel b) when compared with rachitic controls (panel a). When 0.2 m*M* spermine was added to the incubation mixtures, a stimulation of 40S-[^{35}*S*]-Met-tRNA$_f$ binding activity was observed for both samples. However, the extent of stimulation of binding reaction directed by cytosolic initiation factors from rachitic chicks was on the order of 10% whereas the stimulation of the same reaction directed by cytosolic initiation factors from normal chicks was on the order of 70%. This effect was not due to different amounts of endogenous polyamines present in the assays, and it appears quite specific since spermidine was partially active only at much higher concentrations (1 m*M*). Other polyamines such as putrescine, 1,12-diaminododecane up to 3 m*M*, or Mg^{2+} were absolutely ineffective. As far as the differential effect of spermine on the stimulation of the binding reactions is concerned, it can be interpreted as a differential action of the tetramine on a vitamin D-dependent component of the translational machinery in its target tissue.

IV. CONCLUDING REMARKS

Most of the studies on the effect of polyamines on protein synthesis reviewed here have been performed in vitro using cell lysates or reconstituted systems that bear a pale resemblance

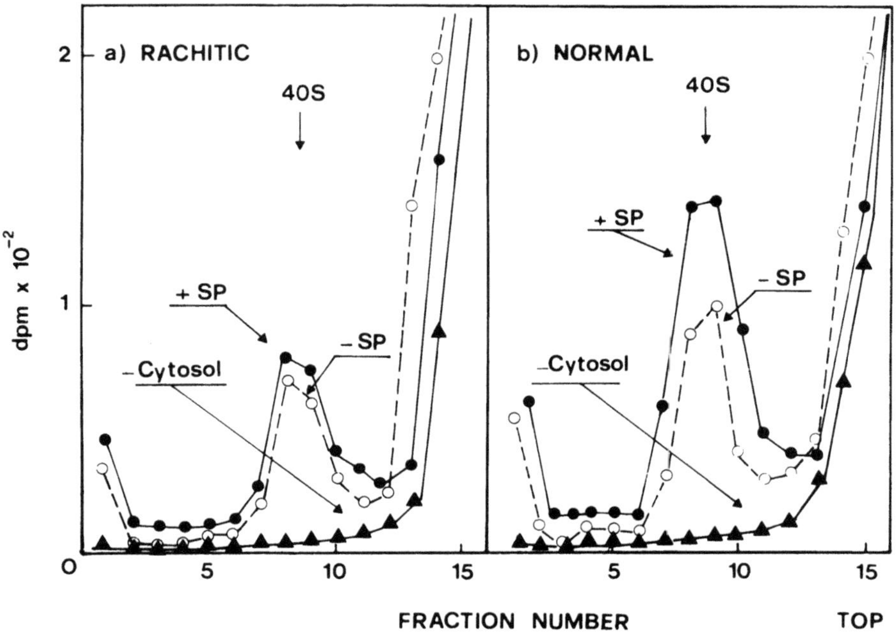

FIGURE 2. Formation of 40S preinitiation complex analyzed by sucrose density gradient centrifugation. Cytosols (350 μg of protein) from intestinal mucosa of rachitic (panel a) or normal chicks (panel b) were studied for their ability to facilitate [^{35}S]-Met-tRNA$_f$ binding to 40S ribosomal subunits either in the absence (○) or presence (●) of 0.2 m*M* spermine. The incubation mixture contained 40 m*M* Tris/HCl, pH 7.5, 50 m*M* KCl, 3 m*M* dithiothreitol, 5 m*M* $MgCl_2$, 2.5 m*M* GTP, 5 μg of AUG, and 0.7 A_{260} unit of 40S ribosomal subunits. The reaction was started by addition of 0.03 A_{260} unit of [^{35}S]-Met-tRNA$_f$ and the mixture was incubated at 30°C for 30 min. 200 μℓ aliquots were layered over a 3.4 mℓ of linear 5 to 20% sucrose gradient in the same buffer. Gradients were centrifuged 1 hr at 2°C and 300,000 x g in the SW 60 Spinco rotor; 0.2-mℓ fractions were collected from the bottom of the gradients and the radioactivity was measured. The arrows indicate the position to which purified 40S ribosomal subunits sedimented in parallel control gradients.

with the physiological condition of the intact cell, whereas less research effort has been addressed to the study of the effect of these polycations in in vivo systems.[53-55] Therefore, much remains to be done to determine whether polyamines actually function in vivo in the regulation of peptide chain initiation. To this aim it is relevant to obtain accurate estimates of the concentrations of free intracellular Mg^{2+} and polyamines available for the protein synthesis machinery, and concentrate research efforts on studies employing whole cells where intracellular polyamines levels can be altered employing specific agents or substances.

However, there is reason to believe that a considerable number of cell-free studies can be safely extrapolated to the intact cell. Indeed, the mechanisms involved in the process of initiation of translation are known to be operative also in cell-free systems which from this point of view do not differ excessively from the intact cell. In addition, the fact that a number of reactions of the initiation process are influenced in vitro by polyamines indicate that these compounds possess the versatility necessary to control them. This fact can be positively associated with the capability of polyamines to modulate a wide range of cellular processes, including cell proliferation and either physiological or pathological protein biosynthesis in tissues or cells responding to hormones.

REFERENCES

1. **Austin, S. A.,** Translational regulation of protein synthesis in eukaryotes, *Essays Biochem.*, 18, 79, 1982.
2. **Jackson, R. J.,** The cytoplasmic control of protein synthesis, in *Protein Biosynthesis in Eukaryotes,* Pérez-Bercoff, R., Ed., Plenum Press, New York, 1982, 363.
3. **Moldave, K.,** Eukaryotic protein synthesis, *Annu. Rev. Biochem.*, 54, 1109, 1985.
4. **Takeda, Y.,** Polyamines and protein synthesis. II. The shift in optimal concentration of Mg^{2+} by polyamines in the MS2 phage RNA-directed polypeptide synthesis, *Biochem. Biophys. Acta,* 179, 232, 1969.
5. **Atkins, J. F., Lewis, J. B., Anderson, C. W., and Gesteland, R. F.,** Enhanced differential synthesis of proteins in a mammalian cell-free system by addition of polyamines, *J. Biol. Chem.*, 250, 5688, 1975.
6. **Igarashi, K., Yabuk, M., Yoshioka, Y., Eguchi, K., and Hirose, S.,** Mechanism of stimulation of polyphenylalanine synthesis by spermidine, *Biochem. Biophys. Res. Commun.*, 75, 163, 1977.
7. **Bachrach, U.,** *Function of Naturally Occurring Polyamines,* Academic Press, New York, 1973, 68.
8. **Tabor, C. W. and Tabor, H.,** Polyamines, *Annu. Rev. Biochem.*, 53, 749, 1984.
9. **Cohen, S. S.,** *Introduction to the Polyamines,* Prentice-Hall, Englewood Cliffs, N.J., 1971.
10. **Russel, D. H.,** *Polyamines in Normal and Neoplastic Growth,* Raven Press, New York, 1973.
11. **Raina, A. and Jänne, J.,** Physiology of the natural polyamines putrescine, spermidine and spermine, *Med. Biol.*, 53, 121, 1975.
12. **Heby, O.,** Role of polyamines in the control of cell proliferation and differentiation, *Differentiation,* 19, 1, 1981.
13. **Algranati, I. D. and Goldemberg, S. H.,** Polyamines and their role in protein synthesis, *Trends Biochem. Sci.*, 2, 272, 1977.
14. **Pegg, A. E.,** Recent advances in the biochemistry of polyamines in eukaroytes, *Biochem. J.*, 234, 249, 1986.
15. **Mezzetti, G., Moruzzi, M. S., Capone, G., and Barbiroli, B.,** Polyamine binding by a cytoplasmic factor in the duodenal mucosa of new-born chick, *Biochem. Biophys. Res. Commun.*, 97, 222, 1980.
16. **Mezzetti, G., Drusiani, F., Montagnani, G., Capone, G., and Barbiroli, B.,** Heat-sensitive spermine-binding factor from peripheral human lymphocyte (41343), *Soc. Exp. Biol. Med.*, 169, 280, 1982.
17. **Mezzetti, G., Loor, R., and Liao, S.,** Androgen-sensitive spermine-binding protein of rat ventral prostate. Purification of the protein and characterization of the hormonal effect, *Biochem. J.*, 184, 431, 1979.
18. **Liao, S., Mezzetti, G., and Chen, C.,** Androgen receptor and early biochemical responses, in *The Cell Nucleus,* Vol. 7, Bush, H., Ed., Academic Press, New York, 1979, 201.
19. **Schreier, A. A. and Schimmel, P. R.,** Interaction of polyamines with fragments and whole molecules of yeast phenylalanine-specific transfer RNA, *J. Mol. Biol.*, 93, 323, 1975.
20. **Sakai, T. T. and Cohen, S. S.,** Effects of polyamines on the structure and reactivity of tRNA, in *Progress in Nucleic Acid Research and Molecular Biology,* Vol. 17, Cohn, W. E., Ed., Academic Press, New York, 1976.
21. **Loftfield, R. B., Eigner, E. A., and Pastuszyn, A.,** Polyamines and protein synthesis, in *Polyamines in Biology and Medicine,* Vol. 8, Morris, D. R. and Marton, L. J., Eds., Marcel Dekker, New York, 1981, 207.
22. **Liang, T., Mezzetti, G., Chen, C., and Liao, S.,** Selective polyamine-binding proteins. Spermine binding by an androgen-sensitive phosphoprotein, *Biochem. Biophys. Acta,* 542, 430, 1978.
23. **McMahon, M. E. and Erdmann, V. A.,** Binding of spermidine to transfer ribonucleic acid, *Biochemistry,* 21, 5280, 1982.
24. **Tropp, J. S. and Redfield, A. G.,** Protein exchange rates in transfer RNA as a function of spermidine and magnesium, *Nucleic Acids Res.*, 11, 2121, 1983.
25. **Takeda, Y. and Igarashi, K.,** Polyamines and protein synthesis. IV. Stimulation of aminoacyl transfer RNA formation by polyamines, *Biochem. Biophys. Res. Commun.*, 37, 917, 1969.
26. **Mezzetti, G.,** Differential polyamine inhibition of Met-tRNA deacylation by prostate enzymes, *Ital. J. Biochem.*, 30, 90, 1981.
27. **Peebles, C. L., Gegenheimer, P., and Abelson, J.,** Precise excision of intervening sequences from precursor tRNAs by a membrane-associated yeast endonuclease, *Cell,* 32, 525, 1983.
28. **Kramer, G., Odom, O. W., and Hardesty, B.,** Polyamines in eukaryotic peptide initiation, *Methods Enzymol.*, 40, 555, 1979.
29. **Konecki, D., Kramer, G., Pinphanichakarn, P., and Hardesty, B.,** Polyamines are necessary for maximum *in vitro* synthesis of globin peptides and play a role in chain initiation, *Arch. Biochem. Biophys.*, 169, 192, 1975.
30. **Sperrazza, J. M. and Spremulli, L. L.,** Quantitation of cation binding to wheat germ ribosomes: influences on subunit association equilibria and ribosome activity, *Nucleic Acids Res.*, 11, 2665, 1983.
31. **Kimes, B. W. and Morris, D. R.,** Cations and ribosome structure. II. Effects on the 50S subunit of substituting polyamines for magnesium ion, *Biochemistry,* 12, 442, 1973.

32. **Hershey, J. W. B.,** The initiation factors, in *Protein Biosynthesis in Eukaryotes,* Pérez-Bercoff, R., Ed., Plenum Press, New York, 1982, 97.
33. **Ochoa, S.,** Regulation of protein synthesis initiation in eukaryotes, *Arch. Biochem. Biophys.,* 223, 325, 1983.
34. **Revel, M. and Groner, Y.,** Post-transcriptional and translational controls of gene expression in eukaryotes, *Annu. Rev. Biochem.,* 47, 1079, 1978.
35. **Gupta, N. K.,** Roles of eukaryotic initiation factor 2 ancillary factors in the regulation of eukaryotic protein synthesis initiation, *Curr. Top. Cell. Regul.,* 21, 1, 1982.
36. **Rannels, D. E., Pegg, A. E., and Rannels, S. R.,** Starvation results in decreased initiation factor in rat skeletal muscle, *Biochem. Biophys. Res. Commun.,* 72, 1481, 1976.
37. **Liang, T., Castañeda, E., and Liao, S.,** Androgen and initiation of protein synthesis in the prostate, *J. Biol. Chem.,* 252, 5692, 1977.
38. **Mezzetti, G., Ferrari, S., Davalli, P., Battini, R., and Corti, A.,** Peptide chain initiation and analysis of *in vitro* translation products in rat heart undergoing hypertrophic growth, *J. Mol. Cell Cardiol.,* 15, 629, 1983.
39. **Mezzetti, G., Moruzzi, M. S., and Barbiroli, B.,** Peptide-chain initiation in duodenal mucosa of rachitic chicks after 1α,25-dihydroxycholecalciferol administration, *Biochem. J.,* 219, 99, 1984.
40. **Mezzetti, G., Moruzzi, M. S., Monti, M. G., and Barbiroli, B.,** Effect of polyamines on peptide chain initation in chick intestine, in *Advances in Polyamines in Biomedical Science,* Calderera, C. M. and Bachrach, U., Eds., CLUEB, Bologna, Italy, 1984, 45.
41. **Igarashi, K., Yabuki, M., Yoshioka, Y., Eguchi, K., and Hirose, S.,** Mechanism of stimulation of polyphenylalanine synthesis by spermidine, *Biochem. Biophys. Res. Commun.,* 75, 163, 1977.
42. **Igarashi, K., Kojima, M., Watanabe, Y., Maeda, K., and Hirose, S.,** Stimulation of polypeptide synthesis by spermidine at the level of initiation in rabbit reticulocyte and wheat germ, *Biochem. Biophys. Res. Commun.,* 97, 480, 1980.
43. **Takemoto, T., Nagamatsu, Y., and Oka, T.,** The study of spermidine-stimulated polypeptide synthesis in cell-free translation of mRNA from lactating mouse mammary gland, *Biochem. Biophys. Acta,* 740, 73, 1983.
44. **Kuroda, Y., Merrick, W. C., and Sharma, R. K.,** Polyamines inhibit the protein kinase 380-catalyzed phosphorylation of eukaryotic initiation factor 2α, *Science,* 215, 415, 1982.
45. **Takeda, Y.,** Polyamines and protein synthesis. III. The effect of polyamines on the binding of phenylalanyl-tRNA to ribosomes, *Biochem. Biophys. Acta,* 182, 258, 1969.
46. **Thomas, A. A. M., Benne, R., and Voorma, H. D.,** Initiation of eukaryotic protein synthesis, *FEBS Lett.,* 128, 177, 1981.
47. **Benne, R. and Hershey, J. W. B.,** The mechanism of action of protein synthesis initiation factors from rabbit reticulocytes, *J. Biol. Chem.,* 253, 3078, 1978.
48. **Park, M. H., Chung, S. I., Cooper, H. L., and Folk, J. E.,** The mammalian hypusine-containing protein, eukaryotic initiation factor 4D, *J. Biol. Chem.,* 259, 4563, 1984.
49. **Torrelio, B. M., Paz, M. A., and Gallop, P. M.,** Cellular proliferation and hypusine synthesis, *Exp. Cell Res.,* 154, 454, 1984.
50. **Chan, L. and O'Malley, B. W.,** Mechanism of action of the sex steroid hormones, *N. Engl. J. Med.,* 294, 1322, 1976.
51. **Liang, T. and Liao, S.,** A very rapid effect of androgen on initiation of protein synthesis in prostate, *Proc. Natl. Acad. Sci. U.S.A.,* 72, 706, 1975.
52. **Hung, S. C., Liang, T., Gluesing, L. M., and Liao, S.,** On the factors affecting the initiation of protein synthesis in the rat ventral prostate: androgens, polyamines and conjugated proteins, *J. Steroid Biochem.,* 7, 1001, 1976.
53. **Gibson, K. and Harris, P.,** The *in-vitro* and *in-vivo* effects of polyamines on cardiac protein biosynthesis, *Cardiovasc. Res.,* 8, 668, 1974.
54. **Rudkin, B. B., Mamont, P. S., and Seiler, N.,** Decreased protein-synthetic activity is in early consequence of spermine depletion in rat hepatoma tissue-culture cells, *Biochem. J.,* 217, 731, 1984.
55. **Luk, G. D.,** Essential role of polyamine metabolism in hepatic regeneration. Inhibition of deoxyribonucleic-acid and protein-synthesis and tissue regeneration by difluoromethylornithine in the rat, *Gastroenterology,* 90, 1261, 1986.

Chapter 9

POLYAMINES AND RNA SYNTHESIS

Samson T. Jacob and Dean A. Stetler

TABLE OF CONTENTS

I. INTRODUCTION

Over the past 2 decades, numerous studies have implicated polyamines in the regulation of gene expression at the level of transcription. The observation that the increased levels of polyamines present during cell growth and development were almost always accompanied by increased RNA polymerase activity and RNA synthesis suggested that polyamines might have a role in controlling RNA metabolism (reviewed in References 1 to 4). Subsequent studies in a number of laboratories established that polyamines were capable of stimulating RNA synthesis in vitro via different mechanisms. These studies were primarily based on measurements of RNA polymerase activity in response to exogenous polyamines. Recent advances in the establishment of cell-free systems that are capable of transcribing specific genes accurately have paved way to understanding the exact role of polyamines in gene expression. This chapter deals with some of the early work as well as the more recent attempts to explore the function of polyamines in RNA synthesis.

II. EFFECT OF POLYAMINES ON RNA SYNTHESIS IN VITRO USING NONDEFINED TEMPLATES

A. Transcription in Prokaryotes

Doerfler et al.[5] were the first to report that addition of polyamines to an in vitro RNA polymerase reaction mixture prevented the characteristic leveling off of RNA synthesis. They attributed this effect to inhibition of the nucleases present in their enzyme preparation of polyamines. However, subsequent investigations in several laboratories demonstrated that the effect of polyamines on in vitro RNA synthesis is mediated by other mechanisms as well. As described below, these include interactions of polyamines with the template, the product, and possibly with RNA polymerase itself.

The effects of polyamines on ribonucleases and on the stability of RNA are both complex and variable as reviewed by Karpetsky et al.[6] Briefly, polyamines (strong organic bases) can form complexes with RNA through hydrogen bonding and electrostatic forces and consequently protect it from nucleolytic digestion. Interestingly, polyamines have been shown to stimulate several prokaryotic and eukaryotic ribonucleases and, in a few cases, to modify their substrate specificity. Thus, it is apparent that polyamines are capable of affecting RNA degradation by interacting with the polyribonucelotide as well as with the nucleases.

Evidence for polyamine-mediated RNA synthesis by other mechanisms was first presented by Krakow[7] who observed stimulation of RNA polymerase activity in preparation from *Azobacter vinlandii* that was essentially free of ribonuclease. Subsequently, other laboratories reported a similar stimulatory effect of polyamines on ribonuclease-free RNA polymerase preparations from different bacterial sources.[8-14] In addition, it was found that polyamines stimulated the bacterial RNA polymerase reaction only when native DNA was used as the template.[7,9,10,12] These observations suggested that polyamines influence transcription reactions at a level other than RNA product stabilization and that these compounds exert an effect on the template. Peterson et al.[10] presented evidence which suggested that polyamines increase the availability of initiation sites on native DNA. Interestingly, the stimulation achieved with spermidine was relatively small with highly purified RNA polymerase, which might be due to the loss of contaminating nucleases during purification. Spermidine has also been shown to increase significantly the degree of asymetric transcription of ΦX 174 RF DNA by *Micrococcus lysodeikticus* RNA polymerase.[11] It is not known whether this was due to an effect of the polyamine on the template or on the enzyme. However, an effect on the enzyme was considered more likely.

Addition of RNA to an in vitro RNA polymerase reaction results in a drastic inhibition of RNA synthesis. Accumulation of RNA product during transcription reaction similarly causes a leveling off of RNA synthesis. Inclusion of polyamines in the reaction reduces this inhibition.[9,12] It was determined that polyamines dissociate RNA polymerase-RNA complexes releasing RNA polymerase for reinitiation on the template.[12] On the contrary, the formation of enzyme-DNA complex was not affected by polyamines. Thus, one of the mechanisms by which polyamines stimulate RNA synthesis, regardless of whether they are added at the beginning of the reaction or later, is by complexing with newly synthesized RNA, resulting in the release of RNA polymerase for additional RNA synthesis.

Other compounds are known to mimic some effects of polyamines. For example, inhibition of RNA synthesis by accumulation of newly synthesized RNA or by addition of RNA to the reaction can be reduced, not only with polyamines, but also with potassium chloride or ammonium sulfate.[12,13] Similarly, like polyamines, magnesium chloride and potassium chloride can increase asymetric transcription of X174 RF DNA.[11] However, the ability of polyamines to alter RNA synthesis is not merely due to their cationic properties since even in the presence of optimal concentrations of magnesium chloride, potassium chloride, or ammonium sulfate polyamines can augment RNA synthesis in vitro.[7,13-15]

Although it is clear that polyamines are capable of stimulating prokaryotic RNA synthesis in vitro through interactions with DNA, the RNA product, and possibly with RNA polymerase itself, the exact mechanisms and other factors involved are not yet fully understood. It is not known whether polyamines can direct the transcription of specific gene sequences by modulating these parameters.

B. Transcription in Eukaryotes

Ballard and Williams-Ashman,[16,17] using enzyme extracted from rat testis, first demonstrated that eukaryotic RNA polymerases can be stimulated in vitro by the addition of polyamines to the reaction mixture. Subsequently, other investigators reported similar results with RNA polymerases from a variety of sources, including beef brain,[18] pig kidney,[19] rat liver,[20-23] rat hepatoma,[22] calf thymus,[24] Ehrlich ascites tumor cells,[25] and mouse spleen.[26] Generally, transcription was stimulated three- to fourfold by the optimal concentration of polyamine. The concentration of polyamine required to achieve maximum stimulation of the three eukaryotic RNA polymerases (I, II, and III) has varied somewhat from laboratory to laboratory. Whether this was due to different tissue sources of the enzyme or to variability in polyamine preparations is not known. However, we have noted significant variations in spermine preparations from different suppliers and even between different lots from the same manufacturer. The concentration of polyamine required to achieve stimulation of RNA synthesis also depends upon the presence or absence of other cations. For example, in the absence of ammonium sulfate, the optimal concentrations of spermine were determined to be 0.31, 0.62, and 2.5 m*M* for transcription of deproteinized DNA by purified rat liver RNA polymerases I, II, and III, respectively.[20] In the presence of the optimal concentration of ammonium sulfate for each polymerase, significantly higher concentrations of spermine were required to produce maximal stimulation (1.25 m*M* for RNA polymerase I and 5 to 10 m*M* for RNA polymerases II and III). Similar optimal concentrations were observed for RNA polymerases I, II, and III purified from mouse spleen.[26] In the presence of ammonium sulfate, rat liver RNA polymerase I was inhibited by spermine concentrations above 1.26 m*M* and was completely inhibited by 5 m*M*.[20] Since RNA polymerases II and III are maximally stimulated at 5 m*M* spermine, this relatively high concentration can be used to selectively inhibit RNA polymerase I activity in unpurified preparations. Mandel and Chambon[24] observed a similar inhibition of calf thymus RNA polymerase I-directed transcription of SV4 DNA by 1 m*M* spermine. However, if the enzyme was allowed to initiate transcription prior to addition of spermine, inhibition was overcome, indicating that the inhibitory effect was at the level of initiation. We have made similar observations with pig kidney RNA polymerases I and II.[19]

As with the prokaryotic system, the effect of polyamines on eukaryotic RNA polymerase activity was shown to be additive to that of divalent cations.[21] Similarly, the stimulatory effect of polyamines with relatively impure preparations of the polymerase was partially due to inhibition of degradation of newly synthesized RNA.[3] Stirpe and Novello[27] reported that spermine stimulated total rat liver RNA polymerases when native DNA, but not denatured DNA, was used as template. Since stimulation of the RNA polymerases by ammonium sulfate was more marked when denatured DNA was used as the template, this result clearly demonstrated that the action of polyamines was more than an ionic effect.

Mandel and Chambon[24] also addressed the importance of the type of DNA template in studying the effect of spermine on transcription. When native calf thymus DNA was used as the template, addition of 0.5 m*M* spermine to transcription reactions catalyzed by purified RNA polymerases I and II (calf thymus) resulted in 1.5- and 3.0-fold enhancement of activity, respectively. In contrast, when SV40 DNA was used, RNA polymerase I was not stimulated by 0.5 m*M* spermine and was markedly inhibited by 1 m*M* spermine. Transcription of SV40 DNA by RNA polymerase II was stimulated to a lesser extent than that observed with calf thymus DNA and concentrations of spermine greater than 0.5 m*M* inhibited the enzyme. Jänne et al.[19] showed that with limiting template concentrations, polyamine concentrations which stimulated RNA synthesis in the presence of excess template actually inhibited RNA polymerase activity. These observations suggest that some of the variability in the optimal concentrations of polyamines required to achieve stimulation of RNA synthesis in vitro could be due to differences in the type and amount of templates used for these studies.

Spermine was shown[22] to have little effect on the activity of either purified RNA polymerase I or II with denatured DNA although the latter template is an efficient template for solubilized RNA polymerase II.[28] In contrast, with native DNA template, spermine stimulated the activity of RNA polymerase II to a level which approached or even exceeded the activity obtained with denatured DNA template. It was suggested that spermine might be replacing or mimicking the effect of factors necessary for the transcription of native templates by RNA polymerase II which were lost during the extraction/purification process.

Jänne et al.[19] demonstrated that partially purified RNA polymerases I and II from pig kidney could be stimulated both by spermine and spermidine which increased the maximal velocity of transcripton with no change in the Km for UTP. The stimulation observed was the same whether spermine was added at the beginning or 5 min after the start of the reaction, suggesting that the effect was at the elongation rather than the initiation step. This was confirmed by analysis of average chain length (UMP/uridine ratio) and size (sucrose density gradient centrifugation) of RNA synthesized in the presence and absence of spermine. Moruzzi et al.[21] also have concluded that the polyamine effect was on the elongation step of RNA synthesis.

Moruzzi et al.[21] observed a biphasic effect of spermine on RNA polymerase II activity in isolated nuclei. RNA synthesis was stimulated at 0.1 and 1 m*M* polyamine. In contrast, only the lower concentration of spermine was effective in enhancing transcription of deproteinized rat liver chromatin by homologous RNA polymerase II. However, when native chromatin was used, rat liver RNA polymerase II was stimulated at both concentrations of the polyamine. This indicated a specific effect of the lower concentration of spermine on the chromatin template. Further experiments,[29] involving spectroscopic titration of actinomycin D binding to various templates as well as UV absorption and circular dichroism spectra, demonstrated that the lower concentration of polyamine is capable of causing changes in the secondary structure of chromatin DNA. These changes were apparently mediated by interaction of the polyamine with chromatin proteins, which results in an increased number of binding sites for actinomycin D and in a better template for homologus RNA polymerase II.

In summary, relatively low (micromolar) concentrations of polyamines are capable of stimulating eukaryotic RNA synthesis through interaction with the chromatin proteins. In addition, higher concentrations of polyamines (0.1 to 5 m*M*) produce a general stimulatory effect by inhibiting degradation of newly synthesized RNA, facilitating release of polymerase from RNA product, and by maximizing the velocity of the elongation step of transcription through interaction with either the enzyme or template. Very high concentrations of polyamines (5 to 10 m*M*) selectively inhibit the polymerases (I, II, and III) at the level of chain initiation. Thus, it appears that polyamines may be one of the mediators by which eukaryotic cells modulate transcription. Whether polyamines also affect selective transcription and to what extent any of the effects are due to modification of the RNA polymerases or other essential transcription factor(s) remains to be determined.

III. PHOSPHORYLATION BY POLYAMINE-MEDIATED NUCLEAR PROTEIN KINASE II AND rRNA SYNTHESIS

A. Characteristics of Polyamine-Mediated Protein Kinase

A major emphasis in recent years has been to explore the probable role of polyamines in the expression of a specific gene. In recent years, a protein kinase which can be modulated by polyamines has been described. Since this kinase can activate RNA polymerases I and II, there have been a few investigations to relate the phosphorylation of the polymerases and transcription in vitro. A few details of these studies are described below.

Nuclear preparations from a variety of eukaryotic cells contain two major cyclic nucleotide-independent protein kinases (for a recent review, see Reference 30). These kinases are designated protein kinases NI and NII based on their order of elution from DEAE-Sephadex or DEAE-cellulose columns.[31] Although these two kinases are similar to casein kinases I and II respectively,[32] it has not been established whether the nuclear kinases and the corresponding casein kinases are identical proteins. Protein kinase NII and casein kinase II contain similar subunits (42/43 and 24/25 kdaltons), utilize both ATP and GTP as substrates, exhibit similar Km, phosphorylate serine and threonine, and respond to similar inhibitors. However, they do exhibit some functional differences. Thus, casein kinase II does not phosphorylate S_2 (M_r 120,000) subunit of RNA polymerase I[33] whereas protein kinase NII phosphorylates S_2, S_3, and S_5 subunits of RNA polymerase I.[34] Further, unlike protein kinase NII which activates RNA polymerase I,[35] casein kinase II is not capable of activating this polymerase.[33] Casein kinase II can phosphorylate ornithine decarboxylase (ODC) whereas protein kinase NII does not phosphorylate or activate ODC.[38] There is also some evidence that the nuclear kinases and cytoplasmic kinases are not structurally identical entities.[36]

In general, the protein kinases purified from isolated nuclei have been found to activate RNA polymerases whereas the cytoplasmic kinases do not activate the polymerases (see Reference 30). Several factors could contribute to the activation of RNA polymerase I by protein kinase NII. First, Mg^{2+}, the divalent metal ion required for the kinase activity, must be present in the assays for RNA polymerase I activity. Second, the salt concentration must be kept below 20 m*M* $(NH_4)_2SO_4$ since higher levels of the salt are known to inactivate the kinase.[34] Third, care must be taken to minimize RNA degradation. Fourth, RNA polymerase I must be subjected to density gradient fractionation to remove the bulk of the free protein kinase. Finally, the polymerase I must be in a relatively dephosphorylated state. The lack of activation of calf thymus RNA polymerase I by homologous casein kinase II might be due to the highly phosphorylated state of the enyzme used as the substrate or to the use of casein kinase II instead of the nuclear kinase.

A unique property of protein kinase NII is the dramatic stimulation of protein phosphorylation by physiological (1 to 5 m*M*) or subphysiological ($<$1 m*M*) concentrations of spermine at optimal Mg^{2+} concentration.[39-41] Spermidine also could stimulate the phosphorylation reactions, albeit at relatively higher concentrations whereas putrescine had only a minimal effect.[40]

Ahmed et al.[42] investigated the effect of polyamines on prostatic chromatin and nonhistone protein-associated kinase reactions using the endogenous proteins as acceptors. These authors postulated that polyamines stimulate phosphorylation by altering conformations of the substrates rather than acting on the kinases themselves. Using highly purified protein kinase NII, we demonstrated that spermine directly interacts with protein acceptor; the K_m with respect to casein in the absence of spermine was eightfold greater than that in the presence of polyamine.[40] Similar conclusions have also been reached by Hara and Endo.[43] That the polyamine exerts its effect on phosphorylation by interaction with the protein acceptor was further demonstrated by phosphorylating casein in the presence of excess protein kinase NII. The reaction reached a plateau within 60 min. At this time point, addition of 5 m*M* spermidine dramatically stimulated (as much as 12-fold) the phosphorylation of casein following further incubation for 2 hr.[41] N^1 or N^8-acetyl spermidine, even at 10 m*M*, had no stimulatory effect whereas 5 m*M* of spermidine augmented phosphorylation reaction three- to fivefold.[41] Similar conclusions have also been reached by Hara and Endo.[43] The spermine-induced increase in phosphorylation was independent of protein kinase NII concentration. The exposure of new phorphorylation sites on protein substrates by spermine was confirmed by demonstrating that the threonine/serine ratio was substantially elevated in presence of the polyamine.[39] Recent studies using extensively purified protein kinase from rat liver and ventral prostate nuclei have also shown that the enhancement of protein kinase reactions by polyamines is primarily due to conformational changes in the acceptor molecule.[44]

B. Probable Relationship of Polyamine-Mediated Phosphorylation of Proteins and rRNA Synthesis

Although polyamines are known to stimulate RNA synthesis in vitro, it has not been established whether the polyamine-induced increase in gene expression is caused by interaction of polyamine with DNA or RNA polymerase I or transcription factor(s) or a combination of effects on these three parameters. We have shown that in addition to the effects on the template, polyamines can affect RNA synthesis via phosphorylation of RNA polymerase I. For these studies, two purine nucleotide analogues, 5′-adenylyl imidodiphosphate and 5′-guanylyl imidodiphosphate (GMP-PMP) were substituted for ATP.[41] Because these analogues do not contain hydrolyzable phosphate at γ position, transfer of γ-P to proteins will not occur thereby preventing phosphorylation. However, RNA synthesis will continue unabated due to hydrolyzable phosphate at the α position.[45] When Mg^{+2} was used as divalent cation, RNA synthesis directed by RNA polymerase I increased two- to sixfold in response to spermine (condition A). However, when the purine analogues were used instead of ATP and GTP, the spermine-induced stimulation of RNA synthesis was reduced considerably (condition B). The difference in the extent of stimulation observed between conditions A and B accounted for approximately 35% of the total polyamine induced increase in RNA synthesis.[41]

It is of interest to note that spermine stimulated phosphorylation of S_2 (120 kdaltons), S_3 (65 kdaltons), S_5 (25 kdaltons), and S_6 (19.5 kdaltons) subunits of RNA polymerase I 50-, 98-, 626-, and 28-fold, respectively.[46] It is likely that the dramatic effect on the phosphorylation of 65-kdalton subunit may be due to highly dephosphorylated state of this polypeptide obtained from this preparation. The relatively smaller degree of stimulation of phosphorylation cf these subunits derived from another preparation lends credence to the above possibility.[40] Nevertheless, the overall stimulation of phosphorylation of RNA polymerase I subunits by spermine is impressive. Although the functions of these polypeptides in rDNA transcription are not known, it is plausible that their phosphorylation by protein kinase NII and the phosphorylation at additional sites by physiological concentrations of polyamines may be functionally relevant.

Kuehn and colleagues[47-49] have demonstrated that in the slime mold, *Physarum polycephalum*, ODC is phosphorylated by a polyamine-dependent protein kinase and the phosphorylated ODC stimulates transcription of rDNA. There are also claims that ODC is a regulatory subunit of RNA polymerase I and may play a role in the synthesis of RNA by acting as an initiation factor for RNA polymerase I.[50-52] Although this kinase appears to be functionally related to protein kinase NII, it consists of one polypeptide of M_r 26,000.[47] In this respect, it resembles nuclear kinase NI which is known to contain only one polypeptide of M_r 28,000 (see Reference 32). A similar protein kinase and a substrate protein of M_r 70,000 have also been identified in Ehrlich ascites tumor cells,[53] bovine spermatozoa,[54] and rat liver.[49] ODC appears to be inactivated by phosphorylation by the kinase in the slime mold.[49] Despite repeated trials, we have been unable to identify a 26 to 28 kdalton polyamine-mediated protein kinase in rat liver nuclei. Incubation of mouse kidney ODC with protein kinase NII from liver did not result in phosphorylation or inactivation of ODC.[37] Further, removal of ODC with a specific antiserum did not abolish the ability to stimulate RNA polymerase I. Antibodies to RNA polymerase I did not interact with ODC and antibodies to ODC did not react with RNA polymerase I.[37] These results indicate that mammalian ODC activity is not regulated by phosphorylation by nuclear protein kinase NII and that the ability of impure ODC preparations to activate RNA polymerase I is due to a contaminating protein other than ODC. Clearly, further studies are required to establish the effect of phosphorylation of RNA polymerase I and/or transcription factor(s) on transcription of rDNA.

IV. EFFECT OF POLYAMINES ON TRANSCRIPTION OF SPECIFIC GENES

Studies by Burnick et al.[55] showed that the ATP analogue 5′-adenylyl imidodiphosphate inhibits initiation of transcription of adenovirus 2 DNA and human globin DNA in whole cell extract. Since this analogue did not inhibit RNA polymerase III-directed transcription, it was concluded that the availability of the β-γ bond of ATP is essential for accurate initiation of transcription by RNA polymerase II. Subsequent studies in the same laboratory demonstrated that another ATP analogue, 5,6-dichloro-1-β-1-ribofuranosylbenzimidazole (DRB) and its mono- and triphosphate derivatives also inhibit accurate transcription of human E-globin gene as well as adenovirus transcription from early region E1a and the major late promoter. DRB was shown to inhibit preferentially casein kinase type II from HeLa cell and calf thymus, suggesting that this kinase may be involved in the RNA polymerase II-directed transcription. Partial restoration of transcription by exogenous purified calf thymus casein kinase II following DRB-induced inhibition supports this conclusion.

Recently, we have demonstrated that DRB or its nucleotide derivative could inhibit transcription of a cloned mouse metallothionein gene (MT-1) in a fractionated nuclear extract containing RNA polymerase II and essential transcription factors and that this inhibition could be reversed by the addition of purified nuclear protein kinase NII.[56] cAMP-dependent protein kinase or protein kinase NI did not markedly restore the DRB-inhibited transcription. Spermine at concentrations as low as 100 μM, enhanced the ability of the exogenous protein kinase NII to restore transcription following DRB-induced inhibition. Although spermine alone stimulated transcription at least fivefold prior to addition of the adenosine analogue, the polyamine had no effect when added alone after addition of the inhibitor. These data indicate that the polyamine mediates initiation of RNA polymerase II-directed transcription via protein kinase NII. It is not known whether (1) the protein kinase exerts its effect by phosphorylating RNA polymerase II or an essential transcription factor(s) and (2) phosphorylation at additional sites on these proteins plays a key role in the initiation of transcription by RNA polymerase II.

ACKNOWLEDGMENTS

The investigations in the authors' laboratory were supported by the U.S. Public Health Services Grants CA25078 and CA31894 (awarded to S. T. Jacob). The authors would like to thank Ms. Carol Buck for assistance in the preparation of this article.

REFERENCES

1. **Bachrach, U.,** *Function of Naturally Occurring Polyamines,* Academic Press, New York, 1973.
2. **Cohen, S. S.,** *Introduction to the Polyamines,* Prentice-Hall, Englewood Cliffs, N.J., 1971.
3. **Raina, A. and Jänne, J.,** Polyamines and the accumulation of RNA in mammalian systems, *Fed. Proc. Fed. Am. Soc. Exp. Biol.,* 29, 1568, 1970.
4. **Russell, D. H.,** Polyamines in growth — normal and neoplastic, in *Polyamines in Normal and Neoplastic Growth,* Russell, D. H., Ed., Raven Press, New York, 1973, 1.
5. **Doerfler, W., Zillig, W., Fuchs, E., and Abers, M.,** Studies on the biosynthesis of proteins. V. The function of nucleic acids in the incorporation of amino acids into proteins in a cell free system in *Escherichia coli, Z. Physiol. Chem.,* 330, 96, 1962.
6. **Karpetsky, T. P., Hieter, P. A., Frank, J. J., and Levy, C. C.,** Polyamines, ribonucleases, and the stability of RNA, *Mol. Cell. Biochem.,* 17, 89, 1977.
7. **Krakow, J. S.,** Ribonucleic acid polymerase of *Azotobacter vinelandii.* III. Effect of polyamines, *Biochim. Biophys. Acta,* 72, 566, 1963.
8. **Fox, C. F. and Weiss, S. B.,** Enzymatic synthesis of ribonucleic acid. II. Properties of the deoxyribonucleic acid-primed reaction with *Micrococcus lysodeikticus* ribonucleic acid polymerase, *J. Biol. Chem.,* 239, 175, 1964.
9. **Abraham, K. A.,** Studies on DNA-dependent RNA polymerase from *E. coli.* I. The mechanism of polyamine induced stimulation of enzyme activity, *Eur. J. Biochem.,* 5, 143, 1968.
10. **Petersen, E. E., Kroger, H., and Hagen, U.,** The influence of spermidine on the reaction of RNA nucleotidyltransferase, *Biochem. Biophys. Acta,* 161, 325, 1968.
11. **Gumport, R. G. and Weiss, S. B.,** Transcriptional and sedimentation properties of ribonucleic acid polymerase from *Micrococcus lysodeikticus, Biochemistry,* 8, 3618, 1969.
12. **Fox, C. F., Gumport, R. I., and Weiss, S. B.,** The enzymatic synthesis of ribonucleic acid. V. The interaction of ribonucleic acid polymerase with nucleic acids, *J. Biol. Chem.,* 240, 2101, 1965.
13. **So, A. G., Davie, E. W., Epstein, R., and Tissieres, A.,** Effects of cations on DNA-dependent RNA polymerase, *Proc. Natl. Acad. Sci. U.S.A.,* 58, 1739, 1967.
14. **Fuchs, E., Millette, R. L., Zillig, W., and Walter, G.,** Influence of salts on RNA synthesis by DNA-dependent RNA-polymerase from *Escherichia coli, Eur. J. Biochem.,* 3, 183, 1967.
15. **Herbst, E. J., Byus, C. V., and Nuss, D. L.,** The stimulation of RNA synthesis by spermidine: studies with Drosophila larvae and RNA polymerase, in *Polyamines in Normal and Neoplastic Growth,* Russell, D. H., Ed., Raven Press, New York, 1973, 71.
16. **Ballard, P. L. and Williams-Ashman, H. G.,** Isolation of a soluble RNA polymerase from rat testis, *Nature (London),* 203, 150, 1964.
17. **Ballard, P. L. and Williams-Ashman, H. G.,** Isolation and properties of a testicular ribonucleic acid polymerase, *J. Biol. Chem.,* 241, 1602, 1966.
18. **Singh, V. K. and Sung, S. B.,** Effect of spermidine on DNA-dependent RNA polymerases from brain cell nuclei, *J. Neurochem.,* 19, 2885, 1972.
19. **Jänne, O., Bardin, C. W., and Jacob, S. T.,** DNA-dependent RNA polymerases I and II from kidney. Effect of polyamines on the *in vitro* transcription of DNA and chromatin, *Biochemistry,* 14, 3589, 1975.
20. **Jacob, S. T. and Rose, K. M.,** Stimulation of RNA polymerases I, II, and III from rat liver by spermine and specific inhibition of RNA polymerase I by higher spermine concentrations, *Biochim. Biophys. Acta,* 425, 125, 1976.
21. **Moruzzi, G., Barbiroli, B., Moruzzi, M. S., and Tadolini, B.,** The effect of spermine on transcription of mammalian deoxyribonucleic acid-dependent ribonucleic acid polymerase, *Biochem. J.,* 146, 697, 1975.
22. **Rose, K. M., Ruch, P. A., Morris, H. P., and Jacob, S. T.,** RNA polymerases from a rat hepatoma. Partial purification and comparison of properties with corresponding liver enzymes, *Biochim. Biophys. Acta,* 432, 60, 1976.
23. **Goldberg, M. I., Perriard, J.-C., and Rutter, W. J.,** A protein cofactor that stimulates the activity of DNA-dependent RNA polymerase I on double-stranded DNA, *Biochemistry,* 16, 1648, 1977.
24. **Mandel, J.-L. and Chambon, P.,** Animal DNA-dependent RNA polymerases. Studies on the reaction parameters of transcription *in vitro* of Simian virus 40 DNA by mammalian RNA polymerases A and B, *Eur. J. Biochem.,* 41, 367, 1974.
25. **Chan, J. Y. H., Loor, R. M., and Wang, T. Y.,** Transcription *in vitro* of Ehrlich ascites tumor DNA and chromatin by purified homologous RNA polymerase II (or B), *Arch. Biochem. Biophys.,* 173, 564, 1976.
26. **Blair, D. G. R.,** Stimulation of RNA polymerases from mouse spleen by polyamines and its modification by ammonium sulfate, *Int. J. Biochem.,* 16, 747, 1984.
27. **Stirpe, F. and Novello, F.,** Factors affecting the activity of ribonucleic acid polymerase solubilized from rat liver nuclei. Effect on ionic strength, spermine and divalent ions with native and denatured deoxyribonucleic acid, *Eur. J. Biochem.,* 15, 505, 1970.

28. **Jacob, S. T.,** Mammalian RNA polymerases, *Prog. Nucleic Acid Res. Mol. Biol.*, 13, 93, 1973.
29. **Barbiroli, B., Masotti, L., Moruzzi, M. S., Monti, M. G., and Moruzzi, G.,** *In vitro* stimulation of transcription by spermine and its interaction with rat liver genome, in *Advances in Polyamine Research,* Campbell, R. A., Morris, D., Bartos, D., Daves, G. D., and Bartos, F., Eds., Raven Press, New York, 1978, 217.
30. **Rose, K. M. and Jacob, S. T.,** Phosphorylation of DNA-dependent RNA polymerases, in *Enzyme Regulation by Reversible Phosphorylation — Further Advances,* Cohen, P., Ed., Elsevier, Amsterdam, 1984, 209.
31. **Desjardins, P. R., Lue, P. F., Liew, C. C., and Gornall, A. G.,** Purification and properties of rat liver nuclear protein kinases, *Can. J. Biochem.*, 50, 1249, 1972.
32. **Hathaway, G. M. and Traugh, J. A.,** Casein kinases — multipotential protein kinases, *Curr. Topics Cell. Regul.*, 21, 109, 1982.
33. **Dahmus, M. E.,** Phosphorylation of eukaryotic DNA-dependent RNA polymerase, *J. Biol. Chem.*, 256, 3332, 1981.
34. **Rose, K. M., Stetler, D. A., and Jacob, S. T.,** Protein kinase activity of RNA polymerase I purified from a rat hepatoma: probable function of Mr 42,000 and 24,600 polypeptides, *Proc. Natl. Acad. Sci. U.S.A.*, 78, 2833, 1981.
35. **Duceman, B. W., Rose, K. M., and Jacob, S. T.,** Activation of purified hepatoma RNA polymerase I by homologous protein kinase NII, *J. Biol. Chem.*, 246, 10755, 1981.
36. **Hoppe, J., Erdmann, H., and Wagner, K. G.,** The acidic peptide-specific type II protein kinases from the nucleus and the cytosol of porcine liver are distinct, *Arch. Biochem. Biophys.*, 245, 504, 1986.
37. **Seely, J. E., Stetler, D. A., Jacob, S. T., and Pegg, A. E.,** Absence of inactivation or phosphorylation of ornithine decarboxylase by nuclear protein kinase NII and of immunological cross-reactivity between RNA polymerase I and ornithine decarboxylase, *Biochem. Biophys. Res. Commun.*, 120, 219, 1984.
38. **Meggio, F., Flamigni, F., Caldarera, C.-M., Guarnieri, C., and Pinna, L. A.,** Phosphorylation of rat heart ornithine dearboxylase by type 2 casein kinase, *Biochem. Biophys. Res. Commun.*, 122, 997, 1984.
39. **Rose, K. M., Bell, L. E., Siefken, D. A., and Jacob, S. T.,** A heparin-sensitive nuclear protein kinase, *J. Biol. Chem.*, 256, 7468, 1981.
40. **Jacob, S. T., Duceman, B. W., and Rose, K. M.,** Spermine-mediated phosphorylation of RNA polymerase I and its effect on transcription, *Med. Biol.*, 59, 381, 1981.
41. **Jacob, S. T., Rose, K. M., and Canellakis, Z. N.,** *Adv. Polyamine Res.*, 4, 631, 1983.
42. **Ahmed, K., Wilson, M. J., and Goueli, S. A.,** Effects of polyamines on prostatic chromatin and non-histone-protein-associated protein kinase reactions, *Biochem. J.*, 176, 739, 1978.
43. **Hara, T. and Endo, H.,** Polyamines alter the substrate preference of nuclear protein kinase NII, *Biochemistry,* 2, 2632, 1982.
44. **Ahmed, K., Goueli, A., and Williams-Ashman, H. G.,** Characteristics of polyamine stimulation of cyclic nucleotide-independent protein kinase reactions, *Biochem. J.*, 232, 767, 1985.
45. **Yount, R. G., Babcock, D., Ballantyne, W., and Ojala, D.,** Adenlyl imidodiphosphate, an adenosine triphosphate analog containing a P-N-P linkage, *Biochemistry,* 10, 2484, 1971.
46. **Jacob, S. T., Stetler, D. A., and Rose, K. M.,** in *Protein Synthesis,* Abraham, A. K., Eikhom, T. S., and Pryme, I. F., Humana Press, Clifton, N.J., 1983, 409.
47. **Daniels, G. R., Atmar, V. J., and Kuehn, G. D.,** Polyamine-activated protein kinase reaction from nuclei and nucleoli of *Physarum polycephalum* which phosphorylates a unique Mr 70,000 nonhistone protein, *Biochemistry,* 20, 2525, 1981.
48. **Atmar, V. J. and Kuehn, G. D.,** Phosphorylation of ornithine decarboxylase by a polyamine-dependent protein kinase, *Proc. Natl. Acad. Sci. U.S.A.*, 78, 5518, 1981.
49. **Kuehn, G. D. and Atmar, V. J.,** Posttranslational control of ornithine decarboxylase by polyamine-dependent protein kinase, *Fed. Proc. Fed. Am. Soc. Exp. Biol.*, 41, 3078, 1982.
50. **Manen, C. A. and Russell, D. H.,** Ornithine decarboxylase may function as an initiation factor for RNA polymerase I, *Science,* 195, 505, 1977.
51. **Russell, D. H.,** Microinjection of purified ornithine decarboxylase into Xenopus oocytes selectively stimulates ribosoma RNA synthesis, *Proc. Natl. Acad. Sci. U.S.A.*, 80, 1318, 1983.
52. **Russell, D. H.,** Ornithine decarboxylase may be a multifunctional protein, *Adv. Enzyme Regul.*, 21, 201, 1983.
53. **Seker, V., Atmar, V. J., Krim, M., and Kuehn, G. D.,** Interferon induction of polyamine-dependent protein kinase activity in Ehrlich ascites tumor cells, *Biochem. Biophys. Res. Commun.*, 106, 305, 1982.
54. **Atmar, V. J., Kuehn, G. D., and Casillas, E. R.,** A polyamine-dependent protein kinase from bovine epididymal spermatozoa, *J. Biol. Chem.*, 256, 8275, 1981.
55. **Burnick, D., Zandomeni, R., Ackerman, S., and Weinmann, R.,** Mechanism of RNA polymerase II-specific initiation of transcription in vitro: ATP requirement and uncapped runoff transcripts, *Cell,* 29, 877, 1982.
56. **Webb, M. L. and Jacob, S. T.,** in preparation.

Chapter 10

EFFECTS OF POLYAMINES AND ANTIBIOTICS ON THE STRUCTURE AND FUNCTION OF RIBOSOMES

Israel D. Algranati and Sara H. Goldemberg

TABLE OF CONTENTS

I. INTRODUCTION

Polyamines are organic bases found at rather high concentrations in all living cells. These substances behave as strongly cationic at physiological pH and together with Mg^{++} they are the main polycations occurring inside the cells.[1-4] The most common polyamines are putrescine, spermidine, and spermine. Prokaryotic cells usually contain only putrescine and spermidine while in eukaryotes the three above-mentioned organic bases are present. Several other less common polyamines have been found in thermophilic organisms,[5-10] methanogenic bacteria,[11] algae,[12] molds,[13] and some plants.[14]

The intracellular amount of each individual polyamine varies widely in different cell types, but the total concentration of polyamines in bacteria (30 to 40 m*M*) and in mammalian cells (2 to 5 m*M*) is of the same order or even higher than Mg^{++} levels.[4,15-18] This fact, together with the polybasic character of these substances and the aliphatic and flexible nature of their structure can explain why polyamines are able to interact with most of the intracellular polyanions such as nucleic acids, phospholipids, and proteins. Although some of these interactions are unspecific and can be related to the positive charges of polyamines, others are rather specific and cannot be accomplished by physiological levels of inorganic cations.

The absolute requirement of polyamines for growth of several microorganisms has been shown as early as 1948 by Herbst and Snell[19] in *Haemophilus parainfluenzae* and a few years later for some strains of *Neisseria perflava*[20] and certain nutritional mutants of *Aspergillus nidulans* obtained by UV irradiation.[21] More recently, a great number of experiments carried out mainly with bacterial polyamine-auxotrophic mutants or specific inhibitors of polyamine biosynthesis in mammalian cells have indicated the essential roles of the organic cations in different steps of DNA replication[22,23] and supercoiling,[24] RNA transcription and processing,[25-27] and protein synthesis.[28-30]

This chapter will deal with the effects of polyamines on the translation machinery, and more specifically on the structure and function of ribosomal particles mainly in bacteria. In addition the effects of several antibiotics on protein synthesis of wild-type bacteria or polyamine-deficient mutants will also be described. Various comprehensive reviews on different aspects of polyamine functions in protein synthesis have been published.[4,31-33]

II. PROTEIN SYNTHESIS IN POLYAMINE-DEFICIENT MUTANTS OF *ESCHERICHIA COLI*

Many reports have described the stimulation of different steps of polypeptide synthesis by the addition of polyamines to bacterial and eukaryotic cell-free systems.[32,34-40] The isolation of *Escherichia coli* mutants blocked in the biosynthesis of polyamines[41-45] has provided a new approach to investigate in vivo and in vitro the effects of polyamine intracellular depletion and the subsequent restoration of normal levels of these substances.

Young and Srinivasan[46] have reported that the addition of polyamines to previously starved bacteria caused an almost immediate increase of protein synthesis and, after a period of 1 to 2 hr, the stimulation of RNA and DNA synthesis. The same authors[47] have found that bacteriophage f_2 grew poorly and that its protein synthesis was markedly reduced under conditions of polyamine depletion. Although these results might be explained in part by a decreased adsorption of phages to polyamine-deficient bacteria as observed by Tabor,[48] they strongly support the idea of a direct role of polyamines in protein synthesis. This conclusion has been confirmed by our in vivo and in vitro studies with polyamine-auxotrophic mutants.[28,49] Previously Morris and co-workers[50,51] had reported that polyamine starvation provoked reduced elongation rates of both polypeptide and RNA chains. Since these in vivo experiments generally cannot distinguish the polyamine effects on translation from those on transcription, we have investigated the polypeptide synthesis directed by poly(U) or natural mRNA in cell-free systems derived from polyamine-depleted and supplemented bacteria. In this way it was possible to study the complete translation process and the structure and functioning of its machinery independently of RNA synthesis.

Recently it has been reported that supplementation of polyamines to previously depleted bacteria was not only able to increase total protein synthesis, but also to induce the preferential in vivo synthesis of some particular proteins such as the ββ′ subunits of RNA polymerase[52] and the polyamine-induced protein.[53] The latter, designated in this way because its synthesis could be induced by polyamines, has been shown to stimulate the in vitro synthesis of certain proteins in the presence of spermidine.[53,54]

III. STRUCTURE AND FUNCTION OF RIBOSOMAL SUBUNITS IN POLYAMINE-REQUIRING BACTERIA

A number of studies dealing with the ribosome cycle during bacterial protein synthesis and the dissociation and association of ribosomal particles have indicated that polyamines are involved in the interactions occurring between 30S and 50S ribosomal subunits.[55-61]

Experiments carried out with ribosomes of *Bacillus stearothermophilus* have shown that spermidine at physiological concentrations was able to promote the in vitro association of both subparticles in a single or two-step reaction.[62-64] During the first step the polyamine was bound to 30S particles giving a complex which was subsequently able to form monomers when 50S subunits were added. Although more spermidine could be bound directly to 50S particles than to 30S subunits, the resulting 50S-polyamine complexes were not able to associate to the small subparticles. On the other hand, two different types of binding sites for polyamines were detected in the 30S particle and only one of them gave an active complex suitable for the formation of 30S-50S couples.[65] The binding of polyamine might stabilize the conformation of the 30S subunit to render it more active for association. Many other studies have confirmed that polyamines, as well as Mg^{++}, are able to promote the association of ribosomal subparticles.[66,67]

The isolation of complexes between polyamines and the macromolecules involved in protein synthesis has indicated that ribosomal particles and different species of tRNA contain polyamines in variable amounts depending upon the experimental conditions used.[57,68,69] However, these findings do not necessarily prove that the above-mentioned complexes indeed occur in vivo, because the organic cations can easily change their distribution during or after cell disruption. In order to overcome this shortcoming, the in vivo distribution of intracellular polyamine has been studied using a nondestructive technique such as nuclear magnetic resonance spectroscopy.[70,71] The results obtained so far have shown that about 50% of the intracellular putrescine was free and the rest was bound only to ribosomes and membrane fractions.

In order to demonstrate the involvement of polyamines in the equilibrium occurring between ribosomal particles and in the formation or assembly of ribosomal subunits, several experiments were performed with *Escherichia coli* mutants blocked in the biosynthesis of polyamines and grown in the absence and presence of these substances. The analyses of ribosomal profiles in cell extracts obtained after slow cooling of growing cultures have shown that polyamine-starved bacteria contained a low amount of 70S monomers with most of the ribosomal particles accumulated as subunits. In contrast, after restoration of normal intracellular levels of polyamines the ribosomal subparticles interacted with each other to give an appreciable amount of 70S ribosomes.[72-74] These results showed that the equilibrium between 70S monomers and ribosomal subunits was strongly shifted towards the dissociation in polyamine-starved cells. Moreover, it was possible to demonstrate that bacteria depleted of polyamines contained a high proportion of defective small ribosomal subunits with a sedimentation coefficient lower than 30S, a decreased affinity for 50S particles, and a low activity for protein synthesis.[49,75] These findings, consistent with the idea that polyamines have a physiological role in the biogenesis or the correct assembly of 30S particles, have been confirmed and extended by Igarashi and co-workers.[76-78] These authors worked with reconstituted 30S subunits formed by different combinations of ribosomal fractions obtained from polyamine-starved or supplemented bacteria and concluded that the small ribosomal particles of polyamine-depleted *E. coli* contained subnormal amounts of S_1 and some other split proteins and undermethylated 16S RNA with a decreased methylation of adenine residues located near the 3′ end of the ribosomal RNA. Igarashi et al.[78] proposed that polyamines are able to stabilize the 23S core particle in the conformation more active for a correct methylation of its 16S RNA. This results in the subsequent stimulation of the binding of split proteins with the concomitant normal assembly of the 30S ribosomal subunit. Although earlier reports by Morris and Jorstad[79] have indicated that polyamine starvation did not affect the methylation of tRNA and rRNA, these findings are not consistent with the above-described results showing the undermethylation of 16S RNA in polyamine-depleted bacteria. On the other hand, the stimulation of bacterial ribosomal RNA enzymatic methylation by polyamines has also been demonstrated.[80]

The stabilization of small ribosomal particles of *Bacillus subtilis* by polyamines has been shown to be due to the inhibition of the 16S RNA degradation.[81]

IV. POLYPEPTIDE SYNTHESIS IN CELL-FREE SYSTEMS DERIVED FROM POLYAMINE-STARVED AND SUPPLEMENTED BACTERIA

It has been shown that the rate of in vivo protein synthesis in polyamine-requiring bacteria supplemented with these substances was severalfold higher than in putrescine-starved cells.[49,82] The polypeptide synthesis of cell-free systems from bacteria with normal intracellular levels of polyamines was also two- to threefold higher than in extracts of putrescine-depleted *E. coli*.[72,74,75,83] Moreover, the in vitro studies of different steps of the translation process have indicated that practically every single step of polypeptide synthesis could be stimulated by polyamines, and that these organic bases were able to partially replace the Mg^{++} requirements of the reactions.[4,31,32]

When polypeptide synthesis was measured in reconstituted systems combining soluble fraction (S_{150}) and ribosomes or purified ribosomal subparticles from polyamine-starved and unstarved cells, it was clear that the amount of polypeptide synthesized depended exclusively on the source of ribosomes and particularly of the 30S subunits. The amino acid incorporation into polypeptides was about the same with both types of supernatant fractions or 50S subparticles, but decreased markedly when 30S subunits from polyamine-depleted bacteria were used instead of the small subparticles obtained from polyamine-supplemented cells.[28]

Detailed studies of translation directed by a natural mRNA and carried out with cell-free extracts from polyamine-deficient mutants of *E. coli* grown in the absence and presence of putrescine have demonstrated that systems obtained from polyamine-supplemented cells showed an increased capacity for polypeptide synthesis initiation as measured by the formation of the initiation complex[82,83] (Table 1). Since some of the 30S ribosomal subunits present in extracts from bacteria grown in the absence of polyamines are defective, these particles could be unable to form active initiation complexes and only the normal small subunits would be involved in protein synthesis. Therefore, in the cell-free system from polyamine-depleted bacteria, a relative decreased number of growing polypeptide chains should be expected. This conclusion was in full agreement with measurements of peptidyl-puromycin formation and the corresponding calculated percentages of active ribosomes in extracts from polyamine-starved and supplemented cells[82] (Table 1).

The rates of elongation and termination of polypeptide synthesis measured by indirect methods were very similar in the systems derived both from polyamine-starved and unstarved

Table 1
INITIATION COMPLEX FORMATION AND PERCENTAGES OF ACTIVE RIBOSOMES IN CELL-FREE SYSTEMS FROM POLYAMINE-DEPLETED AND SUPPLEMENTED BACTERIA PROGRAMMED BY MS_2 PHAGE RNA

System	Initiation complex[a]	Active ribosomes (%)
Polyamine-starved bacteria	7.9	7.6
Polyamine-supplemented bacteria	16.3	19.3

[a] f[^{3}H]met-tRNA-ribosome complex (pmol/mg ribosomes).

From Algranati, I. D. and Goldemberg, S. H., *Biochem. Biophys. Res. Commun.*, 103, 8, 1981. With permission.

bacteria.[82] However, other reports have indicated that polyamines affect polypeptide chain elongation and termination.[84-86]

The results described above, in accordance with Igarashi's findings,[76,78] have shown that polyamines participate in the normal biogenesis, assembly, and stabilization of the 30S particle structure more active for protein synthesis.

V. EFFECTS OF ANTIBIOTICS ON PROTEIN SYNTHESIS AND RIBOSOMAL PARTICLES IN WILD-TYPE BACTERIA AND POLYAMINE-DEFICIENT MUTANTS

Most of the known antibiotics act on different steps of protein biosynthesis. Streptomycin and other aminoglycosides are strongly cationic molecules which bind to the ribosomes, acidic constituents of the cell.[87] Aminoglycosides bear many structural resemblances with polyamines; both are basic substances which interact with negatively charged nucleic acids. These similarities led to the suggestion of a common transport system for streptomycin and polyamines in *E. coli*.[88] Höltje[88] proposed that the antibiotic initial uptake and binding to the ribosomes resulted in the induction of a polyamine transport system which could be utilized by streptomycin itself, thus enhancing the aminoglycoside uptake. However, Campbell and Kadner[89] found several conditions in which the transport of both substrates was affected in opposite ways. They reported that the uptake of putrescine was much less sensitive to anaerobiosis and that some mutants resistant to low levels of streptomycin, as well as putrescine-depleted polyamine auxotrophs, showed decreased initial rates of antibiotic uptake and increased putrescine transport. Moreover, polyamines were unable to induce the uptake of streptomycin[88] or gentamicin.[90] These and other findings argue against the identity of the polyamine and aminoglycoside transport system.[91]

As mentioned above, polyamines are also able to affect the association of ribosomal subunits. In a comparable fashion, streptomycin, which induces a reversible modification in 30S subunit assembly,[92] promotes the association of ribosomal subparticles.[93] It has been shown that the dissociation factor activity is inhibited in vitro by streptomycin, kanamycin, spectinomycin, and neomycin, while the tetracyclines promote ribosomal dissociation. Only antibiotics which interfere with the 30S subunit functions affect the dissociation activity.[94]

The endogenous polyamine content of some *E. coli* mutants seems to modulate the aminoglycoside action on polypeptide synthesis both in vivo and in vitro.[95] Working with polyamine auxotrophs grown in the presence or absence of putrescine, streptomycin caused a marked inhibition of protein synthesis in polyamine-supplemented cells while putrescine-depleted bacteria behaved as phenotypically resistant to the antibiotic (Table 2). Similar

Table 2
STREPTOMYCIN INHIBITION OF IN VIVO AND IN VITRO PROTEIN SYNTHESIS

Conditions	Putrescine[a]	Inhibition (%)
In vivo	−	11.2
	+	71.2
In vitro	−	17.6
	+	51.4

[a] Bacteria and cell-free systems were obtained from cultures in the absence (−) or presence (+) of putrescine.

From Goldemberg, S. H., Fernandez-Velasco, J. G., and Algranati, I. D., *FEBS Lett.*, 142, 275, 1982. With permission.

effects were observed with neomycin, kanamycin, and kasugamycin; spectinomycin, gentamicin, and tetracycline (a nonaminoglycoside antibiotic) showed no differential action dependent on the polyamine content. The aminoglycosides are also known to affect the fidelity of translation, both in vivo and in vitro.[87,96] There seemed to be a relationship between polyamine effect on antibiotic activity and miscoding,[95] since streptomycin, neomycin, and kanamycin, which increase mistranslation, required polyamine for inhibitory activity. Spectinomycin, which does not interfere with fidelity of translation, had the same effect in the presence or absence of putrescine. Gentamicin, even though it increases misreading, was equally inhibitory either in the presence or absence of putrescine, probably because of the multiple sites of action of this antibiotic on both ribosomal particles.[97,98] On the contrary, kasugamycin, which does not affect accuracy of translation,[87,96] was dependent on polyamine for an inhibitory effect. This might be explained by the fact that in polyamine-depleted bacteria the 16S ribosomal RNA is undermethylated,[76] a condition also observed in some kasugamycin-resistant cells.[99]

Under polyamine starvation the defective 30S ribosomal particles formed do not seem to be appropriate targets for streptomycin. Only in polyamine-containing bacteria does the antibiotic promote the small subunit deformation which could be visualized by sucrose gradient centrifugation.[100,101] The binding of dihydrostreptomycin to ribosomes is enhanced by spermine, spermidine, or putrescine at low concentrations of Mg^{++}, but critical levels of this ion, specific for each polyamine, are necessary to observe the stimulatory effect.[102] On the contrary, spermine, and spermidine to a lesser extent, inhibit the binding of erythromycin to ribosomes independently of the Mg^{++} concentration.[103]

The binding of radioactive dihydrostreptomycin to ribosomes of putrescine-starved and unstarved cells of a polyamine-auxotrophic mutant of *E. coli* revealed different types of interaction.[101] Specific binding of the antibiotic with high and low affinity was observed in bacteria cultivated in the presence of putrescine, but only low affinity interaction was evident in polyamine-depleted cells. Streptomycin-resistant strains showed nonspecific binding independently of polyamine content.

It is interesting to remark that some of the antibiotics which interact with the small ribosomal particle contain polyamines or polyamine analogues in their structure. The best known example is the large group of aminoglycosides, with so many similarities to polyamines that Umezawa[104] has called them "antibacterial polyamine compounds".

Negamcycin is a peptide-like antibiotic which may be considered as a hydroxy putrescine substituted in position 1.[105] It inhibits protein synthesis termination and causes miscoding probably acting on the 30S subunit.

Many substances with different antibiotic or antitumoral activities contain spermidine moieties. Edeine is terminated by a spermidine molecule (edeine A_1, used in most studies) or guanylspermidine (edeine B_1). Edeine A_1 interferes with initiation of protein synthesis both in cell-free extracts of *E. coli* and in intact eukaryotic cells.[96,106] Though it binds to both ribosomal subunits, it has a preferential affinity for the small one and stabilizes the association of subparticles in 70S subunits.[107] The substitution of the aminoterminal group of spermidine in edeine A_1 by a guanidino group in edeine B_1 does not seem to affect antibiotic activity.[108]

Many other spermidine containing compounds with antibacterial or cytotoxic action have been isolated, such as bleomycin A5, tallysomycins A and B, glysperins, glycocinnamoyl-spermidines, and spergualin.[104] More complete studies on the mode of action of these substances may help to understand the participation of the polyamine moiety in the antibiotic or antitumoral effect, which in turn may lead to the design of compounds with better specific applications. It may be also relevant to mention the widespread occurrence of polyamine conjugates in plants, and their potential pharmacological importance.[109,110]

VI. POLYAMINES AND FIDELITY OF TRANSLATION

It is well known that translation occurs in vivo with a high level of accuracy. The direct in vivo measurements of valine misincorporation instead of isoleucine in ovalbumin and rabbit globin[111,112] and the replacement of arginine by cysteine in bacterial flagellin[113] gave an average misreading frequency per codon of 1 to 4 $\times$ 10^{-4}. This value is a correct estimation of the mean frequency of translational errors corresponding to various amino acids at different position of the polypeptide chains. However, recent studies have indicated that the error frequency of misreading or nonsense suppression varied widely depending on the particular codon involved and the messenger context[114,115] and in several cases reached values of about 10^{-3} or even higher.[116]

Many intracellular factors contribute to the high fidelity of in vivo translation. Some of them have been studied in cell-free systems using poly(U) as messenger and the relative incorporation of leucine to phenylalanine into polypeptides as a measurement of the error frequency. In this way, it was possible to demonstrate that fidelity of translation depended on several experimental conditions such as temperature, pH, and amino acid levels, as well as on the concentrations of polyvalent cations such as Mg^{++} and polyamines.[117-119] When these organic bases replaced partially the Mg^{++} requirements, the accuracy of protein synthesis was enhanced considerably in bacterial and eukaryotic systems.[117-120] This effect is mainly due to a decrease of tRNA misacylation[121] as well as to an improvement of the discrimination during the initial binding of aminoacyl-tRNA to the ribosomes and the subsequent proofreading steps.[120,122] Based on these results, Jelenc and Kurland[123] have devised a modified in vitro system with a high content of GTP and a mixture of putrescine and spermidine in addition to Mg and Ca salts. Using these conditions, it was possible to increase total protein synthesis and at the same time to reduce the error frequency to the in vivo level.[124,125] Although the addition of polyamines at suboptimal Mg^{++} concentration caused a significant increase of translation accuracy, in some cases, when spermidine was added to a system containing optimal Mg^{++} levels, the error frequency was not altered or even increased.[119,123,126]

Experiments performed in vivo with polyamine-deficient bacteria have shown that in the absence of polyamines polypeptides of high molecular weight were synthesized in smaller amounts than in polyamine-supplemented cells.[95] These results have suggested that the absence of polyamines provoked a more frequent premature termination when certain codons

Table 3
FREQUENCY OF IN VIVO MISREADING IN MS_2 COAT PROTEIN

Strain	Misreading	Error frequency[a] −P	Error frequency[a] +P
Escherichia coli LM27 (speB, speC)	His incorporation	0.10	0.08
	Lys for Asn substitution	7.0	3.4
E. coli LM53 (speA, speB, speC, speD)	His incorporation	0.25	0.22
	Lys for Asn substitution	4.1	2.5

[a] Moles of amino acids misincorporated per 100 moles of coat protein. −P and +P correspond to polyamine-starved and unstarved cultures.

From McMurry, L. M. and Algranati, I. D., *Eur. J. Biochem.*, 155, 383, 1986. With permission.

corresponding to various amino acids were misread as termination signals. On the other hand it has also been reported that polyamines stimulated the suppression of termination codons on several viral messenger RNAs.[127,128] From all these data it was apparent that there was not a complete agreement on whether polyamines were able to enhance or decrease the fidelity of in vivo translation. In order to elucidate this problem, several studies were initiated to measure directly the error frequency of protein synthesis occurring in intact cells. For this purpose, polyamine-deficient mutants cultivated in the absence and presence of putrescine were infected with MS_2 phage and the error of the viral coat protein translation was determined in two different ways: (1) the incorporation of radioactive histidine which is not coded in the coat protein gene[129] and (2) the substitution of an asparagine residue by lysine which produces a modified coat protein with abnormal isoelectric point.[130] These two types of errors were shown to occur at the level of translation since their frequency increased in the presence of streptomycin, known to enhance misreading.[131]

The results obtained with two different polyamine-deficient mutants of *E. coli* have shown that the effect of polyamine depletion on the frequencies of in vivo translation errors depended on the particular codons involved. As shown in Table 3, the errors occurring at low frequency as in the case of histidine incorporation reached the same value of mistranslation level both in polyamine-depleted or supplemented cells. In contrast, for those errors occurring at high frequency, as in the case of lysine for asparagine substitutions, the misreading was about twice as frequent under conditions of polyamine deprivation.[132] The addition of polyamines might presumably contribute to stabilize the U-A base pair in the recognition of the asparagine codon AAU involved,[130] with a concomitant increase of fidelity.

In accordance with the possible dual effect of polyamines enhancing or decreasing the accuracy of in vivo translation under different situations, it is interesting to mention that Tabor et al.[133] have reported that the introduction of a streptomycin-resistance mutation in a polyamine-deficient strain of *E. coli* caused the absolute requirement of polyamines for growth. Since the *rpsL* mutation alters the structure of ribosomal particles in such a way that they translate with higher accuracy,[134] the above-mentioned results might indicate that in this case polyamines would be able to decrease the fidelity of in vivo protein synthesis. On the other hand, the opposite effect might explain the fact that polyamines enhanced the translation efficiency of amber codons in bacteria carrying amber suppressors.[135]

VII. CONCLUDING REMARKS

Several years have gone by since Cohen,[136,137] trying to answer the question "What do the polyamines do?", clearly described the role of these organic cations in maintaining the native and active conformation of tRNA. At the same time, Cohen emphasized that polyamines, like many biochemical substances, most probably play several other roles in living cells. Have we progressed in the understanding of these other functions of polyamines? From the point of view of protein synthesis, we can mention new findings and advance some ideas and promising approaches for future work.

Experiments with polyamine-deficient bacteria have shown that these organic bases are essential for the complete methylation of 16S RNA and the normal biogenesis of active 30S ribosomal particles.[49,75-77] How do polyamines accomplish these roles at the molecular level and why are they so critical in the methylation of 16S RNA and not of other RNAs? We do not yet have definite answers, but the new methods using nondestructive techniques to investigate the in vivo distribution of polyamines and the study with structural analogues of these substances will contribute to our knowledge on the interactions between nucleic acids and the organic cations as well as on the relationship between structure and functions of polyamines.

REFERENCES

1. **Cohen, S. S.,** *Introduction to the Polyamines,* Prentice Hall, Englewood Cliffs, N.J., 1971.
2. **Bachrach, U.,** *Function of Naturally Occurring Polyamines,* Academic Press, New York, 1973.
3. **Tabor, C. W. and Tabor, H.,** Polyamines in microorganisms, in *Microbiol. Rev.*, 49, 81, 1985.
4. **Loftfield, R. B., Eigner, E. A., and Pastuszyn, A.,** Polyamines and protein synthesis, in *Polyamines in Biology and Medicine,* Morris, D. R. and Marton, L. J., Eds., Marcel Dekker, New York, 1981, chap. 11.
5. **De Rosa, M., De Rosa, S., Gambacorta, A., Carteni-Farina, M., and Zappia, V.,** Occurrence and characterization of new polyamines in the extreme termophile *Caldariella acidophila, Biochem. Biophys. Res. Commun.*, 69, 253, 1976.
6. **De Rosa, M., De Rosa, S., Gambacorta, A., Carteni-Farina, M., and Zappia, V.,** The biosynthetic pathway of new polyamines in *Caldariella acidophila, Biochem. J.*, 176, 1, 1978.
7. **Oshima, T.,** Thermine: a new polyamine from an extreme thermophile, *Biochem. Biophys. Res. Commun.*, 63, 1093, 1975.
8. **Oshima, T.,** A new polyamine, thermospermine, 1,12-diamino-4,8-diazadodecane from an extreme thermophile, *J. Biol. Chem.*, 254, 8720, 1979.
9. **Oshima, T.,** A pentaamine is present in an extreme thermophile, *J. Biol. Chem.*, 257, 9913, 1982.
10. **Oshima, T. and Kawahata, S.,** Homocaldopentamine: a new naturally occurring pentaamine, *J. Biochem.*, 93, 1455, 1983.
11. **Scherer, P. and Kneifel, H.,** Distribution of polyamines in methanogenic bacteria, *J. Bacteriol.*, 154, 1315, 1983.
12. **Hamana, K. and Matsuzaki, S.,** Widespread occurrence of norspermidine and norspermine in eukaryotic algae, *J. Biochem.*, 91, 1321, 1982.
13. **Hamana, K. and Matsuzaki, S.,** Unusual polyamines in slime molds *Physarum polycephalum* and *Dictyostelium discoideum, J. Biochem.*, 95, 1105, 1984.
14. **Kuttan, R., Radhakrishnan, A. N., Spande, T., and Witkop, B.,** sym-Homospermidine, a naturally occurring polyamine, *Biochemistry,* 10, 361, 1971.
15. **Herbst, E. J., Weaver, R. H., and Keister, D. L.,** The gram reaction and cell composition: diamines and polyamines, *Arch. Biochem. Biophys.*, 75, 171, 1958.
16. **Lusk, J. E., Williams, R. J. P., and Kennedy, E. P.,** Magnesium and the growth of *Escherichia coli, J. Biol. Chem.*, 243, 2618, 1968.
17. **Lövgren, T. N. E., Pettersson, A., and Loftfield, R. B.,** The mechanism of aminoacylation of transfer ribonucleic acid. The role of magnesium and spermine in the synthesis of isoleucyl-tRNA, *J. Biol. Chem.*, 253, 6702, 1978.
18. **Hurwitz, C. and Rosano, C. L.,** The intracellular concentration of bound and unbound magnesium ions in *Escherichia coli, J. Biol. Chem.*, 242, 3719, 1967.
19. **Herbst, E. J. and Snell, E. E.,** Putrescine as a growth factor for *Haemophilus parainfluenzae, J. Biol. Chem.*, 176, 989, 1948.

20. **Martin, W. H., Jr., Pelczar, M. J., Jr., and Hansen, P. A.,** Putrescine as a growth requirement for *Neisseria, Science,* 116, 483, 1952.
21. **Sneath, P. H. A.,** Putrescine as an essential growth factor for a mutant of *Aspergillus nidulans, Nature,* 175, 818, 1955.
22. **Geiger, L. E. and Morris, D. R.,** Stimulation of deoxyribonucleic acid replication fork movement by spermidine analogs in polyamine-deficient *Escherichia coli, J. Bacteriol.,* 141, 1192, 1980.
23. **Seyfried, C. E. and Morris, D. R.,** Relationship between inhibition of polyamine biosynthesis and DNA replication in activated lymphocytes, *Cancer Res.,* 39, 4861, 1979.
24. **Morris, D. R. and Lockshon, D.,** Polyamines, chromosome conformation and macromolecular synthesis in prokaryotes: a hypothesis, in *Advances in Polyamine Research,* Vol. 3, Caldarera, C., Zappia, V., and Bachrach, U., Eds., Raven Press, New York, 1981, 299.
25. **Abraham, K. A.,** Studies on DNA-dependent RNA polymerase from *Escherichia coli.* I. The mechanism of polyamine induced stimulation of enzyme activity, *Eur. J. Biochem.,* 5, 143, 1968.
26. **Emanuelsson, H., Lowkvist, B., Egyhazi, E., Sjoberg, J., and Heby, O.,** Effects of polyamine depletion on ribonucleotide concentrations and transcriptional activity in early chick embryos, in *Polyamines: Basic and Clinical Aspects,* Imahori, K., Suzuki, F., Suzuki, O., and Bachrach, U., Eds., VNU Science Press, Utrecht, Netherlands, 1985, 119.
27. **Peebles, C. L., Gegenheimer, P., and Abelson, J.,** Precise excision of intervening sequences from precursor tRNAs by a membrane-associated yeast endonuclease, *Cell,* 32, 525, 1983.
28. **Algranati, I. D. and Goldemberg, S. H.,** Polyamines and their role in protein synthesis, *Trends Biochem. Sci.,* 2, 272, 1977.
29. **Atkins, J. F., Lewis, J. B., Anderson, C. W., and Gesteland, R. E.,** Enhanced differential synthesis of proteins in a mammalian cell-free system by addition of polyamines, *J. Biol. Chem.,* 250, 5688, 1975.
30. **Hölttä, E. and Hovi, T.,** Polyamine depletion results in impairment of polyribosome formation and protein synthesis before onset of DNA synthesis in mitogen-activated human lymphocytes, *Eur. J. Biochem.,* 152, 229, 1985.
31. **Igarashi, K. and Hirose, S.,** Mechanism of stimulation of polypeptide synthesis by spermidine, in *Advances in Polyamine Research,* Vol. 1, Campbell, R. A., Morris, D. R., Bartos, D., Daves, G. D., and Bartos, F., Eds., Raven Press, New York, 1978, 243.
32. **Takeda, Y.,** Role of polyamines in aminoacyl transfer-RNA formation, in *Advances in Polyamine Research,* Vol. 1, Campbell, R. A., Morris, D. R., Bartos, D., Daves, G. D., and Bartos, F., Eds., Raven Press, New York, 1978, 255.
33. **Morris, D. R.,** In vivo studies of the roles of putrescine and spermidine in *Escherichia coli,* in *Polyamines in Biology and Medicine,* Morris, D. R. and Marton, L. J., Eds., Marcel Dekker, New York, 1981, chap. 12.
34. **Algranati, I. D. and Lengyel, P.,** Polynucleotide dependent incorporation of amino acids in a cell-free system from thermophilic bacteria, *J. Biol. Chem.,* 241, 1778, 1966.
35. **Changehieri, L. M. and Aronson, J. M.,** Spermidine requirement for *Bacillus thuringiensis* ribosomes in cell-free phenylalanine incorporation, *J. Bacteriol.,* 103, 734, 1970.
36. **Takeda, Y. and Igarashi, K.,** Polyamines and protein synthesis. IV. Stimulation of aminoacyl transfer RNA formation by polyamines, *Biochem. Biophys. Res. Commun.,* 37, 917, 1969.
37. **Igarashi, K., Sugawara, K., Izumi, I., Nagayama, C., and Hirose, S.,** Effect of polyamines on polyphenylalanine synthesis by *Escherichia coli* and rat-liver ribosomes, *Eur. J. Biochem.,* 48, 495, 1974.
38. **Konecki, D., Kramer, G., Pinphanichakarn, P., and Hardesty, B.,** Polyamines are necessary for maximum in vitro synthesis of globin peptide and play a role in chain initiation, *Arch. Biochem. Biophys.,* 169, 192, 1975.
39. **Hunter, A. R., Farrell, P. J., Jackson, R. J., and Hunt, T.,** The role of polyamines in cell-free protein synthesis in the wheat-germ system, *Eur. J. Biochem.,* 75, 149, 1977.
40. **Kramer, G., Odom, O. W., and Hardesty, B.,** Polyamines in eukaryotic peptide initiation, in *Methods in Enzymology,* Vol. 60, Moldave, K. and Grossman, L., Eds., Academic Press, New York, 1979, 555.
41. **Hirschfield, I. N., Rosenfeld, H. J., Leifer, Z., and Maas, W. K.,** Isolation and characterization of a mutant of *Escherichia coli* blocked in the synthesis of putrescine, *J. Bacteriol.,* 101, 725, 1970.
42. **Morris, D. R. and Jorstad, C. M.,** Isolation of conditionally putrescine-deficient mutants of *Escherichia coli, J. Bacteriol.,* 101, 731, 1970.
43. **Maas, W. K., Leifer, Z., and Poindexter, J.,** Studies with mutants blocked in the synthesis of polyamines, *Ann. N.Y. Acad. Sci.,* 171, 957, 1970.
44. **Hafner, E. W., Tabor, C. W., and Tabor, H.,** Mutants of *Escherichia coli* that do not contain 1,4-diaminobutane (putrescine) or spermidine, *J. Biol. Chem.,* 254, 12419, 1979.
45. **Tabor, H., Hafner, E. W., and Tabor, C. W.,** Construction of an *Escherichia coli* strain unable to synthesize putrescine, spermidine or cadaverine: characterization of two genes controlling lysine decarboxylase, *J. Bacteriol.,* 144, 952, 1980.

46. **Young, D. V. and Srinivasan, P. R.,** Regulation of macromolecular synthesis by putrescine in a conditional *Escherichia coli* putrescine auxotroph, *J. Bacteriol.*, 112, 30, 1972.
47. **Young, D. V. and Srinivasan, P. R.,** Growth of ribonucleic acid bacteriophage f_2 in a conditional putrescine auxotroph of *Escherichia coli:* evidence for a polyamine role in translation, *J. Bacteriol.*, 117, 1280, 1974.
48. **Tabor, H.,** Polyamine biosynthesis in *Escherichia coli*: construction of polyamine-deficient mutants, *Med. Biol.*, 59, 389, 1981.
49. **Algranati, I. D., Echandi, G., Garcia-Patrone, M., Gonzalez, N. S., and Goldemberg, S. H.,** Polyamines, equilibrium between ribosomal particles and protein synthesis in bacteria, *Arch. Biol. Med. Exp.*, 10, 49, 1976.
50. **Morris, D. R. and Hansen, M. T.,** Influence of polyamine limitation on the chain growth rates of β-galactosidase and of its messenger ribonucleic acid, *J. Bacteriol.*, 116, 588, 1973.
51. **Jorstad, C. M. and Morris, D. R.,** Polyamine limitation of growth slows the rate of polypeptide chain elongation in *Escherichia coli, J. Bacteriol.*, 119, 857, 1974.
52. **Kazuhiro, M., Ohnishi, R., Hirose, S., and Igarashi, K.,** Necessity of polyamines for maximum *in vivo* synthesis of ββ′ subunits of RNA polymerase, *Biochim. Biophys. Res. Commun.*, 123, 528, 1984.
53. **Kazuhiro, M., Igarashi, K., Kakegawa, T., and Hirose, S.,** Preferential stimulation of the *in vivo* synthesis of a protein by polyamines in *Escherichia coli*: purification and properties of the specific protein, *Biochemistry,* 23, 2679, 1984.
54. **Igarashi, K., Kazuhiro, M., Kakegawa, T., Ohnishi, R., and Hirose, S.,** Preferential stimulation of the *in vivo* synthesis of Mr 62K protein and ββ′ subunits of RNA polymerase by polyamines, in *Polyamines: Basic and Clinical Aspects,* Imahori, K., Suzuki, F., Suzuki, O., and Bachrach, U., Eds., VNU Science Press, Utrecht, Netherlands, 1985, 127.
55. **Garcia-Patrone, M., Gonzalez, N. S., and Algranati, I. D.,** Association factor of ribosomal subunits from *Bacillus stearothermophilus, Proc. Natl. Acad. Sci. U.S.A.*, 68, 2822, 1972.
56. **Hardy, S. J. S. and Turnock, G.,** Stabilization of 70S ribosomes by spermidine, *Nature New Biol.*, 229, 17, 1971.
57. **Cohen, S. S. and Lichtenstein, J.,** Polyamimes and ribosome structure, *J. Biol. Chem.*, 235, 2112, 1960.
58. **Azzam, M. E., Perazzolo, C. A., and Algranati, I. D.,** Polypeptide synthesis in *Escherichia coli* extracts: effect of spermidine on the exchange of ribosomal subunits, *FEBS Lett.*, 21, 165, 1972.
59. **Perazzolo, C. A., Azzam, M. E., and Algranati, I. D.,** Termination of polypeptide synthesis in bacteria and the exchange of ribosomal subunits, *Eur. J. Biochem.*, 34, 467, 1973.
60. **Norton, J. W., Erdmann, V. A., and Herbst, E. J.,** Polyamine-inorganic cation interaction with ribosomes of *Escherichia coli, Biochim. Biophys. Acta,* 155, 293, 1968.
61. **Silman, N., Artman, M., and Engelberg, H.,** Effect of magnesium and spermine on the aggregation of bacterial and mammalian ribosomes, *Biochim. Biophys. Acta,* 103, 231, 1965.
62. **Garcia-Patrone, M., Gonzalez, N. S., and Algranati, I. D.,** Association of ribosomal subunits: mechanism of the reaction induced by association factor, *FEBS Lett.*, 24, 126, 1972.
63. **Garcia-Patrone, M., Gonzalez, N. S., and Algranati, I. D.,** Association of ribosomal subunits. III. Properties of the 30S-50S couples produced *in vitro* by the association factor from *Bacillus stearothermophilus, Biochim. Biophys. Acta,* 299, 452, 1973.
64. **Garcia-Patrone, M., Gonzalez, N. S., and Algranati, I. D.,** Association of ribosomal subunits. IV. Polyamines as active components of the association factor from *Bacillus stearothermophilus, Biochim. Biophys. Acta,* 395, 373, 1975.
65. **Garcia-Patrone, M. and Algranati, I. D.,** Association of ribosomal subunits: studies on the binding of polyamines to 30S and 50S particles, *FEBS Lett.*, 66, 39, 1976.
66. **Sperrazza, J. M. and Spremulli, L. L.,** Quantitation of cation binding to wheat germ ribosomes: influences on subunit association equilibria and ribosome activity, *Nucleic Acids Res.*, 11, 2665, 1983.
67. **Rosano, C. L., Bunce, S. C., and Hurwitz, C.,** Localization of polyamine enhancement of protein synthesis to subcellular components of *Escherichia coli* and *Pseudomonas* sp. strain Kim, *J. Bacteriol.*, 153, 326, 1983.
68. **Tabor, C. W. and Kellog, P. D.,** The effect of isolation conditions on the polyamine content of *Escherichia coli* ribosomes, *J. Biol. Chem.*, 242, 1044, 1967.
69. **Quigley, G. J., Teeter, M. M., and Rich, A.,** Structural analysis of spermine and magnesium ion binding to yeast phenylalanine transfer RNA, *Proc. Natl. Acad. Sci. U.S.A.*, 75, 64, 1978.
70. **Frydman, B., Frydman, R. B., De Los Santos, C., Alonso-Garrido, D., Goldemberg, S. H., and Algranati, I. D.,** Putrescine distribution in *Escherichia coli* studied *in vivo* by C^{13} nuclear magnetic resonance, *Biochim. Biophys. Acta,* 805, 337, 1984.
71. **Frydman, B., Frydman, R. B., De Los Santos, C., Alonso-Garrido, D., Goldemberg, S. H., and Algranati, I. D.,** Putrescine-macromolecule complexes detected *in vivo* by carbon-13 nuclear magnetic resonance, in *Polyamines: Basic and Clinical Aspects,* Imahori, K., Suzuki, F., Suzuki, O., and Bachrach, U., Eds., VNU Science Press, Utrecht, Netherlands, 1985, 187.

72. **Echandi, G. and Algranati, I. D.,** Protein synthesis and ribosomal distribution in a polyamine auxotroph of *Escherichia coli*: studies in cell-free systems, *Biochem. Biophys. Res. Commun.*, 62, 313, 1975.
73. **Algranati, I. D., Echandi, G., Goldemberg, S. H., Cunningham-Rundles, S., and Maas, W. K.,** Ribosomal distribution in a polyamine-auxotroph of *E. coli, J. Bacteriol.*, 124, 1122, 1975.
74. **Goldemberg, S. H. and Algranati, I. D.,** Polyamines and protein synthesis: studies in various polyamine-requiring mutants of *Escherichia coli, Mol. Cell. Biochem.*, 16, 71, 1977.
75. **Echandi, G. and Algranati, I. D.,** Defective 30S ribosomal particles in a polyamine auxotroph of *Escherichia coli, Biochem. Biophys. Res. Commun.*, 67, 1185, 1975.
76. **Igarashi, K., Kashiwagi, K., Kishida, K., Watanabe, Y., Kogo, A., and Hirose, S.,** Defect in the split proteins of 30S ribosomal subunits and under-methylation of 16-S ribosomal RNA in a polyamine-requiring mutant of *Escherichia coli* grown in the absence of polyamines, *Eur. J. Biochem.*, 93, 345, 1979.
77. **Igarashi, K., Kashiwagi, K., Kishida, K., Kakegawa, T., and Hirose, S.,** Decrease in the S_1 protein of 30-S ribosomal subunits in polyamine-requiring mutants of *Escherichia coli* grown in the absence of polyamines, *Eur. J. Biochem.*, 114, 127, 1981.
78. **Igarashi, K., Kishida, K., Kashiwagi, K., Tatokoro, I., Kakegawa, T., and Hirose, S.,** Relationship between methylation of adenine near the 3′ end of 16S ribosomal RNA and the activity of 30S ribosomal subunits, *Eur. J. Biochem.*, 113, 587, 1981.
79. **Morris, D. R. and Jorstad, C. M.,** Growth and macromolecular composition of a mutant of *Escherichia coli* during polyamine limitation, *J. Bacteriol.*, 113, 271, 1973.
80. **Igarashi, K., Kishida, K., and Hirose, S.,** Stimulation by polyamines of enzymatic methylation of two adjacent adenines near the 3′ end of 16S ribosomal RNA of *Escherichia coli, Biochem. Biophys. Res. Commun.*, 96, 678, 1980.
81. **Igarahsi, K., Kakegawa, T., and Hirose, S.,** Stabilization of 30S ribosomal subunits of *Bacillus subtilis* W168 by spermidine and magnesium ions, *Biochim. Biophys. Acta*, 697, 185, 1982.
82. **Algranati, I. D. and Goldemberg, S. H.,** Initiation, elongation and termination of polypeptide synthesis in cell-free systems from polyamine-deficient bacteria, *Biochim. Biophys. Res. Commun.*, 103, 8, 1981.
83. **Algranati, I. D. and Goldemberg, S. H.,** Translation of natural mRNA in cell-free systems from a polyamine-requiring mutant of *Escherichia coli, Biochem. Biophys. Res. Commun.*, 75, 1045, 1977.
84. **Sander, G., Parlato, G., Crechet, J. B., Nagel, K., and Parmeggiani, A.,** Regulation of turnover GTPase activity of elongation factor G: the 30S-coupled and 30S-uncoupled reactions. Coordinated effects of cations, pH and polyamines, *Eur. J. Biochim.*, 86, 555, 1978.
85. **Takemoto, T., Nagamatsu, Y., and Oka, T.,** The study of spermidine-stimulated polypeptide synthesis in cell-free translation of mRNA from lactating mouse mammary gland, *Biochim. Biophys. Acta*, 740, 73, 1983.
86. **Hryniewicz, M. M. and Vonder Haar, R. A.,** Polyamines enhance readthrough of the UGA termination codon in a mammalian messenger RNA, *Mol. Gen. Genet.*, 190, 336, 1983.
87. **Vazquez, D.,** *Inhibitors of Protein Biosynthesis*, Springer-Verlag, Berlin, 1979, 38.
88. **Höltje, J.-V.,** Streptomycin uptake via an inducible polyamine transport system in *Escherichia coli, Eur. J. Biochem.*, 86, 345, 1978.
89. **Campbell, B. D. and Kadner, R. J.,** Relation of aerobiosis and ionic strength to the uptake of dihydrostreptomycin in *Escherichia coli, Biochim. Biophys. Acta*, 593, 1, 1980.
90. **Bryan, L. E., Nicas, T., Holloway, B. W., and Crowther, C.,** Aminoglycoside-resistant mutation of *Pseudomonas aeruginosa* defective in cytochrome C_{552} and nitrate reductase, *Antimicrob. Agents Chemother.*, 17, 71, 1980.
91. **Bryan, L. E. and Kwan, S.,** Roles of ribosomal binding, membrane potential, and electron transport in bacterial uptake of streptomycin and gentamicin, *Antimicrob. Agents Chemother.*, 28, 835, 1983.
92. **Garvin, R. T., Rosset, R., and Gorini, L.,** Ribosomal assembly influenced by growth in the presence of streptomycin, *Proc. Natl. Acad. Sci. U.S.A.*, 70, 2762, 1973.
93. **Kaempfer, R.,** Initiation factor IF-3: a specific inhibitor of ribosomal subunit association, *J. Mol. Biol.*, 71, 583, 1972.
94. **Garcia-Patrone, M., Perazzolo, C. A., Baralle, F., Gonzalez, N. S., and Algranati, I. D.,** Studies on dissociation factor of bacterial ribosomes: effects of antibiotics, *Biochim. Biophys. Acta*, 246, 291, 1971.
95. **Goldemberg, S. H. and Algranati, I. D.,** Polyamine requirement for streptomycin action on protein synthesis in bacteria, *Eur. J. Biochem.*, 117, 251, 1981.
96. **Pestka, S.,** Inhibitors of protein synthesis, in *Molecular Mechanisms of Protein Biosynthesis*, Weissbach, H. and Pestka, S., Eds., Academic Press, New York, 1977, 467.
97. **Kuhberger, R., Piepersberg, W., Petzet, A., Buckel, P., and Bock, A.,** Alteration of ribosomal protein L6 in gentamicin-resistant strains of *Escherichia coli*. Effects on fidelity of protein biosynthesis, *Biochemistry*, 18, 187, 1979.
98. **Tai, P. C. and Davis, B. D.,** Triphasic concentration effects of gentamicin on activity and misreading in protein synthesis, *Biochemistry*, 18, 193, 1979.

99. **Helser, T. L., Davies, J. E., and Dahlberg, J. E.,** Change in methylation of 16S ribosomal RNA associated with mutation to kasugamycin resistance in *Escherichia coli, Nature New Biol.*, 233, 12, 1971.
100. **Modolell, J. and Davis, B. D.,** Mechanism of inhibition of ribosomes by streptomycin, *Nature*, 224, 345, 1969.
101. **Goldemberg, S. H., Fernandez-Velasco, J. G., and Algranati, I. D.,** Differential binding of streptomycin to ribosomes of polyamine-deficient bacteria grown in the absence and presence of putrescine, *FEBS Lett.*, 142, 275, 1982.
102. **Teraoka, H. and Tanaka, K.,** Effect of polyamines on the binding of dihydrostreptomycin and *N*-acetylphenylalanyl-tRNA to ribosomes from *Escherichia coli, Eur. J. Biochem.*, 40, 423, 1973.
103. **Teraoka, H. and Tanaka, K.,** Effect of spermine on the binding of erythromycin to *Escherichia coli* ribosomes and the peptidyl transfer reaction, *Eur. J. Biochem.*, 33, 578, 1973.
104. **Umezawa, H.,** Roles of amines in the action of antibiotics — bleomycin, spergualin and aminoglycosides, in *Advances in Polyamine Research*, Vol. 4, Bachrach, U., Kaye, A., and Chayen, R., Eds., Raven Press, New York, 1983, 1.
105. **Uehara, Y., Hori, M., and Umezawa, H.,** Specific inhibition of the termination process of protein synthesis by negamycin, *Biochim. Biophys. Acta*, 442, 251, 1976.
106. **Vazquez, D.,** *Inhibitors of Protein Biosynthesis*, Springer-Verlag, Berlin, 1979, 31.
107. **Szer, W. and Kurylo-Borowska, Z.,** Interactions of edeine with bacterial ribosomal subunits. Selective inhibition of aminoacyl-tRNA binding sites, *Biochim. Biophys. Acta*, 259, 357, 1972.
108. **Hettinger, T. P. and Craig, L. C.,** Edeine. IV. Structures of the antibiotic peptides edeines A_1 and B_1, *Biochemistry*, 9, 1224, 1970.
109. **Smith, T. A., Negrel, J., and Bird, C. R.,** The cinnamic acid amides of the di- and polyamines, in *Advances in Polyamine Research*, Vol. 4, Bachrach, U., Kaye, A., and Chayen, R., Eds., Raven Press, New York, 1983, 347.
110. **Smith, T. A.,** Polyamines, in *Annual Review of Plant Physiology*, Vol. 36, Briggs, W. R., Jones, R. L., and Walbot, V., Eds., Annual Reviews, Palo Alto, Calif., 1985, 117.
111. **Loftfield, R. B.,** The frequency of errors in protein biosynthesis, *Biochem. J.*, 89, 82, 1963.
112. **Loftfield, R. B. and Vanderjagt, D.,** The frequency of errors in protein biosynthesis, *Biochem. J.*, 128, 1353, 1972.
113. **Edelmann, P. and Gallant, J.,** Mistranslation in *E. coli, Cell*, 10, 131, 1977.
114. **Bossi, L. and Roth, J. R.,** The influence of codon context on genetic code translation, *Nature*, 286, 123, 1980.
115. **Andersson, D. I., Bohman, K., Isaksson, L. A., and Kurland, C. G.,** Translation rates and misreading characteristics of *rpsD* mutants in *Escherichia coli, Mol. Gen. Genet.*, 187, 467, 1982.
116. **Bouadloun, F., Donner, D., and Kurland, C. G.,** Codon-specific missense errors *in vivo, EMBO J.*, 2, 1351, 1983.
117. **Abraham, A. K.,** The fidelity of translation, in *Progress in Nucleic Acid Research and Molecular Biology*, Vol. 28, Davidson, J. N. and Cohn, W. E., Eds., Academic Press, New York, 1983, 81.
118. **Abraham, A. K., Olsnes, S., and Pihl, A.,** Fidelity of protein synthesis *in vitro* is increased in the presence of spermidine, *FEBS Lett.*, 101, 93, 1979.
119. **Igarashi, K., Kashiwagi, K., Aoki, R., Kojima, M., and Hirose, S.,** Comparative studies on the increase by polyamines of fidelity of protein synthesis in *Escherichia coli* and wheat germ cell-free systems, *Biochem. Biophys. Res. Commun.*, 91, 440, 1979.
120. **Igarashi, K., Hashimoto, S., Miyake, A., Kashiwagi, K., and Hirose, S.,** Increase of fidelity of polypeptide synthesis by spermidine in eukaryotic cell-free systems, *Eur. J. Biochem.*, 128, 597, 1982.
121. **Loftfield, R. B., Eignar, E. A., and Pastuszyn, A.,** The role of spermine in preventing misacylation by phenylalanyl-tRNA synthetase, *J. Biol. Chem.*, 256, 6729, 1981.
122. **Thompson, R. C. and Karim, A. M.,** The accuracy of protein biosynthesis is limited by its speed: high fidelity selection by ribosomes of aminoacyl-tRNA ternary complexes containing GTP[γS], *Proc. Natl. Acad. Sci. U.S.A.*, 79, 4922, 1982.
123. **Jelenc, P. C. and Kurland, C. G.,** Nucleoside triphosphate regeneration decreases the frequency of translation errors, *Proc. Natl. Acad. Sci. U.S.A.*, 76, 3174, 1979.
124. **Patterson, I. and Kurland, C. G.,** Ribosomal protein L7/L12 is required for optimal translation, *Proc. Natl. Acad. Sci. U.S.A.*, 77, 4007, 1980.
125. **Wagner, G. H., Jelenc, P. C., Ehremberg, M., and Kurland, C. G.,** Rate of elongation of polyphenylalanine *in vitro, Eur. J. Biochem.*, 122, 193, 1982.
126. **Thompson, R. C., Dix, D. B., Gerson, R. B., and Karim, A. M.,** Effect of Mg^{++} concentration, polyamines, streptomycin, and mutations in ribosomal proteins on the accuracy of the two-step selection of aminoacyl-tRNAs in protein biosynthesis, *J. Biol. Chem.*, 256, 6676, 1981.
127. **Hirashima, A., Harigai, H., and Watunabe, I.,** Enhancing effect of magnesium ion on cell-free synthesis of read-through protein of bacteriophage Q_β, *Biochem. Biophys. Res. Commun.*, 88, 1046, 1979.

128. **Morch, M. D. and Benicourt, C.,** Polyamines stimulate suppression of amber termination codons *in vitro* by normal tRNAs, *Eur. J. Biochem.*, 105, 445, 1980.
129. **Min Jon, W., Haegeman, G., Ysebaert, M., and Fiers, W.,** Nucleotide sequence of the gene coding for the bacteriophage MS_2 coat protein, *Nature*, 237, 82, 1972.
130. **Parker, J., Johnston, T. C., Borgia, P. T., Holtz, G., Remant, E., and Fiers, W.,** Codon usage and mistranslation. *In vivo* basal level misreading of the MS_2 coat protein message, *J. Biol. Chem.*, 258, 10007, 1983.
131. **Gorini, L.,** Streptomycin and misreading of the genetic code, in *Ribosomes*, Nomura, M., Tissieres, A., and Lengyel, P., Eds., Cold Spring Harbor Laboratory, Cold Spring Harbor, N.Y., 1974, 791.
132. **McMurry, L. M. and Algranati, I. D.,** Effect of polyamines on translation fidelity *in vivo, Eur. J. Biochem.*, 155, 383, 1986.
133. **Tabor, H., Tabor, C. W., Cohn, M. S., and Hafner, E. W.,** Streptomycin resistance (*rpsL*) produces an absolute requirement for polyamines for growth of an *Escherichia coli* strain unable to synthesize putrescine and spermidine *[Δ(speA-speB)ΔspeC], J. Bacteriol.*, 147, 702, 1981.
134. **Gorini, L.,** The contrasting role of *strA* and *ram* gene products in ribosomal functioning, *Cold Spring Harbor Symp. Quant. Biol.*, 34, 101, 1969.
135. **Tabor, H. and Tabor, C. W.,** Polyamine requirement for efficient translation of amber codons *in vivo, Proc. Natl. Acad. Sci. U.S.A.*, 79, 7087, 1982.
136. **Cohen, S. S.,** What do the polyamines do?, *Nature*, 274, 209, 1978.
137. **Cohen, S. S.,** The functions of the polyamines, in *Advances in Polyamine Research*, Vol. 1, Campbell, R. A., Morris, D. R., Bartos, D., Daves, G. D., and Bartos, F., Eds., Raven Press, New York, 1978, 1.

Section C.
Polyamine Metabolism

Chapter 11

ACETYLATION AND INTERCONVERSION OF THE POLYAMINES

Nikolaus Seiler

TABLE OF CONTENTS

I. INTRODUCTION

The enzymes involved in polyamine biosynthesis are exclusively intracellularly localized. Catabolic activities exist both in the circulation and within the tissues. This distribution reflects different functions of the catabolic activities within cells and in extracellular compartments.

One type of polyamine catabolism comprises oxidative deaminations by copper containing, semicarbazide-sensitive amine oxidases. Among these a number of serum enzymes, especially the first amine oxidase recognized to be specific for the polyamines,[1] played historically the greatest role.[2] Since the terminal amino groups of the polyamines are removed and compounds are formed which cannot be reconverted into polyamines, the copper amine oxidase-catalyzed reactions have been summarized under the term terminal polyamine catabolism.[3] The physiological significance of these reactions is still obscure.[4]

The other major pathway of polyamine catabolism was termed interconversion pathway[5] for the following reason: administration of either labeled putrescine, spermidine, or spermine to an animal results in the rapid labeling of the other two amines. For example, 2 hr after administration of [tetramethylene-1,4-^{14}C]spermine·4HCl to rats, 4 ± 0.8% of the total radioactivity in the liver was recovered as spermidine, 1 ± 0.3% as putrescine, and 51 ± 6% as spermine.[6] This and related studies[7-12] showed that the metabolic interconversion of the polyamines is not restricted to liver or a certain species, but seems to be a general phenomenon at least in higher organisms. Prokaryotes, yeasts, fungi, and plants seem not to have been studied yet with regard to the existence of an analogous interconversion pathway, as is existing in vertebrates.

The reactions involved in polyamine interconversion and the potential significance of this pathway in eukaryotic cells will be discussed in this chapter.

II. EARLY OBSERVATIONS

A. Evidence for the Physiological Functioning of Polyamine Interconversion

Administration of a labeled polyamine and observation of its metabolic transformations does not produce convincing evidence for a physiological role of the processes involved. One cannot exclude the possibility that the injected (labeled) compounds are exposed to enzymes which have no access to the endogenous amine pools and are metabolized to a significant extent before they have entered physiological polyamine compartments. However, the following observations could only be understood by assuming that polyamine interconversion was a continuous physiological process, active in various organs of vertebrates.

At 28 days after a single intraperitoneal dose of [2,3-^{3}H]putrescine, measurable amounts of the radioactivity which corresponded in its electrophoretic mobility to putrescine were still found in the organs of mice.[13] Using a different methodology this observation was later confirmed.[14] Moreover, injections of [1,4-^{14}C]putrescine in trout showed that after an initial rapid decline of the specific radioactivity of putrescine, its specific radioactivity paralleled that of spermidine and spermine, labeled putrescine being detectable over a period of more than 70 days.[15] This was strong evidence for the fact that putrescine is continuously formed from spermidine under physiological conditions. The specific activities of spermidine and spermine became identical both in trout and mouse brain some time after the administration of the labeled putrescine,[14,15] although there is no doubt that spermidine and spermine are labeled at different rates if putrescine or ornithine are used as precursors.[10,16-19]

B. Reactions Involved in Polyamine Interconversion

Having demonstrated that polyamine interconversion is a physiological event, the next question to be answered was which reactions are involved in the transformation of spermine into spermidine and of spermidine into putrescine, and which are the enzymes catalyzing these reactions.

The biosynthetic reactions and enzymes are, of course, part of the polyamine interconversion system. They have been well characterized.[20-22] It seemed likely and it was later shown[23] that the aminopropyltransferase reactions are not reversible; therefore, they cannot play a direct role in polyamine catabolism.

Serum amine oxidase initiates the conversion of spermine into spermidine, and that of spermidine into putrescine; it has been demonstrated that the aldehydes which are formed by oxidative deamination of the polyamines spontaneously eliminate acrolein.[24] However, there were arguments against the notion that serum amine oxidase is involved in physiological polyamine interconversion:

1. The activity of serum amine oxidase varies enormously in different species.[25] As a consequence, one would expect a great interspecies variation of the rate of polyamine interconversion.
2. From the localization of the amine oxidase in serum, one would have to conclude that not only exogenous but also intracellular polyamines are entering the circulation where their catabolism could be achieved; in addition, one would have to suggest that polyamine interconversion in tissues takes place only in those selected organs, such as small intestine, which are known to have the capacity of oxidatively deaminating the polyamines.[26]

Since polyamine interconversion had been demonstrated to exist under physiological conditions in vertebrate brains[14,15] which have very low, if any, serum amine oxidase-like activity,[26] the latter assumption is not valid.

Polyamine interconversion within the circulation cannot be excluded a priori to exist. However, cellular spermine is presumably entering the circulation only after its displacement from a dead cell; that is, if the above assumption was correct, polyamine interconversion would be a process suited for the metabolic transformation of tissue polyamines after cell death and of exogenous polyamines into putrescine, but not a process involved in the regulation of intracellular polyamine content.

The key to the reaction sequence involved in polyamine interconversion was the purification and characterization of polyamine oxidase (PAO) from rat liver by Hölttä.[27] PAO has the capacity to oxidatively split in vitro spermine into spermidine and 3-aminopropanal; the analogous reaction with spermidine was considerably slower,[28] but also detectable.[27] PAO was therefore a logical candidate for participation in the interconversion pathway. When it turned out that monoacetyl derivatives of the polyamines, N^1-acetylspermidine and N^1-acetylspermine were far better substrates of PAO than the polyamines, a reaction scheme was proposed which included as first step N^1-acetylation, and as second step the oxidative cleavage of the acetyl derivatives.[28,29]

III. ACETYLATION OF POLYAMINES

A. AcetylCoA:Spermidine/Spermine N^1-Acetyltransferase

A cytosolic acetyltransferase capable of acetylating spermidine was first described by Matsui and Pegg.[30] This enzyme has been purified to homogeneity from livers of rats treated with carbon tetrachloride.[31] It has an apparent molecular weight of about 115,000. The K_M for acetylCoA was 1.5 μM, for spermidine 130 μM, and for spermine 35 μM.[31] With spermidine as substrate exclusively, N^1-acetylspermidine and no N^8-acetylspermidine is formed, whereas both primary amino groups of spermine can be acetylated by spermidine/spermine N^1-acetyltranferase (SAT). However, neither putrescine[31,32] nor histones[33] are substrates.

Purified SAT permitted the preparation of a monospecific antiserum which could be used to study the regulation of the enzyme.[34] Immunoblotting techniques revealed that the antiserum reacted with a protein in liver extracts having a molecular weight of 18,000, which is identical with the subunit of the purified enzyme.[34]

B. Induction of Spermidine/Spermine N[1]-Acetyltransferase and Biological Consequences

SAT activity is low in normal tissues. Specific antiserum against SAT had little effect on the polyamine acetylating capacity of crude cytosolic extracts indicating that most of the basal acetylating capacity is not due to SAT.[34] There are, however, a great number of stimuli which enhance SAT activity dramatically: hepatotoxins (carbon tetrachloride,[30,33,35] thioacetamide,[36] nitrosamines[37]), certain hormones,[36,38,39] folic acid,[40] 3-isobutyl-1-methylxanthine (IBMX),[41] methylglyoxal-bis(guanylhydrazone),[42] and also spermidine.[43] As a physiological stimulus fasting deserves especial interest.[38,44] The increase of SAT activity was in all cases studied due to the enhanced amount of enzyme protein.[34,42,43,45] N[1]-Acetylspermidine concentration was regularly increased in the tissues with elevated SAT levels[6,39-41,46,47] and the urinary excretion of N[1]-acetylspermidine was also enhanced,[6,40] but in agreement with the site-specificity of SAT, no changes were observed in N[8]-acetylspermidine metabolism. N[1]-Acetylspermidine concentrations are usually unmeasurably low, but after induction of SAT by treatment with carbon tetrachloride, which increases the amount of liver enzyme about 250-fold,[45] it could be determined.[46]

Enhanced formation of putrescine from labeled spermidine in the liver of carbon tetrachloride-treated rats had been previously demonstrated.[48] It is apparent that an increase in the rate of conversion of one polyamine into another, the enhancement of SAT, and the increase of N[1]-acetylpolyamine concentrations are strictly interdependent phenomena.[6,30,33,35,36,46] This does not necessarily imply increased tissue polyamine levels: the decrease in spermidine and putrescine concentrations in the liver of fasting animals was also correlated with an enhanced SAT activity and an elevated level of N[1]-acetylspermidine.[44]

Inhibition of protein biosynthesis by cycloheximide showed that the induced enzyme has a biological half-life as low as 15 min[34,49] indicating that SAT is suitable as a regulatory protein.

C. Formation of N[8]-Acetylspermidine

The existence of an enzyme capable of monoacetylating putrescine was first shown in microsomal and purified nuclear fractions of liver and brain, but not in the cytosol.[50] Blankenship and Walle demonstrated subsequently that this enzyme was localized in chromatin.[51] With spermidine as substrate, N[8]-acetylspermidine is its major reaction product.[33,51,52] It is also capable of acetylating spermine and, according to Libby,[53] histones.

The nuclear enzyme was not affected by the monospecific antibody, indicating that it is not related to the cytosolic SAT.[52]

None of the mentioned stimuli which are inducing SAT has had a measureable effect on acetylCoA:spermidine N[8]-acetyltransferase, and since N[8]-acetylspermidine is not a substrate of PAO[28] it has no role in polyamine interconversion.

It has recently been demonstrated that spermidine N[8]-acetylation is active in cultured rat hepatoma (HTC) cells,[54] and it can be assumed that it is also active in various tissues under physiological conditions because N[8]-acetylspermidine is a normal urinary constituent;[55] functions of spermidine N[8]-acetylation are at present a matter of speculation.

IV. POLYAMINE OXIDASE

A. Occurrence and Characteristics

PAO is found in virtually all tissues. Its activity in rat organs ranges between 380 (pancreas)

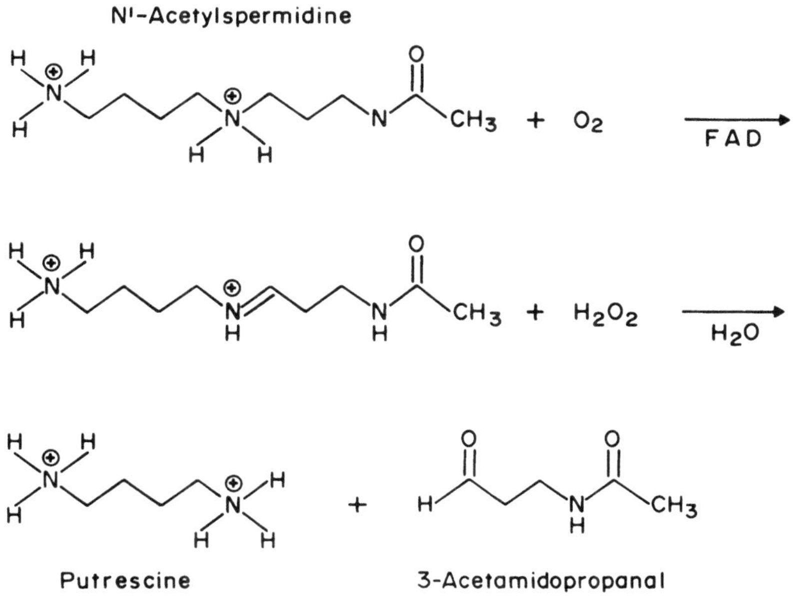

FIGURE 1. PAO-catalyzed formation of putrescine from N[1]-acetylspermidine.

and 10 (skeletal muscle) nmol of N^1,N^{12}-diacetylspermine oxidized per minute per gram, i.e., PAO activity is high in comparison with endogenous polyamine contents, which are in the order of 0.2 (skeletal muscle) to 2 (liver) μmol/g tissue.

The enzyme has been purified from rat liver. Its molecular weight is around 60,000. It contains tightly bound flavin adenine dinucleotide (FAD) as prosthetic group, and there is evidence that Fe^{2+} may be a co-factor.[57] Isoenzymes are not known. Unlike monoamine oxidase (MAO), with which it has a number of common features, PAO is not localized in mitochondria, but in peroxisomes, at least in liver and kidney;[27,58] other organs have not yet been studied. Morgan[59] has described an amine oxidase in human pregnancy serum with characteristics that are compatible with the assumption that it is PAO.

PAO activity was not induced in rat liver by partial hepatectomy or adminstration of carbon tetrachloride, thioacetamide, and growth hormone,[27] i.e., by stimuli which have been demonstrated to cause enhanced polyamine interconversion.[30,35,36,48] Precise data on PAO turnover do not exist. From experiments with cycloheximide[56] and from the rate of recovery of active enzyme after its inactivation with a specific, irreversible inhibitor,[60] a biological half life of at least 3 to 4 days must be assumed, but it could be considerably longer.

PAO needs molecular oxygen as electron acceptor and forms H_2O_2.[27] With a K_M of 0.6 μM, N^1-acetylspermine is the substrate with the highest affinity, followed by N^1-acetylspermidine (K_M = 14 μM). N^8-Acetylspermidine is not attacked by PAO.[28]

The reaction with N^1-acetylspermidine is formulated in Figure 1. The aminopropyl moiety which originates from methionine is split off in the form of 3-acetamidopropanal.[38] In vivo, the aldehyde is further oxidized to *N*-acetyl-β-alanine, which then is hydrolyzed, presumably by an enzyme[61] for which previously no functional role was known. In situations of very active oxidation of N^1-acetylspermidine, β-alanine accumulation in tissue can be observed.[39,41]

B. Inhibitors of Polyamine Oxidase

Iron chelators and FAD analogues, such as quinacrine have been found to inhibit PAO;[27] however, since they lack specificity they were of little use.

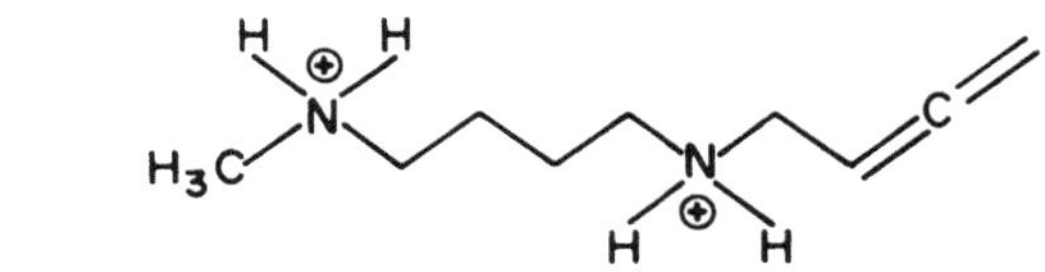

N¹-Methyl, N⁴-(2,3-butadienyl)-1,4-diaminobutane

MDL 72521

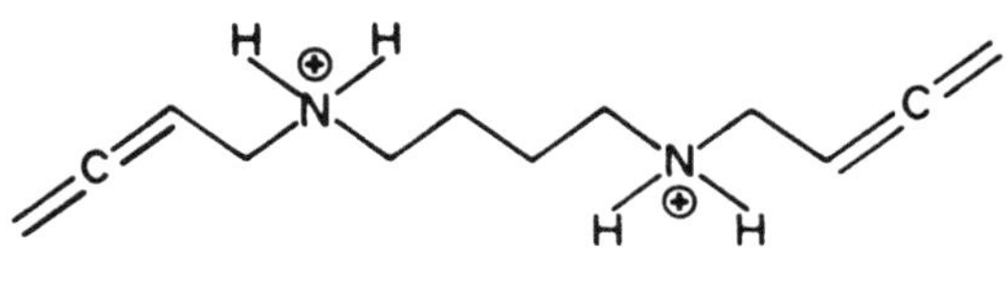

N¹, N⁴-Bis-(2,3-butadienyl)-1,4-diaminobutane

MDL 72527

FIGURE 2. Irreversible inhibitors of PAO.

From the analogies between PAO and MAO — both are FAD dependent and have assumedly the same reaction mechanism — it was presumed that irreversible inactivators of these two enzymes would require similar structural features, and differ mainly in that part of the molecule which provides affinity to the active site of the enzyme by close structural analogy to natural substrates. The fact that pargyline (*N*-methyl-*N*-benzyl-propargylamine) is a weak inhibitor of LPAO[27] despite considerable structural differences between pargyline and the substrates of PAO was in support of this concept. In a first attempt a number of substituted putrescines were synthesized.[62] In the series of derivatives having 2-propynyl, 2-propenyl, and 2,3-butadienyl residues on one or both nitrogen atoms, the 2,3-butadienyl derivatives were most potent. N^1-methyl,N^4-(2,3-butadienyl)-1,4-butanediamine (MDL 72521) and N^1,N^4-bis(2,3-butadienyl)-1,4-butanediamine (MDL 72527) (Figure 2) were investigated further in some detail.

Both compounds irreversibly inactivate PAO in a dose and time-dependent manner.[62] Mechanistic studies have not yet been carried out, but their properties are compatible with the assumption that these compounds are enzyme-activated irreversible inhibitors.[63] They are supposed to bind to FAD in the course of the enzymic reaction, as was suggested for MAO inhibitors of similar structure.[64] MDL 72521 and 72527 are specific. They do not inhibit any other enzyme significantly at concentrations which are sufficient to inactivate PAO in vitro and in vivo completely,[62] and they do not exhibit toxic effects in experimental animals.[60]

MDL 72527 seems to penetrate the blood brain barrier more effectively than MDL 72521,[60] therefore it is the more advantageous compound for in vivo studies.

C. Evidence for N^1-Acetylspermidine and N^1-Acetylspermine Being the Natural Substrates of Polyamine Oxidase

Oxidative splitting of labeled spermine and spermidine, and formation of 3-aminopropanal by PAO has previously been demonstrated.[27,65] However, there is no evidence that this reaction is of any significance in vertebrates. On the contrary, all pertinent observations indicate that in vivo only N^1-acetylspermidine and N^1-acetylspermine serve as substrates of PAO:

1. In HTC cells, depleted of their putrescine and spermidine content by use of D,L-α-difluoromethylornithine (DFMO), normal intracellular spermidine concentrations can be restored by addition of N[1]-acetylspermine, or N[1]-acetylspermidine, but not by addition of spermine to the culture medium.[66] There is no evidence that N[1]-acetylspermidine was cleaved hydrolytically in this experiment, as would be the case with N[8]-acetylspermidine.[67] If PAO was inactivated by 10 μ*M* N[1]-methyl, N[4]-allenylputrescine (MDL 72521) before N[1]-acetylspermine was added to the culture medium, the cellular spermidine content was not restored.[54]
2. From the inactivation of PAO in vivo over an extended period of time, one expects an increase of the concentration of its substrates. Addition of MDL 72521 to the culture medium of growing HTC cells in doses greater than 10 μ*M* caused a gradual increase of N[1]-acetylspermidine and N[1]-acetylspermine to reach levels of 90 and 75 pmol per 10^6 cells, respectively, on day 3 of culture.[41] No significant changes of spermidine or spermine concentrations were observed.
3. Single and multiple doses of MDL 72527 produced an increase in N[1]-acetylspermidine and N[1]-acetylspermine levels in various organs of mice.[60] Spermidine concentrations were in some organs — brain and testes — significantly decreased after long-term administration of the PAO inhibitor[60] but not increased, as one would expect it, if PAO was attacking spermidine directly. As was mentioned previously, administration of carbon tetrachloride enhances putrescine formation from spermidine, that is, one observes a dramatic increase of putrescine concentration in liver with a first maximum around 6 hr,[48] and a concomitant decrease of spermidine concentration. Inactivation of PAO enhanced acetylpolyamine accumulation, but did not prevent the decrease of spermidine.[60] The fact that inactivation of PAO in carbon tetrachloride-treated mice prevented at the same time the accumulation of putrescine in liver suggested that the early formation of putrescine in carbon tetrachloride intoxication is entirely due to spermidine degradation via the interconversion reactions.
4. Rodents excrete the major portion of polyamines in the form of putrescine, but significant amounts of free spermidine and of N[1]-acetylspermidine and N[8]-acetylspermidine are also found in urine,[68] and, in addition, amino acids and amino acid derivatives which are formed from the polyamines by oxidative deamination.[69-71] When PAO was inactivated, the average daily N[1]-acetylspermidine excretion increased from 0.24 ± 0.03 to 3.1 ± 0.4 μmol/24 hr per rat, while spermidine (and N[8]-acetylspermidine) excretion did not change. Putrescine excretion decreased under these conditions from 2.4 ± 0.3 to 1.8 μmol/24 hr, demonstrating that a considerable portion of the urinary putrescine is formed from spermidine, and is not the direct product of ornithine decarboxylation.[3]

V. FUNCTIONAL ASPECTS OF POLYAMINE INTERCONVERSION

A. The Metabolic Scheme of the Polyamines

Based on the present knowledge of polyamine synthesis, and the catabolic reactions, which have in part been discussed above, it is possible to draw a scheme which demonstrates the major features of polyamine metabolism (Figure 3).

The interconversion pathway as it appears in this scheme is essentially a sequence of reactions which produces spermidine and spermine from putrescine and degrades the polyamines again in the inverse direction by N[1]-acetylation and oxidative splitting of the acetyl derivatives. The putrescine moiety is conserved within this metabolic cycle, but the aminopropyl moieties of spermidine and spermine are irreversibly eliminated and finally converted into β-alanine. One complete turn of the cycle requires two molecules of methionine and forms two molecules of 5′-methylthioadenosine and β-alanine.

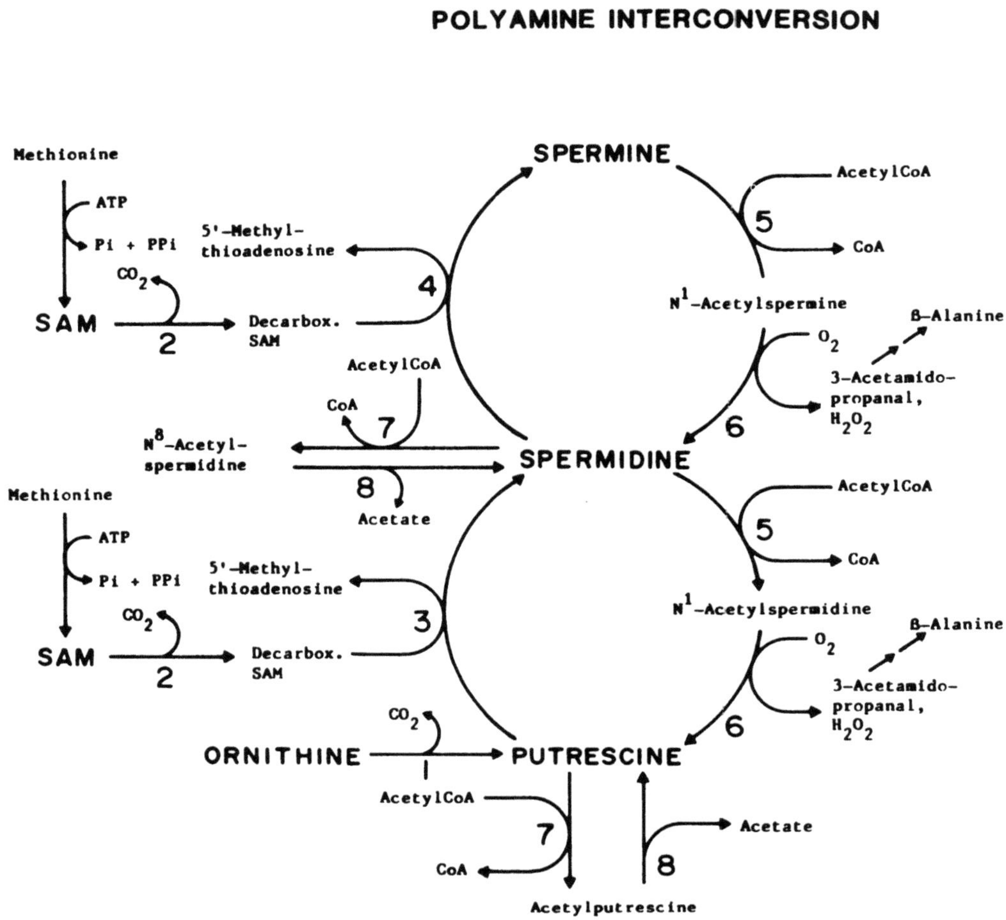

A

FIGURE 3. (A) Metabolic scheme of the polyamines. 1 = ornithine decarboxylase (ODC) (EC 4.1.1.17), 2 = *S*-Adenosyl-L-methionine decarboxylase (SAM-DC) (EC 4.1.1.50), 3 = spermidine synthase, 4 = spermine synthase, 5 = acetylCoA:spermidine/spermine N^1-acetyltransferase (cytosolic) (SAT), 6 = polyamine oxidase (PAO), 7 = acetylCoA:spermidine N^8-acetyltransferase (nuclear), and 8 = N^8-Acetylspermidine acetylhydrolase. (B) The two horizontal arrows in the scheme of terminal polyamine catabolism indicate transformations catalyzed by copper amine oxidases and an aldehyde dehydrogenase (or another enzyme capable of oxidizing the reaction products of the amine oxidases to a carboxylic acid).

In a steady-state situation ornithine decarboxylation has the function to produce putrescine in order to substitute the irreversible losses of the putrescine moiety. Irreversible losses may occur by two different ways: (1) terminal polyamine catabolism and (2) transport, i.e., in vivo by excretion with the urine. All intermediates of the interconversion pathway are substrates of copper amine oxidases,[4,72] and most of the compounds, which are mentioned in the scheme of terminal polyamine catabolism, have been identified in urine.[69-71]

ODC is usually considered to be the rate-limiting enzyme of polyamine biosynthesis. Although, this may be essentially true in many instances, especially in certain tumor cells which are known to have slow rates of polyamine degradation[73] and excretion,[74,75] it appears from the metabolic scheme that the intracellular rate of spermidine synthesis is limited by the availability of *S*-adenosyl-5′-deoxy-(5′)-3-methylthiopropylamine (decarboxylated SAM),

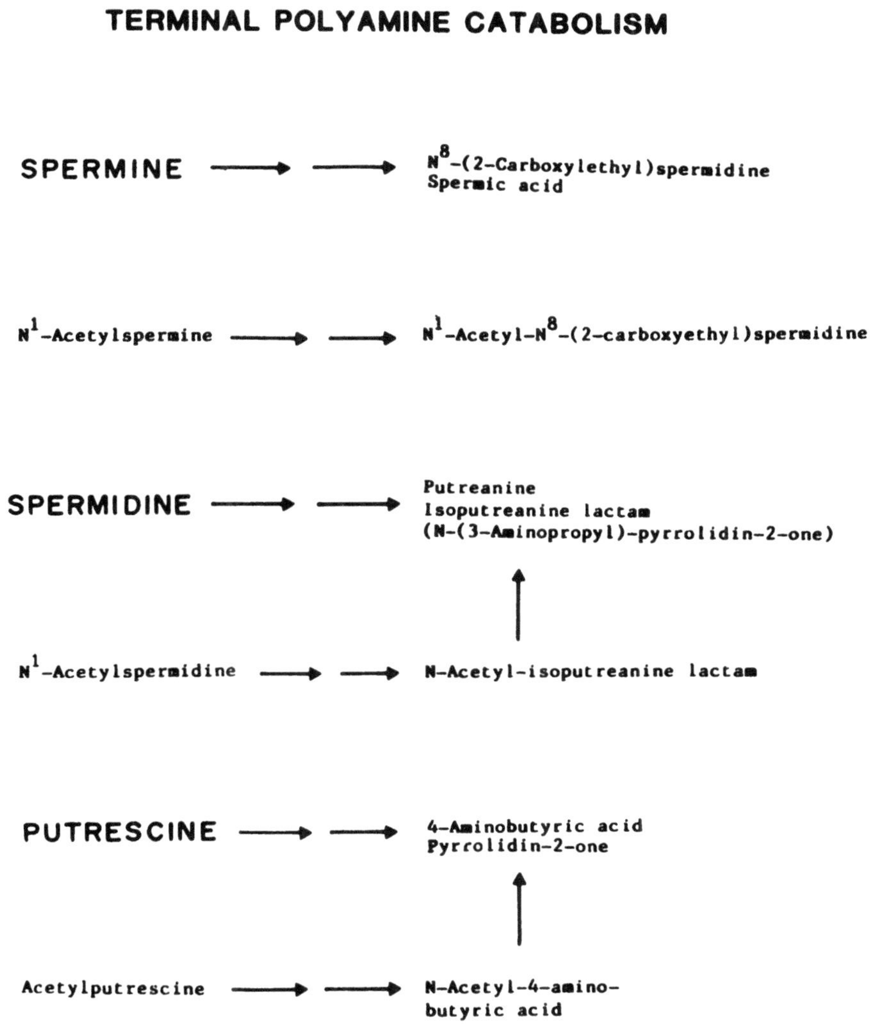

FIGURE 3B.

together with that of putrescine, assuming that spermidine synthase is usually present in excess.[76] The amount of putrescine formed per minute within a given cell is not only a function of ODC activity, but at the same time a function of the rate of putrescine formation from spermidine, which is dependent on the rate of N^1-acetylation. The picture is further complicated by the fact that the size of the intracellular putrescine pool may in addition be considerably affected by transport — both inward and outward transport[75] — and by diamine oxidase-catalyzed putrescine catabolism.

In contrast with putrescine, there is only a single source for decarboxylated SAM, that is, the actual rate of its formation is an absolute limit for spermidine and spermine formation. Since this rate is affected by the intracellular putrescine[77] and spermidine[78] pools, there is an intimate interdependence between one regulatory site and the other. We must assume that the relative importance of the factors which are involved in the control of cellular pools of putrescine and decarboxylated SAM is greatly varying in different tissues and with their physiopathological state.

An example is the mammalian brain during maturation. In immature rat brain the formation of decarboxylated SAM is presumably much slower than the overall rate of putrescine formation,[79] even if we assume that SAM-DC is fully activated, due to the high putrescine concentration. Therefore, it is likely that SAM-DC activity limits the rate of spermidine and spermine formation in this physiological situation. In general, only a careful analysis of a given physiological and pathological status will allow us to define rate-limiting steps of polyamine metabolism.

B. Why has Acetylation Been Chosen by Nature?

Cellular acetylpolyamine concentrations are usually very low, even under conditions of enhanced polyamine catabolism[5,6,44,46,47] or after inactivation of PAO,[41,60] and in no case was it possible to ascribe a biochemical or pharmacological effect to elevated acetylpolyamine tissue levels. The fact that 1 m*M* concentrations of N^1-acetylspermidine in the culture medium were reported to inhibit the growth of BHK-21/C13 cells[75] is not a contradiction to this statement, because in no case has such an enormous accumulation of N^1-acetylspermidine been observed. Thus, it appears that acetylation of polyamines has only regulatory and metabolic functions. The products of acetylation seem to be without any specific significance.

Due to their polycationic nature it has long been recognized that polyamines bind to a variety of polyanions,[80] among which polynucleotides have been most extensively studied.[81-83] Polyamine binding is compatible with early observations on their effects on the melting temperature of DNA. Putrescine is least effective in protecting DNA against thermal denaturation, and spermine is most effective, due to the increased number of positive charges.[84] It follows that the concentrations of free spermidine and spermine within the various cellular compartments are very low,[85] although their binding (at least to DNA) is such that the mobility of the molecules is effectively independent of that of the macromolecule.[86]

Acetylation is an obvious means to reduce the number of positive charges of the polyamines and thus reduce their interaction with polyanions.[51] The monoacetylspermidines are indeed much less effective in protecting DNA against thermal denaturation than spermidine, and are comparable in this regard with putrescine,[87] as is expected of molecules with only two positive charges.

Taking the cellular localization of SAT into account,[45] N^1-acetylation of spermidine and spermine is compatible with the suggestion that acetylation is a means to terminate physiologically significant interactions of the polyamines with polyanionic binding sites within the cytoplasm. N^8-Acetylation of spermidine and monoacetylation of spermine by the same nuclear acetyltransferase could fulfil the analogous role in the cell nucleus. Since the deacetylase, which is responsible for the hydrolytic cleavage of N^8-acetylspermidine, is a cytoplasmic enzyme,[51] N^8-acetylation may serve to transport polyamines from the cell nucleus into the cytoplasm.

Canellakis et al.[88] suggested that acetylation could permit the directional flow of polyamines through lipophilic membranes from one hydrophilic region (intracellular compartment) to another (extracellular space). In their scheme, they proposed for putrescine a stepwise acetylation of both amino groups on one surface. The partial deacetylation of the lipophilic diacetylputrescine at the opposite membrane surface is presumed to lead to the reverse orientation of monoacetylputrescine in the lipid membrane, and complete deacetylation to the transport of putrescine through the membrane.

There is presently no evidence for putrescine acetylation on cell membrane surfaces or of bis-acetylations of the polyamines, except that N^1-acetylspermine can in vitro be converted to N^1, N^{12}-diacetylspermine by SAT.[31] However, monoacetylputrescine, in contrast with putrescine, and N^1, N^{12}-diacetylspermine, in contrast with N^1-acetylspermine and spermine, do not seem to be taken up well by HTC cells,[66] a finding which is compatible with the unidirectional transport at least of certain polyamine derivatives. The repeatedly mentioned fact that acetylated polyamines are a major excretory form, both of mammalian cells[75] and of vertebrates,[68] may also be interpreted in terms of facilitation of polyamine transport by acetylation, although acetylation does not seem to be an absolute requirement for polyamine excretion.

C. Polyamine Reutilization and Turnover

The metabolic scheme of the polyamines (Figure 3) suggests that spermidine and putrescine, which are formed from spermine and spermidine, respectively, may be reutilized for polyamine synthesis, and it has been mentioned that HTC cells deficient of active ODC, putrescine, and spermidine are capable of restoring their normal polyamine complement from N^1-acetylspermidine, and N^1-acetylspermine, respectively.[66] The question is, however, whether polyamine reutilization is a physiological phenomenon. First positive evidence to this question has been obtained from the comparison of the biological half-lives of the putrescine and aminopropyl moieties of spermidine and spermine, by double-labeling experiments with [^{14}C]putrescine and [^{3}H]methionine. It turned out that the apparent half-life of the putrescine part of the polyamines was considerably longer than that of the aminopropyl moieties.[14] This observation led to the concept of polyamine reutilization and the formulation of metabolic cycles for the polyamines, analogous to that shown in Figure 3. With the advent of specific inhibitors of PAO, it is now possible to determine polyamine reutilization in closed metabolic systems. Two examples are given in the following sections.

1. Polyamine Reutilization and Turnover in Mouse Brain

The mammalian brain can be considered as a closed metabolic system for those compounds which are not penetrating the blood-brain barrier. Polyamines[89] and N^1-acetylspermidine[79] have been shown to belong to this category; therefore, it seems allowed, at least in a first approximation, to calculate reaction rates from time-dependent concentration changes of these compounds.

Inactivation of PAO by administration of 50 mg/kg N^1, N^4-bis-(2,3-butadienyl)-1,4-butanediamine to mice causes a linear increase of N^1-acetylspermidine and N^1-acetylspermine concentrations with time (1,1 nmol/g brain per hour and 0.05 nmol/g brain per hour, respectively).[90] Since the PAO inhibitor did not have any toxic effects even after prolonged administration, one can assume that the accumulation of the acetylated polyamines reflects the physiological turnover at steady-state conditions.

In agreement with the expectations, putrescine concentration decreased concomitantly with the increase of N^1-acetylspermidine. After about 8 hr this decrease leveled off to reach at around 24 hr a minimum value of about 30% of that of controls. Even repeated administration of the PAO inhibitor did not lower this value further.[90]

Under the premises that elimination of the acetylpolyamines from brain by transport and metabolism is slow, as compared with formation, their rate of accumulation is a direct measure of polyamine turnover.[90] The biological half-life ($\tau_{1/2}$) of spermidine in brain, which was calculated from its accumulation rate is in reasonable agreement with the $\tau_{1/2}$ obtained by labeling with [^{3}H] methionine, and determination of the decrease of the specific radioactivity with time, but $\tau_{1/2}$ for spermine was considerably longer than in the previous study, indicating perhaps that significant amounts of N^1-acetylspermine are lost either by metabolism and/or via circulation.

The observed decrease of putrescine concentration is a direct measure of the proportion of putrescine which is formed by the interconversion pathway. In the adult mouse brain, about 70% of the putrescine is the product of spermidine degradation, and only 30% is formed by decarboxylation of ornithine. The spermidine-derived putrescine is normally reutilized for spermidine biosynthesis. This is prevented by inactivation of PAO, and consequently one observes a small but significant decrease of spermidine concentration.[90]

The proportion of putrescine formed along the interconversion pathway is changing during brain development. Within the first 5 days of postnatal life of rats, it is below the detection limit of the method and reaches adult values around day 20.[79]

2. Polyamine Reutilization by an Intact Rat

Average polyamine excretion by an untreated Sprague-Dawley rat of about 240 g is 3.1 ± 0.4 μmol putrescine equivalents per 24 hr, i.e., the sum of excreted putrescine, spermidine, N^1-acetylspermidine, and N^8-acetylspermidine. Spermine excretion in rats is usually negligibly low. After inactivation of PAO, the urinary output of putrescine equivalents increases to 5.2 ± 1.2 μmol/24 hr. The additional excretion occurs in the form of N^1-acetylspermidine.[3] This increase of polyamine excretion by about 2 μmol/24 hr is that amount of putrescine which has been formed from spermidine and is normally reutilized for spermidine biosynthesis. Extensive inhibition of ODC by administration of DFMO has practically no effect on the amount of putrescine which is reutilized.[3]

From corresponding data obtained after complete inhibition of the terminal polyamine catabolism by aminoguanidine, it appears that the amount of reutilized putrescine per rat per day is significant, namely, more than 25% of the normally excreted amount of polyamines. Unfortunately, we have presently no data available, which would allow us to calculate the total putrescine amount which must be provided per day by decarboxylation of ornithine, in order to maintain a steady-state situation in an adult rat. Thus, it is not possible to compare *de novo* synthesis of putrescine directly with putrescine reutilization.

3. Physiological Significance of Polyamine Reutilization

A hypothesis about the functional significance of polyamine reutilization has previously been presented.[14,93] The following suggestions and considerations are based on the same concept, but have been updated to take into account our present knowledge of polyamine interconversion.

Rapidly proliferating cells are geared to maximum rates of *de novo* polyamine biosynthesis. They usually have high ODC activities and consequently high putrescine concentrations, which ensure full activation of SAM-DC. In such a system further enhancement of polyamine formation can only be achieved by a considerable increase of the concentration of the substrates of spermidine and spermine synthase, that is, if ODC activity is increased, first spermidine formation will be enhanced, and if spermidine concentration has increased by *de novo* synthesis, spermine formation will also be enhanced according to the rules of the Michaelis-Menten kinetics. This is the reason why rapidly growing tissues frequently exhibit high putrescine concentrations and a high spermidine/spermine ratio. The latter has been suggested as indicator of rapid cell growth,[91] although it is not always relevant because some nongrowing tissues, such as myelin-rich parts of the nervous system also exhibit very high spermidine/spermine ratios,[92] presumably due to a specific function of spermidine in myelin.

In nongrowing organs, the enzyme pattern is opposite to that in fast-growing tissues. One finds usually higher SAM-DC than ODC activities, and spermidine and spermine synthase activities are higher by an order of magnitude or more than the activities of the two decarboxylases.[76] Due to the low putrescine concentration, SAM-DC is assumedly not fully activated. (For brain we have evidence that this is the case.[77]) In contrast with dividing cells which share their polyamines among the daughter cells, diminution of the polyamine content in nondividing cells can only occur by degradation and/or excretion. Since spermine does not seem to be excreted by intact cells, and since terminal polyamine catabolism is not as ubiquitous as the interconversion pathway, the basal activity of SAT is suggested to dictate the rate of spermine turnover under steady-state conditions. Spermidine turnover rate may also be limited in many instances by SAT activity. The mature mammalian brain is presumably the best example in this regard, because neither polyamines nor acetylpolyamines penetrate the blood brain barrier very well,[79,89] and the activity of enzymes of the terminal catabolic reactions are low.[26] In other tissues excretion and/or terminal catabolism reactions may play a greater role than in brain and thus contribute to a greater extent to the rate of polyamine turnover.

The turnover rate is limiting the rate with which a metabolic system can be switched from one steady-state situation to another, according to the physiological needs, by changing the activity of rate-limiting enzymes. Since polyamine turnover in nongrowing tissues is of the order of many days, the regulation of polyamine formation by mass action of the substrates of the biosynthetic enzymes, as has been described above for embryonal and tumor cells, would only permit very slow changes. However, the pattern of enzyme activities present in nongrowing tissues, together with the observed polyamine reutilization suggest a different type of regulation. This suggestion is based on the assumption that ODC and SAT are part of a regulatory circuit, the interconversion cycle. The activities of both enzymes can be independently modulated in response to appropriate signals. Furthermore, it is a necessary prerequisite that cells have the ability to control the extent of polyamine reutilization, i.e., they can change according to their physiological needs the elimination of products of the interconversion cycle by transport and/or the rate of terminal polyamine catabolism.

In order to illustrate the expected metabolic changes in this system, we discuss briefly three cases.

a. Induction of Ornithine Decarboxylase

The fact that spermidine is continuously degraded to putrescine, which can be reutilized for spermidine formation, ensures basal polyamine turnover at very low ODC activity. Induction of ODC will enhance cellular putrescine concentration, which results in the activation of SAM-DC. Enhanced formation of decarboxylated SAM will increase the rate of both spermidine and spermine formation at the same time. This is in sharp contrast to the mentioned sequential increase of spermidine and spermine concentrations in embryonal and tumor cells. Since the spermidine concentration is "normal", enhancement of spermine formation will be directly proportional to the increase of SAM-DC activity. In contrast, spermidine formation is limited by the availability of putrescine as substrate.

Whether spermidine and spermine synthase are competing for the same pool of decarboxylated SAM, or whether separate pools of decarboxylated SAM exist for the two synthases, is not known.

If SAT activity is not changed, a net increase of cellular spermidine and spermine concentrations will be observed.

As soon as ODC has returned to steady-state level, the newly formed polyamines will be submitted to gradual elimination processes, which are initiated by SAT.

b. Induction of Spermidine/Spermine N^1-Acetyltransferase

Several observations have been reported which suggest that SAT is induced whenever cells must get rid of excessive polyamines, whereby the excess indicates not necessarily a supernormal concentration, but only an amount above that which is physiologically needed under specific circumstances.[72] According to our scheme (Figure 3), induction of SAT, as well as induction of ODC, is expected to enhance putrescine formation, and thus activate the formation of spermidine and spermine. As a result, polyamine turnover will be enhanced. In order to achieve a decrease of cellular polyamine levels it is necessary that a certain portion of the putrescine and/or spermidine, which are formed as products of the catabolic reactions of the interconversion cycle, are eliminated from the cell by transport, or transformed into terminal catabolic products. One can expect, therefore, that outward transport is usually coupled with an elevated SAT activity, whereas uptake of polyamines from the environment is preferentially coupled with an enhanced ODC activity and with cell growth.[94,95]

Spermine and monoacetylspermine have not yet been identified as substrates of polyamine export, but all other products of the interconversion pathway (spermidine, N^1-acetylspermidine, putrescine, and monoacetylputrescine) may be excreted by one cell type, or another.

c. Induction of Ornithine Decarboxylase and Spermidine/Spermine N^1-Acetyltransferase

Proportional induction of both ODC and SAT is a means to increase polyamine turnover rate, without changing equilibrium concentrations of the polyamines.

The most conspicuous advantage of the interconversion cycle as a regulatory circuit is that it provides a means for the rapid, short term increase of spermidine and spermine concentrations by small changes of ODC activity, and a decrease of polyamine concentration by induction of SAT.

That equilibrium concentrations of putrescine are directly correlated with spermidine turnover rates has been proposed.[90] Inhibition of ODC by DFMO, and the consequent decrease of putrescine concentration was followed by a concomitant decrease of spermidine turnover rate in mouse brain. In contrast, induction of ODC by treatment with inhibitors of 4-aminobutyrate:2-oxoglutarate aminotransferase and elevation of putrescine levels enhanced brain spermidine turnover. These and previously mentioned findings suggest, therefore, the brain as the most likely candidate for the sophisticated type of polyamine regulation that has been described above. It is tempting to speculate that it is the regulatory system of all highly differentiated cell types.

VI. POLYAMINE ACETYLATION IN NORMAL AND TRANSFORMED CELLS

It is well documented that transformed cells incorporate exogenous putrescine more actively and synthesize spermidine at a faster rate than do normal cells. This has been demonstrated for phytohemagglutinin-stimulated human lymphocytes,[96] Rous sarcoma virus-transformed chick embryo fibroblasts,[95] and polyoma virus-transformed baby-hamster-kidney (BHK-21) cells,[97] findings which are in agreement with the above-mentioned characteristics of tumor cells and the role of polyamines in cell growth. Surprisingly, in all three cases,[95-97] polyamine acetylation has been found to be enhanced. This seems surprising in view of the finding that rapidly growing cells were found to have less catabolic activity[73] and excrete less polyamines than slowly growing cells.[74,75] Since acetylase activity reaches a peak early in the cell cycle, it was suggested that in growing cells acetylation and subsequent oxidation of the acetylated polyamines may provide putrescine.[98] The fact that inhibition of PAO, i.e., prevention of putrescine formation in HTC cells along this pathway, had no effect on cell proliferation rate[41] argues against the notion that conversion of spermidine into putrescine is essential for growth of tumor cells.

From what has been said above about the presumed consequences of induction of SAT, it may be suggested that polyamine turnover is enhanced in transformed cells, if SAM-DC is not fully activated by the putrescine which deirves from ornithine. This assumption is even more likely if we postulate that SAM-DC is compartmentalized within the cells, and cellular putrescine is not uniformly distributed. However, the few isolated observations do not allow us to draw conclusions. It will be an important task for the future to establish whether the enhancement of polyamine acetylation is a general characteristic of neoplastic growth. Only a more profound knowledge of the changes that occur during dedifferentiation and differentiation of a variety of cell types can be expected to form the basis for the solution of the riddle that is hiding behind the apparently paradoxical enhancement of polyamine acetylation in transformed cells.

VII. CONCLUSIONS

Cellular polyamine concentrations are regulated by at least three inducible enzymes, ODC,

SAM-DC, and SAT, and in addition by transport. In certain cell types, terminal polyamine catabolism by copper-dependent amine oxidases, such as diamine oxidase, in addition limit polyamine concentrations. It is suggested that acetylation is not only a means to transform polyamines into an excretable form, but that it is an inherent part of a complex regulatory circuit, which is assumed to control and adjust the levels of the individual polyamines according to the physiological needs of the cell. This regulatory system is supposed to be capable of producing even relatively rapid changes of spermidine and spermine concentrations, in spite of a slow basal turnover rate of the polyamines.

REFERENCES

1. **Hirsch, J. G.,** Spermine oxidase: an amine oxidase with specificity for spermine and spermidine, *J. Exp. Med.*, 97, 345, 1953.
2. **Morgan, D. M. L.,** Polyamine oxidases, in *Polyamines in Biomedical Research,* Gaugas, J. M., Ed., John Wiley & Sons, Chichester, England, 1980, 285.
3. **Seiler, N., Bolkenius, F. N., and Knödgen, B.,** The influence of catabolic reactions on polyamine excretion, *Biochem. J.*, 225, 219, 1985.
4. **Seiler, N.,** Inhibition of enzymes oxidizing polyamines, in *Inhibition of Polyamine Synthesis: Biological Significance and Basis for New Therapies,* McCann, P. P., Pegg, A. E., and Sjoerdsma, A., Eds., Academic Press, New York, 1987, 49.
5. **Seiler, N., Bolkenius, F. N., and Rennert, O. M.,** Interconversion, catabolism and elimination of the polyamines, *Med. Biol.*, 59, 334, 1981.
6. **Seiler, N., Bolkenius, F. N., and Knödgen, B.,** Acetylation of spermidine in polyamine catabolism, *Biochim. Biophys. Acta,* 633, 181, 1980.
7. **Siimes, M.,** Studies on the metabolism of 1,4-^{14}C-spermidine and 1,4-^{14}C-spermine in the rat, *Acta Physiol. Scand. Suppl.*, 298, 1, 1967.
8. **Seiler, N.,** Polyamine metabolism in the brain, in *Polyamines in Normal and Neoplastic Growth,* Russell, D. H., Ed., Raven Press, New York, 1973, 137.
9. **Sturman, J. A., Ingoglia, N. A., and Lindquist, T. D.,** Interconversion of putrescine, spermidine and spermine in gold fish and rat retina, *Life Sci.*, 19, 719, 1976.
10. **Halliday, C. A. and Shaw, G. G.,** The distribution and metabolism of putrescine, spermidine and spermine injected into the cerebral ventricles of rabbits, *J. Neurochem.*, 26, 1199, 1976.
11. **Hietala, O. A., Pajunen, A. E. I., Lapinjoki, S. P., and Piha, R. S.,** Interconversion of polyamines in mouse brain, *Acta Univ. Oul. A97: Biochem.*, 29, 49, 1980.
12. **Pegg, A. E., Hibasami, H., Matsui, I., and Bethell, D. R.,** Formation and interconversion of putrescine and spermidine in mammalian cells, *Adv. Enzyme Regul.*, 19, 427, 1981.
13. **Fischer, H. A., Korr, H., Seiler, N., and Werner, G.,** Interrelationships between polyamines and nucleic acids. III. Metabolic and autoradiographic studies of putrescine in mouse brain, *Brain Res.*, 39, 197, 1972.
14. **Antrup, H. and Seiler, N.,** On the turnover of polyamines spermidine and spermine in mouse brain and other organs, *Neurochem. Res.*, 5, 123, 1980.
15. **Seiler, N., Al-Therib, M. J., Fischer, H. A., and Erdmann, G.,** Dynamic and regional aspects of polyamine metabolism in the brain of trout (*Salmo irideus,* Gibb), *Int. J. Biochem.*, 10, 961, 1979.
16. **Snyder, S. H., Kreutz, D. S., Medina, V. J., and Russell, D. H.,** Polyamine synthesis and turnover in rapidly growing tissue, *Ann. N.Y. Acad. Sci.*, 171, 749, 1970.
17. **Russell, D. H., Medina, V. J., and Snyder, S. H.,** The dynamics of synthesis and degradation of polyamines in normal and regenerating rat liver and brain, *J. Biol. Chem.*, 245, 6732, 1970.
18. **Nakajima, T.,** Formation of putreanine, *N*-(4-aminobutyl)-3-aminopropionic acid from spermidine in rat brain and liver, *J. Neurochem.*, 20, 735, 1973.
19. **Shaskan, E. G. and Snyder, S. H.,** Polyamine turnover in different regions of rat brain, *J. Neurochem.*, 20, 1453, 1973.
20. **Raina, A. and Jänne, J.,** Physiology of the natural polyamines putrescine, spermidine and spermine, *Med. Biol.*, 53, 121, 1975.
21. **Pegg, A. E. and Williams-Ashman, H. G.,** Biosynthesis of putrescine, in *Polyamines in Biology and Medicine,* Morris, D. R. and Marton, L. J., Eds., Marcel Dekker, New York, 1981, 3.
22. **Williams-Ashman, H. G. and Pegg, A. E.,** Aminopropyl group transfers in polyamine biosynthesis, in *Polyamines in Biology and Medicine,* Morris, D. R. and Marton, L. J., Eds., Marcel Dekker, New York, 1981, 43.

23. **Pegg, A. E., Seely, J. E., Pösö, H., Della Ragione, F., and Zagon, I. S.,** Polyamine biosynthesis and interconversion in rodent tissues, *Fed. Proc. Fed. Am. Soc. Exp. Biol.*, 41, 3065, 1982.
24. **Tabor, C. W., Tabor, H., and Bachrach, U.,** Identification of the aminoaldehydes produced by the oxidation of spermine and spermidine in purified plasma amine oxidase, *J. Biol. Chem.*, 239, 2194, 1964.
25. **Blaschko, H.,** The amine oxidases of mammalian blood plasma, *Adv. Comp. Physiol. Biochem.*, 1, 67, 1962.
26. **Seiler, N., Knödgen, B., Bink, G., Sarhan, S., and Bolkenius, F.,** Diamine oxidase and polyamine catabolism, *Adv. Polyamine Res.*, 4, 135, 1983.
27. **Hölttä, E.,** Oxidation of spermidine and spermine in rat liver: purification and properties of polyamine oxidase, *Biochemistry*, 16, 91, 1977.
28. **Bolkenius, F. N. and Seiler, N.,** Acetylderivatives as intermediates in polyamine catabolism, *Int. J. Biochem.*, 13, 287, 1981.
29. **Seiler, N.,** Amide-bond-forming reactions of polyamines, in *Polyamines in Biology and Medicine*, Morris, D. R. and Marton, L. J., Eds., Marcel Dekker, New York, 1981, 127.
30. **Matsui, I. and Pegg, A. E.,** Increase in acetylation of spermidine in rat liver extracts brought about by treatment with carbon tetrachloride, *Biochem. Biophys. Res. Commun.*, 92, 1009, 1980.
31. **Della Ragione, F. and Pegg, A. E.,** Purification and characterization of spermidine/spermine N^1-acetyltransferase from rat liver, *Biochemistry*, 24, 6152, 1982.
32. **Della Ragione, F. and Pegg, A. E.,** Studies on the specificity and kinetics of rat liver spermidine/spermine N^1-acetyltransferase, *Biochem. J.*, 213, 701, 1983.
33. **Matsui, I., Wiegand, L., and Pegg, A. E.,** Properties of spermidine *N*-acetyltransferase from livers of rats treated with carbon tetrachloride and its role in the conversion of spermidine into putrescine, *J. Biol. Chem.*, 256, 2454, 1981.
34. **Persson, L. and Pegg, A. E.,** Studies of the induction of spermidine/spermine N^1-acetyltransferase using a specific antiserum, *J. Biol. Chem.*, 259, 12364, 1984.
35. **Pösö, H. and Pegg, A. E.,** Effect of carbon tetrachloride on polyamine metabolism in rodent liver, *Arch. Biochem. Biophys.*, 217, 730, 1982.
36. **Matsui, I. and Pegg, A. E.,** Effect of thioacetamide, growth hormone or partial hepatectomy on spermidine acetylase activity of rat liver cytosol, *Biochim. Biophys. Acta*, 633, 87, 1980.
37. **Matsui, I. and Pegg, A. E.,** Induction of spermidine N^1-acetyltransferase by dialkylnitrosamines, *Cancer Res.*, 42, 2990, 1982.
38. **Grillo, M. A.,** Acetylation of spermidine in chicken tissue: effect of starvation, insulin and glucagon, *Adv. Polyamine Res.*, 4, 321, 1983.
39. **Danzin, C., Bolkenius, F. N., Claverie, N., Wagner, J., and Grove, J.,** Secretin-induced accumulation of N^1-acetylspermidine and putrescine in the rat pancreas, *Biochem. Biophys. Res. Commun.*, 109, 1234, 1982.
40. **Matsui, I. and Pegg, A. E.,** Induction of spermidine N^1-acetyltransferase in rat kidney by treatment with folic acid, *FEBS Lett.*, 139, 205, 1982.
41. **Seiler, N., Bolkenius, F. N., Bey, P., Mamont, P. S., and Danzin, C.,** Biochemical significance of inhibition of polyamine oxidase, in *Recent Progress of Polyamine Research*, Selmeci, L., Brosnan, M. E., and Seiler, N., Eds., Akademiai Kiado, Budapest, 1985, 305.
42. **Pegg, A. E., Erwin, B. G., and Persson, L.,** Induction of spermidine/spermine N^1-acetyltransferase by methylglyoxal bis(guanylhydrazone), *Biochim. Biophys. Acta*, 842, 111, 1985.
43. **Pegg, A. E. and Erwin, B. G.,** Induction of spermidine/spermine N^1-acetyltransferase in rat tissues by polyamines, *Biochem. J.*, 231, 285, 1985.
44. **Seiler, N., Bolkenius, F. N., and Sarhan, S.,** Formation of acetylpolyamines in the liver of fasting animals, *Int. J. Biochem.*, 13, 1205, 1981.
45. **Persson, L., Erwin, B. G., and Pegg, A. E.,** Spermidine/spermine N^1-acetyltransferase: studies using a specific antiserum, in *Recent Progress in Polyamine Research*, Selmeci, L., Brosnan, M. E., and Seiler, N., Eds., Akademiai Kiado, Budapest, 1985, 287.
46. **Seiler, N., Bolkenius, F. N., Knödgen, B., and Heagele, K.,** The determination of N^1-acetylspermine in mouse liver, *Biochim. Biophys. Acta*, 676, 1, 1981.
47. **Abdel-Monem, M. M. and Merdink, J. L.,** Formation of N^1-acetylspermidine in rat liver after treatment with carbon tetrachloride, *Life Sci.*, 28, 2017, 1981.
48. **Hölttä, E., Sinervirta, R., and Jänne, J.,** Synthesis and accumulation of polyamines in rat liver, regenerating after treatment with carbon tetrachloride, *Biochem. Biophys. Res. Commun.*, 54, 350, 1973.
49. **Matsui, I. and Pegg, A. E.,** Effect of inhibitors of protein synthesis on rat liver spermidine N^1-acetyltransferase, *Biochim. Biophys. Acta*, 675, 373, 1981.
50. **Seiler, N. and Al-Therib, M. J.,** AcetylCoA:1,4-diaminobutane *N*-acetyltransferase. Occurrence in vertebrate organs and subcellular localization, *Biochim. Biophys. Acta*, 354, 206, 1974.
51. **Blankenship, J. and Walle, T.,** In vitro studies of enzymatic synthesis and metabolism of N-acetylated polyamines, *Adv. Polyamine Res.*, 2, 97, 1978.

52. **Erwin, B. G., Persson, L., and Pegg, A. E.,** Differential inhibition of histone and polyamine acetylases by multisubstrate analogues, *Biochemistry,* 23, 4250, 1984.
53. **Libby, P. R.,** Rat liver nuclear *N*-acetyltransferases: separation of two enzymes with both histone and spermidine acetyltransferase activity, *Arch. Biochem. Biophys.,* 203, 384, 1980.
54. **Mamont, P., Bolkenius, F., Seiler, N., Bey, P., and Kolb, M.,** Biochemical consequences of polyamine oxidase and N^8-acetylspermidine deacetylase inhibition in cultured rat hepatoma (HTC) cells, Abstr. No. 6, Int. Conf. on Polyamines, Budapest, 1984.
55. **Abdel-Monem, M. M. and Ohno, K.,** Polyamine metabolism. II. *N*-(Monoaminoalkyl)- and *N*-(polyaminoalkyl)acetamides in human urine, *J. Pharm. Sci.,* 66, 1195, 1977.
56. **Seiler, N., Bolkenius, F. N., Knödgen, B., and Mamont, P.,** Polyamine oxidase in rat tissues, *Biochem. Biophys. Acta,* 615, 480, 1980.
57. **Hölttä, E.,** Polyamine oxidase, *Meth. Enzymol.,* 94, 306, 1983.
58. **Beard, M. E., Baker, R., Conomas, P., Pugatch, D., and Holtzman, E.,** Oxidation of oxalate and polyamines by rat peroxisomes, *J. Histochem. Cytochem.,* 33, 460, 1985.
59. **Morgan, D. M. L.,** Human pregnancy-associated polyamine oxidase: partial purification and properties, *Biochem. Soc. Trans.,* 13, 351, 1985.
60. **Bolkenius, F. N., Bey, P., and Seiler, N.,** Specific inhibition of polyamine oxidase in vivo is a method for the elucidation of its physiological role, *Biochim. Biophys. Acta,* 838, 69, 1985.
61. **Fujimoto, D., Kayama, T., and Tamiya, N.,** *N*-Acetyl-β-alanine deacetylase in hog kidney, *Biochim. Biophys. Acta,* 167, 407, 1968.
62. **Bey, P., Bolkenius, F. N., Seiler, N., and Casara, P.,** *N*-2,3-Butadienyl-1,4-butanediamine derivatives: potent irreversible inactivators of mammalian polyamine oxidase, *J. Med. Chem.,* 28, 1, 1985.
63. **Seiler, N., Jung, M. J., and Koch-Weser, J., Eds.,** *Enzyme Activated Irreversible Inhibitors,* Elsevier/North-Holland Biomedical Press, Amsterdam, 1978.
64. **Krantz, A., Kobel, B., Sachdeva, Y. P., Salach, J., Claesson, A., and Sahlberg, C.,** Allenic amines as inactivators of mitochondrial monoamine oxidase, in *Drug Action and Design: Mechanism-Based Enzyme Inhibitors,* Kalman, T., Ed., Elsevier/North-Holland Biomedical Press, Amsterdam, 1979, 145.
65. **Quash, G. and Taylor, D. R.,** Serum β-aminopropionaldehyde: identification and origin, *Clin. Chim. Acta,* 30, 17, 1970.
66. **Mamont, P. S., Seiler, N., Siat, M., Joder-Ohlenbusch, A.-M., and Knödgen, B.,** Metabolism of acetyl derivatives of polyamines in cultured polyamine-deficient rat hepatoma cells, *Med. Biol.,* 59, 347, 1981.
67. **Mamont, P., Joder-Ohlenbusch, A.-M., and Siat, M.,** Metabolism of N^8-monoacetylspermidine in rat hepatoma cells. Investigation of its effect on the activity of L-ornithine decarboxylase, *Eur. J. Biochem.,* 133, 613, 1983.
68. **Seiler, N., Koch-Weser, J., Knödgen, B., Richards, W., Tardif, C., Bolkenius, F. N., Schechter, P., Tell, G., Mamont, P., Fozard, J., Bachrach, U., and Grosshans, E.,** The significance of acetylation in the urinary excretion of polyamines, *Adv. Polyamine Res.,* 3, 197, 1981.
69. **Nakajima, T., Noto, T., and Kato, N.,** Isolation and identification of polyamine metabolites in urine of animals, *Physiol. Chem. Phys.,* 12, 401, 1980.
70. **Seiler, N. and Knödgen, B.,** *N*-(3-Aminopropyl)pyrrolidin-2-one: a physiological excretory product deriving from spermidine, *Int. J. Biochem.,* 15, 907, 1983.
71. **Van den Berg, G. A., Elzinga, H., Nagel, G. T., Kingsma, A. W., and Muskiet, F. A. J.,** Catabolism of polyamines in the rat. Polyamines and their non-α-amino-acid metabolites, *Biochim. Biophys. Acta,* 802, 175, 1984.
72. **Seiler, N.,** Acetylpolyamines as substrates of amine oxidases, in *Structure and Function of Amine Oxidases,* Mondovi, B., Ed., CRC Press, Boca Raton, Fla., 1985, 21.
73. **McCormick, F.,** Kinetics of polyamine synthesis and turnover in mouse fibroblasts, *Biochem. J.,* 174, 427, 1978.
74. **Melvin, M. A. L., Wallace, H. M., and Keir, H. M.,** Conjugation of polyamines in mammalian cells in culture, *Physiol. Chem. Phys.,* 12, 431, 1980.
75. **Wallace, H. M. and Keir, H. M.,** Polyamine acetylation and excretion by mammalian cells in culture, in *Recent Progress in Polyamine Research,* Selmeci, L., Brosnan, M. E., and Seiler, N., Eds., Akademiai Kiado, Budapest, 1985, 297.
76. **Raina, A., Pajula, R.-L., Eloranta, T., and Tuomi, K.,** Synthesis of polyamines and *S*-adenosylmethionine in rat tissues and tumor cells: effect of D,L-α-hydrazino-δ-aminovaleric acid on cell proliferation, *Adv. Polyamine Res.,* 1, 75, 1978.
77. **Ientile, R., Ruggeri, P., Russo, P., and Macaione, S.,** Effect of intraventricular putrescine on adenosylmethionine decarboxylase in rat hypothalamus and caudate nucleus, *J. Neurochem.,* 44, 1315, 1985.
78. **Mamont, S., Danzin, C., Wagner, J., Siat, M., Joder-Ohlenbusch, A. -M., and Claverie, N.,** Accumulation of decarboxylated *S*-adenosyl-L-methionine in mammalian cells as a consequence of the inhibition of putrescine biosynthesis, *Eur. J. Biochem.,* 123, 499, 1982.

79. **Bolkenius, F. N. and Seiler, N.,** Developmental aspects of polyamine interconversion in rat brain, *Int. J. Dev. Neurosci.*, 4, 217, 1985.
80. **Cohen, S. S.,** *Introduction to the Polyamines,* Prentice-Hall, Englewood Cliffs, N.J., 1971.
81. **Bloomfield, V. A. and Wilson, R. W.,** Interaction of polyamines with polynucleotides, in *Polyamines in Biology and Medicine,* Morris, D. R. and Marton, L. J., Eds., Marcel Dekker, New York, 1981, 183.
82. **Abraham, K. A. and Pihl, A.,** Role of polyamines in macromolecular synthesis, *Trends Biochem. Sci.*, 6, 106, 1981.
83. **Braunlin, W. H., Strick, T. J., and Record, M. T., Jr.,** Equilibrium dialysis studies of polyamine binding to DNA, *Biopolymers,* 21, 1301, 1982.
84. **Stevens, L.,** Studies on the interaction of homologues of spermine with deoxyribonucleic acid and with bacterial protoplasts, *Biochem. J.*, 103, 811, 1967.
85. **Brosnan, M. E. and Hu, Y. -W.,** Synthesis and function of polyamines in mammary gland during lactation, in *Recent Progress in Polyamine Research,* Selmeci, L., Brosnan, M. E., and Seiler, N., Eds., Akademiai Kiado, Budapest, 1985, 169.
86. **Wemmer, D. E., Srivenugopal, K. S., Reid, B. R., and Morris, D.,** Nuclear magnetic resonance studies of polyamine binding to a defined DNA sequence, *J. Mol. Biol.*, 185, 457, 1985.
87. **Thomas, T. J., Bloomfield, V. A., and Canellakis, Z. N.,** Differential effects on the B-to-Z transition of poly(dG-me^{5}dC) produced by N^1- and N^8-acetylspermidine, *Biopolymers,* 24, 725, 1985.
88. **Canellakis, E. S., Viceps-Madore, D., Kyriakidis, D. A., and Heller, J. S.,** The regulation and function of ornithine decarboxylase and of the polyamines, *Curr. Topics Cell. Regul.*, 15, 155, 1979.
89. **Shin, W. W., Fong, W.-F., and Wong, P. C.-L.,** Limited blood-brain barrier transport of polyamines, *J. Neurochem.*, 44, 1056, 1985.
90. **Seiler, N. and Bolkenius, F. N.,** Polyamine reutilization and turnover in brain, *Neurochem. Res.*, 10, 529, 1985.
91. **Russell, D. H.,** Polyamines as biochemical markers of tumor growth parameters, in *Biochemical Markers for Cancer,* Ming Chu, T., Ed., Marcel Dekker, New York, 1982, 241.
92. **Seiler, N. and Schmidt-Glenewinkel, T.,** Regional distribution of putrescine, spermidine and spermine in relation to the distribution of RNA and DNA in the rat nervous system, *J. Neurochem.*, 24, 791, 1975.
93. **Seiler, N.,** Turnover of polyamines, in *Polyamines in Biology and Medicine,* Morris, D. R. and Marton, L. J., Eds., Marcel Dekker, New York, 1981, 169.
94. **Wallace, H. M. and Keir, H. M.,** Uptake and excretion of polyamines from baby hamster kidney cells (BHK-21/C13). The effect of serum on confluent cell cultures, *Biochim. Biophys. Acta,* 676, 25, 1981.
95. **Bachrach, U. and Seiler, N.,** Formation of acetylpolyamines and putrescine from spermidine by normal and transformed chick embryo fibroblasts, *Cancer Res.*, 41, 1205, 1981.
96. **Bachrach, U., Menashe, M., Faber, J., Desser, H., and Seiler, N.,** Polyamine biosynthesis and metabolism in transformed human lymphocytes, *Adv. Polyamine Res.*, 4, 259, 1981.
97. **Wallace, H. M. and Keir, H. M.,** A comparison of polyamine metabolism in normal and transformed baby-hamster-kidney cells, *Biochem. J.*, 202, 785, 1982.
98. **Wallace, H. M., MacGowan, S. H., and Keir, H. M.,** Polyamine metabolism in mammalian cells in culture, *Biochem. Soc. Trans.*, 13, 329, 1985.

Chapter 12

OXIDATION OF DIAMINES AND POLYAMINES

Bruno Mondovì, Pierluigi Riccio, Enzo Agostinelli, and Giordana Marcozzi

TABLE OF CONTENTS

I. INTRODUCTION

The metabolic pathways responsible for the inactivation of all biogenic amines including di- and polyamines are relatively simple and essentially the same in all living organisms. The oxidation of polyamines plays a role of particular importance in polyamine metabolism in that not only is it responsible for the production of catabolytes to be excreted, but above all it represents the key reaction by means of which diamines and polyamines, basic molecules, are irreversibly converted into compounds having different physicochemical properties and therefore different physiological functions.

For the sake of brevity, the term "polyamines" is often used in this chapter instead of the expression "diamines and polyamines", thus including putrescine and cadaverine.

Oxidized polyamines contain one or more aldehydic functional groups in the molecule. Oxidation is accomplished either by the removal of one primary amino group or by the cleavage of the molecule containing one or two secondary amino groups at one of the secondary nitrogen atoms at a time, and oxidation of the terminal carbon of the remaining molecule to an aldehyde group during the same catalytic cycle. Oxidative deamination is operated by a class of enzymes, the amine oxidases (AOs), which represent the main topic concerned in this chapter.

$$R\text{–}CH_2\text{–}NH_2 + H_2O + O_2 \rightarrow R\text{–}CHO + NH_3 + H_2O_2$$

Cleavage at the primary amino group (operated by both

Cu- and FAD-dependent AOs) (1)

$$R\text{–}CH_2\text{–}NH\text{–}CH_2\text{–}R + H_2O + O_2 \rightarrow R\text{–}CHO + H_2N\text{–}CH_2\text{–}R + H_2O_2$$

Cleavage at the secondary amino group (operated by the FAD-AOs) (2)

Aldehydes derived from polyamine oxidation generally undergo further modification: either they can be oxidized to carboxylic acids in an enzymatic reaction involving aldehyde dehydrogenase or, alternatively, they may be reduced to alcohols by alcohol dehydrogenase.

Quantitatively, a minor amount of the intracellular polyamines is oxidatively deaminated by AOs after being conjugated with acetic acid to yield the aminoacetyl derivatives of the parent compounds in a reaction involving the amine, acetyl-CoA, and a specific acetyltransferase. In the central nervous system the latter is the principal metabolic pathway for the breakdown of putrescine.

As the research on polyamine oxidation has been carried out with a heterogenous variety of living organisms, this chapter contains information on a large number of AOs from various sources; special emphasis has been given to the functional characteristics of the enzymes (i.e., substrate specificity and mode of oxidation) rather than to their structural features. In order to make the reading of the text as steadfast as possible and to avoid misunderstanding concerning the identity of some enzymes, we have adopted the nomenclature and abbreviations proposed for AOs in a recent book on this subject.[1] Table 1 reports, in alphabetical order, the abbreviation and corresponding appropriate name of each of the enzymes appearing in this chapter. The oxidative pathways followed by polyamines are summarized at the end of this chapter.

II. OCCURRENCE OF AMINE OXIDASES IN LIVING ORGANISMS

AOs are ubiquitous enzymes that have been found practically in all living organisms

Table 1
ABBREVIATION AND CORRESPONDING NAME OF SOME AMINE OXIDASES

Abbreviation	Name
ANAO	*Aspergillus niger* amine oxidase (Cu)
AO(s)	Amine oxidase(s)
ATPAO	*Aspergillus terreus* polyamine oxidase (FAD)
BSAO	Bovine serum amine oxidase (Cu)
DAO	Diamine oxidase (Cu)
ELAO	*Euphorbia* latex amine oxidase (Cu)
HPDAO	Human placental diamine oxidase (Cu)
LSAO	Lentil seedlings amine oxidase (Cu)
MAO	Monoamine oxidase (FAD)
MRAO	*Micrococcus rubens* amine oxidase (Cu)
MSPAO	Maize seedlings polyamine oxidase (FAD)
PAO	Polyamine oxidase (FAD)
PCAO	*Pencillium chrysogenum* amine oxidase (Cu)
PCPAO	*P. chrysogenum* polyamine oxidase (FAD)
PSAO	Pea seedlings amine oxidase (Cu)
RLPAO	Rat liver polyamine oxidase (FAD)
SKDAO	Swine kidney diamine oxidase (Cu)
SSAO	Swine serum amine oxidase (Cu)
VLAO	*Vicia* leaves amine oxidase (Cu)

Table 2
COMPARATIVE TABLE OF THE MAIN SUBSTRATES OF SOME AMINE OXIDASES

	BSAO	SKDAO	RLPAO	MAO
Pu		***		
AcPu		**		*
Ca		***		
AcCa		**		*
Spd	***		*	
N¹AcSpd		**	***	
N⁸AcSpd	***	*		*
Spm	***		**	
N¹AcSpm		*	***	
N¹,N¹²diAcSpm			***	
His		***		

Note: *: very weak oxidation, **: weak oxidation, and ***: good oxidation. Pu: putrescine, AcPu: monoactylputrescine, Ca: cadaverine, AcCa: monoacetylcadaverine, spd: spermidine, N^1AcSpd: N^1-acetylspermidine, N^8AcSpd: N^8-acetylspermidine, Spm: spermine, N^1AcSpm: N^1-acetylspermine, N^1,N^{12}diAcSpm: N^1,N^{12}-diacetylspermine, and His: histamine.

including animals, plants, and microorganisms. To this class belong a large numer of enzymic proteins showing not only heterogeneity in structure, but also differences in substrate and inhibitor specificity as well as different subcellular localization.

AOs catalyze the oxidative deamination of diamines, polyamines, and their acetyl derivatives as well as of monoamines and of some products of previous polyamine oxidation, namely aminoaldehydes. Table 2 reports the substrate specificity of some AOs.

A first tentative classification of AOs was made according to their substrate specificity, i.e., mono-, di-, and polyamine oxidase (PAO). Unfortunately, such a classification appeared unsatisfactory as it did not provide adequate information in the precise identification of these enzymes. As an example, some Cu-containing serum AOs oxidize monoamines as well as spermidine and spermine, these latter being also good substrates of the intracellular flavin-adenin-dinucleotide (FAD)-dependent PAO.

The best criterion of classification seems that of dividing AOs into two groups: the "flavin" AOs containing FAD as a cofactor, and the Cu AOs having, in addition to Cu, another organic prosthetic group, the nature of which will be discussed later.

A. Microorganisms

Both Cu- and FAD-dependent AOs have been purified and studied in a variety of microorganisms including bacteria and fungi. The characteristics and properties of some of them are briefly reported here.

1. Serratia marcescens

An enzyme which oxidizes spermidine was purified from *Serratia marcescens*.[2] The enzyme is a monomeric protein, weighing approximately 76,000 daltons, which shows a major absorption peak at 414 nm. This enzyme is quite unusual for an AO in that it contains, in addition to FAD, a heme moiety, iron-protoporphyrin IX, as a cofactor which is responsible for the 414-nm peak.[2] Moreover, the enzyme requires, unlike the other AOs, an electron carrier, the natural acceptor being probably cytochrome C.[2] Since it is incapable of utilizing dioxygen directly, it should be classified as a dehydrogenase.

The *Serratia* enzyme cleaves spermidine at the secondary nitrogen as follows:

$$\underset{\text{spermidine}}{H_2N\text{–}(CH_2)_4\text{–}NH\text{–}(CH_2)_3\text{–}NH_2} + H_2O + FAD \rightarrow$$

$$\underset{\gamma\text{-aminobutyraldehyde}}{H_2N\text{–}(CH_2)_3\text{–}CHO} + \underset{\text{1,3-diaminopropane}}{H_2N\text{–}(CH_2)_3\text{–}NH_2} + FADH_2 \qquad (3)$$

In addition to spermidine, for which the K_m value is lower than 10^{-8} *M*, spermine and N^1-acetylspermine, as well as several other polyamines, are oxidized by spermidine dehydrogenase, the better substrates being those compounds containing a 4-aminobutylimino moiety.[3] N^8-acetylspermidine and diamines are not oxidized by *Serratia* spermidine dehydrogenase.

2. Micrococcus rubens

A FAD-dependent AO was purified from a strain of *Micrococcus rubens* which oxidized putrescine, cadaverine, and spermidine.[4,5] MRAO has a molecular weight of 82,000 daltons and contains 1 mol of FAD per mole of enzyme.[6,7]

3. Aspergillus niger

A Cu-containing AO was found to be inducibly formed in mycelia of *Aspergillus niger* grown on a monoamine- or diamine-containing medium as the sole nitrogen source.[8,9] The enzyme is a dimeric protein having a molecular weight of about 250,000 daltons, composed of two identical subunits, and containing two Cu atoms per enzyme molecule.[10] Titration of the carbonyl cofactor suggested the presence of two organic prosthetic groups per dimer.[11] ANAO shows optical properties which resemble those of plant and animal Cu-dependent AOs, with an absorbance maximum at 480 nm. The enzyme oxidizes preferentially a variety of monoamines, but diamines of C_4 to C_6 and agmatine, as well as histamine, are also utilized as substrates, although in lower rates with respect to the aliphatic monoamines.[12]

4. Penicillum chrysogenum

Agmatine, putrescine, or spermidine addition to the medium as nitrogen or carbon sources stimulate the synthesis of an inducible Cu-dependent AO from *Pencillum chrysogenum.*[13] The enzyme weighs about 160,000 daltons, and is composed of two identical subunits. The Cu content has been determined as two atoms per dimer.[14] This AO catalyzes the oxidative deamination of agmatine, which is its best substrate, but also C_3-C_5 aliphatic diamines and histamine are slowly oxidized. Monoamines and polyamines are not substrates of this Cu-AO.[14]

This same fungus, when grown on a spermine- or spermidine-containing medium as nitrogen sources, produces an FAD-dependent PAO weighing 160,000 daltons.[15] This PAO contains two moles of FAD per mole of enzyme and is composed of two identical subunits.[16] Spermine and spermidine, but not mono- and diamines, are oxidized at the secondary nitrogen to yield putrescine and 3-aminopropionaldehyde as final products.[16] The reaction stoichiometry is as follows:

$$\underset{\text{spermine}}{H_2N{-}(CH_2)_3{-}NH{-}(CH_2)_4{-}NH{-}(CH_2)_3{-}NH_2} + 2H_2O + 2O_2 \rightarrow$$

$$\underset{\text{putrescine}}{H_2N{-}(CH_2)_4{-}NH_2} + \underset{\text{3-aminopropionaldehyde}}{2H_2N{-}(CH_2)_2{-}CHO} + 2H_2O_2 \qquad (4)$$

5. Aspergillus terreus

This fungus produces a PAO similar to that of *P. chrysogenum.* The enzyme has a molecular weight of 130,000 daltons, is composed of two identical subunits, and contains two molecules of FAD per enzyme molecule. ATPAO oxidizes spermidine and spermine at the same site as PCPAO.[17]

B. Plants

Both Cu- and FAD-dependent AOs have been found in plants. Although considerable information is available on the former, much less is known on the FAD enzymes.

Cu-AOs have been investigated from many vegetable sources, but only four of them have been highly purified and characterized in detail. Their molecular and spectroscopic features are similar to those of the animal Cu-AOs and, like the latter, they catalyze the oxidative deamination of primary amino groups. All of them are diamine oxidases.

Vegetable FAD-AOs have been found in members of the Gramineae family and cleave polyamines at the secondary nitrogen. The only plant PAO in which FAD has been extracted and identified is that of the maize seedlings, whereas for those from oats and barley seedlings the involvement of FAD as a cofactor has been claimed on the basis of indirect evidence.

1. Pisum sativum

The Cu-dependent AO purified from pea seedlings is a dimeric protein weighing about 180,000 daltons.[18-20] It oxidizes preferentially diamines, but spermidine, agmatine, and histamine are also good substrates. Spermine is oxidized as well, but at a considerably lower rate.[21] The reaction stoichiometry with putrescine as substrate, identical to the other plant Cu-AO-catalyzed reactions, is the following:

$$H_2N{-}(CH_2)_4{-}NH_2 + H_2O + O_2 \rightarrow H_2N{-}(CH_2)_3{-}CHO + NH_3 + H_2O_2 \qquad (5)$$

2. Lens esculenta

The AO purified from lentil seedlings has a molecular weight of about 180,000 daltons and is composed of two identical subunits. It contains two Cu atoms per enzyme molecule and oxidizes only diamines and agmatine.[22]

3. Euphorbia characias

A Cu-containing AO was purified from the latex of *Euphorbia,* a perennial mediterranean shrub.[23] The molecular weight of the enzyme, as determined by analytical ultracentrifugation, is comparable to those of PSAO and LSAO, about 180,000 daltons,[24] whereas gel chromatography gives a smaller value. ELAO is a dimeric enzyme which oxidizes diamines and agmatine, but not polyamines.[23]

4. Vicia faba

The leaves of *Vicia faba* (broad bean) contain a Cu-AO which appears to be a dimeric protein weighing about 130,000 daltons.[25] Like the other plant Cu-AOs, VLAO oxidatively deaminates putrescine and cadaverine, but agmatine and spermidine also show a good affinity for this enzyme. Histamine and spermine are substrates as well, but are oxidized much more slowly.[26]

5. Zea mays

PAO from maize seedlings is a monomeric, FAD-dependent enzyme weighing 65,000 daltons.[27,28] MSPAO, like the corresponding enzymes from oats and barley, are similar to the bacterial FAD-PAOs in that they attack the secondary amino group of polyamines to produce 1,3-diaminopropane.[29] Spermidine and spermine are best oxidized by plant PAOs at a slightly acidic pH around 6.[28]

C. Animals

The most numerous reports on AOs are available on the animal enzymes. Under physiological conditions, some amine oxidizing activity is present in practically all animal tissues, as well as in certain biological fluids such as the blood plasma and seminal fluid.

Both Cu- and FAD-dependent animal AOs have been investigated in detail; the former are either extra- or intracellular enzymes, the latter mainly intracellular. Among those having intracellular localization, it has been observed that some of them are associated with the particulate fraction, such as the Cu-SKDAO inside the microsomal vescicles and the FAD-MAO with the outer mitochondrial membrane. Other enzymes, such as the human intestinal DAO, are dissolved in the cytosol.

Due to the vast literature dealing with AOs from a large number of animal species and, within a single species, from various tissues and organs, it is practically impossible to draw a comprehensive picture accounting for even the differences in substrate specificity between each other. Nevertheless, it is necessary to supply some fundamental data on the animal enzymes which, obviously, are the most interesting ones for the physiology of polyamines under a perspective of possible clinical application (i.e., pharmacological treatment). Thus, only the most significant and best-characterized enzymes from serum, kidney, placenta, and liver of various animal species will be briefly discussed.

1. Bovine and Swine Serum Amine Oxidases

Among the serum AOs, bovine serum amine oxidase (BSAO) and swine serum amine oxidase (SSAO) are the best characterized.[30-43] Their molecular weights correspond, respectively, to about 180,000 and 190,000 daltons.[36,37,42] Both enzymes are dimeric proteins containing two Cu atoms and one phenylhydrazine-titrable carbonyl cofactor per molecule.[35,36,39-42] Their amino acid composition is similar,[30,38] and they both catalyze the oxidative deamination of primary amino groups of several aliphatic and aromatic monoamines, but while BSAO oxidizes spermidine and spermine, SSAO does not, being capable of deaminating histamine.[44] BSAO oxidizes spermine as follows:

$$H_2N\text{-}(CH_2)_3\text{-}NH\text{-}(CH_2)_4\text{-}NH\text{-}(CH_2)_3\text{-}NH_2 + 2H_2O + 2O_2 \rightarrow$$
spermine

$$OHC\text{-}(CH_2)_2\text{-}NH\text{-}(CH_2)_4\text{-}NH\text{-}(CH_2)_2\text{-}CHO + 2NH_3 + 2H_2O_2 \quad (6)$$
dioxidized spermine

2. Swine Kidney Diamine Oxidase

SKDAO is a dimeric enzyme weighing about 180,000 daltons and resembling the serum AOs as for both Cu and organic cofactor content.[45-49] This enzyme is particularly active towards C_4-C_7 aliphatic diamines and histamine.[50] Despite the demonstration that a single enzyme was responsible for both histamine and diamine oxidizing activities that has been made several years ago, some recent papers on DAO still indicate this enzyme as ''histaminase''.

3. Human Placental Diamine Oxidase

HPDAO appears to be different from the classical Cu-AOs so far described in that it has been claimed to be a monomer weighing about 80,000 daltons. HPDAO oxidizes several diamines and histamine.[51,52] It is worthwhile to stress its biochemical and immunological similarity with the DAO present in the effusions of cancer patients and tumor extracts.[53,54]

4. Rat Liver Polyamine Oxidase

PAO from rat liver is a FAD-containing protein consisting of a single polypeptide chain and weighing approximately 60,000 daltons.[55] It catalyzes the cleavage of spermidine and spermine at the secondary nitrogen with production of 3-aminopropionaldehyde and, respectively, putrescine and spermidine.[55]

III. STRUCTURE AND MECHANISM OF ACTION OF AMINE OXIDASES

Evident differences in the structure can be observed between the Cu- and FAD-dependent AOs; therefore, these two kinds of oxidases must be dealt separately. Because of their role of primary importance in polyamine metabolism, and because they are the most numerous enzymes of this class that have been described, the Cu-AOs will be discussed in much greater detail than the FAD enzymes.

A. Structure of Copper Amine Oxidases

The structure of Cu-AOs resemble a basic pattern common to all enzymes of this group. The active molecule of these enzymes generally consists of a dimeric protein composed of two apparently identical subunits. Two Cu atoms and one or two carbonyl-reactive groups belonging to the organic cofactor (see below) appear to be present per active dimer. All enzymes contain a carbohydrate moiety.[56]

We unified the information as far as possible, avoiding details of limited interest for the kind of presentation intended to give in this chapter. At any rate, whenever significant differences of particular interest exist regarding some specific enzyme, they will be mentioned.

1. Structure of the Protein Molecule

The molecular weight ranges from about 130,000 (VLAO) to 250,000 daltons (ANAO). The average weight is approximately 180,000 daltons for most Cu-AOs. For some of the enzymes, the molecular weight has been controversial for a long time since they show a marked tendency to undergo association-dissociation phenomena varying the enzyme concentration and the environmental conditions.[48]

As already mentioned, Cu-AOs appear to be active as dimeric enzymes, one important exception being represented by HPDAO which has been claimed to function as a monomer. Association of the subunits appears to be due to noncovalent interaction.

The carbohydrate residues composing the molecule represent about 7 to 14% of the total molecular weight; this glycidic portion allows affinity chromatographic separation on concanavalin-A Sepharose.[33,34,36] Isozymes of the BSAO have been claimed to differ only in their carbohydrate composition.[32,57]

For some of the enzymes, the amino acid composition has been reported.[20,22,23,25,30,38]

2. Metal Cofactor

The presence of Cu as a cofactor has been established unambiguously, but its exact role during the catalytic cycle has not yet been clarified. Either it could have a structural significance, thus being essential in the maintenance of the correct tertiary structure of the protein, or it could play a functional role, being involved in the redox reaction during catalysis. Possibly, it could have both structural and functional value. In any case, the removal of metal from the enzyme molecule results in a complete loss of activity which can be restored upon reconstitution with suitable amounts of Cu but not with other metals.[35,42,46,58] As the presence of Cu is essential, its possible role in dioxygen activation has been postulated.[59] Actually, nearly all the known Cu proteins are oxidoreductases; therefore, the addition of substrate (or substrate analogues) under anaerobic conditions is likely to reduce the Cu. Contrary to expectations, the divalent state of the metal ion is not appreciably affected by the previous treatment.[46,60-63] Evidence of the functional role comes from the observation that azide simultaneously temporarily modifies the EPR spectrum and inhibits the activity of the Cu-AOs (see below).[42,64]

AOs generally contain two ions of cupric Cu, both EPR detectable, per enzyme molecule; enzymes from various sources show similar spectral characteristics. The binding of Cu for AOs is quite different from that of the blue oxidases in that the metal ion is present in a complex of tetragonal symmetry only slightly distorted from square planar and of considerably ionic character, in the sense of a low delocalization degree of the cupric ion-unpaired hole.[46,60,61] The g value and binding parameters account for the presence of more than one nitrogen atom at each of the metal centers as the Cu ligands.[35,46] Further information is provided by NMR spectroscopy demonstrating the presence of water molecules coordinated to Cu.[65]

The two Cu sites appear to be not identical, as demonstrated on SSAO by 35 GHz EPR spectroscopy,[42,63] which showed that the two cupric ions are affected differently by the presence of substrate; this one, however, does not bind directly to Cu as demonstrated by EPR spectroscopy employing ^{15}N-putrescine.[46] Azide or cyanide binding to the metal ion abolishes the 35 GHz EPR spectral heterogeneity.[42] Indirect evidence supporting the nonequivalency of the two Cu sites is provided by two observations which conflict if the two sites are assumed to be identical; one, on the basis of NMR studies, that azide and cyanide binding displaces the water molecule coordinated in equatorial position,[42] the other, based on the pH dependance of these same inhibitors binding, that in this case is displaced a protein ligand having a pK_a close to 9 in the displaced state.[66] In addition, it has been demonstrated that the reaction with phenylhydrazine (PH) modifies the EPR spectrum of approximately half of the Cu.[24]

3. Organic Cofactor

The problem of identifying the organic cofactor of Cu-AOs has probably been the most controversial point for a long time. Its identity with pyridoxal phosphate (PLP) has long been claimed on the basis of several indirect evidences.[38,67-72] Only in recent times has this intriguing problem seemed to have been solved with the isolation from purified BSAO of

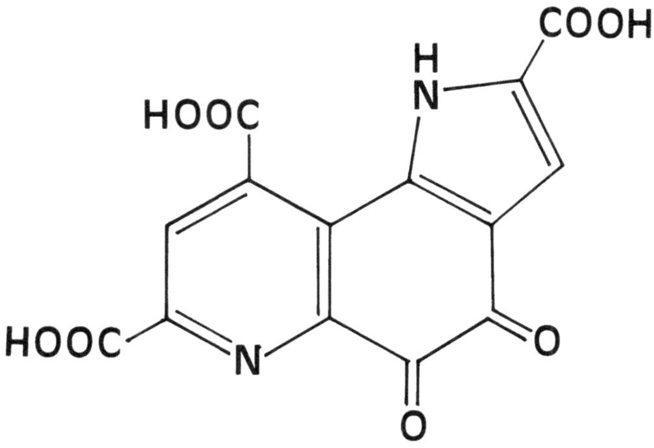

FIGURE 1. Structure of pyrroloquinoline quinone (PQQ).

pyrroloquinoline quinone (PQQ),[73] a new coenzyme for an eukaryotic protein since it had been previously described only in some bacterial dehydrogenases. Figure 1 reports the structure PQQ.

The new coenzyme was first identified a few years ago in methanol dehydrogenases of methylotrophic bacteria. The recent discovery of PQQ, as well as the fact that it is covalently bound to the enzyme molecule and that proteolytic treatment of the enzyme seems to affect PQQ integrity, account for the long time required for its identification. Moreover, its carbonyl-group reactivity explain why PLP was long thought to be the best candidate. Further research is necessary, being at present the involvement of PLP not yet definitively ruled out, to shed light on some ambiguous observations, such as the decrease of AO activity in animals on a vitamin B_6-deficient diet.[38]

4. Spectroscopic Studies

Some pieces of information on EPR and NMR studies have already been reported describing the AO Cu.

Purified preparations of Cu-AOs show a pink color for which the organic cofactor appears to be responsible.[67,74-76] Evidence of the organic chromophore contribution to the visible spectrum of Cu-AOs is supported by the observation that disappearance of the typical 500-nm absorption band by substrate addition is not associated to equally significative changes in the EPR spectrum.[46,60-63] Moreover, procedures which cause a marked alteration of the EPR spectrum, such as azide or cyanide treatment, do not modify the optical spectrum.[64,72,77] In addition, Cu removal under nonreducing conditions produces metal-free enzymes with unmodified visible absorption features.[76] Reduction of the visible chromophore of the Cu-depleted enzyme with dithionite results in the disappearance of the 500-nm absorption band which can be restored not only by Cu(II), but also by Co(II) or Ni(II) reconstitution.[58] The obvious conclusion is that the Cu cofactor, although not directly involved in the genesis of the 500-nm absorption band, is essential for the reoxidation of reduced chromophore.

The broad absorption band which peaks around 500 nm (460 to 510 nm depending on the enzyme) is the result of several electronic transitions demonstrable by CD spectroscopy. The CD bands at 350 (+), 450 (−), and 540 (−) nm observed in purified BSAO,[58] at first attributed to charge-transfer bands from hypothetical sulfur ligands to Cu, should be considered in the light of the recent acquirements on the organic cofactor. The same enzyme CD bands peaking at 660 (+) and 810 (−) nm are more likely to be generated by the Cu d-d transitions.

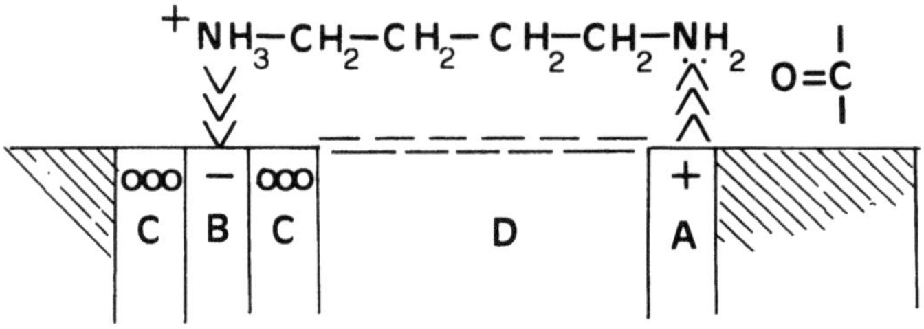

FIGURE 2. Schematic structure of the substrate binding site of SKDAO. Putrescine is represented. A: protonated site for the binding of the substrate amino group to be removed, B: anionic site for the binding of the second amino group of diamines in the protonated form, 6 to 9 Å distant from site A, C: hydrophilic region surrounding site B, and D: hydrophobic region for the binding of the hydrocarbon chain of substrates. O=C is the reactive carbonyl of the organic cofactor.

5. Structure of the Active Site

The attempt to describe the structure of the active site is mainly based on kinetic studies carried out on SKDAO with a number of compounds derived from chemical modification of substrates.

The whole substrate molecule is responsible for the enzyme-substrate interaction, but conflicting opinions have been expressed with respect to the exact details of such interaction. The most reliable hypothesis is reported here. Figure 2 represents schematically the substrate binding site of SKDAO. AOs having different substrate specificity should have differently structured sites. The amino group of the substrate to be removed binds as a nucleophile, thus in the unprotonated state, to a protonated enzymic group, close to the reactive carbonyl function of the organic cofactor, having a pK_a of about 8.8.[78,79] Actually, substrate binding by a Schiff base requires the unprotonated amino group. In diamine oxidizing enzymes, a second negatively-charged enzymic group should be located 6 to 9 Å distant from the former site for the binding of the second amino group of diamines in the protonated form.[80] The incapability of oxidizing those compounds containing a trimethylammonium group suggests the presence of a hydrophilic region surrounding the negatively-charged site. Between the two sites, a hydrophobic region for the binding of the hydrocarbon chain of substrates should be present.[80]

B. Structure of FAD Amine Oxidases

The fact that the most widely investigated enzymes of this group, namely the mitochondrial MAOs, are primarily involved in the metabolism of the biogenic monoamine neurotransmitters, and therefore play a role of secondary importance only in the oxidation of acetylated polyamines, also accounts for their less extensive description in this chapter.

The mitochondrial enzymes, located in the outer membrane of mitochondria, have been classified into two forms, A and B, on the basis of their selective inhibition by clorgyline and deprenyl.[81] The amino acid composition has been reported for the bovine and swine liver enzymes, showing that about 40% of the residues are hydrophobic.[82,83] Chemical modification of MAO-B revealed the presence of one essential cysteine residue and one essential histidine residue per monomer,[84-86] whose molecular weight corresponds to about 53,000 daltons. Kinetic studies suggested the involvement of the essential cysteine and histidine residues in, respectively, binding and cleavage of substrate.[85]

Rat liver PAO appears to be localized in the peroxisomes, but the presence of PAO activity in the cytosolic fraction raises the question on whether an isozyme exists or rather leakage of PAO out of the peroxisomes may have occurred during the fractionation procedure.[55]

Both MAO and PAO have been proved to contain FAD as the cofactor.[55,87]

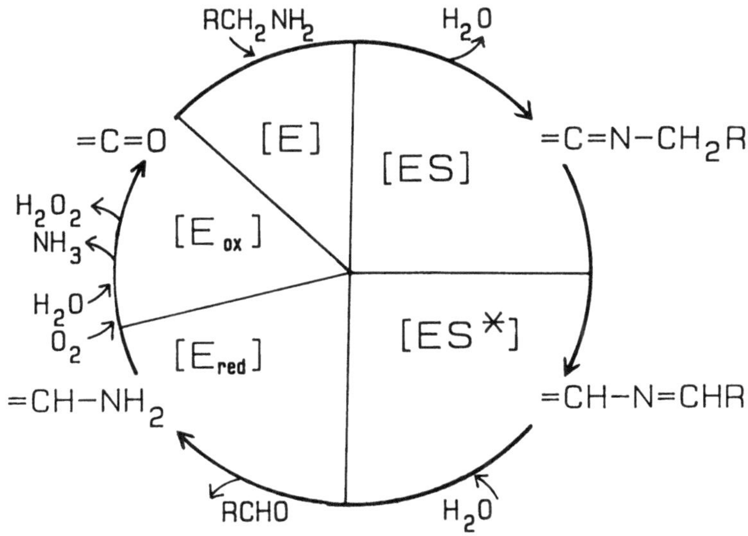

FIGURE 3. Transaminase-type mechanism of action of copper AOs.

C. Mechanism of Action of Copper Amine Oxidases

Cu-AOs oxidatively deminate those polyamines having primary amino groups in a reaction involving the consumption of three substrates (the amine, dioxygen, and water) and the release of three products (ammonia, hydrogen peroxide, and the corresponding aldhehyde). These enzymes operate by an enzyme-substitution (double displacement or ping-pong bi-ter) mechanism.[62,88-90]

As already mentioned, titration with PH revealed the presence of one carbonyl cofactor (and therefore one active site) per dimer,[24,34,41] the only exception being the enzyme from *Aspergillus niger*.[11] However, recent experiments on LSAO seem to indicate the presence of two active sites per enzyme molecule.[24,91] At present, with no more experimental evidence, we can only postulate hypothetical mechanisms to reconcile the PH titration experiments with the possible presence of two active sites. The two subunits may be structured in such a way that they can function one at a time, and the enzyme would therefore function through a "flip-flop" mechanism.[92] This could explain the half-of-the-sites reactivity observed with PH.

AOs follow a transaminase-type mechanism of action. Figure 3 summarizes the catalytic steps described in the text. Substrate [S] "recognizes" the active site of free enzyme [E] and interacts with it by electrostatic and van der Waals forces. Before the reduced enzyme [E_{red}] is formed, at least two distinct intermediate stages are detectable by stopped-flow kinetics.[78,93] At the [ES] stage, substrate binds covalently to the carbonyl cofactor by the formation of a Schiff base. The tautomerization of the Schiff base (refer to Figure 3) is probably responsible for the appearance of a 360 nm absorbing species [ES*] which, together with the 500-nm absorption band, is bleached as the aldehyde product (but not ammonia and hydrogen peroxide) is released leaving the reduced enzyme [E_{red}].

The reaction so far described develops anaerobically. The other products are released in a second time, and the reaction is dioxygen dependent. The step leading to reconstitution of the "resting" enzyme [E] from [E_{red}] is accomplished in two distinguishable stages:[89,90,94] during the first one, reaction with dioxygen produces the release of NH_3 and H_2O_2 and reappearance of the 500-nm absorption band.[94,95] In the second stage the reoxidized enzyme [E_{ox}] is converted into the resting enzyme [E]. It has been hypothesized that inhibition of

AOs by excess of substrate should occur because of the formation of a catalytically inactive "abortive complex" between the reoxidized enzyme [E_{ox}] and substrate rather than to the presence of an inhibitory binding site.[43,62,94,96]

IV. FUNCTIONAL ROLE OF AMINE OXIDASES

The functional role played by AOs is closely related to the physiological relevance of polyamines, which has already been widely dealt with throughout this book; therefore, repeating of many important aspects on this matter will be avoided.

The importance of polyamines in such processes as cellular growth and differentiation has been clearly established, but an equally relevant function appears to be played by the oxidation products of polyamines, aminoaldehydes. In some cases these latter, which represent the primary product of polyamine oxidation, are not ulteriorly modified, but generally the aldehyde function undergoes further oxidation to be converted into a carboxylic group by aldehyde dehydrogenase. As an example, putrescine is converted into γ-aminobutyric acid, which is a neurotransmitter and an inhibitor of protein synthesis as well.[97] Another example of polyamine conversion into an amino acid is represented by the β-alanine formed from aminopropionaldehyde derived from spermidine and spermine oxidation by PAO.[98]

Aldehydes produced by polyamine oxidation are toxic compounds, being able to impair cell growth and division acting as inhibitors of nucleic acids and protein syntheses.[99,100] A toxic product of subsequent degradation of aldehydes is acrolein, whose mode of action differs from that of aldehydes.[101,102] Aldehydes act as inhibitors of cell proliferation probably by means of their interaction with nucleotides and DNA.[103,104] It has been observed that immobilized SKDAO injected intraperitoneally in Ehrlich ascite tumor-bearing mice inhibits tumor growth probably through the release of either toxic aldehydes or H_2O_2 derived from oxidation of endogenous polyamines.[105]

Another possible biological function of oxidized polyamines may be the modulation of the immune response. Reduction of DNA synthesis and impairment of cytolysis was observed in vitro with murine lymphocytes incubated with irradiated DAB/2 spleen cells in the presence of polyamines and bovine serum which presents AO activity.[106]

Of particular interest is the finding that the maternal part of placenta shows a high DAO activity, whereas the fetal part does not, suggesting again the possible role of oxidized polyamines in immunity allowing the survival of the fetus and placenta in spite of their different immunological specificity with respect to the mother.[51,107] The presence of DAO activity in the serum of pregnant women in addition to that of the placenta also indicates a possible protecting role against the release of polyamines from the fetoplacental unit.[108]

Several other functions played by AOs in animal cell physiology will become evident dealing with the amine oxidation in pathophysiological conditions.

It is worthwhile to remember a role of AOs in plant physiology in the synthesis of compounds of pharmacological interest containing cyclic derivatives of polyamine oxidation. The incorporation of radioactive putrescine into nicotine, strychnine, and atropine has been reported.[109-111] Moreover, radiolabeled cadaverine has been proved to be involved in the biosynthesis of several other cyclic compounds containing piperidine rings.[112-116]

Not only aldehydes, but also H_2O_2 produced during AO catalysis may have a functional role: its involvement in human platelet aggregation has very recently been investigated.[117] Moreover, H_2O_2 seems to possess other dynamic properties; its insulin-like effect on adipocyte, as well as a stimulating effect on lymphocyte blastization were also reported.[117] In addition, a correlation between H_2O_2 and Ca^{2+} flux through the biological membranes has been established.[117]

A. Balance Between Biosynthetic and Degradative Enzymes in the Control of Intracellular Diamine and Polyamine Levels

It is not easy to provide a comprehensive description of all the reactions involved in the regulation of the activity of those enzymes involved in polyamine metabolism, primarily because of our ignorance of many aspects of the molecular mechanisms by means of which this control is exerted particularly in the case of AOs. We would like to focus on the important problem of how the intracellular polyamine concentration is adjusted and maintained at a level optimal for the cells at a certain time, especially in consideration of the ever-increasing number of functions which polyamines appear to play in cell metabolism.

The intracellular polyamine concentration is controlled either by regulating the activity levels of both the biosynthetic (ODC, SAMD) and degradative (AOs) enzymes, or by controlling the net direction of the synthetic and hydrolytic reactions involving acetic acid and polyamines.

Since a thorough discussion on the regulation of the biosynthetic decarboxylase activity is already available in other parts of this book (see Chapters 15 to 20), a few data will be referred only when strictly necessary. Similarly, the interconversion of acetylpolyamines and polyamines is dealt extensively in Chapter 11.

Despite the large number of studies carried out to determine the changes in AO activity occurring under several pathophysiological conditions (see below), little is known about the mechanisms by means of which the regulation of intracellular AO activity is carried out.

Significative variations of the ODC to SAMD ratio were observed as measured in different experimental systems and at different times for the same sample,[118-120] and the same behavior could perhaps be observed with the amine oxidizing enzymes having different substrate specificity; unfortunately, for AOs too little is known on this matter, being the determinations generally carried out on a single enzyme species, i.e., usually DAO employing putrescine as substrate. Actually, DAO plays a pivotal role in polyamine metabolism; by controlling the intracellular putrescine level, it indirectly regulates spermidine and spermine concentrations, which are synthesized by adding aminopropyl moieties to putrescine. For the sake of simplicity we will simply use the expressions "biosynthetic" and "degradative" referred to polyamine-metabolizing enzymes.

Several observations demonstrate that the activity levels of both biosynthetic and degradative enzymes are correlated, in that a variation of ODC and SAMD activities usually corresponds to a modification of AO activity.[118-127] Some representative examples of this behavior are reported here.

Experiments carried out on regenerating rat liver after partial hepatectomy reveal a large increase in putrescine concentration in response to the induction of ODC.[121] Under the same experimental conditions, a significative increase in DAO activity is observed.[125]

Several other evidences demonstrating the increase in polyamine biosynthetic activity in rapid growing tissues were obtained directly (by measuring the activity of ODC and SAMD) or indirectly (by evaluating the concentration of putrescine, spermidine and spermine). Many of the systems employed to determine the biosynthetic decarboxylases activity have also been investigated to establish AO activity. Embryonic tissues,[128,129] rat kidney during compensatory hypertrophy after unilateral nephrectomy,[130] and human and animal tumors showed,[125,127,131-133] together with an increased polyamine biosynthesis, an enhanced amine oxidizing activity that correlated well with the intracellular polyamine level.

Samples of human tumors of the central nervous system obtained after surgical treatment showed even more direct evidence of the correlation existing between polyamine biosynthetic and degradative activities on one hand, and rates of polyamine metabolism and tumor growth on the other. Both rates of polyamine biosynthesis and oxidation of putrescine were in fact proportional to the degree of histological malignancy of the tumors.[118,127]

Under physiological conditions, a high polyamine biosynthetic activity is generally induced in response to certain external stimuli such as growth hormones, whereas the increase of AO activity appears to be a consequence of enhanced intracellular polyamine concentration. Actually, the enhancement of DAO has been recently observed to follow the increase in putrescine concentration,[130,134,135] suggesting that induction of other AOs may also be related to the intracellular polyamine level.

In order to clarify to what extent the observed increase in DAO activity was imputable to synthesis of new protein, activation of preexisting inactive enzyme or slowing of its turnover, protein and RNA syntheses inhibitors were injected into both partially hepatectomized and unilateral nephrectomized rats.[130,134] Results clearly demonstrated that administration of either cycloheximide or actinomycin D completely prevented the enhancement of DAO activity when given at the time of operation, or previous to the large increase in polyamine concentration, revealing at the same time that the half-life of this enzyme remained the same (about 14 to 16 hr) both in pathological and in control animals. Thus, the rise of DAO activity should be attributed entirely to *de novo* synthesis of mRNA codifying for the enzyme.[130,134,135]

It is not easy to draw conclusions on the role played by AOs in the regulation of polyamine metabolism, being the functions of this class of enzymes not yet completely clarified. Nevertheless, when considering the behavior of DAO, acting mainly on putrescine which is the precursor of the other polyamines, we can hypothesize that, among the others, it is very likely to take part in the precise regulation of intracellular putrescine level and, consequently, of the total polyamines, cooperating with the biosynthetic enzymes in maintaining an optimal concentration at any time.

B. Amine Oxidation in Physiological Conditions

The early stages of development of a living organism are characterized by modifications of several metabolic parameters with respect to the normal adult condition. Particularly marked are those differences regarding the patterns of polyamines and their related enzymes. Fetal tissues synthesize and oxidize polyamines at different rates according to their respective metabolic activities. Oxidative deamination not only occurs as a simple homeostatic mechanism in the control of intracellular polyamine levels, but may possibly supply the fetus with biologically active compounds, too.

In guinea pig fetuses, liver DAO activity, and therefore the ability to oxidize diamines, increases rapidly from the middle of gestation and reaches its maximum 3 days after delivery. During the following week DAO activity decreases to the normal adult level.[128,129] In humans also liver DAO activity develops early, its average value being significantly higher around the middle of gestation than that found in adults.[136]

Intestinal DAO levels show a different behavior: throughout the chick embryo development, its activity is considerably lower than the adult value, whereas the guinea pig intestine, at the beginning of the 3rd month of gestation, shows about one third of the adult value. After birth, DAO activity rapidly increases in chick, guinea pig, and rat intestine.[137-139] This behavior relates to the proliferation of the intestinal mucosa as postnatal food intake begins. To confirm this, it is known that the rate of enterocyte proliferation depends on feeding; as food intake activates and starvation lowers, ODC activity in the intestinal mucosa and, as mentioned above, ODC and DAO activities are surprisingly parallel.[138,142]

DAO activity is particularly elevated in kidney; its role in the regulation of polyamine concentration in blood and urine is suggested by the localization of the enzyme in the basal portion of the tubule cells and towards the lumen.[143]

Pregnancy also is characterized by a modification of the polyamine metabolism of the mother in response to the fetal development. In the pregnant woman, serum DAO activity increase begins between 9 and 28 days after ovulation and continues until about the 20th week of gestation, when it reaches a plateau. Before delivery, serum DAO level has been reported to be more than 500 times higher than in nonpregnant subjects. After delivery or artificial abortion serum DAO activity returns to its basal value within 24 hr.[108,144-146]

As already mentioned, pregnancy serum DAO is most probably of placental origin, being particularly abundant in the maternal part of placenta.[147,148] The most convincing evidence of this, apart from the biochemical and immunological similarity shared by serum and placental enzymes, is provided by the finding that in animals with nondeciduate placenta, such as hog and horse, serum DAO activity could not be demonstrated.[149,150] Diffusion of DAO also occurs between placenta and amniotic fluid, where the enzyme activity pattern resembles that of the serum.[151]

C. Amine Oxidation in Pathological Conditions

Investigation of DAO activity in pathological conditions is of noteworthy interest in that it provides useful information on the state of activity of polyamine metabolism and, possibly, the development of sensitive and reliable routine assay methods may be of clinical use in monitoring certain diseases.

DAO activity was assayed in some experimental conditions reproducing possible pathological situations in order to provide further information on the biology of cell response against some alterations of the metabolism of the living organism.

In the regenerating rat liver after partial hepatectomy, DAO was shown to increase with a peak of activity between 16 and 48 hr. The maximum activity was about 3 to 15 times higher than in normal liver, where this enzyme activity is low. The normal values were approached about 7 days after the operation.[134]

DAO activity was observed to follow a biphasic increase in the rat kidney undergoing compensatory hypertrophy after unilateral nephrectomy, reaching three to four times that observed in sham-nephrectomized rats.[130]

In the rat heart, the development of isoprotenerol-induced cardiac hypertrophy caused a two- to fourfold increase of DAO activity. After interruption of drug administration, DAO activity regressed, together with the cardiac hypertrophy, to the basal value.[135] In addition, the hypertrophic heart of spontaneous hypertensive rats showed a DAO activity about 2.5 times higher than that of normotensive rats.[135]

Regarding the research on human neoplasms, a high DAO activity was found in a human endocrine tumor, medullary thyroid carcinoma. The activity was enhanced not only in neoplastic tissue, but, for some patients with this type of pathology and particularly those having metastases, also in blood.[132,152] Subsequently, another human tumor, small cell carcinoma of the lung, which frequently shows endocrine cell properties, was demonstrated to have an enhanced DAO activity.[131,153] Other lung cancers of different histologic types also had increased DAO levels.[133,154]

An increased DAO activity was also detected in ascites and pleural fluids of patients having various other types of cancer. Among the others, we will mention ovarian carcinoma, which appeared to be the source of the DAO present in the ascites fluid.[155]

The tumor tissue and cerebrospinal fluid of patients with malignant brain tumors were able to oxidize putrescine much faster than the corresponding tissue and liquor of nonmalignant tumor patients.[127] In some cases putrescine oxidase activity was also present, and was sometimes even higher than in the neoplastic tissue of some patients.[127] It is not completely clear which kind of enzyme is responsible for the variation leading to enhanced putrescine oxidation since in the brain acetylation precedes oxidative deamination of diamines by mitochondrial MAO.[156]

V. INHIBITION OF POLYAMINE OXIDATION

As already mentioned, oxidation of polyamines not only serves as a means to control their intracellular level, but it also provides metabolites which may be involved in the regulation of some physiological processes as well as precursors of other metabolic pathways. Therefore, it is to be expected that substances which interfere with AO activity may be able to influence both such processes and those functions proper to polyamines themselves. The current interest in the development and study of the effect of polyamine-metabolizing enzymes inhibitors should then be considered of particular importance in view of their possible use in biomedical, industrial, and agricultural fields. From this point of view, special interest arouses the in vivo effects of AO inhibitor administration rather than the kinetic studies carried out with isolated enzymes in vitro. Before reporting a few data on the effect of in vivo inhibition of amine oxidation, some information is given about the most representative AO inhibitors.

A clear distinction between FAD- and Cu-dependent AO inhibitors is often hard to make; actually, compounds such as pargyline, nialamide, iproniazide, and isoniazide, although considered classical MAO inhibitors, are able to impair some Cu-AO activity.[157] Similarly, PH, a well-known carbonyl-group reagent, inhibits MAO in addition to the Cu enzymes.[158]

The compounds inhibiting AO activity can be classified into different types: substrate analogues, suicide substrates, time-dependent inactivators, Cu-chelating agents, and sometimes, at high concentrations, substrates.

Those compounds having structural features similar to the amine substrate can strongly inhibit AOs, but their effect can be reversed by dialysis. Among them, sulfonium and trimethylammonium compounds competitively inhibit AOs, whereas noncompetitive inhibition occurs with other bis-onium compounds.[80,159-161] The keto analogues of diamines are potent inhibitors of plant and animal DAOs[162]

Suicide substrates (very often called also suicide inhibitors) produce a time-dependent inhibition because they must be altered by the enzyme catalysis before they can react covalently with the enzyme in order to produce an inactivation. Compounds of this kind, able to inhibit SKDAO, are aminoacetonitrile, propargylamine, allylamine, *cis*-imidazolylallylamine, and 1,4-diamino-2-butyne.[163-168]

Time-dependent inactivators should practically be considered irreversible inhibitors because they form a covalent bond with an essential group of the enzyme. Partial recovery of activity, if any, occurs very slowly. Among these compounds, members of particular importance are the carbonyl group reagents PH, semicarbazide, hydroxylamine, and aminoguanidine, well-known inhibitors of the Cu-AOs.[169-171]

Metal-chelating agents, such as diethyldithiocarbamate, inhibit the Cu-AOs by removing the enzyme-bond Cu.[172-174] Partial or total recovery of the enzyme activity occurs after reconstitution of the enzyme with suitable amounts of cupric ions. Azide and cyanide function as inhibitors through their binding to Cu which causes the displacement of either a water molecule or a protein ligand coordinating the metal ion (see above).[42,64]

It has been observed that some substrates, at high concentration, inhibit the activity of some Cu-AOs. Inhibition by excess of substrate is particularly pronounced with those substrates having, like histamine, π electrons.[50,175,176]

Inhibition of polyamine catabolism in vivo can be obtained by using compounds which block specific oxidative pathways. Administration of such substances separately or simultaneously may help to quantify the extent of each catabolic route followed by a particular substrate.

Aminoguanidine (AG) and MDL 72521 are potent inhibitors of, respectively, Cu-AOs, particularly DAO, and PAO. Administration of the former to rats makes it possible to evaluate, by means of quantitative and qualitative determination of urinary polyamine excretion, that only about 40% of the total polyamines destined for elimination are normally present in the urine, the remaining 60% being oxidized by Cu-AOs. Combined AG and MDL 72521 treatment further enhances polyamine excretion in the form of N^1-acetylspermidine. The use of these AO inhibitors proves an important means to study the interconversion of polyamines.[177,178]

Oral administration of AG to rats in order to evaluate histamine elimination evidenced, together with a decrease of AO activity, stimulation of the appetite of the animal and increase in body weight.[179] A possible explanation of this phenomenon may be the enhancement of protein and nucleic acids syntheses due to either increased polyamine or decreased aldehyde levels.

The in vivo inhibition of mouse mitochondrial MAO by daily administration of pargyline or nialamide results in a noteworthy increase in brain acylcadaverines.[180]

From a clinical point of view, it should be remembered that the use of drugs normally administered for purposes other than the impairment of polyamine catabolism may cause undesired effects due to amine oxidation inhibition. The antimalarial drugs amodiaquine, quinacrine, and chloroquine inhibit putrescine catabolism in the rat.[181] Inhibition of intraperitoneally injected radioactive putrescine oxidation was monitored in vivo by measuring the decrease, after the drug administration, of $^{14}CO_2$ formed after a series of conversions through γ-aminobutyrate to succinate leading to ultimate oxidation in the cytric acid cycle. The ability of antimalarial drugs to inhibit DAO was also confirmed by using partially purified rat intestinal enzyme in vitro.[181]

Another example is provided by a recent study suggesting the possible responsibility of serum Cu-AO inhibition in the appearance of secondary effects in patients treated with a prolonged therapy of the antiparkinson drugs carbidopa and benserazide.[182]

Moreover, levamisole, an anthelmintic drug also known as an inhibitor of sperm motility, was proved to be a potent inhibitor of both seminal and placental DAOs.[183] A number of other DAO inhibitors were also found to be effective in reducing rabbit sperm motility.[183]

The first active intermediate of gyromitrin, the poison of the mushroom false morel, namely *N*-methyl-*N*-formylhydrazine, acts as a noncompetitive inhibitor of human intestinal DAO.[184] DAOs from five other sources are also inhibited by the metabolite of the poison. The results of a study suggests the possibility that inhibition of intestinal DAO may be involved in mushroom poisoning.[184]

ACKNOWLEDGMENT

This work was in part supported by the Italian National Research Council, special project on "Oncology", contract no. 8400685.44 and the "Ministero della Pubblica Istruzione".

APPENDIX

Below are summarized the oxidative pathways of the main diamines and polyamines.

Putrescine

$$H_2N{-}(CH_2)_4{-}NH_2 \xrightarrow{\text{DAO}} H_2N{-}(CH_2)_3{-}CHO$$

Acetylputrescine

$$H_3C{-}\overset{\overset{\displaystyle O}{\|}}{C}{-}NH{-}(CH_2)_4{-}NH_2 \xrightarrow{\text{MAO}} H_3C{-}\overset{\overset{\displaystyle O}{\|}}{C}{-}NH{-}(CH_2)_3{-}CHO$$

Cadaverine

$$H_2N{-}(CH_2)_5{-}NH_2 \xrightarrow{\text{DAO}} H_2N{-}(CH_2)_4{-}CHO$$

Acetylcadaverine

$$H_3C\text{–}C(=O)\text{–}NH\text{–}(CH_2)_5\text{–}NH_2 \xrightarrow{\text{MAO}} H_3C\text{–}C(=O)\text{–}NH\text{–}(CH_2)_4\text{–}CHO$$

Spermidine

$$\xrightarrow{\text{microorganism or vegetable PAOs}} H_2N\text{–}(CH_2)_3\text{–}NH_2 + H_2N\text{–}(CH_2)_3\text{–}CHO$$

$$H_2N\text{–}(CH_2)_3\text{–}NH\text{–}(CH_2)_4\text{–}NH_2 \xrightarrow{\text{serum AO}} H_2N\text{–}(CH_2)_4\text{–}NH(CH_2)_2\text{–}CHO$$

$$\xrightarrow{\text{animal PAO}} H_2N\text{–}(CH_2)_2\text{–}CHO + H_2N\text{–}(CH_2)_4\text{–}NH_2$$

N[1] Acetylspermidine

$$\xrightarrow{\text{DAO}} H_3C\text{–}C(=O)\text{–}NH\text{–}(CH_2)_3\text{–}NH\text{–}(CH_2)_3\text{–}CHO$$

$$H_3C\text{–}C(=O)\text{–}NH\text{–}(CH_2)_3\text{–}NH\text{–}(CH_2)_4\text{–}NH_2$$

$$\xrightarrow{\text{PAO}} H_3C\text{–}C(=O)\text{–}NH\text{–}(CH_2)_2\text{–}CHO + H_2N\text{–}(CH_2)_4\text{–}NH_2$$

N[8] Acetylspermidine

$$H_2N\text{–}(CH_2)_3\text{–}NH\text{–}(CH_2)_4\text{–}NH\text{–}C(=O)\text{–}CH_3 \xrightarrow{\text{serum AO, MAO}}$$

$$OHC\text{–}(CH_2)_2\text{–}NH\text{–}(CH_2)_4\text{–}NH\text{–}C(=O)\text{–}CH_3$$

Spermine

$$\xrightarrow{\text{PAO}} H_2N\text{–}(CH_2)_3\text{–}NH\text{–}(CH_2)_4\text{–}NH_2 + H_2N\text{–}(CH_2)_2\text{–}CHO$$

$$H_2N\text{–}(CH_2)_3\text{–}NH\text{–}(CH_2)_4\text{–}NH\text{–}(CH_2)_3\text{–}NH_2 \xrightarrow{\text{serum AO}}$$

$$H_2N\text{–}(CH_2)_3\text{–}NH\text{–}(CH_2)_4\text{–}NH\text{–}(CH_2)_2\text{–}CHO$$

$$\downarrow \text{serum AO}$$

$$OHC\text{–}(CH_2)_2\text{–}NH\text{–}(CH_2)_4\text{–}NH\text{–}(CH_2)_2\text{–}CHO$$

N[1] Acetylspermine

$$H_3C{-}\overset{\displaystyle O}{\overset{\|}{C}}{-}NH{-}(CH_2)_3{-}NH{-}(CH_2)_4{-}NH{-}(CH_2)_3{-}NH_2 \xrightarrow{\text{serum AO}}$$

$$H_3C{-}\overset{\displaystyle O}{\overset{\|}{C}}{-}NH{-}(CH_2)_3{-}NH{-}(CH_2)_4{-}NH{-}(CH_2)_2{-}CH$$

REFERENCES

1. **Mondovì, B.,** Preface, in *Amine Oxidases: Structure and Function,* Mondovì, B., Ed., CRC Press, Boca Raton, Fla., 1986.
2. **Tabor, C. W. and Kellogg, P. D.,** Identification of flavin adenine dinucleotide and heme in a homogeneous spermidine dehydrogenase from *Serratia marcescens, J. Biol. Chem.*, 245, 5424, 1970.
3. **Okada, M., Kawashima, S., and Imahori, K.,** Substrate binding characteristics of the active site of spermidine dehydrogenase from *Serratia marcescens, J. Biochem.*, 85, 1235, 1979.
4. **Adachi, O., Yamada, H., and Ogata, K.,** Purification and properties of putrescine oxidase of *Micrococcus rubens, Agric. Biol. Chem.*, 30, 1202, 1966.
5. **Okada, M., Kawashima, S., and Imahori, K.,** Affinity chromatography of putrescine oxidase from *Micrococcus rubens* and spermidine dehydrogenase from *Serratia marcescens, J. Biochem.*, 85, 1225, 1979.
6. **DeSa, R. J.,** Putrescine oxidase from *Micrococcus rubens*. Purification and properties of the enzyme, *J. Biol. Chem.*, 247, 5527, 1972.
7. **Yamada, H., Adachi, O., and Ogata, K.,** Putrescine oxidase, a diamine oxidase requiring flavin adenine dinucleotide, *Agric. Biol. Chem.*, 29, 1148, 1965.
8. **Yamada, H., Adachi, O., and Ogata, K.,** Amine oxidase of microorganisms. I. Formation of amine oxidase by fungi, *Agric. Biol. Chem.*, 29, 117, 1965.
9. **Adachi, O. and Yamada, H.,** Amine oxidase of microorganisms. VII. An improved purification procedure and further properties of amine oxidase of *Aspergillus niger, Agric. Biol. Chem.*, 33, 1707, 1969.
10. **Kumagai, H., Kishimoto, N., and Yamada, H.,** Subunit structure of amine oxidase from *Aspergillus niger, Agric. Biol. Chem.*, 42, 893, 1978.
11. **Suzuki, H., Ogura, Y., and Yamada, H.,** Stoichiometry of the reaction by amine oxidase from *Aspergillus niger, J. Biochem.*, 69, 1065, 1971.,
12. **Yamada, H., Adachi, O., and Ogata, K.,** Amine oxidases of microorganisms. III. Properties of amine oxidase of *Aspergillus niger, Agric. Biol. Chem.*, 29, 864, 1965.
13. **Isobe, K., Tani, Y., and Yamada, H.,** A new enzyme, agmatine oxidase, from fungi, *Agric. Biol. Chem.*, 46, 1345, 1982.
14. **Isobe, K., Tani, Y., and Yamada, H.,** Crystallization and characterization of agmatine oxidase from *Pencillum chrysogenum, Agric. Biol. Chem.*, 46, 1353, 1982.
15. **Yamada, H., Isobe, K., and Tani, Y.,** Oxidation of polyamines by fungal enzymes, *Agric. Biol. Chem.*, 44, 2469, 1980.
16. **Isobe, K., Tani, Y., and Yamada, H.,** Crystallization and characterization of polyamine oxidase from *Pencillum chrysogenum, Agric. Biol. Chem.*, 44, 2651, 1980.
17. **Isobe, K., Tani, Y., and Yamada, H.,** Crystallization and characterization of polyamine oxidase from *Aspergillus terreus, Agric. Biol. Chem.*, 44, 2749, 1980.
18. **Srivastava, S. K. and Prakash, V.,** Purification and properties of pea cotyledon and embryo diamine oxidase, *Phytochemistry*, 16, 189, 1977.
19. **Macholan, L. and Haubrova, J.,** Isolation and some characteristics of diamine oxidase from etiolated pea seedlings, *Collect. Czech. Chem. Commun.*, 41, 2987, 1976.
20. **Yanagisawa, H., Hirasawa, E., and Suzuki, Y.,** Purification and properties of diamine oxidase from pea epicotyls, *Phytochemistry*, 20, 2105, 1981.
21. **Kenten, R. H. and Mann, P. J. G.,** The oxidation of amines by extracts of pea seedlings, *Biochem. J.*, 50, 360, 1962.

22. **Floris, G., Giartosio, A., and Rinaldi, A.,** Diamine oxidase from *Lens esculenta* seedlings: purification and properties, *Phytochemistry,* 22, 1871, 1983.
23. **Rinaldi, A., Floris, G., and Finazzi-Agrò, A.,** Purification and properties of diamine oxidase from *Euphorbia latex, Eur. J. Biochem.,* 127, 417, 1982.
24. **Rinaldi, A., Floris, G., Sabatini, S., Finazzi-Agrò, A., Giartosio, A., Rotilio, G., and Mondovì, B.,** Reaction of beef plasma and lentil seedlings Cu-amine oxidases with phenylhydrazine, *Biochem. Biophys. Res. Commun.,* 115, 841, 1983.
25. **Matsuda, H. and Suzuki, Y.,** Purification and properties of the diamine oxidase from *Vicia faba* leaves, *Plant Cell Physiol.,* 22, 737, 1981.
26. **Matsuda, H. and Suzuki, Y.,** Some properties of the amine oxidase in *Vicia faba* seedlings, *Plant Cell Physiol.,* 18, 1131, 1977.
27. **Hirasawa, E. and Suzuki, Y.,** Purification and properties of a polyamine oxidase from *Zea mays, Phytochemistry,* 14, 99, 1975.
28. **Suzuki, Y. and Yanagisawa, H.,** Purification and properties of maize polyamine oxidase: a flavoprotein, *Plant Cell Physiol.,* 21, 1085, 1980.
29. **Smith, T. A.,** Purification and properties of the polyamine oxidase of barley plants, *Phytochemistry,* 11, 899, 1972.
30. **Yamada, H., Gee, P., Ebata, M., and Yasunobu, K. T.,** Monoamine oxidase. VI. Physicochemical properties of plasma monoamine oxidase, *Biochem. Biophys. Acta,* 81, 165, 1964.
31. **Achee, F. M., Chervenka, C., Smith, R. A., and Yasunobu, K. T.,** Amine oxidase. XII. The association and dissociation, and number of subunits of beef plasma amine oxidase, *Biochemistry,* 7, 4329, 1968.
32. **Yasunobu, K. J., Ishizaki, H., and Minamiura, N.,** Molecular, mechanistic and immunological properties of amine oxidases, *Mol. Cell Biochem.,* 13, 3, 1976.
33. **Watanabe, K. and Yasunobu, K. T.,** Carbohydrate content of bovine plasma amine oxidase and isolation of a carbohydrate-containing fragment attached to asparagine, *J. Biol. Chem.,* 245, 4612, 1972.
34. **Ishizaki, H. and Yasunobu, K. T.,** Bovine plasma amine oxidase interactions with concanavalin A in solution and with concanavalin A sepharose, *Biochem. Biophys. Acta,* 611, 27, 1980.
35. **Suzuki, S., Sakurai, T., Nakahara, A., Oda, O., Manabe, T., and Okuyana, T.,** Spectroscopic aspects of copper binding site in bovine serum amine oxidase, *FEBS Lett.,* 116, 17, 1980.
36. **Turini, P., Sabatini, S., Befani, O., Chimenti, F., Casanova, C., Riccio, P. L., and Mondovì, B.,** Purification of bovine plasma amine oxidase, *Anal. Biochem.,* 125, 294, 1982.
37. **Buffoni, F. and Blaschko, H.,** Benzylamine oxidase and histaminase: purification and crystallisation of an enzyme from pig plasma, *Proc. R. Soc. London,* 161, 153, 1964.
38. **Blaschko, H. and Buffoni, F.,** Pyridoxal phosphate as constituent of the histaminase (benzylamine oxidase) of pig plasma, *Proc. R. Soc. London B,* 163, 45, 1965.
39. **Lindstrom, A. and Pettersson, G.,** Active-site titration of pig-plasma benzylamine oxidase with hydrazine derivatives, *Eur. J. Biochem.,* 34, 564, 1973.
40. **Buffoni, F. and Ignesti, G.,** Active site titration of pig-plasma benzylamine oxidase with phenyl hydrazine, *Biochem. J.,* 145, 369, 1975.
41. **Lindstrom, A. and Pettersson, G.,** Active site titration of pig-plasma benzylamine oxidase, *Eur. J. Biochem.,* 83, 131, 1978.
42. **Barker, R., Boden, N., Cayley, G., Charlton, S. C., Henson, R., Holmes, M. C., Kelly, I. D., and Knowles, P. F.,** Properties of cupric ions in benzylamine oxidase from pig plasma as studied by magnetic resonance and kinetic methods, *Biochem. J.,* 177, 289, 1979.
43. **Kelly, I. D., Knowles, P. F., Yadav, K. D. S., Bardlsey, W. G., Leff, P., and Waight, R. D.,** Steady state kinetic studies on benzylamine oxidase from pig plasma, *Eur. J. Biochem.,* 114, 133, 1981.
44. **Buffoni, F.,** Histaminase and related amine oxidases, *Pharmacol. Rev.,* 18, 1163, 1966.
45. **Mondovì, B., Rotilio, G., Finazzi-Agrò, A., and Scioscia-Santoro, A.,** Purification of pig-kidney diamine oxidase and its identity with histaminase, *Biochem. J.,* 91, 408, 1964.
46. **Mondovì, B., Rotilio, G., Costa, M. T., Finazzi, Agrò, A., Chiancone, E., Hansen, R., and Beinert, H.,** Diamine oxidase from pig kidney. Improved purification and properties, *J. Biol. Chem.,* 242, 1160, 1967.
47. **Yamada, H., Kumagai, H., Kawasa Ki, H., Matsui, H., and Ogata, K.,** Crystallization and properties of diamine oxidase from pig kidney, *Biochem. Biophys. Res. Commun.,* 29, 723, 1967.
48. **Pionetti, J. M.,** Analytical-band centrifugation of the active form of pig kidney diamine oxidase, *Biochem. Biophys. Res. Commun.,* 58, 495, 1974.
49. **Kluetz, M. D. and Schmidt, P. G.,** Diamine oxidase: molecular weight and subunit analysis, *Biochem. Biophys. Res. Commun.,* 76, 40, 1977.
50. **Hill, C. M. and Bardsley, W. G.,** Histamine and related compounds as substrates of diamine oxidase (histaminase), *Biochem. Pharmacol.,* 24, 253, 1975.
51. **Bardsley, W. G., Crabbe, J. C., and Scott, I. V.,** The amine oxidases of human placenta and pregnancy plasma, *Biochem. J.,* 139, 169, 1974.

52. **Crabbe, M. G. C., Waight, R. D., Bardsley, W. G., Barker, R. W., Kelly, I. D., and Knowles, P. F.**, Human placental diamine oxidase, *Biochem. J.*, 155, 679, 1976.
53. **Lin, C. W. and Kirley, S. D.**, Histaminase as an oncoplacental enzyme: biochemical and immunological evidence, in *Onco-Developmental Gene Expression*, Fishman, W. H. and Sella, S., Eds., Academic Press, New York, 1976, 469.
54. **Lin, C. W., Inglis, N. R., Rule, A. H., Turksoy, R. N., Chapman, L. M., Kirley, S. D., and Stolbach, L. L.**, Histaminase and other tumor markers in malignant effusion fluids, *Cancer Res.*, 39, 4894, 1979.
55. **Hölttä, E.**, Oxidation of spermidine and spermine in rat liver; purification and properties of polyamine oxidase, *Biochemistry*, 16, 91, 1977.
56. **Knowles, P. F. and Yadav, K. D. S.**, Amine oxidases, in *Copper Proteins and Copper Enzymes*, Vol. 2, Lontie, R., Ed., CRC Press, Boca Raton, Fla., 1984, chap. 4.
57. **Falk, M. C., Staton, A. J., and Williams, T. J.**, Heterogenetity of pig plasma amine oxidase: molecular and catalytic properties of chromatographically isolated forms, *Biochemistry*, 22, 3746, 1983.
58. **Suzuki, S., Sakurai, T., Nakahara, A., Manabe, T., and Okuyama, T.**, Effect of metal substitution on the chromophore of bovine serum amine oxidase, *Biochemistry*, 22, 1630, 1983.
59. **Mondovì, B., Finazzi-Agrò, A., Rotilio, G., and Sabatini, S.**, Redox titration and reversible removal of copper from Cu-AO, in *Frontiers in Bioinorganic Chemistry*, Xavier, A. V., Ed., VHC, 1986, 604.
60. **Yamada, H. and Yasunobu, K. T.**, Mono amine oxidase. II. Copper, one of the prosthetic groups of plasma monoamine oxidase, *J. Biol. Chem.*, 237, 3077, 1962.
61. **Yamada, H., Adachi, O., and Yamano, T.**, Electron paramagnetic resonance spectra of amine oxidase from *Aspergillus niger*, *Biochem. Biophys. Acta*, 191, 751, 1969.
62. **Lindstrom, A., Olsson, B., and Pettersson, G.**, Kinetics of the interaction between pig-plasma benzylamine oxidase and substrate, *Eur. J. Biochem.*, 35, 70, 1973.
63. **Grant, J., Kelly, I., Knowles, P., Olsson, J., and Pettersson, G.**, Changes in the copper centers of benzylamine oxidase from pig plasma during the catalytic cycle, *Biochem. Biophys. Res. Commun.*, 83, 1216, 1978.
64. **Lindstrom, A., Olsson, B., and Pettersson, G.**, Effect of azide on some spectral and kinetic properties of pig-plasma benzylamine oxidase, *Eur. J. Biochem.*, 48, 237, 1974.
65. **Boden, N., Holmes, M. C., and Knowles, P. F.**, Binding of water to type I and II Cu^{2+} in proteins, *Biochem. Biophys. Res. Commun.*, 57, 845, 1974.
66. **Olsson, B., Olsson, J., and Pettersson, G.**, Effect on enzyme activity of ligand-binding to copper in pig-plasma benzylamine oxidase, *Eur. J. Biochem.*, 87, 1, 1978.
67. **Yamada, H. and Yasunobu, K. T.**, Monoamine oxidase. IV. Nature of the second prosthetic group of plasma monoamine oxidase, *J. Biol. Chem.*, 238, 2669, 1963.
68. **Buffoni, F.**, Pyridoxal catalysis in pig plasma benzylamine oxidase (histaminase), in *Pyridoxal Catalysis: Enzymes and Model Systems*, Snell, E. G., Braunstein, A. E., Severin, E. S., and Torchinsky, Y. M., Eds., Wiley Interscience, New York, 1968, 363.
69. **Hamilton, G. A.**, A possible model reaction for pyridoxal phosphate dependent amine oxidases, in *Pyridoxal Catalysis: Enzymes and Model Systems*, Snell, E. G., Braunstein, A. E., Severin, E. S., and Torchinsky, Y. M., Eds., Wiley Interscience, New York, 1968, 375.
70. **Mondovì, B., Costa, M. T., Finazzi-Agrò, A., and Rotilio, G.**, Further investigations on the structure of pig kidney diamine oxidase in *Pyridoxal Catalysis: Enzymes and Model Systems*, Snell, E. G., Braunstein, A. E., Severin, E. S., and Torchinsky, Y. M., Eds., Wiley Interscience, New York, 1968, 403.
71. **Buffoni, F. and Della Corte, L.**, Pig plasma benzymaline oxidase, *Adv. Biochem. Psychopharmacol.*, 5, 133, 1972.
72. **Finazzi-Agrò, A., Guerrieri, P., Costa, M. T., and Mondovì, B.**, On the nature of chromophore in pig kidney diamine oxidase, *Eur. J. Biochem.*, 74, 435, 1977.
73. **Lobenstein-Verbeek, C. L., Jongejan, J. A., Frank, J., and Duine, J. A.**, Bovine serum amine oxidase: a mammalian enzyme having covalently bound PQQ as prosthetic group, *FEBS Lett.*, 170, 305, 1984.
74. **Hill, J. M. and Mann, P. J. G.**, Further properties of the diamine oxidase of pea seedlings, *Biochem. J.*, 91, 171, 1966.
75. **Kluetz, M. D., Adamsons, K., and Flynn, J. R., Jr.**, Cryoenzymology and spectrophotometry of pea seedling diamine oxidase, *Biochemistry*, 19, 1617, 1980.
76. **Rinaldi, A., Giartosio, A., Floris, G., Medda, R., and Finazzi-Agrò, A.**, Lentil seedlings amine oxidase: preparation and properties of the copper-free enzyme, *Biochem. Biophys. Res. Commun.*, 120, 242, 1984.
77. **Dooley, D. M. and Golnik, K. C.**, Spectroscopic and kinetic studies of the inhibition of pig kidney diamine oxidase by anions, *J. Biol. Chem.*, 258, 4245, 1983.
78. **Lindstrom, A., Olsson, B., Olsson, J., and Pettersson, G.**, Effect of pH on the transient reduction of pig plasma benzylamine oxidase by benzylamine derivatives, *Eur. J. Biochem.*, 64, 321, 1976.
79. **Lindstrom, A., Olsson, B., Pettersson, G., and Szymanska, J.**, Kinetics of the interaction between pig-plasma benzylamine oxidase and various monoamines, *Eur. J. Biochem.*, 47, 99, 1974.

80. **Bardsley, W. G., Ashford, J. S., and Hill, C. M.,** Synthesis and oxidation of aminoalkyl-onium compounds by pig kidney diamine oxidase, *Biochem. J.*, 122, 557, 1971.
81. **Squires, R. F.,** Multiple forms of monoamine oxidase in intact mitochondria as characterized by selective inhibitors and thermal stability: a comparison of eight mammalian species, *Adv. Biochem. Psychopharmacol.*, 5, 355, 1972.
82. **Oreland, L.,** Purification and properties of pig liver monoamine oxidase, *Arch. Biochem. Biophys.*, 146, 410, 1971.
83. **Minamiura, N. and Yasunobu, K. T.,** Bovine liver monoamine oxidase, *Arch. Biochem. Biophys.*, 189, 481, 1978.
84. **Hiramatsu, A., Tsurushiin, S., and Yasunobu, K. T.,** Evidence for essential cysteine residues in bovine liver mitochondrial monoamine oxidase, *Eur. J. Biochem.*, 57, 587, 1975.
85. **Oi, S., Yasunobu, K. T., and Westley, J.,** The effect of pH on the kinetic parameters and mechanism of beef liver monoamine oxidase, *Arch. Biochem. Biophys.*, 145, 557, 1971.
86. **Zeidan, H., Watanabe, K., Piette, L. H., and Yasunobu, K. T.,** ESR studies of spin labeled bovine liver monoamine oxidase B., in *Frontiers in Protein Chemistry*, Liu, T. Y., Mamiya, G., and Yasunobu, K. T., Eds., Elsevier/North-Holland, New York, 1980, 131.
87. **Nara, S., Igaue, I., Gomes, B., and Yasunobu, K. T.,** The prosthetic groups of animal amine oxidases, *Biochem. Biophys. Res. Commun.*, 23, 342, 1966.
88. **Finazzi-Agrò, A., Rotilio, G., Costa, M. T., and Mondovì, B.,** Evidence for a ping-pong mechanism in the diamine oxidase reaction, *FEBS Lett.*, 14, 31, 1969.
89. **Oi, S., Inamasu, M., and Yasunobu, K. T.,** Mechanistic studies of beef plasma amine oxidase, *Biochemistry*, 9, 3378, 1970.
90. **Taylor, C. E., Taylor, R. S., Rasmussen, L., and Knowles, P. F.,** A catalytic mechanism for the enzyme benzylamine oxidase from pig plasma, *Biochem. J.*, 130, 713, 1972.
91. **Berg, K. A. and Abeles, R. H.,** Mechanism of action of plasma amine oxidase: products released under anaerobic conditions, *Biochemistry*, 19, 3186, 1980.
92. **Falk, M. C.,** Stoichiometry of phenylhydrazine inactivation of pig plasma amine oxidase, *Biochemistry*, 22, 3740, 1983.
93. **Olsson, B., Olsson, J., and Pettersson, G.,** Stopped flow spectrophotometric characterisation of enzymic reaction intermediates in the anaerobic reduction of pig plasma benzylamine oxidase by amine substrates, *Eur. J. Biochem.*, 71, 375, 1976.
94. **Olsson, B., Olsson, J., and Pettersson, G.,** The kinetics of re-oxidation of reduced benzylamine oxidase, *Eur. J. Biochem.*, 74, 329, 1977.
95. **Lindstrom, A. and Pettersson, G.,** The order of substrate addition and product release during the catalytic action of pig-plasma benzylamine oxidase, *Eur. J. Biochem.*, 84, 479, 1978.
96. **Lindstrom, A., Olsson, B., and Pettersson, G.,** Transient kinetics of benzaldehyde formation during the catalytic action of pig plasma benzylamine oxidase, *Eur. J. Biochem.*, 42, 377, 1974.
97. **Seiler, N.,** On the role of GABA in vertebrate polyamine metabolism, *Physiol. Chem. Phys.*, 12, 411, 1980.
98. **Razin, S., Bachrach, U., and Gery, I.,** Formation of β-alanine from spermine and spermindine by *Pseudomonas aeruginosa*, *Nature*, 181, 700, 1958.
99. **Bachrach, U., Abzug, S., and Bekierkunst, A.,** Cytotoxic effect of oxidized spermine on Ehrlich arcites cells, *Biochim. Biophys. Acta*, 134, 174, 1967.
100. **Bachrach, U.,** Oxidized polyamines, *Ann. N.Y. Acad. Sci.*, 171, 939, 1970.
101. **Alacorn, R. A.,** Acroelin. IV. Evidence for the formation of the cytotoxic aldehyde acrolein from enzymatically oxidized spermine or spermidine, *Arch. Biochem. Biophys.*, 137, 365, 1970.
102. **Bachrach, U., Don, S., and Weiner, H.,** Antivirus action of acrolein, glutaraldehyde and oxidized spermine, *J. Gen. Virol.*, 13, 415, 1971.
103. **Eilon, G. and Bachrach, U.,** Interaction of oxidized polaymines with DNA. III. Association with nucleosides, mono and polynucleotides, *Biochim. Biophys. Acta*, 179, 464, 1969.
104. **Bachrach, U. and Eilon, G.,** Interaction of oxidized polyamines with DNA. I. Evidence for the formation of cross-links, *Biochim. Biophys. Acta*, 145, 418, 1967.
105. **Mondovì, B., Gerosa, P., and Cavaliere, R.,** Studies on the effect of polyamines and their products on Ehrlich ascites tumours, *Agents Actions*, 12, 450, 1982.
106. **Byrd, W. J., Jacobs, D. M., and Amoss, M. S.,** Synthetic polyamines added to cultures containing bovine sera, reversibly inhibit *in vitro* parameters of immunity, *Nature*, 267, 621, 1977.
107. **Gaugas, J. M. and Curzen, P.,** Polyamine interaction with pregnancy serum in suppression of lymphocyte transformation, *Lancet*, 1, 18, 1978.
108. **Morgan, D. M. L.,** Polyamine oxidation and human pregnancy, in *Advances in Polyamine Research*, Vol. 4, Bachrach, U., Kaye, A., and Chayen, R., Eds., Raven Press, New York, 1983, 155.
109. **Leete, E.,** The biogenesis of nicotine. V. New precursors of the pyrrolidine ring, *J. Am. Chem. Soc.*, 80, 2162, 1958.

110. **Cromwell, B. T.,** The role of putrescine in the synthesis of hyoscyamine, *Biochem. J.*, 37, 722, 1943.
111. **Diaper, D. G. M., Kirkwood, S., and Marion, L.,** The biogenesis of alkaloids. III. A study of hyoscyamine biosynthsis using isotopic putrescine, *Can. J. Chem.*, 29, 964, 1951.
112. **Schütte, H. R.,** Die Biosynthese von Alkaloiden. III. Uber die gegenseitige Umwandelbarkeit von Lupinin und Spartein sowie uber deren Biogeneseort in Lupinus Lutens, *Atompraxis*, 7, 91, 1961.
113. **Schütte, H. R., Bohlmann, F., and Reusche, W.,** Die Biosynthese des Sparteins, *Arch. Pharm.*, 294, 610, 1961.
114. **Schütte, H. R.,** Zur Biosynthese der Lupinenalkaloide, *Arch. Pharm.*, 293, 1006, 1960.
115. **Schütte, H. R., Nowacki, E., and Schaffer, C.,** Die Bildung von Lupanin und Hydroxylupanin aus Cedaverin (1,5-^{14}C)-dihydrochloride in Lupinus angustifolius, *Arch. Pharm.*, 295, 20, 1962.
116. **Hasse, K. and Berg, P.,** Bildung von Anabasin aus Cadaverin in Genenwart von Ptlanzenextrakten, *Biochem. J.*, 331, 1959.
117. **Del Principe, D., Menichelli, A., De Matteis, W., Di Corpo, M. L., Di Giulio, S., and Finazzi-Agrò, A.,** Hydrogen peroxide has a role in the aggregation of human platelets, *FEBS Lett.*, 185, 142, 1985.
118. **Scalabrino, G., Ferioli, M. E., Modena, D., Puerari, M., and Luccarelli, G.,** Levels of activity of the polyamine biosynthetic decarboxylases as indicators of the degree of malignancy of human brain tumors, *Ital. J. Biochem.*, 31, 60, 1982.
119. **Williams-Ashaman, H. G., Coppoc, G. L., and Weber, G.,** Imbalance in ornithine metabolism in hepatomas of different growth rates as expressed in formation of putrescine, spermidine, and spermine, *Cancer Res.*, 32, 1924, 1972.
120. **O'Brien, T. G., Simsiman, R. C., and Boutwell, R. K.,** Induction of the polyamine-biosynthetic enzymes in mouse epidermis and their specificity for tumor promotion, *Cancer Res.*, 35, 2426, 1975.
121. **Russell, D. H. and Snyder, S. H.,** Amine synthesis in rapidly growing tissues: ornithine decarboxylase activity in regenerating rat liver, chick embryo, and various tumors, *Proc. Natl. Acad. Sci. U.S.A.*, 60, 1420, 1968.
122. **Jänne, J. and Raina, A.,** Stimulation of spermidine synthesis in the regenerating rat liver: relation to increased ornithine decarboxylase activity, *Acta Chem. Scand.*, 22, 1349, 1968.
123. **Scalabrino, G., Pigatto, P., Ferioli, M. E., Modena, D., Puerari, M., and Carù, A.,** Levels of activity of the polyamine biosynthetic decarboxylases as indicators of degree of malignancy of human cutaneous epitheliomas, *J. Invest. Dermatol.*, 74, 122, 1980.
124. **Quash, G., Keolouangkhot, T., Gazzolo, L., Ripoli, H., and Saez, S.,** Diamine oxidase and polyamine oxidase activities in normal and transformed cells, *Biochem. J.*, 177, 275, 1979.
125. **Sessa, A., Desiderio, M. A., Baizini, M., and Perin, A.,** Diamine oxidase activity in regenerating rat liver and in 4-dimethyl-aminoazobenzene-induced and Yoshida AH 130 hepatomas, *Cancer Res.*, 41, 1929, 1981.
126. **Perin, A., Sessa, A., and Desiderio, M. A.,** Induction of diamine oxidase activity in some processes of growth, in *Advances in Polyamine Research*, Vol. 4, Bachrach, U., Kaye, A., and Chayen, R., Eds., Raven Press, New York, 1983, 175.
127. **Mondovì, B., Riccio, P., Riccio, A., and Marcozzi, G. S.,** Amine oxidase activity in malignant human brain tumors, in *Advances in Polyamine Research*, Vol. 4, Bachrach, U., Kaye, A., and Chayen, R., Eds., Raven Press, New York, 1983, 183.
128. **Fogel, W. A., Bieganski, T., Bechtold, E., and Maslinski, C.,** Diamine oxidase activity and imidazoleatic acid formation in the fetal and maternal guinea pig liver, *Agents Actions*, 12, 49, 1982.
129. **Fogel, W. A., Bieganski, T., and Maslinski, C.,** Gamma-aminobutyric acid/GABA formation from putrescine in guinea pig liver during ontogenesis, *Comp. Biochem. Physiol. C*, 73, 431, 1982.
130. **Desiderio, M. A., Sessa, A., and Perin, A.,** Induction of diamine oxidase activity in rat kidney during compensatory hypertrophy, *Biochim. Biophys. Acta*, 714, 243, 1982.
131. **Baylin, S. B., Abeloff, M. D., Tomford, J. W., Wieman, K. C., and Ettinger, D. S.,** Elevated histaminase (diamine oxidase) activity in small cell carcinoma of the lung, *N. Engl. J. Med.*, 293, 1286, 1975.
132. **Baylin, S. B., Beaven, M. A., Engelman, K., and Sjoerdsma, A.,** Elevated histaminase activity in medullary carcinoma of the thyroid gland, *N. Engl. J. Med.*, 283, 1239, 1970.
133. **Baylin, S. B., Abeloff, M. D., Goodwin, G., Carney, D. N., and Gazdar, A. F.,** Activities of L-dopa decarboxylase and diamine oxidase (histaminase) in human lung cancers and decarboxylase as a marker for small (oat) cell cancer in cell culture, *Cancer Res.*, 40, 1990, 1980.
134. **Sessa, A., Desiderio, M. A., and Perin, A.,** Diamine oxidase activity induction in regenerating rat liver, *Biochem. Biophys. Acta*, 698, 11, 1982.
135. **Perin, A., Sessa, A., and Desiderio, M. A.,** Polyamine levels and dimaine oxidase activity in hypertrophic heart of spontaneously hypertensive rats and of rats treated with isoproterenol, *Biochim. Biophys. Acta*, 755, 344, 1983.

136. **Kusche, J., Bieganski, T., Hesterberg, R., Stahlknecht, C. D., Feussner, K. D., Stahlenberg, I., and Lorenz, W.**, The influence of carcinoma growth on diamine oxidase activity in human gastrointestinal tract, *Agents Actions*, 10, 110, 1980.
137. **Jänne, J. and Hölttä, E.**, Putrescine metabolizing enzyme activities in some rat tissues during postanatal development, *Acta Chem. Scand.*, 27, 2399, 1973.
138. **Luk, G. D., Bayless, T. M., and Baylin, S. B.**, Diamine oxidase (histaminase), a circulating marker for rat intestinal mucosal maturation and integrity, *J. Clin. Invest.*, 66, 66, 1980.
139. **Luk, G. D., Marton, L. J. M., and Baylin, S. B.**, Ornithine decarboxylase is important in intestinal mucosal maturation and recovery from injury in rats, *Science*, 210, 195, 1980.
140. **Maudsley, D. V., Leif, J., and Kobayashi, Y.**, Ornithine decarboxylase in rat small intestine: stimulation with food or insulin, *Am. J. Physiol.*, 231, 1557, 1976.
141. **Smith, M. W. and Jarvis, L. G.**, Growth and cell replacement in the newborn pig intestine, *Proc. R. Soc. London. B*, 203, 69, 1978.
142. **Stanley, B. A. and Kazarinoff, M. N.**, Induction of ornithine decarboxylase in colon and liver by starvation and refeeding: a comparison of effects on total and holoenzyme, *Biochem. Biophys. Res. Commun.*, 105, 773, 1982.
143. **Argento-Cerù, M. P., Oratore, A., Mondovì, B., and Finazzi-Agrò, A.**, Localization of diamine oxidase in pig kidney: immunofluorescence method, *Cell Mol. Biol.*, 27, 359, 1981.
144. **Kobayashi, Y.**, Blood levels of diamine oxidase as a test for pregnancy, *J. Lab. Clin. Med.*, 62, 669, 1963.
145. **Elmfors, B. and Tryding, N.**, Date of confinement prediction from serum diamine oxidase determination in early pregnancy, *Br. J. Obstet. Gyanaecol.*, 83, 6, 1976.
146. **Maslinski, C., Niedzielski, A., and Redziejowska, B.**, The histaminase activity in pregnant women and after the artificial abortion, *Gynaecologia*, 157, 43, 1964.
147. **Weisburger, W. R., Mendelsohn, G., Eggleston, J. C., and Baylin, S. B.**, Immunohistochemical localization of histaminase (diamine oxidase) in decidual cells of huamn placenta, *Lab. Invest.*, 38, 703, 1978.
148. **Lin, C. W., Chapman, C. M., De Lellis, R. A., and Kirley, S.**, Immunofluorescent staining of histaminase (diamine oxidase) in human placenta, *J. Histochem. Cytochem.*, 26, 1021, 1978.
149. **Ahlmark, A.**, Studies on the histaminolytic power of plasma with special reference to pregnancy, *Acta Physiol. Scand.*, 9 (Suppl. 28), 1, 1938.
150. **Swanberg, H.**, Histaminase in pregnancy. With special reference to its origin and formation, *Acta Physiol. Scand.*, 23 (Suppl. 79), 1, 1950.
151. **Ward, H., Whyley, G. A., and Millar, M. D.**, A comparative study of diamine oxidase in amniotic fluid and serum in normal pregnancy and in pregnancy complicated by rhesus iso-immunization, *J. Obstet. Gynaecol.*, 80, 525, 1973.
152. **Baylin, S. B., Beaven, M. A., Buja, L. M., and Keiser, H. R.**, Histaminase activity: a biochemcial marker for medullary carcinoma of the thyroid., *Am. J. Med.*, 53, 723, 1972.
153. **Baylin, S. B., Weisburger, W. R., Eggleston, J. C., Mendelsohn, G., Beaven, M. A., Abeloff, M. D., and Ettinger, D. S.**, Variable content of histaminase L-dopa decarboxylase and calcitonin in small cell carcinoma of the lung, *N. Engl. J. Med.*, 299, 105, 1978.
154. **Berger, C. L., Goodwin, G., Mendelsohn, G., Eggleston, J. C., Abeloff, M. D., Aisner, S., and Baylin, S. B.**, Endocrine related biochemistry in the spectrum of human lung carcinoma, *J. Clin. Endocrinol. Metab.*, 53, 422, 1981.
155. **Ettinger, D. S., Rosenshein, N. B., Parmley, T. H., Mendelsohn, G., and Baylin, S. B.**, Tumor cell origin of histaminase activity in ascites fluid from patients with ovarian carcinoma, *Cancer*, 45, 2568, 1980.
156. **Seiler, N. and Al-Therib, M. J.**, Putrescine catabolism in mammalian brain, *Biochem. J.*, 144, 29, 1974.
157. **Crabbe, M. J. C. and Bardsley, W. G.**, Monoamine oxidase inhibitors and other drugs as inhibitors of diamine oxidase from human placenta and pig kidney, *Biochem. Pharmacol.*, 23, 2983, 1974.
158. **Roth, J. A., Eddy, B. J., Pearce, L. B., and Mulder, K. M.**, Phenylhydrazine: selective inhibition of human type B monoamine oxidase, *Biochem. Pharmacol.*, 30, 945, 1981.
159. **Bardsley, W. G. and Ashford, J. S.**, Inhibition of pig kidney diamine oxidase by substrate analogues, *Biochem. J.*, 128, 253, 1972.
160. **Crabbe, M. J. C. and Bardsley, W. G.**, The inhibition of human placental diamine oxidase by substrate analogues, *Biochem. J.*, 139, 183, 1974.
161. **Shindler, J. S. and Bardsley, W. G.**, Human kidney diamine oxidase: inhibition studies, *Biochem. Pharmacol.*, 25, 2689, 1976.
162. **Macholan, L.**, Selective and reversible inhibition of diamine oxidase by 1,5-diamino-3-pentanone, *Coll. Czech. Chem. Commun.*, 39, 653, 1974.
163. **Rando, R. J. and De Mairena, J.**, Propargyl amine-induced irreversible inhibition of non-flavin-linked amine oxidase, *Biochem. Pharmacol.*, 23, 463, 1974.

164. **Hevey, R. C., Babson, J., Maycock, A. L., and Abeles, R. H.,** Highly specific enzyme inhibitors. Inhibition of plasma amine oxidase, *J. Am. Chem. Soc.*, 95(18), 6125, 1973.
165. **Maycock, A. L., Suva, R. H., and Abeles, R. H.,** Novel inactivators of plasma amine oxidase, *J. Am. Chem. Soc.*, 97(19), 5613, 1975.
166. **Kagan, H. M., Soucy, D. M., Zoski, C. G., Resnick, R. J., and Tang, S. S.,** Multiple modes of catalysis-dependent inhibition and inactivation of aortic lysyl oxidase, *Arch. Biochem. Biophys.*, 221, 158, 1983.
167. **Tang, S. S., Trackman, P. C., and Kagan, H. M.,** Reactions of aortic lysyl oxidase with β-aminopropionitrile, *J. Biol. Chem.*, 258, 4331, 1983.
168. **Tobes, M. C., Griffith, R. K., Di Pietro, R. A., and Zoorob, H. H.,** Design and evaluation of mechanism-based inhibitors of diamine oxidase, *Fed. Proc. Fed. Am. Soc. Exp. Biol.*, 42, 2042, 1983.
169. **Child, R. E. and Bardsley, W. G.,** Time dependent inhibition of enzymes by active-site directed reagents. A theoretical treatment of the kinetics of affinity labelling, *J. Theor. Biol.*, 53, 381, 1975.
170. **Crabbe, M. J. C., Childs, R. E., and Bardsley, W. G.,** Time-dependent inhibition of diamine oxidase by carbonyl-group reagents and urea, *Eur. J. Biochem.*, 60, 325, 1975.
171. **Bieganski, T., Osinska, Z., and Maslinski, C.,** Inhibition of plant and mammalian diamine oxidases by hydrazine and guanidine compounds, *Int. J. Biochem.*, 14, 949, 1982.
172. **Mondovì, B., Rotilio, G., Finazzi-Agrò, A., and Antonini, E.,** Amine oxidases: a new class of copper oxidases, in *Magnetic Resonances in Biological Research*, Franconi, C., Ed., Gordon & Breach Scientific, New York, 1971, 233.
173. **Bardsley, W. G., Childs, R. E., and Crabbe, M. J. C.,** Inhbiition of enzymes by metal ion-chelating reagents. The action of copper-chelating reagents on diamine oxidase, *Biochem. J.*, 137, 61, 1974.
174. **Lindström, A. and Pettersson, G.,** The mechanism of inhibition of pig-plasma benzylamine oxidase by the copper-chelating reagent Cuprizone, *Eur. J. Biochem.*, 48, 229, 1974.
175. **Costa, M. T., Rotilio, G., Finazzi-Agrò, A., Vallogini, M. P., and Mondovì, B.,** On the active site of diamine oxidase: kinetic studies, *Arch. Biochem. Biophys.*, 147, 8, 1971.
176. **Bardsley, W. G., Solano-Munoz, F., Wright, A. J., and McGinlay, P. B.,** Inhibition of enzyme-catalysed reaction by excess substrate. A theoretical and Monte Carlo study of turning points in v(S) graphs, *J. Mol. Biol.*, 169, 597, 1983.
177. **Seiler, N., Bolkenius, F. N., and Knödgen, B.,** The influence of catabolic reactions on polyamine excretion, *Biochem. J.*, 225, 219, 1985.
178. **Bolkenius, F. N., Bey, P., and Seiler, N.,** Specific inhibition of polyamine *in vivo* is a method for the elucidation of its physiological role, *Biochim. Biophys. Acta*, 838, 69, 1985.
179. **Baylin, S., Horakova, Z., and Beaven, M. A.,** Increase in food consumption and growth after treatment with aminoguanidine, *Experientia*, 31, 562, 1975.
180. **Dolezalova, H., Stepita-Klauco, M., Kucera, J., Uchimura, H., and Hirano, M.,** Blood concentrations of monoacylcadaverines in schizophrenia, in *Mass Spectrometry in Drug Metabolism*, Frigerio, A. and Ghisalberti, E. L., Eds., Plenum Press, New York, 1977, 201.
181. **Ma, K. and Sourks, T.,** Inhibition of diamine oxidase by antimalarial drugs, *Agents Actions*, 10(5), 395, 1980.
182. **Coelho, H., Azevedo, M., and Manso, C.,** Inhibitory effect of drugs used in the treatment of Parkinson's disease on plasma monoamine oxidase activity, *J. Neural. Trans.*, 61, 271, 1985.
183. **Kavanagh, J. P., Brightwell, R., Bardsley, W. G., and Schnieden, H.,** Levamisole and other diamine oxidase inhibitors as inhibitors of sperm motility, *Arch. Androl.*, 7, 51, 1981.
184. **Bieganski, T., Braun, R., and Kusche, J.,** *N*-Methyl-*N*-formylhydrazine: a toxic and mutagenic inhibitor of the intestinal diamine oxidase, *Agents Actions*, 14, 351, 1984.

Chapter 13

POLYAMINE OXIDASES AND OXIDIZED POLYAMINES

David M. L. Morgan

TABLE OF CONTENTS

"This tale grew in the telling until it became a history . . . and included many glimpses of the yet more ancient history that preceded it."

J.R.R. Tolkien
Foreword to *The Lord of the Rings*

I. NOMENCLATURE OF THE POLYAMINE OXIDASES

The nomenclature of the amine oxidases is confused (for a guide to the earlier literature see References 1 to 9) and there are many instances where so-called mono- and diamine oxidases have been shown to act on polyamines.[10,11] Indeed, the latest edition of the *Recommendations of the Commission on Biological Nomenclature*[12] lists just two amine oxidases: EC 1.4.3.4 amine:oxygen oxidoreductase (deaminating, flavin-containing) for which the recommended name is amine oxidase (flavin-containing) and EC 1.4.3.6 amine:oxygen oxidoreductase (deaminating, copper-containing) or amine oxidase (copper-containing). The first is defined as a flavoprotein acting on primary amines, and usually also on secondary and tertiary amines with small substituents; the second as a group of enzymes that oxidize primary monoamines, diamines, and histamine. In both cases the enzyme is considered to act on a CH_2-NH_2 group of the donor and oxygen is the acceptor. EC 1.5.3.3, spermine oxidase, is now deleted. Spermidine dehydrogenase,[13] EC 1.5.99.6, is also listed; however, for this enzyme the acceptor is a compound other than oxygen, so it will not be further considered here.

An alternative, and perhaps more logical, grouping would be to divide the polyamine oxidases into those that act at the primary amino groups, and those (the majority) that act at the secondary amino group(s) of the aminopropyl moieties of spermine or spermidine. Polyamine oxidases that act at the secondary amino group would be further subdivided according to whether diaminopropane or aminopropionaldehyde were among the products (Figure 1). The problems inherent in any attempt to derive a coherent classification for the amine oxidases have been discussed by Bachrach.[15] For the purposes of this review amine oxidases able to utilize the polyamines spermine or spermidine as substrates will be considered as polyamine oxidases, whether or not they can also act on mono- or diamines.

The terminology used here for N-substituted derivatives of spermine and spermidine is that of Tabor et al.;[14] the order of the numbering

$$\begin{matrix} 1 & 2 & 3 & 4 & 5 & 6 & 7 & 8 \\ \end{matrix}$$
$$NH_2CH_2CH_2CH_2NH_2CH_2CH_2CH_2CH_2NH_2$$

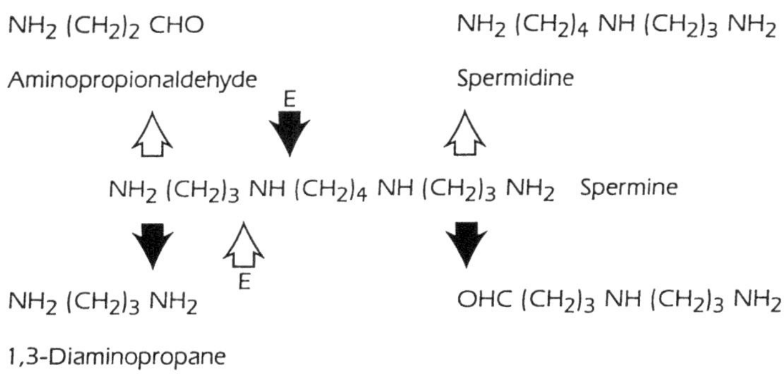

FIGURE 1. Alternative sites of cleavage of spermine by polyamine oxidases from various sources. The substrate is cleaved at a secondary amino group with the formation of amino-propionaldehyde or diaminopropane as one of the products. Spermidine is attacked in a similar manner. (Reprinted by permission from *Essays Biochem.*, 23, 94, copyright © 1987, The Biochemical Society, London.)

being determined by the convention that the secondary amino group should have the lowest possible number. In the chemical literature spermine is variously referred to as *N,N'*-bis(3-aminopropyl)-1,4-butanediamine, *N,N'*-bis(3-aminopropyl)-1,4-diaminobutane, or 4,9-diazadodecane-1,12-diamine (Chemical Abstracts Registry number 71-44-3); similarly spermidine may be referred to as *N*-(3-aminopropyl)-1,4-butanediamine or 4-azaoctane-1,8-diamine (CA 124-20-9); putrescine is 1,4-butanediamine or diaminobutane (CA 110-60-1).

II. BOVINE PLASMA POLYAMINE OXIDASE

A. Properties

In the 30 years since its discovery the bovine plasma polyamine oxidase has been the subject of extensive study.[16] A number of procedures for the isolation of the enzyme have been described.[10,17-24] The enzyme has been purified to apparent homogeneity, as shown by electrophoresis and ultracentrifugation, and crystallized.[10] However, electrophoresis in polyacrylamide disc gels, with or without sodium dodecyl sulfate, revealed the presence of minor ($<5\%$) components that also possessed amine oxidizing activity[17,23,24] and appear to share a common mechanism.[21]

The enzyme has been shown to act on spermine and spermidine in accordance with Scheme I.

The aldehydes formed were reduced with sodium borohydride, extracted with butanol, and shown to comigrate with the corresponding synthetic amino alcohols by paper chromatography in nine solvent systems.[25] Identical i.r. spectra were obtained from the borohydride-reduced product of spermine oxidation and authentic *N,N'*-bis(3-hydroxypropyl)-1,4-diaminobutane. The enzyme has a molecular weight, determined by gel filtration and ultracentrifugation, of about 180,000;[23,26] electrophoresis in the presence of sodium dodecyl sulfate and ultracentrifugation of the guanidine-treated enzyme indicated the presence of two noncovalently linked subunits of molecular weight (MW) 90,000. Association-dissociation between monomer-dimer-trimer forms has been noted,[26] providing an explanation for reported molecular weights of the order of 265,000.[27-29] Each subunit contains a disulfide bridge, and there are two copper atoms and one carbonyl group per molecule. Only one of the copper atoms appears to be essential for activity,[30] as is the carbonyl group, since the enzyme is inhibited by carbonyl reagents (see below). Early reports suggested pyridoxal phosphate (2 mol/25,400 MW) as the carbonyl-containing cofactor[19,31] but its presence was never directly demonstrated and later work cast doubt on this conclusion;[20,32,33] indeed, Suva

$$NH_2(CH_2)_3NH(CH_2)_4NH(CH_2)_3NH_2 + 2O_2 + 2H_2O \rightarrow$$
spermine

$$OCH(CH_2)_2NH(CH_2)_4NH(CH_2)_2CHO + 2NH_3 + 2H_2O_2$$
N,N'-bis(3-propionaldehyde)-
1,4-diaminobutane

$$NH_2(CH_2)_4NH(CH_2)_3NH_2 + O_2 + H_2O \rightarrow$$
spermidine

$$NH_2(CH_2)_4NH(CH_2)_2CHO + NH_3 + H_2O_2$$
N-(4-aminobutyl)-3-
aminopropionaldehyde

SCHEME I. Oxidation of spermine and spermidine by bovine plasma amine oxidase.

and Abeles[34] stated categorically that pyridoxal phosphate was not a cofactor for this enzyme. Recently it has been shown that the prosthetic group is covalently bound pyroloquinoline quinone,[35] a cofactor for a number of dehydrogenases.[36] The presence of this compound was demonstrated by isolation of the quinone adduct from a proteolytic digest of the dinitrophenylhydrazine-treated enzyme and comparison with the authentic compound by HPLC and spectroscopy.

The enzyme is a glycoprotein,[37] containing 7 to 8% carbohydrate made up of galactose (18; numbers are residues per mole of enzyme), mannose (32), sialic acid (10), and *N*-acetylglucosamine (24).[20] The carbohydrate moiety, which is attached by a glycosylamine linkage between *N*-acetylglucosamine and an asparagine residue in the protein,[37] is not essential for activity and is remote from the active site.[22] Three forms of the enzyme have been reported[10,20,21,23,24] but data on the minor components is sparse. Yasunobu et al.[20] have suggested that the multiple forms of the enzyme may be isoenzymes that differ only in carbohydrate content but this has yet to be established. When pure, the enzyme is pink in color and in solution shows absorption maxima at 280 and 480 nm.[10,23] The latter peak disappears after addition of sodium dithionite, or substrate, under anaerobic conditions but reappears on admission of oxygen.

B. Substrates and Kinetics

The enzyme oxidizes polyamines containing primary amino groups, those forming part of an aminopropyl moiety being the more readily attacked,[38] and some primary amines including benzylamine.[10,17] Molecular oxygen is required, since spermine was not degraded by the enzyme in a helium atmosphere;[17] also, the yield of aminoaldehyde, measured by assessing cytotoxicity, was increased in an oxygen-rich atmosphere.[39] The relative rates of oxidation of spermine, spermidine, and benzylamine have been variously reported as 125:56:100 and 164:164:100, both based on manometric measurement of oxygen consumption;[10,17] 250:250:100 from polarographic determinations of oxygen uptake,[24] and 238:294:100 (*o*-dianisidine/peroxidase assay of hydrogen peroxide production)[29] with Michaelis constants of 2, 2.7, and 0.67 μM (V_{max} 18.7, 11.5, and 6.9 nmol of O_2 per minute per milligram), respectively.[40] Gahl and Pitot[29] have reported oxidation rates (recalculated to the same relative scale) for N^8-acetylspermidine (198; K_m 44 μM), N^1-acetylspermidine (59), putrescine (15; K_m 2.3 mM) and N^1-acetylputrescine (15); these workers also reported K_m values of 33 μM for spermine and 48 μM for spermidine. In contrast, Mamont et al.,[41] reported that the amine oxidase in newborn calf serum did not attack N^1-acetylspermidine, although the N^8-isomer was rapidly degraded to *N*-acetylputrescine.[29] This enzyme also acted on N^1-acetylspermine to form N^1-acetylspermidine.[41] The presence of spermidine or N^8-acetylspermidine as contaminants in the preparation of N^1-acetylspermidine used by Gahl and Pitot has been suggested[42] to explain the discrepancy, but this remains to be resolved. Oi et al.[43] found K_m values of 13.5 μM for oxygen with benzylamine as substrate (air-equilibrated buffers at pH 7 contain 2 to 3 mM oxygen[44]) and 1.4 mM for benzylamine, in good agreement with other workers (1.4 mM[17] and 1.3 mM[29]). Cadaverine (1,5-diaminopentane), [lysine]vasopressin, the peptide Pro-Lys-Gly-NH_2, and elastin are also oxidized slowly.[28] *N,N'*-bis(3-aminopropyl)ethane-1,2-diamine, *N,N'*-(3-aminopropyl)hexane-1,6-diamine,[45] and *N,N'*-(3-aminopropyl)nonane-1,9-diamine[46] were also oxidized. Immobilization of the enzyme on Sepharose 4B increased the K_m for spermine and spermidine to 0.11 and 0.5 mM, respectively.[47] Putrescine is also a substrate for this enzyme[29,47,48] and Gahl and Pitot[29] have reported the presence in fetal calf serum of a form of the enzyme showing a greater activity with putrescine than that of the enzyme in adult bovine serum. This could indicate a change in isoenzyme pattern between fetal and adult life. These workers were able to achieve a partial separation of the putrescine oxidizing activity from spermidine oxidase activity in fetal but not in adult bovine sera.[29,48]

C. Inhibitors

Bovine serum polyamine oxidase is inhibited by isoniazid (isonicotinic acid hydrazide, 4-pyridine carboxylic acid hydrazide), iproniazid [isonicotinic acid 2-isopropylhydrazide, 4-pyridine carboxylic acid 2-(1-methylethyl)hydrazide], and carbonyl-binding agents such as semicarbazide, hydrazine, and hydroxylamine, with 95 to 100% inhibition occurring at concentrations of 1 mM or less.[17] Other inhibitors include sulfhydryl reagents[49] (K_i for *p*-chloromercuribenzoate, a noncompetitive inhibitor, 1.66 mM), copper chelators such as cuprizone,[50] aminoguanidine, and methylglyoxal bis(guanylhydrazone),[29] which also inhibits *S*-adenosylmethionine decarboxylase (K_i 70 and 20 μM, respectively), 3-hydroxybenzyloxyamine,[51] β-aminoproprionitrile, penicillamine, and pargyline (*N*-methyl-*N*-2-propynylbenzylamine, and *N*-methyl N^1-propynylbenzenemethanamide).[28] Benzoic acid and benzyl alcohol are noncompetitive inhibitors[49] (K_i 30 and 33.6 mM, respectively), as is benzaldehyde (K_i 1.1 mM).[43] The enzyme was irreversibly inactivated by 2-bromoethylamine, binding 2 mol of inhibitor per mole of enzyme.[52] The effects of inhibitors on copper-containing oxidases have been reviewed by Bardsley.[53]

D. Mechanism

The mechanism of the enzyme has been investigated by several groups[21,34,43,52,54] and appears to be of the ping-pong or double-displacement type with suggested involvement of lysine,[33] histidine,[55] cysteine[34,54] (in contrast to an earlier report[49]), and copper[20,30] in the active site. It has been suggested that the amine oxidizing site and the copper binding site are far apart[54] but the study referred to[56] was of hog kidney diamine oxidase. A formal reaction mechanism has been proposed.[43]

All of these studies have been carried out with benzylamine or a similar aromatic monoamine as the substrate, largely for technical reasons since the oxidative deamination of benzylamine to form benzaldehyde can conveniently be followed by monitoring the change in absorbance at 250 nm. However, benzylamine is a monoamine and a poor substrate for this enzyme, also an unnatural one, and mechanistic interpretations based on its use fail to explain the preferential oxidation of polyamines containing an aminopropyl moiety.[40] Also germane to considerations of mechanism are investigations on the protonation of spermine and spermidine[57-62] and the prochiral hydrogens at C-1' of spermidine.[63] Dissociation constants of 11.50, 10.95, 9.79, and 8.90 for the amino groups of spermine and 11.56, 10.80, and 9.52 for spermidine have been reported.[61]

After more than 30 years investigation of this enzyme by many laboratories it is salutory

I

$NH_2(CH_2)_3COOH$
4-amino-*n*-butyric acid
γ-aminobutyric acid

II

$HOOC(CH_2)_2NH(CH_2)_3COOH$
N-(2-carboxyethyl)-4-amino-*n*-butyric acid

$HOOC(CH_2)_2NH(CH_2)_4NH_2$
N-(2-carboxyethyl)-1,4-diaminobutane
N-(4-aminobutyl)-3-aminopropionic acid
putreanine

$NH_2(CH_2)_3NH(CH_2)_3COOH$
N-(3-aminopropyl)-4-amino-*n*-butyric acid
isoputreanine

III

$NH_2(CH_2)_3NH(CH_2)_4NH(CH_2)_2COOH$
N-(3-aminopropyl)-*N'*-(2-carboxyethyl)-1,4-diaminobutane
N^8-(2-carboxyethyl)spermidine

$HOOC(CH_2)_2NH(CH_2)_4NH(CH_2)_2COOH$
N,N'-bis(2-carboxyethyl)-1,4-diaminobutane
spermic acid

SCHEME II. Amino acids derived from putrescine (I), spermidine (II), and spermine. All have been found in normal human and rat urine.[69-72]

to list what we do not yet know about it. We do not know the amino acid composition of sequence, although a preliminary composition of one form was published in 1976.[20] Neither do we know the mechanism by which the enzyme oxidizes spermine or spermidine, preferentially attacking the aminopropyl end and we are not even certain if these are the natural substrates of this enzyme. We have no information on the rate of secretion of the enzyme, its half-life in the circulation, tissue of origin, or its function. Nor do we know why enzymes similar to that in bovine serum have been found in the sera of all other ruminants examined,[1,64] and in some connective tissues,[8,28] but not, apparently, in sera from nonruminants. The method used in these surveys (Warburg manometry), however, lacks sensitivity.

Seiler et al.[65,66] have presented indirect evidence for the presence in murine sera and tissues of such an enzyme following the intraperitoneal administration of large doses of spermine or spermidine. The demonstration of *N*-(3-aminopropyl)pyrrolidin-2-one[67,68] and a variety of amino acids derived from spermine or spermidine[69-72] in normal human and rat urine (see Scheme II) has also been interpreted as indicating the presence in theses species of an enzyme able to oxidatively deaminate the terminal aminopropyl groups of spermine and spermidine

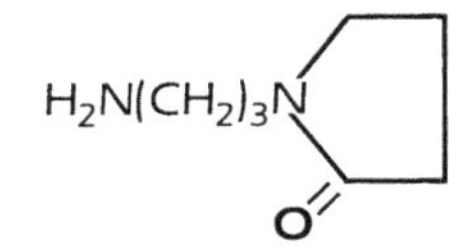

N-(3-aminopropyl)pyrrolidin-2-one

Human seminal plasma may contain a similar enzyme[73] (see also Section IV).

$NH_2(CH_2)_3NH(CH_2)_4NH(CH_2)_3NH_2 + O_2 + H_2O \rightarrow$
spermine

$NH_2(CH_2)_3NH(CH_2)_4NH_2 + NH_2(CH_2)_2CHO + H_2O_2$
spermidine — 3-aminopropionaldehyde

$NH_2(CH_2)_3NH(CH_2)_4NH_2 + O_2 + H_2O \rightarrow$
spermidine

$NH_2(CH_2)_2CHO + NH_2(CH_2)_4NH_2 + H_2O_2$
3-aminopropionaldehyde — putrescine

SCHEME III. Oxidation of spermine and spermidine by rat liver polyamine oxidase.

III. RAT LIVER POLYAMINE OXIDASE

A. Properties

Hölttä[74,75] isolated and characterized an oxidase from rat liver which catalyzed the oxidation of spermine, spermidine, and derivatives formed by the acetylation of their aminopropyl-moieties; benzylamine was not oxidized. The highest specific activity was found in the peroxisomal fraction, where its presence has been confirmed histochemically in both rat liver and kidney,[76] although polyamine oxidase was not included in a recent review of peroxisomal oxidases.[77] The greatest proportion of the total activity was, however, in the cytosol.[74,75] The enzyme, which has been purified 4000-fold to electrophoretic homogeneity, is a single polypeptide of MW 60,000, containing tightly-bound flavin adenine dinucleotide (FAD) as the prosthetic group, and possibly iron.[75] It acts on the secondary amino groups of spermine or spermidine with the production of 3-aminopropionaldehyde as shown in Scheme III — oxidation of spermine and spermidine by rat liver polyamine oxidase.

Aminopropionaldehyde was identified as a reaction product by ion-exchange chromatography followed by (1) reduction with sodium borohydride to an alcohol that comigrated with authentic 3-aminopropanol on electrophoresis and on paper chromatography in four solvent systems, (2) demonstration of the co-identity of the reaction product with 3-aminopropionaldehyde prepared from 3-aminopropanol using horse-liver alcohol dehydrogenase, and (3) oxidation with dichromate that gave a product that comigrated with β-alanine on electrophoresis and paper chromatography. Confirmation of this reaction scheme was provided by Bokenius and Seiler[78] who demonstrated the formation of *N*-acetyl-3-aminopropionadehyde, N^1-acetylspermidine, and putrescine from N^1,N^{12}-diacetylspermine by this enzyme. Thus, rat liver polyamine oxidase preferentially attacks the aminopropyl moieties of free or acetylated spermine or spermidine.

B. Substrates and Kinetics

An unusual feature of this enzyme is the stimulatory effect of aldehydes, particularly benzaldehyde, on the reaction rate. Stimulation by benzaldehyde was much greater with spermidine (40- to 100-fold, depending on pH) than with spermine (only about tenfold) as the substrate. A number of nonaldehydic compounds structurally related to benzaldehyde were without effect.[74] Thus, the presence of the aldehyde group is necessary for stimulation, which probably results from Schiff base formation between the aldehyde and the primary amino groups of the substrate, since oxidation of N^1-acetylspermidine is unaffected by benzaldehyde, and benzaldehyde is not bound to the enzyme.[75] Oxygen appears to be the sole electron acceptor.

Apparent Michaelis constants of 20 and 50 μM for spermine and spermidine have been reported which were reduced to 5 and 15 μM in the presence of 5 mM benzaldehyde.[74]

Bolkenius and Seiler[78] reported K_m and V_{max} values for N^1-acetylspermidine (14 μM; 2.5 nmol H_2O_2 per minute per milligram protein), N^1-acetylspermine (0.6 μM; 2.8), and N^1,N^{12}-diacetylspermine (5 μM; 4.0) for a partially purified preparation of the enzyme from rat liver cytosol (K_m for spermine 40 μM, V_{max} 0.1). N^1-Acetyl-1,3-diaminopropane, N^1-acetylputrescine, N^8-acetylspermidine, N^1,N^8-diacetylspermidine, putrescine, cadaverine, and monoamines were not attacked. Spermine (K_i 18 μM), spermidine (22 μM), putrescine (270 μM), and N^1-acetylputrescine were competitive inhibitors of N^1-acetylspermine oxidation; inhibition by N^8-acetylspermidine (K_i 11 μM) and *N*-(3-aminopropyl)-1,3-diaminopropane (60 μM) was noncompetitive.[78] A crude rat liver mitochondrial preparation was reported[79] to oxidize acetylpolyamines at the following relative rates: N^1-acetylspermine, 100; N^1,N^{12}-diacetylspermine, 35; N^1-acetylspermidine, 69; N^8-acetylspermidine, 32; N^1-acetylcadaverine, 80; N^1-acetylputrescine, 682; and N^1-acetyldiaminopropane, 10. Whether the striking discrepancies between these two sets of results are due to differing enzyme preparations or to differences in methodology remains to be resolved, although both groups used fluorescence assay of hydrogen peroxide formation to measure rates of oxidation.

C. Inhibitors

Sulfhydryl reagents (*p*-hydroxymercuribenzoate, iodoacetamide, and *N*-ethylmaleimide) inhibited enzyme activity, which was increased in the presence of 5 mM dithiothreitol or mercaptoethanol. Carbonyl reagents, such as phenylhydrazine, hydroxylamine, semicarbazide, and 4-bromo-3-hydroxybenzyloxyamine were strongly inhibitory at concentrations of 10 mM.[74] Thus, the enzyme appears to contain both sulfhydryl and carbonyl groups that are essential for activity.

Typical inhibitors of pyridoxal phosphate enzymes, such as Isoniazid and canaline [2-amino-4-(aminooxy)butyric acid], or copper chelators (such as thiourea and diethyl-dithiocarbamate) were without effect. Iron chelators 2,2′-dipyridyl, 8-hydroxyquinoline, and *o*-phenanthroline at 1 mM concentrations, showed 12, 40, and 52% inhibition, respectively. Treatment of the partially purified enzyme with *o*-phenanthroline and cysteine for 12 hr resulted in a 72% loss of activity which was partially (70%) restored by the addition of ferrous ions. Quinacrine [N^4-(6-chloro-2-methoxy-9-acridinyl)-*N,N*′-diethyl-1,4-pentanediamine; 6-chloro-9[(4-diethylaminol)-1-methylbutyl]amino-2-methoxyacridine)] a flavoprotein inhibitor, which does not inhibit bovine serum polyamine oxidase, strongly inhibited (80% at 0.1 mM) the rat liver enzyme. Pargyline, at the same concentration, had no effect.[74]

IV. HUMAN PREGNANCY-ASSOCIATED POLYAMINE OXIDASE

An association between maternal blood amine oxidase activity and pregnancy has been recognized for more than 40 years (there is much literature on this topic, hence only selected references will be cited here; for reviews see References 4 and 80 to 82). Histaminase,[83] diamine (putrescine) oxidase,[84] spermidine oxidase,[85] and polyamine (spermine) oxidase[86,87] all show a progressive increase in activity with increase in gestational age to 21 weeks or beyond, and decline to vanishingly low levels 3 to 4 days after delivery. Only traces of similar activities are found in male sera or in sera from nonpregnant women of reproductive age.[83,86,88] The placenta has long been regarded as a source of these pregnancy-associated enzymes but there is increasing evidence that they may also originate from the decidua.[89-91]

A. Placental Enzymes

Smith[92] partially purified (500-fold) an enzyme from human placenta that oxidized cadaverine (K_m 0.2 mM), putrescine (0.07 mM), diaminopropane, histamine, and benzylamine.

Relative rates of oxidation, assessed by measurement of concomitantly formed hydrogen peroxide using indigodisulfonate reduction, were 100, 54, 23, 23, and 5, respectively. The enzyme was completely inhibited by 1 m*M* aminoguanidine, diethyldithiocarbamate, semicarbazide, and hydroxylamine.

Paolucci and co-workers[93] obtained a 9600-fold purification of a placental amine oxidase. Gel filtration gave four active fractions with molecular masses that were multiples of 125,000 ± 5000. The enzyme oxidized putrescine (K_m 330 μ*M*), histamine (6 μ*M*), cadaverine (3.3 μ*M*), diaminopropane, and spermine with relative rates of 100, 85, 69, 13, and 6, also assessed by indigodisulfonate reduction; benzylamine was not attacked. Aminoguanidine and semicarbazide inhibited the enzyme (100 and 27% at 0.1 m*M*); hydroxylamine did not.

The most extensive study of placental amine oxidases is that of Bardsley and co-workers[11,94-98] who isolated two amine oxidases from homogenates of human placenta. One, described as a diamine oxidase, was a single polypeptide of MW 70,000 containing 1-g atom each of copper and manganese[98] that oxidized *o*-bis(aminomethyl)benzene, cadaverine, putrescine, spermine, and histamine in the ratio 214:147:100:32:30, but not spermidine or benzylamine, [8-arginine]-vasopressin or [8-lysine]vasopressin (K_m 18 μ*M*); collagen and tropocollagen were also oxidized. The enzyme was completely inhibited by aminoguanidine (0.01 m*M*) and hydroxylamine (0.05 m*M*) and to a lesser extent (50%) by semicarbazide (0.05 m*M*) but not by sulfhydryl reagents. Inhibition by β-aminopropionitrile was time dependent.[99] The second enzyme, described as a monoamine oxidase, was inhibited by sulfhydryl reagents (0.05 m*M* iodoacetamide or *N*-ethylmalimide) but not by aminoguanidine; relative rates of oxidation of a number of substrates were adrenaline: 1300, putrescine: 100, cadaverine: 84, spermine: 50, spermidine: 140, histamine: 49, and benzylamine: 67. Measurement of oxidation of substrates was by coupled oxidation of *o*-dianisidine by peroxidase and the hydrogen peroxide formed. Both enzymes oxidized *p*-dimethylaminomethylbenzylamine (K_m 0.1 and 0.14 m*M*, respectively). Placental amine oxidases have been discussed in a recent review by Mondovì and Riccio.[100]

B. Human Serum Polyamine Oxidases

Gahl et al.[85] were unable to separate the putrescine and spermidine oxidizing activities in human pregnancy serum and obtained K_m values of 2.5 and 10.9 μ*M*, respectively. Both activities were competitively inhibited by aminoguanidine (K_i 3 μ*M* with putrescine and 0.8 μ*M* with spermidine). Relative rates of oxidation, determined by the *o*-dianisidine-peroxidase assay, were spermidine: 100, spermine: 102, putrescine: 112, N^1-acetylputrescine: 135, N^1-acetylspermidine: 157, N^8-acetylspermidine: 101, cadaverine: 88, histamine: 58, 1,3-diaminopropane: 140, and benzylamine: 46.[101]

A polyamine oxidase of about 67,000 MW (determined by gel filtration and gradient gel electrophoresis) was partially purified[82,102] from human retroplacental blood serum, known to be a particularly rich source of this enzyme.[86] Gradient gel electrophoresis showed a major band associated with enzyme activity, accounting for 80% of the protein, and three minor bands that appeared to be without enzyme activity. This preparation contained 0.4-g atoms of iron and 0.09-g atoms of copper with 67,000 MW, determined by electrothermal atomic absorption. The optical absorption spectrum showed peaks at 280 and 405 nm unchanged by addition of spermidine or sodium dithionite. Oxidation of spermine and spermidine resulted in the production of aldehydes in the molar ratios of 2:1 and 1:1, respectively, but neither ammonia nor superoxide formation could be detected. The reaction requires molecular oxygen as anoxic conditions resulted in a decrease in enzyme activity. Oxidation of [^{14}C]-labeled spermine (*N,N'*-bis(3-aminopropyl)-[^{14}C]tetramethylene-1,4-damine) or spermidine (*N*-3-aminopropyl)-[1,4-^{14}C]tetramethylene-1,4-diamine) gave products that comigrated with authentic spermidine or putrescine, respectively, on paper electrophoresis, ion-exchange chromatography, and thin-layer chromatography of the dansyl

$$CH_3CONH(CH_2)_3NH(CH_2)_4NH(CH_2)_3NHOCH_3 + O_2 + H_2O \rightarrow$$
N^1,N^{12}-diacetylspermine

$$CH_3CONH(CH_2)_3NH(CH_2)_4NH_2 + OCH(CH_2)_2NHOCH_3 + H_2O_2$$
N^1-acetylspermidine *N*-acetylaminopropionaldehyde

SCHEME IV. Oxidation of N^1,N^{22}-diacetylspermine by human polyamine oxidase.

derivatives. Oxidation of N^1N^{12}-diacetylspermine gave N^1-acetylspermidine, demonstrated by thin-layer chromatography of the dansylated reaction products, in accordance with Scheme IV.

Thus, the mode of action of the enzyme resembles that of the rat liver polyamine oxidase and differs from that of the bovine serum enzyme. Relative rates of oxidation, determined kinetically from fluorescent measurement of the rate of hydrogen peroxide formation, were spermine: 1, spermidine: 4, N^1-acetylspermine: 10, N^1-acetylspermidine: 10, N^8-acetylspermidine: 4, N^1,N^{12}-diacetylspermine: 2, and putrescine: 20. Addition of benzaldehyde (5 m*M*) had no effect on the rate of oxidation of N^1-acetylspermidine, N^1,N^{12}-diacetylspermine, or putrescine; the rate for N^8-acetylspermidine was increased to about that of the N^1-isomer; spermine and spermidine oxidation rates were increased fourfold, a feature in common with rat liver polyamine oxidase.

The enzyme was completely inhibited by Quinacrine and to a lesser extent by β-aminopropionitrile (60%, both at 0.1 m*M*),[103] aminoguanidine, and methylglyoxal bis(guanylhydrazone) but not by pargyline or tranylcypromine (*trans*-phenylcyclopropylamine).[82,102]

C. Functions of the Amine Oxidases in Pregnancy

Again 40 or more years of investigation have yielded only meager knowledge of the function(s) of the amine oxidases in pregnancy. It seems probable that in human tissues there are, to quote Crabbe,[104] "a range of catalytic proteins with amine oxidising activity, some preferring monoamines to diamines, some vice-versa, and some polyamines or protein-bound lysyl groups." In pregnancy one or more of these enzymes is released by the placenta, and possibly the decidua, and appears in the circulation in quantities that, when compared with the tissue levels and half-life of the enzyme,[91] suggest active secretion, in contrast to the more passive release of some other pregnancy-associated proteins.[105] It is of interest to note that a lysyl oxidase (MW 30,000 by gel filtration in 6 *M* urea or sodium dodecyl sulfate-polyacrylamide gel electrophoresis) has been purified recently from human placenta,[106] but its ability to oxidize polyamines has not been tested. For a recent review of lysyl oxidase see Reference 107.

What, if any, the extracellular (or intracellular) functions of polyamine oxidase in pregnancy are is not at all clear. In a preliminary study Illei and Morgan[108] found that serum polyamine oxidase activity was significantly lower in 38 cases of spontaneous abortion compared to normal pregnancies of the same gestational age. An immunosuppressive role has been proposed,[104,109-115] based on observations that cells involved in the immune responses were more susceptible to the effects of oxidized polyamines than were other cells such as fibroblasts. However, this may be merely a special adaptation of a much more general function.

V. POLYAMINE OXIDASE ACTIVITY IN MILK

Human and bovine milk contain polyamine oxidase activity.[116] Preliminary experiments indicated that the human milk enzyme was not inhibited by Quinacrine, aminoguanidine, or 3-hydroxybenzyloxyamine (at 10 μ*M*; inhibition of the serum enzyme, with the same concentrations, was 50, 100, and 15%, respectively), hence this enzyme appears to differ from the serum (placental ?) enzyme. Furthermore, since activity was still present in milk 10 days after delivery, the enzyme activity in milk would appear to be independent of that in blood. The level of polyamine oxidase activity in samples of bovine milk, from animals in full lactation, was found to be remarkably constant. Inhibition by aminoguanidine and 3-hydroxybenzyloxyamine was similar in both bovine serum and milk; Quinacrine was ineffective in both. Blaschko[1] was unable to detect spermine oxidase activity in goat colostrum or whey, probably because of the relative insensitivity of the manometric methods in use at the time.

VI. HUMAN SEMINAL PLASMA AMINE OXIDASE

An amine oxidase with an apparent molecular weight of 182,000 (by ultracentrifugation) has been purified 1700-fold from human seminal plasma[73] and shown to be homogenous by polyacrylamide gel electrophoresis. Relative rates of oxidation, assessed by hydrogen peroxide formation, were putrescine: 100 (K_m 13 μ*M*), spermidine: 53 (0.56 m*M*), spermine: 42 (0.1 m*M*), and histamine: 32. The enzyme was inhibited by carbonyl reagents, such as Isoniazid, semicarbazide, and canaline (1,4-diaminooxybutyric acid), and activity could not be restored by the addition of exogenous pyridoxal-5′-phosphate. It was suggested that the enzyme acts on the primary amino groups of spermine and spermidine in the same way as the bovine serum amine oxidase, but the evidence is not conclusive.

VII. HUMAN LIVER POLYAMINE OXIDASE

Homogenates of liver from human cadavers oxidized N^1-acetylspermine, N^1,N^{12}-diacetylspermine, and N^1-acetylspermidine with relative rates of 100, 72, and 47, respectively,[117] in the presence of 0.1 m*M* pargyline and 1 m*M* semicarbazide. N^8-acetylspermidine, N^1-acetylcadaverine, N^1-acetylputrescine, spermine, spermidine, cadaverine, and putrescine showed either no or negligible activity.

VIII. TISSUE DISTRIBUTION OF POLYAMINE OXIDASE ACTIVITY IN MAMMALS

Polyamine oxidase activity is high in most tissues. It is present at levels comparable to those of spermine and spermidine synthase and greatly exceed that of spermine/spermidine N^1-acetyltransferase,[118] hence tissue levels of acetylated polyamines verge on the undetectable.[119] The enzyme is present in all mammalian tissues, and in all cultured mammalian cells examined so far. However, data on the distribution of the enzyme in tissues is only available for two species, the rat,[120] and man[91,117,121] (Table 1). Relative activities have been calculated from each set of data, as the use of different substrates and assay conditions makes direct comparison of results exceedingly difficult.

IX. FUNGAL POLYAMINE OXIDASES

A. *Aspergillus* and *Penicillium* Enzymes

Polyamine oxidase activity has been found in the mycelia of fungi belonging to the genera *Aspergillus, Mucor, Penicillium, Rhizopus, Cylindrocarpon, Fusarium,* and *Gibberella* when grown on media containing spermine or spermidine as the sole nitrogen source.[122] In some cases only spermine was oxidized, in other both spermine and spermidine were substrates. Polyamine oxidases have been purified to crystallisation from *P. chrysogenum* and *A. ter-*

Table 1
RELATIVE DISTRIBUTION OF POLYAMINE OXIDASE IN MAMMALIAN TISSUES

	Rat[120]	Human[117a]	Human[91,121]
Pancreas	38.1	1	—
Liver	20.2	25.0	—
Spleen	15.2	8.8	—
Kidney	14.9	16.5	—
Small intestine	10.8	2.6	—
Testes	9.8	21.3	—
Thymus	9.3	—	—
Prostate (ventral)	6.9	—	—
Brain	6.3	2.2	—
Lung	4.3	1.9	—
Heart	2.0	2.4	—
Skeletal muscle	1	—	—
Placenta	—	—	1
Decidua	—	—	9.1
Amnion	—	—	5.7
Chorion	—	—	3.1
Substrate	N^1,N^{12}-Diacetylspermine	N^1-Acetylspermine	[^{14}C]-spermine
Assay	TLC of dansyl derivatives	H_2O_2 production	[^{14}C]-spermidine

[a] Cadaver tissues.

reus.[123,124] Both enzymes (MW 160,000 and 130,000, respectively, by gel filtration) were flavoproteins containing 2 mol of FAD per mole of enzyme, and both consisted of two subunits. Absorption spectra showed maxima at 275, 375, and 450 nm and 275, 380, and 457, respectively. The *Penicillium* enzyme oxidized spermine rapidly and spermidine slightly (relative rates 100:7) with apparent K_m values of 22 and 10μ*M* (V_{max} 3.37 and 0.2 mmol H_2O_2 per minute per milligram protein), respectively. Cadaverine and putrescine were not attacked. Hydroxylamine, semicarbazide, isoniazid, iproniazid, 2,2′-bipyridyl, *o*-phenanthroline, or *p*-chloromercuribenzoate were all inhibitory at 1.4 m*M*.

The enzyme from *Aspergillus* oxidized spermidine and spermine in the ratio 2:1 (apparent K_m 1.2 and 0.54 μ*M*, respectively; V_{max} 0.15 and 6.2 mmol H_2O_2 per minute per milligram protein) but not diamines or monoamines. Hydroxyquinoline, *o*-phenanthroline, phenylhydrazine, or iproniazid inhibited the enzyme at 1.4 m*M*.

Both enzymes were identified as being of the rat liver type by demonstration (using an amino acid analyzer) of the sequential appearance of spermidine and putrescine among the products of spermine oxidation.

B. Extracellular Fungal Polyamine Oxidase

A *Penicillium* sp. (designated sp no. PO-1) has been isolated which, when grown in medium containing propane-1,3-diamine, secretes an extracellular polyamine oxidase.[126] The enzyme (MW 135,000 by gel filtration), was homogenous on polyacrylamide gel electrophoresis and ultracentrifugation. Sodium dodecyl sulfate-gel electrophoresis showed that the enzyme consisted of two identical subunits. Two moles of FAD were found per mole of enzyme. The absorption spectra showed maxima at 276, 370, and 458 nm. The secondary amino groups of spermine or spermidine were attacked, with the formation of spermidine or putrescine, 3-aminopropionaldehyde, and hydrogen peroxide. Phenylhydrazine was found to inhibit the enzyme, but sulfhydryl reagents such as iodoacetate or *p*-chloromercuribenzoate, did not.[127] The relative rates of oxidation and Michaelis constants determined by these

$$[CH_3(CH_2)_3]_2NH + O_2 + H_2O \rightarrow$$

$$CH_3(CH_2)_3NH_2 + CH_3(CH_2)_2CHO + H_2O_2$$

SCHEME V. Oxidation of di-*n*-butylamine by polyamine oxidase from *C. boidinii.*

workers were spermine: 100 (K_m 0.5 μM, V_{max} 10.3 μmol H_2O_2 per minute per milligram protein), spermidine: 76 (0.2 μM, 7.4), N^1-acetylspermidine: 44 (32.3 μM, 3.1), N^8-acetylspermidine: 25 (26.3 μM, 1.4), N^1-acetylspermine: 14 (1.6 μM, 5.4) and N^1,N^{12}-diacetylspermine: 4.3 (6.4 μM, 1.5). Putrescine, cadaverine, and their monoacetyl derivatives were not oxidized. With N^8-acetylspermidine as the substrate N^1-acetylputrescine was formed, and N^1-acetylspermine gave N^1-acetylspermidine;[128] thus, the oxidation of acetylpolyamines by this enzyme must proceed by a route different from that used by the rat liver enzyme.

C. *Candida boidinii* Polyamine Oxidase

A peroxisomal polyamine oxidase (MW in the range 80,000 to 100,000, by gel filtration) has been partially purified from the yeast *Candida boidinii*, grown on spermidine as the sole nitrogen source.[129] Substrates oxidized by the enzyme included spermine (K_m 0.2 mM, V_{max} 4.1 μmol O_2 per minute per milligram protein), spermidine (0.28 mM 0.46), N^1-acetylspermidine (0.054 mM, 11.7), *N-n*-butylpropylamine (6.89 mM, 0.25), di-*n*-butylamine (0.31 mM, 0.36), and di-*n*-hexylamine (0.05 mM, 0.05). The enzyme was strongly inhibited by Quinacrine, *trans*-phenylcyclopropylamine, and 2,2′-bipyridyl, but not by semicarbazide. Oxidation of di-*n*-butylamine led to the formation of stoichiometric amounts of *n*-buytlamine, butyraldehyde, and hydrogen peroxide in accord with Scheme V.

Spermidine was oxidized to putrescine and 3-aminopropionaldehyde; ammonia was not produced in significant amounts. Thus, this enzyme acts on secondary amino groups.

Polyamine oxidase activity has been found also in the yeasts *C. nagoyaensis, Hansenula polymorphia*, and *Trichospora melibiosacseum* when grown on spermidine. No activity could be detected in *C. steatolytica, Sporopachydermia cereana*, or *Pichia pastoris* when these organisms were grown on spermidine.

D. *Pichia pastoria* Polyamine Oxidase

An amine oxidase that acts only on primary amino groups has been isolated from *P. pastoris* grown on butylamine as the sole nitrogen source.[130] This enzyme (MW 265,000, subunit MW 116,000 by polyacrylamide gel electrophoresis) catalyzed the oxidation of spermine, spermidine, putrescine, lysine, ornithine, 1,2-diaminoethane, and a number of monoamines. Semicarbazide, hydroxylamine, cyanide, Isoniazid, *trans*-2-phenylcyclopropylamine, and *p*-hydroxymercuribenzoate strongly inhibited the enzyme at concentrations of 3 mM or less.[131] Examination of the stoichiometry of the formation of ammonia, hydrogen peroxide, and aldehyde suggested that both primary amino groups of spermine and spermidine were oxidized. The mode of action on spermine thus resembles that of the bovine plasma enzyme. In the case of spermidine, however, the reaction postulated (Scheme VI) differs from that for any other amine oxidase so far described. The possibility that putrescine could be formed in this reaction was excluded on the grounds that no 1-pyrroline (the cyclization product of 4-aminobutyraldehyde, a product of putrescine oxidation) could be detected among the reaction products.

X. BACTERIAL POLYAMINE OXIDASES

In comparison with the interest shown in fungal enzymes, there appear to have been no major advances in the field of bacterial polyamine oxidases since these were last reviewed;[8,132] indeed, a recent review on polyamines in microorganisms[133] does not mention them.

$$NH_2(CH_2)_3NH(CH_2)_4NH_2 + 2O_2 + 2H_2O \rightarrow$$

$$OCH(CH_2)_2NH(CH_2)_3CHO + 2NH_3 + 2H_2O$$

SCHEME VI. Oxidation of spermidine by *P. pastoris* polyamine oxidase.

XI. PLANT POLYAMINE OXIDASES

These have been reviewed recently by Smith[134] and Rinaldi et al.,[135] and hence will not be discussed here.

XII. METHODS FOR ASSAYING POLYAMINE OXIDASE ACTIVITY

Enzyme activity has been assessed by measurement of the conversion of substrate to product, by quantitating concomitant formation of hydrogen peroxide, or by determination of oxygen consumption.

Tabor et al.[17] established the use of benzylamine as a substrate for assaying bovine plasma amine oxidase. The assay, which is based on the difference between the molar absorption coefficients of benzylamine and benzaldehyde at 250 nM (ϵ = 12,500 to 12,800 M^{-1} cm^{-1}),[30,52] is simple to perform and has been widely used. Bardsley et al.[136,137] substituted *p*-dimethylaminomethylbenzylamine (ϵ = 11,000 M^{-1} cm^{-1} at 250 nm for the aldehyde) for benzylamine as a substrate for the direct, continuous spectrophotometric assay of placental[11] and other amine oxidases. Colorimetric determination of the aminoaldehydes has also been used.[138] Monitoring the formation of radiolabeled products from the oxidation of [^{14}C]-labeled spermine or spermidine by a polyamine oxidase involves separation of products from substrate, either by paper electrophoresis[68,139] or ion-exchange chromatography[48,140,141] but provides a more specific assay. Dansylation of the amines in the reaction mixture and subsequent separation by thin-layer chromatography followed by quantitation of the emitted fluorescence[120,142] has proved a very sensitive, albeit elaborate, technique. Hydrogen peroxide formation has been assessed using indigodisulfonate reduction,[92,93] the peroxidase-coupled *o*-dianisidine reaction,[29,74,85] peroxidase coupled to 2,2′-azino-di-(3-ethylbenzthiazoline-6-sulfonic acid), ABTS;[98,129,137,143] the molar absorption coefficient of ABTS has been variously reported as 36,000 M^{-1} cm^{-1} at 414 nm,[143] 32,400,[137] and 18,410 at 405 nm;[29] and fluorimetrically following oxidation of homovanillic acid (4-hydroxy-3-methoxyphenylacetic acid) by hydrogen peroxide-peroxidase to form the highly fluorescent condensation product 2,2′-dihydroxy-3,3′-dimethoxybiphenyl-5,5′-diacetic acid.[28,78,117,144]

Oxygen consumption has been assessed by Warburg manometry,[1,10,17,64] or polarographically using an oxygen electrode.[24,40]

Curiously, no method for the assay of polyamine oxidases is included in the third edition of Bergmeyer's treatise.[145]

XIII. POLYAMINE OXIDASES AS REAGENTS

The possibility of using polyamine oxidases as reagents with which to measure polyamines has attracted the attention of several groups over the years.[15] An assay for spermidine[146,147] utilized the spermidine-cleaving oxidase of *Serratia marcescens* with colorimetric determination of the 1-pyrroline formed. Spermine and spermidine have been determined, as total polyamines, by colorimetric estimation of the aldehydes formed following oxidation by bovine plasma amine oxidase.[138,148] A method for the differential determination of spermine and spermidine has been reported,[47] using columns of bovine plasma amine oxidase immobilized on Sepharose 4 and quantitation of the hydrogen peroxide formed. Differential

$(NH_2CH_2)_3NH(CH_2)_nNH{-}R$
(n must be greater than 3)

SCHEME VII. Structural requirements for a polyamine to be a substrate of bovine plasma polyamine oxidase.

determination of spermine and spermidine has also been carried out using polyamine oxidases isolated from *Pencillium chrysogenum* and *Aspergillus terreus* in sequence.[125] A somewhat similar technique using bovine plasma amine oxidase and pea seedling amine oxidase forms the basis of a recently developed diagnostic kit for the determination of diamines and polyamines.[149] The sensitivity of enzymic methods is in the nanomolar range for spermine and spermidine. Dansylation of the reaction products followed by separation using thin-layer chromatography and direct quantitation of the fluorescent dansyl derivatives will permit the measurement of picomole amounts of polyamines.[142] The use of chemiluminescence, however, is reported to offer a 50-fold increase in sensitivity over dansylation.[149]

XIV. BIOLOGICAL EFFECTS OF OXIDIZED POLYAMINES

A. In Vitro Cytotoxicity

The cytotoxic or cytostatic effects of oxidized polyamines in vitro have been well documented for a wide variety of cell types including bacteria, trypanosomes, mammalian spermatozoa, bacterial, plant, and animal viruses, tumor cells, and a number of mammalian cell lines (for reviews see Bachrach[150,151] and Morgan[8]). These observations were made using calf serum (which contains a polyamine oxidase), or oxidized polyamines prepared with the bovine plasma enzyme, or in the presence of bovine serum albumin (Cohn fraction V) in which polyamine oxidase is present as a contaminant,[152,153] even in the crystalline material.[153a] Addition of inhibitors of polyamine oxidase abolished the effects.[154-156] The structural requirements[157] for a substrate of the enzyme to give rise to cytotoxic products appear to be (1) a primary amino group separated from a secondary amino by three CH_2 units and (2) that any additional amino groups be separated from the secondary group by a further three or more CH_2 units (Scheme VII).

In the first experiments to make use of an all-human system, Morgan and Illei[112] showed that human retroplacental blood serum in the presence of exogenous spermine significantly suppressed the in vitro incorporation of [^{3}H]thymidine by spontaneously growing human lymphocytes thus demonstrating that aminopropionaldehyde, a product of the human enzyme, was also cytotoxic.

The finding that oxidized spermine or spermidine can kill intraerythrocytic parasites without perceptible damage to the host cell[158,159] has led to renewed interest in the effects of oxidized polyamines on parasites. Bovine serum amine oxidase plus spermine has been shown to be toxic or inhibitory to *Babesia rodhani* and *Plasmodium yoelii*,[158,159] *P. falciparum*,[160,162] and a variety of trypanosomes.[163,164] Treatment with human retroplacental serum polyamine oxidase plus spermine, or 3-aminopropionaldehyde, is also cytotoxic for trypanosomes and *P. falciparum*.[162,165] Other recently demonstrated effects of oxidized polyamines are inhibition of neutrophil locomotion[166] and of the neutrophil respiratory burst;[167] in vitro toxicity towards schistosomes and similar helminths has also been demonstrated.[168]

The mechanism by which the aminoaldehydes exert their cytotoxicity is unknown. It is probable that the effect is mediated through some change in membrane components which prevents ingress of nutrients, or egress of toxic metabolites, or both. The killing of intracellular parasites by the polyamine-polyamine oxidase system without apparent damage to the host cell supports this hypothesis. It has been reported that the dialdehyde resulting from spermine oxidation by bovine plasma polyamine oxidase can cross-link cells,[115] but this requires confirmation.

Table 2
INCORPORATION OF [^{3}H]LEUCINE BY HUMAN UMBILICAL VEIN ENDOTHELIAL CELLS AND NORMAL SKIN FIBROBLASTS AFTER EXPOSURE TO SPERMINE (100 *M*) FOR VARYING PERIODS IN THE PRESENCE OF BOVINE SERUM

	Control	1 hr	2 hr	4 hr	8 hr
HUVEC	100 ± 5	69 ± 16	67 ± 7	5 ± 0.6	4 ± 0.7
NSFB	100 ± 13	89 ± 6	72 ± 5	4 ± 1	—

Note: HUVEC = human umbilical vein endothelial cells and NSFB = normal skin fibroblasts. Cells were seeded into 96-well microculture plates at 10,000 cells per well in 200 μℓ medium, 199 containing 20% bovine serum (HUVEC) or DMEM with 8% bovine serum (NSFB). After 24 hr spermine (10 μℓ) was added to each well to give a final concentration of 100 μ*M*. After the times shown above the medium was removed, the cell sheet rinsed with serum-free medium, and fresh medium applied. After 24 hr [^{3}H]leucine (1 Ci/mmol, 1 mCi/mℓ; 1 ℓ) was added to each well and the amount of label incorporated into macromolecules was determined 24 hr later. Results are expressed as percentage of control values, and are mean ± SD (n = 10 to 12); control cpm were 8080 ± 530 (HUVEC) and 6753 ± 918 (NSFB). (D. M. L. Morgan, unpublished data).

B. Reversibility of the Cytotoxic Effect

The claims[109,110,169] that the inhibitory effects of oxidized polyamines on cell proliferation are reversible have been based largely on results obtained from lymphocytes which were treated with polyamines in the presence of calf serum before being exposed to a mitogen. Lymphocytes exposed to oxidized polyamines after stimulation by mitogen showed a reduction in uridine incorporation and protein synthesis that was reversible if the treatment was removed after 4 hr exposure, but not if it was continued for 24 hr.[169] However, in each case the experimental design is such that interpretation of the data is difficult. Exposure of human fibroblasts or vascular endothelial cells to spermine in the presence of bovine serum resulted in irreversible damage after 2 hr (Table 2), demonstrated by a marked reduction in [^{3}H]leucine incorporation following incubation of the cells for 48 hr in fresh medium.

It has been reported that oxidized spermine arrests cells in late G1 phase or at the G1/S interface of the cell cycle;[40,140] however, nonproliferating lymphocytes are normally in G0. It is therefore highly probable that the polyamine-polyamine oxidase system will, as suggested by Byrd et al.,[109,110] have little effect on nonproliferating cells. More work is necessary to clarify this point.

C. Identity of the Cytoactive Agent(s)

Aminoaldehydes, ammonia, and hydrogen peroxide are all produced by the action of bovine polyamine oxidase on spermine or spermidine. Initial attempts to identify the aldehyde products were hampered by their apparent instability. The half-lives of the dialdehyde from spermine and the monoaldehyde from spermidine have been shown to be of the order of 40 and 140 min,[170] respectively, at pH 7 and 37°C; Gaugas and Dewey[140] reported a similar value for the dialdehyde in tissue culture medium.

Identification of the products formed by the action of purified bovine serum amine oxidase on spermine and spermidine as *N,N'*-bis(3-propionaldehyde)-1,4-diaminobutane and *N*-(4-

aminobutyl)-3-aminopropionaldehyde, respectively,[25] together with the demonstrated cytotoxicity of synthetic *N,N'*-bis(3-propionaldehyde)-1,4-diaminobutane[45] appeared to provide conclusive evidence for the role of these compounds as the cytotoxic agents. Any contribution by concomitantly formed hydrogen peroxide was ruled out by the finding that addition of exogenous catalase, or prior separation of the aldehydes from the other reaction products before use, did not significantly diminish the effects observed.[45,171,172] Doubt was cast on this conclusion by the work of Alarcon,[173,174] and Kimes and Morris,[175] who reported that acrolein, also highly cytotoxic, was formed in significant amounts from enzymically oxidized spermine or spermidine, probably by β-elimination from the unstable aminoaldehydes. However, it is possible that the acrolein detected was artefactual, as Israel et al.[46] were unable to detect acrolein formation when the synthetic dialdehyde from spermine, prepared by unambiguous synthesis, was allowed to break down under physiological conditions. These workers concluded that acrolein was not produced from the dialdehyde by spontaneous generation, by unimolecular decomposition at neutral pH and temperatures up to 37°C, by the action of bovine polyamine oxidase, or by interaction of the dialdehyde with components of the cell culture medium for up to 72 hr. Furthermore, synthetic 4,9-diazadodecanedialdehyde (a homologue of the spermine dialdehyde) which can be formed from *N,N'*-bis(3-aminopropyl)nonane-1,9-diamine by the action of bovine amine oxidase, was equally cytotoxic and also failed to form acrolein under physiological conditions.[46]

Although some of the observed cytotoxicity of oxidized polyamines may be attributable to the formation of acrolein in long-term incubations, more recent experiments support the concept that the cytotoxic effects are due to aminoaldehyde formation.[40,45,140,161,164,169,172,176,177] Recently, Morgan et al.[162] separated the aldehydes formed from the oxidation of spermine and spermidine by the bovine serum enzyme and demonstrated their cytotoxicity under conditions where no ammonia or hydrogen peroxide were formed; aminopropionaldehyde, the product of the human enzyme, was also cytotoxic under these conditions. Oxidation of spermine or spermidine by human polyamine oxidase also results in the formation of products that are cytotoxic (Table 3). Hydrogen peroxide, formed concomitantly with aminoaldehydes, was not responsible for this effect, as oxidation of acetyl polyamines, which give rise to similar amounts of hydrogen peroxide, did not result in cytotoxicity (Table 3).

It should be noted that recent reports that the structure of the product of chemical or enzymic oxidation of spermine is an *N,N*-acetal[178,179] refer to oxidation by a plant enzyme that removes the primary amino group from the aminobutyl moiety of spermidine,[134] although this is not clear from the texts.

D. Oxidation of Acetylpolyamines

The first, and rate-limiting, step in intracellular polyamine degradation in mammalian cells is acetylation of the aminopropyl nitrogens by spermidine/spermine N1-acetyltransferase.[118,180,181] The discovery of similar enzymes in a bacterium[182] and a number of fungi[183] suggest that those organisms in which a polyamine oxidase is present may also possess this enzyme and hence a common pathway of polyamine degradation. In general the activity of the transferase, although readily inducible[184,185] is much less than that of the oxidase. Acetylpolyamines are also metabolized more rapidly by intracellular polyamine oxidase than the nonacetylated compounds, so tissue levels of acetylated polyamine derivatives are very low. The metabolism of the acetylpolyamines has been reviewed recently by Seiler.[42] Oxidation of N^1-acetylspermidine or N^1,N^{12}-diacetylspermine by bovine plasma amine oxidase or human pregnancy-associated polyamine oxidase does not result in the formation of cytotoxic products (Table 3), although both substrates are oxidized at a greater rate than the nonacetylated polyamines. Oxidation of N^8-acetylspermidine or N^1-acetylspermine by either enzyme, however, gives rise to cytotoxic products. An inhibitor of transforming growth factor-

Table 3
THE EFFECTS OF VARIOUS POLYAMINES ON THE INCORPORATION OF [^{3}H]LEUCINE BY HUMAN UMBILICAL VEIN ENDOTHELIAL CELLS AND NORMAL SKIN FIBROBLASTS IN THE PRESENCE OF FETAL CALF SERUM OR HUMAN RETROPLACENTAL SERUM

	HUVEC		NSFB, 10% FCS
	20% FCS	10% RPS	
Control	100 ± 6	100 ± 11	100 ± 5
Spermine	2 ± 0.6	36 ± 7	3 ± 0.7
Spermidine	2 ± 0.4	43 ± 9	—
Putrescine	99 ± 11	100 ± 11	—
N^1-Acetylspermidine	95 ± 7	93 ± 9	97 ± 9
N^8-Acetylspermidine	6 ± 2	46 ± 7	24 ± 3
N^1-Acetylspermine	2 ± 0.4	53 ± 8	7 ± 3
N^1,N^{12}-Diacetylspermine	81 ± 6	90 ± 7	85 ± 4
N^1-Acetylputrescine	90 ± 6	95 ± 10	—

Note: HUVEC = human umbilical vein endothelial cells, NSFB = normal skin fibroblasts, FCS = fetal calf serum, and RPS = retroplacental serum. Experimental conditions were as described in Table 2, with the exception that the label was added 24 hr after the polyamines. Variations in the effect of a given compound in the different experiments are due to variation in the enzyme activity in the sera used as sources of polyamine oxidases. Results are expressed as percentages of control values and are means ± SD (n = 9 to 12); control cpm were 7115 ± 434, 8467 ± 955, and 4022 ± 186, respectively. (D. M. L. Morgan, unpublished data.)

induced cell growth has been isolated from human placenta and shown to be spermine;[186] acetylation of the inhibitor abolished the effect. Listed below are structures of acetylated spermine and spermidine.

$$CH_3COHN(CH_2)_3NH(CH_2)_4NH_2$$

N^1-acetylspermidine

$$NH_2(CH_2)_3NH(CH_2)_4NHCOCH_3$$

N^8-acetylspermidine

$$CH_3CONH(CH_2)_3NH(CH_2)_4NH(CH_2)_3NH_2$$

N^1-acetylspermine

$$CH_3CONH(CH_2)_3NH(CH_2)_4NH(CH_2)_3NHCOCH_3$$

N^1,N^{12}-diacetylspermine

For the human enzyme these findings are in accord with the reaction sequence described earlier (Scheme III); acetylation of the aminopropyl nitrogens would result in the formation of *N*-acetylaminopropionaldehyde on oxidation. Loss of the cytocidal effect may be due to the reduction of the charge on the nitrogen resulting from acylation. Acetylation of the aminobutyl nitrogen in the parent compound does not modify the resulting aldehyde and hence would not be expected to alter the biological effect. The results obtained with the bovine serum enzyme are more difficult to interpret, since acetylation of the primary amino groups of spermine or spermidine should have no effect on the structure of the aldehyde formed, if the reaction proceeds according to Scheme I. However, it has been reported[41] that N^1-acetylspermidine is not metabolizied by the calf serum enzyme. Hence, this point requires further clarification. The finding that aminopropylacetylation protects against subsequent cytotoxicity must call in question the relevance to the in vivo situation of many of the in vitro studies of aminoaldehyde toxicity. How this protective function is effected is not known but merits investigation.

E. Cytotoxicity of Oxidized Putrescine

There are conflicting reports of the effects of the products of putrescine oxidation on cell proliferation. Gaugas and Curzen[111] found that human pregnancy sera, later shown to contain a polyamine oxidase,[86] were able to convert putrescine (80 to 160 μ*M*) to an inhibitor of lymphocyte transformation, an observation supported by Allen et al.[40] Spontaneously dividing lymphocytes were not affected. In contrast, Byrd et al.[110] reported that putrescine (up to 150 μ*M*) in the presence of bovine plasma oxidase was not inhibitory. The ability of rat kidney fibroblasts to form colonies in soft agar when treated with epidermal growth factor was not inhibited by putrescine (up to 0.8 m*M*) in the presence of calf serum. Hog kidney diamine oxidase has been reported to act on putrescine (>100 μ*M*) to inhibit the proliferation of spontaneously growing lymphocytes, fibroblasts, and melanoma cells.[187] However, addition of putrescine at concentrations up to 160 μ*M* to cultures of human umbilical vein endothelial cells and normal skin fibroblasts in the presence of either bovine serum or human retroplacental serum failed to inhibit leucine incorporation (Table 3) to any great extent in conditions where both spermine and spermidine were effective. Clearly further work is needed to settle this point.

XV. BIOLOGICAL CONSEQUENCES OF POLYAMINE OXIDATION

What is the role of the polyamine oxidases in cellular function? The widespread occurrence and similarities (see Table 4 for examples) of the polyamine-cleaving oxidases (Figure 1) in higher plants,[134,135] bacteria,[8] fungi, protozoa,[188] macrophages,[139] lymphocytes,[176] fibroblasts, vascular endothelial and smooth muscle cells, human trophoblasts,[153a] and all mammalian tissues examined[91,117,120,121,189] indicate that these enzymes have been well conserved in evolutionary terms. This in turn suggests that the enzymes or their products play a fundamental role in cell function.

The discovery that acetylation of the aminopropyl groups prevents the products of oxidation from being cytotoxic may provide the answer to a puzzling question: Why should polyamine degradation apparently proceed by a pathway that results in the production of such potentially hazardous compounds as the aminoaldehydes? Acetylation of the primary amino groups prevented the uptake of polyamines by cultured L1210 leukemia cells[190,191] and this, together with the occurrence of acetylated polyamines in normal urine,[192,193] suggests that acetylation may be an important step in polyamine elimination[194] as well as in degradation via the polyamine oxidase pathway.[195] Indeed, it has been suggested[42] that acetylation may be a means whereby biologically active polyamines are inactivated.

Several attempts have been made to construct comprehensive pathways of polyamine metabolism.[119,192,195] A difficulty with this approach lies in the need to include oxidases that act on primary amino groups and those that act on secondary amino groups (although one may be intracellular and another extracellular) in order to account for the products formed. Two new enzyme-activated, irreversible inhibitors, N^1-methyl-N^2-(2,3-butadienyl)-1,4-butanediamine and N^1,N^2-bis(2,3-butadienyl)-1,4-butanediamine,[196] which appear to be specific for polyamine oxidase, have been used to explore the effects of inhibition of this enzyme

Table 4
SIMILARITIES IN THE PROPERTIES OF POLYAMINE OXIDASES FROM VARIOUS SOURCES

Source	MW[a]	Subunits	FAD	Products[b] include	Ref.
Pencillium chrysogenum	160,000	2	+	Aminopropionaldehyde	123
Aspergillus terreus	130,000	2	+	Aminopropionaldehyde	124
Micrococcus rubens	80,000	1	+	Diaminopropane	205,206
Zea mays	65,000	1	+	Diaminopropane	207
Avena sativa	85,000	1	?	Diaminopropane	208
Rat liver	60,000	1	+	Aminopropionaldehyde	74
Human pregnancy serum	67,000	1(?)	?	Aminopropionaldehyde	102

[a] MW values determined by gel filtration.
[b] Spermine or spermidine as substrate.

on polyamine metabolism in vivo.[197-199] In each case there was an increased accumulation and/or excretion of acetyl polyamines following treatment with the inhibitor. If these compounds become widely available they may prove to be powerful tools with which to investigate polyamine degradation, in vitro and in vivo, and may, perhaps, help to provide answers to some of the questions posed here.

The published values for molecular weights of chalones (the tissue-specific, species-nonspecific regulators of tissue growth) have steadily decreased over the years.[200,201] This, and other evidence indicating a polyamine involvement[202] led to the suggestion[8,202] that chalones may be oxidized polyamines that readily, and reversibly, bind to proteins by virtue of possessing both positively charged amino groups and a carbonyl group in a relatively small molecule. Further evidence for the possible involvement of polyamine oxidases, or oxidized polyamines, in processes regulating cell division is provided by the work of Quash et al.[189] who found that polyamine oxidase activity was three- to fivefold lower in transformed than in normal cells. A five- to tenfold decrease in activity as compared to controls was found in 9,10-dimethyl-[α]-anthracene-induced mammary tumors in rats, and in human esophageal tumors. The authors of a review[203] concluded that the polyamines and their oxidized derivatives form part of an integrated biochemical system for the regulation of cell proliferation and growth, by means of stimulatory substances (the polyamines) and inhibitory substances (the oxidized polyamines or derivatives). However, the finding that the oxidation products of aminopropyl-acetylated polyamines are not cytostatic or cytotoxic indicates that further investigation is necessary to determine the validity of the latter part of this conclusion.

There are still many questions to be answered concerning the role of polyamine oxidases in the functioning of cell and organism. Do they play a part in the control of cell growth or proliferation? Are they part of the regulatory mechanisms of the immune system? The biological importance of the site of polyamine cleavage (Figure 1) is still an open question. As noted by Jänne et al.[204] in 1978, the full physiological significance of the polyamine oxidases remains to be discovered, but their ubiquity, and the potency of some of their products, suggests that their role(s) may be important.

REFERENCES

1. **Blaschko, H.,** The amine oxidases of mammalian blood plasma, *Adv. Comp. Physiol. Biochem.*, 1, 67, 1962.
2. **Blaschko, H.,** Amine oxidase, in *The Enzymes,* Vol. 8, 2nd ed., Boyer, P. D., Lardy, H., and Myrbäch, K., Eds., Academic Press, New York, 1963, 337.

3. **Zeller, E. A.,** Diamine oxidases, in *The Enzymes,* Vol. 8, 2nd ed., Boyer, P. D., Lardy, H., and Myrbäch, K., Eds., Academic Press, New York, 1963, 313.
4. **Buffoni, E.,** Histaminase and related amine oxidases, *Pharmacol. Rev.,* 18, 1163, 1966.
5. **Kapeller-Adler, R.,** *Amine Oxidases and Methods for Their Study,* Wiley-Interscience, New York, 1970.
6. **Tabor, H. and Tabor, C. W.,** Biosynthesis and metabolism of 1,4-diaminobutane, spermidine, spermine and related amines, *Adv. Enzymol.,* 36, 203, 1972.
7. **Malmström, B. G., Andreasson, L.-E., and Reinhammer, B.,** Copper-containing oxidases and superoxide dismutase, in *The Enzymes,* Vol. 12, 3rd ed., Boyer, P. D. Ed., Academic Press, New York, 1975, 507.
8. **Morgan, D. M. L.,** Polyamine oxidases, in *Polyamines in Biomedical Research,* Gaugas, J. M., Ed., John Wiley & Sons, Chichester, England, 1980, 286.
9. **Morgan, D. M. L.,** Polyamine oxidases, *Biochem. Soc. Trans.,* 13, 322, 1985.
10. **Yamada, H. and Yasunobu, K. T.,** Monoamine oxidase. I. Purification, crystallisation, and properties of plasma monoamine oxidase, *J. Biol. Chem.,* 237, 1511, 1962.
11. **Bardsley, W. G., Crabbe, M. J. C., and Scot, I. V.,** The amine oxidases of human placenta and pregnancy plasma, *Biochem. J.,* 139, 169, 1974.
12. Committee of the International Union of Biochemistry, *Enzyme Nonmenclature,* Academic Press, New York, 1984.
13. **Tabor, C. W.,** Spermidine dehydrogenase from *Serratia marcesens*: a heme-flavin enzyme, in *Structure and Functions of Amine Oxidases,* Mondovi, B., Ed., CRC Press, Boca Raton, Fla., 1985, 45.
14. **Tabor, H., Tabor, C. W., and De Meis, L.,** Chemical synthesis of N^1-acetyl-1,4-diaminobutane, N^1-acetylspermidine, and N^8-acetylspermidine, *Meth. Enzymol.,* 17B, 829, 1971.
15. **Bachrach, U.,** Copper amine oxidases and amines as regulators of cellular processes, in *Structure and Functions of Amine Oxidases,* Mondovi, B., Ed., CRC Press, Boca Raton, Fla., 1985, 5.
16. **Pettersson, G.,** Plasma amine oxidase, in *Structure and Functions of Amine Oxidases,* Mondovi, B., Ed., CRC Press, Boca Raton, Fla., 1985, 105.
17. **Tabor, C. W., Tabor, H., and Rosenthal, S. M.,** Purification of amine oxidase from beef plasma, *J. Biol. Chem.,* 208, 645, 1954.
18. **Tabor, C. W., Tabor, H., and Rosenthal, S. M.,** Amine oxidases. A. Amine oxidase from steer plasma, *Meth. Enzymol.,* 2, 390, 1955.
19. **Yasunobu, K. T. and Smith, R. A.,** Amine oxidase (beef plasma), *Meth. Enzymol.,* 17B, 698, 1971.
20. **Yasunobu, K. T., Ishizaki, H., and Minamura, N.,** The molecular, mechanistic and immunological properties of amine oxidases, *Mol. Cell. Biochem.,* 13, 3, 1976.
21. **Summers, M. C., Markovic, R., and Klinman, J. P.,** Stereochemistry and kinetic isotope effects in the bovine plasma amine oxidase catalysed oxidation of dopamine, *Biochemistry,* 18, 1969, 1979.
22. **Ishizaki, H. and Yasunobu, K. T.,** Bovine plasma amine oxidase interactions with concanavalin A in solution and with concanavalin A-Sepharose, *Biochim. Biophys. Acta,* 611, 27, 1980.
23. **Turini, P., Sabatini, S., Befani, O., Chimenti, F., Casanova, C., Riccio, P. L., and Mondovi, B.,** Purification of bovine plasma amine oxidase, *Anal. Biochem.,* 125, 294, 1982.
24. **Mondovi, B., Turini, P., Befani, O., and Sabatini, S.,** Purification of bovine plasma amine oxidase, *Meth. Enzymol.,* 94, 314, 1983.
25. **Tabor, C. W., Tabor, H., and Bachrach, H.,** Identification of the aminoaldehydes produced by the oxidation of spermine and spermidine with purified plasma amine oxidase, *J. Biol. Chem.,* 239, 2194, 1964.
26. **Achee, F., Chervenka, C. H., Smith, R. A., and Yasunobu, K. T.,** Amine oxidase. XII. The association and dissociation, and number of subunits of beef plasma amine oxidase, *Biochemistry,* 7, 4329, 1968.
27. **Yamada, H., Gee, P., Ebata, M., and Yasunobu, K. T.,** Monamine oxidase. VI. Physicochemical properties of plasma monoamine oxidase, *Biochim. Biophys. Acta,* 81, 165, 1964.
28. **Nakano, G., Harada, M., and Nagatsu, T.,** Purification and properties of an amine oxidase in bovine dental pulp and its comparison with serum amine oxidase, *Biochim. Biophys. Acta,* 341, 366, 1974.
29. **Gahl, W. A. and Pitot, H. C.,** Polyamine degradation in foetal and adult bovine serum, *Biochem. J.,* 202, 603, 1982.
30. **Mondovi, B. B., Sabatini, S., and Befani, O.,** Recent studies on the active site of beef plasma amine oxidase, in *Advances in Polyamines in Biomedical Science,* Caldarera, C. M. and Bachrach, U., Eds., CLUEB, Bologna, Italy, 1984, 75.
31. **Yamada, H. and Yasunobu, K. T.,** Monoamine oxidase. IV. Nature of the second prosthetic group of plasma monoamine oxidase, *J. Biol. Chem.,* 238, 2669, 1963.
32. **Watanabe, K., Smith, R. A., Inamatsu, M., and Yasunobu, K. T.,** Recent investigations on the prosthetic group of beef plasma amine oxidase, *Adv. Biochem. Psychopharmacol.,* 5, 107, 1972.
33. **Inamatsu, M., Yasunobu, K. T., and Koenig, W. A.,** Cofactor investigation of bovine plasma amine oxidase. NaBH4 reduction of enzyme-substrate mixture, *J. Biol. Chem.,* 249, 5265, 1974.

34. **Suva, R. H. and Abeles, R. H.**, Studies on the mechanism of action of plasma amine oxidase, *Biochemistry*, 17, 3538, 1978.
35. **Lobenstein-Verbeek, C. L., Jongjan, J. A., Frank, J., and Duine, J. A.**, Bovine serum amine oxidase: a mammalian enzyme having covalently bound PQQ as prosthetic group, *FEBS Lett.*, 170, 305, 1984.
36. **Duine, J. A. and Franke, J.**, Quinoproteins: a novel class of dehydrogenases, *Trends Biochem. Sci.*, 6, 278, 1981.
37. **Watanabe, K. and Yasunobu, K. T.**, Carbohydrate content of bovine plasma amine oxidase and isolation of a carbohydrate containing fragment attached to asparagine, *J. Biol. Chem.*, 245, 4612, 1970.
38. **Israel, M., Rosenfield, J. S., and Modest, E. J.**, Analogs of spermine and spermidine. I. Synthesis of polymethylenepolyamines by reduction of cyanoethylated α,ω-alklenediamines, *J. Med. Chem.*, 7, 710, 1964.
39. **Hegre, O. D., Marshall, S., and Hickey, G. E.**, Spermidine cytotoxicity in vitro: effect of serum and oxygen tension, *In Vitro*, 20, 198, 1984.
40. **Allen, J. C., Smith, C. J., Hussain, J. I., Thomas, J. M., and Gaugas, J. M.**, Inhibition of lymphocyte proliferation by polyamines requires ruminant-plasma polyamine oxidase, *Eur. J. Biochem.*, 102, 153, 1979.
41. **Mamont, P. C., Seiler, N., Siat, M., Joder-Ohlenbusch, A.-M., and Knödgen, B.**, Metabolism of acetyl derivatives of polyamines in cultured polyamine-deficient rat hepatoma cells, *Med Biol.*, 59, 347, 1981.
42. **Seiler, N.**, Acetylpolyamines as substrates of amine oxidases, in *Structure and Functions of Amine Oxidases*, Mondovi, B., Ed., CRC Press, Boca Raton, Fla., 1985, 21.
43. **Oi, S., Inamasu, M., and Yasunobu, K. T.**, Mechanistic studies of beef plasma amine oxidase, *Biochemistry*, 9, 3378, 1970.
44. **Smith, T. A. and Bickley, D. A.**, Further properties of the polyamine oxidase of barley leaves, *Phytochemistry*, 13, 2437, 1974.
45. **Fukami, H., Tomida, I., Morino, T., Yamada, H., Oki, R., Kawasaki, M., and Ogata, K.**, Phagocidal action of oxidised spermine and its analogs, *Biochem. Biophys. Res. Commun.*, 28, 19, 1967.
46. **Israel, M., Zoll, E. C., Muhammad, N., and Modest, E. J.**, Synthesis and antitumor evaluation of the presumed cytotoxic metabolites of spermine and *N,N'*-bis(3-aminopropyl)nonane-1,9-diamine, *J. Med. Chem.*, 16, 1, 1973.
47. **Isobe, K., Tani, Y., Yamada, H., and Hiromi, K.**, Determination of polyamines with immobilised beef plasma amine oxidase, *Agric. Biol. Chem.*, 44, 615, 1980.
48. **Gahl, W. A., Vale, A. M., and Pitot, H. C.**, Separation of putrescine oxidase and spermidine oxidase in foetal bovine serum with the aid of a specific radioactive assay of spermidine oxidase, *Biochem. J.*, 187, 197, 1980.
49. **Wang, T.-M., Achee, F. M., and Yasunobu, K. T.**, Amine oxidase. XI. Beef plasma amine oxidase, a sulphhydryl protein, *Arch. Biochem. Biophys.*, 128, 106, 1968.
50. **Yamada, H. and Yasanobu, K. T.**, Monoamine oxidase. II. Copper, one of the prosthetic groups of plasma monoamine oxidase, *J. Biol. Chem.*, 237, 3077, 1962.
51. **Gaugas, J. M. and Dewey, D. L.**, Oxygen-dependent free radicals in spermine oxidation cytostasis and chemiluminescence and the role of superoxide dismutase, *Br. J. Cancer*, 41, 946, 1980.
52. **Neumann, R., Hevey, R., and Abeles, R. H.**, The action of plasma amine oxidase on β-haloamines. Evidence for proton abstraction in the oxidative reaction, *J. Biol. Chem.*, 250, 6362, 1975.
53. **Bardsley, W. G.**, Inhibitors of copper amine oxidases, in *Structure and Function of Amine Oxidases*, Mondovi, B., Ed., CRC Press, Boca Raton, Fla., 1985, 135.
54. **Zeidan, H., Watanabe, K., Piette, L. H., and Yasunobu, K. T.**, Electron spin resonance studies of bovine plasma amine oxidase. Probing of the environment about the substrate-liberated sulfhydryl groups in the active site, *J. Biol. Chem.*, 255, 7621. 1980.
55. **Tsurushiin, S., Hiramatsu, A., and Yasunobu, K. T.**, The essential histidine residues of bovine plama amine oxidase, *Biochim. Biophys. Acta*, 400, 451, 1975.
56. **Kleutz, M. D. and Schmidt, P. G.**, Proton relaxation study of the hog kidney diamine oxidase active center, *Biochemistry*, 9, 3378, 1977.
57. **Gold, M. and Powell, H. K. J.**, Thermodynamic study of the stepwise protonation of 3-azaheptane-1,7-diamine and 4-azaoctane-1,8-diamine, and of complex formation between copper (II) and 3-azaheptane-1,7-diamine and 4,9-diazadodecane-1,12-diamine, *J. Chem. Soc. Dalton Trans.*, 230, 1976.
58. **Kimberley, M. M. and Goldstein, J. H.**, Determination of pK_a values and total proton distribution pattern of spermidine by carbon-13 nuclear magnetic resonance titrations, *Anal. Chem.*, 53, 789, 1981.
59. **Aikens, D. A., Bunce, S. C., Quasch, O. F., Schwartz, H. M., and Hurwitz, C.**, Two-dimensional N. M. R. investigation of the protonation sequence in spermidine, *J. Chem. Soc. Chem. Commun.*, 43, 1983.

60. **Aikens, D. A., Bunce, S., Onasch, F., Parker, R., Hurwitz, C., and Clemans, S.,** The interactions between nucleic acids and polyamines. II. Protonation constants and ^{13}C-NMR chemical shift assignments of spermidine, spermine, and homologs, *Biophys. Chem.*, 17, 67, 1983.
61. **Takeda, Y., Samejuma, K., Nagano, K., Watanabe, M., Sugeta, H., and Kyogoku, Y.,** Determination of protonation sites in thermospermine and in some other polyamines by ^{15}N and ^{13}C nuclear magnetic resonance spectroscopy, *Eur. J. Biochem.*, 130, 383, 1983.
62. **Onasch, F., Aikens, D., Bunce, S., Schwartz, H., Nairn, D., and Hurwitz, C.,** The interactions between nucleic acids and polyamines. III. Microscopic protonation constants of spermidine, *Biophys. Chem.*, 19, 245, 1984.
63. **Pontoni, G. and Coward, J. K.,** Stereochemical studies of enzyme-catalyzed alkyl-transfer reactions. An NMR method for distinguishing between the two prochiral hydrogens at C-1′ of spermidine, *Tetrahedron Lett.*, 24, 151, 1983.
64. **Blaschko, H. and Bonney, R.,** Spermine oxidase and benzylamine oxidase. Distribution, development and substrate specificity, *Proc. Roy. Soc. B*, 156, 268, 1962.
65. **Seiler, N., Knödgen, B., Gittos, M. W., Chan, W. Y., Griesmann, G., and Rennert, O. M.,** On the formation of amino acids deriving from spermidine and spermine, *Biochem. J.*, 200, 123, 1981.
66. **Seiler, N., Knödgen, B., Bink, G., Sarhan, S., and Bolkenius, F.,** Diamine oxidase and polyamine catabolism, *Adv. Polyamine Res.*, 4, 135, 1983.
67. **Seiler, N., Knödgen, B., and Haegele, K.,** *N*-(3-Aminopropyl)pyrrolidon-2-one, a product of spermidine catabolism in vivo, *Biochem. J.*, 208, 189, 1982.
68. **Seiler, N. and Knodgen, B.,** *N*-(3-Aminopropyl)pyrrolidin-2-one: a physiological excretory product deriving from spermidine, *Int. J. Biochem.*, 15, 907, 1983.
69. **Asatoor, A. M.,** Isolation and characterisation of a new urinary metabolite derived from spermidine, *Biochim. Biophys. Acta*, 586, 55, 1979.
70. **Noto, T., Tanaka, T., and Nakajima, T.,** Urinary metabolites of polyamines in rats, *J. Biochem.*, 83, 543, 1978.
71. **Nakajima, T., Matsuoka, Y., and Akazawa, S.,** Putreanine excretion in human urine, *Biochim. Biophys. Acta*, 222, 405, 1970.
72. **van den Berg, G. A., Elzinga, H., Nagel, G. T., Kingma, A. W., and Muskiet, F. A. J.,** Catabolism of polyamines in the rat. Polyamines and their non-alpha-amino acid metabolites, *Biochim. Biophys. Acta*, 802, 175, 1984.
73. **Hölttä, E., Pulkkinen, P., Elfving, K., and Jänne, J.,** Oxidation of polyamines by diamine oxidase from human seminal plasma, *Biochem. J.*, 145, 373, 1975.
74. **Hölttä, E.,** Oxidation of spermidine and spermine in rat liver: purification and properties of polyamine oxidase, *Biochemistry*, 16, 91, 1977.
75. **Hölttä, E.,** Polyamine oxidase (rat liver), *Meth. Enzymol.*, 94, 306, 1983.
76. **Beard, M. E., Baker, R., Conomos, P., Pugatch, D. R., and Holtzman, E.,** Oxidation of oxalate and polyamines by rat peroxisomes, *J. Histochem. Cytochem.*, 33, 460, 1985.
77. **Hamilton, G.,** Peroxisomal oxidases and suggestions for the mechanism of action of insulin and other hormones, *Adv. Enzymol.*, 57, 85, 1985.
78. **Bolkenius, F. N. and Seiler, N.,** Acetylderivatives as intermediates in polyamine catabolism, *Int. J. Biochem.*, 13, 287, 1981.
79. **Suzuki, O., Matsumoto, T., Oya, M., and Katsumata, Y.,** Metabolism of acetylpolyamines by monoamine oxidase, diamine oxidase and polyamine oxidase, *Biochim. Biophys. Acta*, 677, 190, 1981.
80. **Tabor, H.,** Metabolic studies on histidine, histamine, and related imidazoles, *Pharmacol. Rev.*, 6, 299, 1954.
81. **Morgan, D. M. L.,** Commentary: amine oxidases and pregnancy, *Br. J. Obstet. Gynaecol.*, 89, 177, 1982.
82. **Morgan, D. M. L.,** Polyamine oxidation and human pregnancy, *Adv. Polyamine Res.*, 4, 155, 1983.
83. **Ahlmark, A.,** Studies on the histaminolytic power of plasma with special reference to pregnancy, *Acta Physiol. Scand.*, 9 (Suppl. 28), 1, 1944.
84. **Southren, A. L., Kobayashi, Y., Carmody, N. C., and Weingold, A. B.,** Serial measurements of plasma diamine oxidase (DAO) during normal human pregnancy by an improved method, *Am. J. Obstet. Gynaecol.*, 95, 615, 1966.
85. **Gahl, W. A., Vale, A. M., and Pitot, H. C.,** Spermidine oxidase in human pregnancy serum. Probable identify with diamine oxidase, *Biochem. J.*, 201, 161, 1982.
86. **Illei, G. and Morgan, D. M. L.,** Polyamine oxidase activity in human pregnancy serum, *Br. J. Obstet. Gynaecol.*, 86, 878, 1979.
87. **Morgan, D. M. L., Illei, G., and Royston, J. P.,** Serum polyamine oxidase activity in normal pregnancy, *Br. J. Obstet. Gynaecol.*, 90, 1194, 1983.
88. **Weingold, A. B. and Southren, A. L.,** Diamine oxidase as an index of the fetoplacental unit, *Obstet. Gynaecol.*, 32, 593, 1968.

89. **Lin, C.-W., Chapman, C. M., DeLellis, R. A., and Kinsey, S.,** Immunofluorescent staining of histaminase (diamine oxidase) in human placenta, *J. Histochem. Cytochem.*, 26, 1021, 1978.
90. **Weisburger, W. R., Mendelsohn, G., Eggleston, J. C., and Baylin, S. B.,** Immunohistochemical localization of histaminase (diamine oxidase) in decidual cells of human placenta, *Lab. Invest.*, 38, 703, 1978.
91. **Illei, G. and Morgan, D. M. L.,** The distribution of polyamine oxidase activity in the fetomaternal compartments, *Br. J. Obstet. Gynaecol.*, 86, 873, 1979.
92. **Smith, J. K.,** The purification and properties of placental histaminase, *Biochem. J.*, 103, 110, 1967.
93. **Paolucci, F., Cronenberger, L., Plan, R., and Pachsko, H.,** Purification and properties of the diamine:oxygene oxydoreductase of human placenta, *Biochimie*, 53, 735, 1971.
94. **Bardsley, W. G. and Crabbe, M. J. C.,** Properties of purified human placental amine oxidase and a possible role of this enzyme in normal pregnancy, *Biochem. Soc. Trans.*, 1, 494, 1973.
95. **Crabbe, M. J. C. and Bardsley, W. G.,** The inhibition of placental diamine oxidase by substrate analogs, *Biochem. J.*, 139, 183, 1974.
96. **Crabbe, M. J. C. and Bardsley, W. G.,** Monoamine oxidase inhibitors and other drugs as inhibitors of diamine oxidase from human placenta and pig kidney, *Biochem. Pharmacol.*, 23, 2983, 1974.
97. **Crabbe, M. J. C., Childs, R. E., and Bardsley, W. G.,** Time dependent inhibition of diamine oxidase by carbonyl group reagents and urea, *Eur. J. Biochem.*, 60, 325, 1975.
98. **Crabbe, M. J. C., Waight, R. D., Bardsley, W. G., Barker, R. W., Kelley, I. D., and Knowles, P. F.,** Human placental diamine oxidase. Improved purification and characterization of a copper- and manganese-containing amine oxidase with novel substrate specificity, *Biochem. J.*, 155, 679, 1976.
99. **Crabbe, M. J. C.,** Histaminases in human placenta and seminal fluid and their possible similarities to Lysyl oxidase, *Agents Actions*, 9, 41, 1979.
100. **Mondovì, B. and Riccio, P.,** Animal intracellular amine oxidases, in *Structure and Functions of Amine Oxidases*, Mondovì, B., Ed., CRC Press, Boca Raton, Fla., 1985, 63.
101. **Gahl, W. A. and Pitot, H. C.,** Acetylated polyamines as substrates for human pregnancy serum diamine oxidase, *Life Sci.*, 29, 2177, 1981.
102. **Morgan, D. M. L.,** Human pregnancy-associated polyamine oxidase: partial purification and properties, *Biochem. Soc. Trans.*, 13, 351, 1985.
103. **Morgan, D. M. L.,** Polyamine oxidases and cellular interactions, *Adv. Polyamine Res.*, 3, 65, 1981.
104. **Crabbe, M. J. C.,** Summary of an international workshop on diamine oxidase held at the Department of Biogenic Amines, Polish Academy of Sciences, Lodz, Poland 7-8 June, 1970, *Agents Actions*, 11, 3, 1981.
105. **Grudzinskas, J. G., Teisner, B., and Seppala, M.,** *Pregnancy Proteins*, Academic Press, Sydney, 1982.
106. **Kuivaniemi, H., Savolainen, E.-R., and Kivirikko, K. I.,** Human placental lysyl oxidase. Purification, partial characterisation, and preparation of two specific antisera to the enzyme, *J. Biol. Chem.*, 259, 6996, 1984.
107. **Buffoni, F.,** Lysyl oxidase (characterisation and clinical importance), in *Structure and Function of Amine Oxidases*, Mondovì, B., Ed., CRC Press, Boca Raton, Fla., 1985, 77.
108. **Illei, G. and Morgan, D. M. L.,** Serum polyamine oxidase activity in spontaneous abortion, *Br. J. Obstet. Gynaecol.*, 89, 199, 1982.
109. **Byrd, W. J., Jacobs, D. M., and Amos, M. S.,** Synthetic polyamines added to cultures containing bovine sera reversibly inhibit in vitro parameters of immunity, *Nature*, 267, 621, 1977.
110. **Byrd, W. J., Jacobs, D. M., and Amos, M. S.,** Influence of synthetic polyamines on in vitro responses of immunocompetent cells, *Adv. Polyamine Res.*, 2, 71, 1978.
111. **Gaugas, J. M. and Curzen, P.,** Polyamine interaction with pregnancy serum in suppression of lymphocyte transformation, *Lancet*, i, 18, 1978.
112. **Morgan, D. M. L. and Illei, G.,** Polyamine-polyamine oxidase interaction: part of maternal protective mechanism against fetal rejection, *Br. Med. J.*, 280, 1295, 1980.
113. **Morgan, D. M. L.,** Polyamine oxidase in human pregnancy: a possible immunoregulatory agent, *Biochem. Soc. Trans.*, 9, 400, 1981.
114. **Labib, R. S. and Tomasi, T. B.,** Enzymatic oxidation of polyamines. Relationship to immunosuppressive properties, *Eur. J. Immunol.*, 11, 266, 1981.
115. **Khomenko, A. K., Dyachok, F. I., Gridina, N. Y., Burlaka, D. P., Negrey, G. Z., and Shylakhovenko, V. A.,** Influence of serum polyamine oxidase on normal and tumor cells, *Eksp. Onkol.*, 6, 63, 1984.
116. **Morgan, D. M. L. and Toothill, V. J.,** Polyamine oxidase activity in human and bovine milk, *Biochem. Soc. Trans.*, 13, 352, 1985.
117. **Suzuki, O., Matsumoto, T., and Katsumata, Y.,** Determination of polyamine oxidase activities in human tissues, *Experientia*, 40, 838, 1984.
118. **Pegg, A. E.,** Recent advances in the biochemistry of polyamines in eukaryotes, *Biochem. J.*, 234, 249, 1986.
119. **Seiler, N., Bolkenius, F. N., and Rennert, O. M.,** Interconversion, catabolism and elimination of the polyamines, *Med. Biol.*, 59, 334, 1981.

120. **Seiler, N., Bolkenius, F. N., Knödgen, B., and Mamont, P.,** Polyamine oxidase in rat tissues, *Biochim. Biophys. Acta,* 615, 480, 1980.
121. **Illei, G. and Morgan, D. M. L.,** Polyamine oxidase activity in amniotic fluid and fetal membranes, *Br. J. Obstet. Gynaecol.,* 87, 413, 1980.
122. **Yamada, H., Isobe, K., and Tani, Y.,** Oxidation of polyamines by fungal enzymes, *Agric. Biol. Chem.,* 44, 2469, 1980.
123. **Isobe, K., Tani, Y., and Yamada, H.,** Crystallization and characterization of polyamine oxidase from *Penicillium chrysogenum, Agric. Biol. Chem.,* 44, 2651, 1980.
124. **Isobe, K., Tani, Y., and Yamada, H.,** Crystallization and characterization of polyamine oxidase from *Aspergillus terreus, Agric. Biol. Chem.,* 44, 2749, 1980.
125. **Isobe, K., Tani, Y., and Yamada, H.,** Kinetic properties of fungal polyamine oxidases and their application to differential determination of spermine and spermidine, *Agric. Biol. Chem.,* 44, 2955, 1980.
126. **Kobayashi, Y. and Horikoshi, K.,** Production of extracellular polyamine oxidase by *Penicillium* sp. No. PO-1, *Agric. Biol. Chem.,* 45, 2943, 1981.
127. **Kobayashi, Y. and Horikoshi, K.,** Purification and characterisation of extracellular polyamine oxidase produced by *Penicillium* sp. No. PO-1, *Biochim. Biophys. Acta,* 705, 133, 1982.
128. **Kobayashi, Y., Higashi, T., Michida, H., Iwasaki, S., and Horikoshi, K.,** Oxidation of acetylpolyamines by extracellular polyamine oxidase produced by *Penicillium* sp. No. PO-1, *Biochim. Biophys. Acta,* 743, 431, 1983.
129. **Haywood, G. W. and Large, P. J.,** Partial purification of a peroxisomal polyamine oxidase from *Candida boidinii* and its role in growth on spermidine as sole nitrogen source, *J. Gen. Microbiol.,* 130, 1123, 1984.
130. **Green, J., Haywood, G. W., and Large, P. J.,** Serological differences between the multiple amine oxidases of yeasts and comparison of the specificities of the purified enzymes from *Candida utilis* and *Pichia pastoris, Biochem. J.,* 211, 481, 1983.
131. **Haywood, G. W. and Large, P. J.,** Microbial oxidation of amines. Distribution, purification and properties of two primary-amine oxidases from the yeast *Candida boidinii* grown on amines as sole nitrogen source, *Biochem. J.,* 199, 187, 1981.
132. **Kumagai, H. and Yamada, H.,** Bacterial and fungal amine oxidases, in *Structure and Functions of Amine Oxidases,* Mondovì, B., Ed., CRC Press, Boca Raton, Fla., 1985, 37.
133. **Tabor, C. W. and Tabor, H.,** Polyamines in microorganisms, *Microbiol. Rev.,* 49, 31, 1985.
134. **Smith, T. A.,** The di- and polyamine oxidases of higher plants, *Biochem. Soc. Trans.,* 13, 319, 1985.
135. **Rinaldi, A., Floris, G., and Giartosio, A.,** Plant amine oxidases, in *Structure and Functions of Amine Oxidases,* Mondovì, B., Ed., CRC Press, Boca Raton, Fla., 1985, 51.
136. **Bardsley, W. G., Crabbe, M. J. C., Schindler, J. S., and Ashford, J. S.,** Oxidation of *p*-dimethylaminomethylbenzylamine by pig kidney diamine oxidase. A new method for spectrophotometric assay, *Biochem. J.,* 127, 875, 1972.
137. **Bardsley, W. G., Crabbe, M. J. C., and Scott, I. V.,** A method for determining diamine oxidase (histaminase) in biological fluids, *Biochem. Med.,* 11, 138, 1974.
138. **Bachrach, U. and Reches, B.,** Enzymic assay for spermine and spermidine, *Anal. Biochem.,* 17, 38, 1966.
139. **Morgan, D. M. L., Ferluga, J., and Allison, A. C.,** Polyamine oxidase and macrophage function, in *Polyamines in Biomedical Research,* Gaugas, J. M., Ed., John Wiley & Sons, Chichester, England, 1980, 303.
140. **Gaugas, J. M. and Dewey, D. L.,** Evidence for serum binding of oxidised spermine and its potent G1-phase inhibition of cell proliferation, *Br. J. Cancer,* 39, 548, 1979.
141. **Morgan, D. M. L. and Illei, G.,** Radiochemical estimation of serum polyamine oxidase activity in human pregnancy, *Med. Lab. Sci.,* 38, 49, 1980b.
142. **Seiler, N. and Knödgen, B.,** Determination of the naturally occurring monoacetyl derivatives of di- and polyamines, *J. Chromatogr.,* 164, 155, 1979.
143. **Childs, R. E. and Bardsley, W. G.,** The steady-state kinetics of peroxidase with 2,2′-azino-di-(3-ethylbenzthiazoline-6-sulphonic acid) as chromogen, *Biochem. J.,* 145, 93, 1975.
144. **Suzuki, O., Matsumoto, T., Oya, M., Katsumata, Y., and Samejima, K.,** A new fluorimetric assay for spermidine synthase, *Anal. Biochem.,* 115, 72, 1981.
145. **Bergmeyer, H. U., Ed.,** *Oxidoreductases and Transferases,* (Methods of Enzymatic Analysis, Vol. 3, 3rd ed.), Verlag Chemie, Weinham, West Germany, 1983.
146. **Bachrach, U. and Oser, I. S.,** Enzymatic assay for spermidine, *J. Biol. Chem.,* 238, 2098, 1963.
147. **Bachrach, U.,** Spermidine, in *Methods of Enzymatic Analysis,* Bergmeyer, H. U., Ed., Vol. 4., 2nd ed., Verlag Chemie, Weinheim, West Germany, 1974, 1740.
148. **Bachrach, U.,** Spermine and spermidine, in *Methods of Enzymatic Analysis,* Bergmeyer, H. U., Ed., Vol. 4., 2nd ed., Verlag Chemie, Weinheim, West Germany, 1974, 1744.
149. **Bachrach, U. and Plesser, Y. M.,** A sensitive, rapid, chemiluminescence-based method for the determination of diamines and polyamines, *Anal. Biochem.,* 152, 423, 1986.

150. **Bachrach, U.**, Oxidized polyamines, *Ann. N.Y. Acad. Sci.*, 171, 939, 1970.
151. **Bachrach, U.**, Oxidation of polyamines and diamines, in *Polyamines in Biology and Medicine*, Morris, D. R. and Marton, L. J., Eds., Marcel Dekker, New York, 1981, 151.
152. **Katsuta, H., Takaoka, T., Nose, K., and Nagai, Y.**, Effects of polyamines on the proliferation of mammalian cells in tissue culture, *Jpn. J. Exp. Med.*, 45, 345, 1975.
153. **Pulkkinen, P., Piik, K., Koso, P., and Jänne, J.**, Effect of polyamines and their oxidised derivatives on the metabolism of rat epidydimal spermatozoa, *Acta Endocrinol.*, 87, 845, 1978.
153a. **Morgan, D. M. L.**, unpublished data.
154. **Gahl, W. A. and Pitot, H. C.**, Reversal by aminoguanidine of the inhibition of proliferation of human fibroblasts by spermidine and spermine, *Chem. Biol. Interact.*, 22, 91, 1978.
155. **Maschler, R., Smith, C. J., Allen, J. C., and Maurer, H. R.**, Elucidation of the different effects of polyamines and other naturally-occurring inhibitors of cell proliferation (chalones) and T-lymphocyte and granulocyte colony growth in vitro, *Z. Naturforsch.*, 38c, 74, 1983.
156. **Ali-Osman, F. and Maurer, H. R. C.**, Stimulation of clonal tumor cell growth in vitro by inhibiting the serum polyamine oxidase activity, *J. Cancer Res. Clin. Oncol.*, 106, 17, 1983.
157. **Israel, M. and Modest, E. J.**, Synthesis of aminoethyl derivatives of α,ω-alkylenediamines and structure-activity relationships for the polamine-bovine plasma amine oxidase system, *J. Med. Chem.*, 14, 1042, 1971.
158. **Morgan, D. M. L., Christensen, J. R., and Allison, A. C.**, Polyamine oxidase and the killing of intracellullar parasites, *Biochem. Soc. Trans.*, 9, 563, 1981.
159. **Morgan, D. M. L. and Christensen, J. R.**, Polyamine oxidation and the killing of intracellular parasites, *Adv. Polyamine Res.*, 4, 169, 1983.
160. **Ferrante, A., Rzepczyk, C., and Allison, A. C.**, Polyamine oxidase mediates intra-erythrocytic death of *Plasmodium falciparum, Trans R. Soc. Trop. Med. Hyg.*, 77, 789, 1983.
161. **Rzepczyk, C. K., Saul, A. J., and Ferrante, A.**, Polyamine oxidase-mediated intraerythrocytic killing of *Plasmodium flaciparum:* evidence against the role of reactive oxygen metabolites, *Infect. Immunol.*, 43, 238, 1984.
162. **Morgan, D. M. L., Bachrach, U., Assaraf, Y., Harari, E., and Golenser, J.**, The effect of purified aminoaldehydes produced by polyamine oxidation on the in vitro development of *Plasmodium falciparum* in normal and glucose-6-phosphate-dehydrogenase deficient erythrocytes, *Biochem. J.*, 236, 97, 1986.
163. **Ferrante, A., Allison, A. C., and Hirumi, H.**, Polyamine oxidase-mediated killing of African trypanosomes, *Parasite Immunol.*, 4, 349, 1982.
164. **Ferrante, A., Rzepczyk, C. M., and Saul, A. J.**, Polyamine oxidase-mediated trypanosome killing: the role of hydrogen peroxide and aldehydes, *J. Immunol.*, 133, 2157, 1984.
165. **Egan, J. E., Haynes, J. D., Brown, N. D., and Eismann, C. S.**, Polyamine oxidase in human retroplacental serum inhibits the growth of *Plasmodium falciparum, Am. J. Trop. Med. Hyg.*, 35, 890, 1986.
166. **Ferrante, A.**, Inhibition of human neutrophil locomotion by the polyamine oxidase-polyamine system, *Immunology*, 54, 785, 1985.
167. **Ferrante, A., Maxwell, G. M., Rencis, V. O., Allison, A. C., and Morgan, D. M. L.**, Inhibition of the respiratory burst of human neutrophils by the polyamine oxidase-polyamine system, *Int. J. Immunopharmacol.*, 8, 411, 1986.
168. **Farrante, A., Ljungstrom, I., Rzepczyk, C., and Morgan, D. M. L.**, Differences in sensitivity of *Schistosoma mansoni schistosomula, Dirofilaria immitis microfilariae,* and *Nematospiroides dubius* third stage larvae to damage by the polyamine oxidase-polyamine system, *Infect. Immun.*, 53, 606, 1986.
169. **Hussain, J. I., Smith, C. J., and Allen, J. C.**, Polyamine-mediated inhibition of in-vitro cell proliferation is not due to acrolein, *Cell Tissue Kinet.*, 16, 583, 1983.
170. **Kimes, B. W. and Morris, D. R.**, Preparation and stability of oxidised polyamines, *Biochim. Biophys. Acta*, 228, 223, 1971.
171. **Bachrach, U., Abzug, S., and Bekierkunst, A.**, Cytotoxic effect of oxidised spermine on Ehrlich ascites cells, *Biochim. Biophys. Acta*, 134, 174, 1967.
172. **Bachrach, U. and Rosenkovitch, E.**, Effect of oxidised spermine and other aldehydes on the infectivity of vaccinia virus, *Appl. Microbiol.*, 23, 232, 1972.
173. **Alarcon, R. A.**, Isolation of acrolein from incubated mixtures of spermine with calf serum and its effect on mammalian cells, *Arch. Biochem. Biophys.*, 106, 240, 1964.
174. **Alarcon, R. A.**, Acrolein. IV. Evidence for the formation of the cytotoxic aldehyde acrolein from enzymatically oxidised spermine or spermidine, *Arch. Biochem. Biophys.*, 137, 365, 1970.
175. **Kimes, B. W. and Morris, D. R.**, Inhibition of nucleic acid and protein synthesis in *Escherichia coli* by oxidised polyamines and acrolein, *Biochim. Biophys. Acta*, 228, 235, 1971.
176. **Smith, C. J., Maschler, R., Maurer, H. R., and Allen, J. C.**, Inhibition of cells in culture by polyamines does not depend on the presence of ruminant serum, *Cell Tissue Kinet.*, 16, 269, 1983.
177. **Smith, C. J., Hussain, J. I., and Allen, J. C.**, Inhibition of cell proliferation by polyamines does not depend on the cytotoxicity of acholein, *Biochem. Soc. Trans.*, 13, 326, 1985.

178. **Croker, S. J., Loeffler, R. S. T., and Smith, T. A.,** 1,5-diazabicyclo[4.3.0.]nonane, the oxidation product of spermine, *Tetrahedron Lett.*, 24, 1559, 1983.
179. **Brandange, S. and Eriksson, L.-H.,** Ring-chain tautomerism of an *N,N*-acetal formed by enzymatic oxidation of spermine or spermidine, *Acta Chem. Scand. Ser. B*, 38, 526, 1984.
180. **Della Ragione, F. and Pegg, A. E.,** Purification and characterization of spermidine/spermine N^1-acetyltransferase from rat liver, *Biochemistry*, 21, 6152, 1982.
181. **Della Ragione, F. and Pegg, A. E.,** Role of spermidine/spermine N^1-acetyltransferase in polyamine metabolism, in *Advances in Polyamines in Biochemical Science*, Calderera, C. M. and Bachrach, U., Ed., CLUEB, Bologna, Italy, 1984, 97.
182. **Matsui, I., Kamei, M., Otani, S., Morisawa, S., and Pegg, A. E.,** Occurrence and induction of spermidine N^1-acetyltransferase in *Escherichia coli, Biochem. Biophys. Res. Commun.*, 106, 1155, 1982.
183. **Haywood, G. W. and Large, P. J.,** The occurrence, subcellular localisation and partial purification of diamine acetyltransferase in the yeast *Candida boidinii* grown on spermidine or putrescine as sole nitrogen source, *Eur. J. Biochem.*, 148, 277, 1985.
184. **Pegg, A. E. and Erwin, B. G.,** Induction of spermidine/spermine N^1-acetyltransferase in rat tissues by polyamines, *Biochem. J.*, 148, 277, 1985.
185. **Matsui-Yuasa, I., Otani, S., Morisawa, S., Takigawa, M., Enomoto, M., and Suzuki, F.,** Induction of spermidine/spermine N^1-acetyltransferase by parathyroid hormone in rabbit costal chondrocytes in culture, *J. Biochem.*, 97, 387, 1985.
187. **Gaugas, J. M. and Dewey, D. L.,** Hog kidney diamine oxidase conversion of biogenic diamines to inhibitors of cell proliferation, *J. Pathol.*, 134, 243, 1981.
188. **Mach, M., Kersten, H., and Kersten, W.,** Regulation of tRNA methyltransferase activities by spermidine and putrescine. Inhibition of polyamine synthesis and tRNA methylation by a-methylornithine or 1,3-diaminopropan-2-ol in *Dictyostelium discoideum, Biochem. J.*, 202, 153, 1982.
189. **Quash, G., Keolouankhot, T., Gazzolo, L., Ripoll, H., and Saez, S.,** Diamine and polyamine oxidase activities in normal and transformed cells, *Biochem. J.*, 177, 275, 1979.
190. **Porter, C. W., Bergeron, R. J., and Stolowich, N. J.,** Biological properties of N^4-spermidine derivatives and their potential in anticancer chemotherapy, *Cancer Res.*, 42, 4072, 1982.
191. **Porter, C. W., Cavanaugh, P. F., Stolowich, N., Ganis, B., Kelly, E., and Bergeron, R. J.,** Biological properties of N^4- and N^1,N^8-spermidine derivatives in cultured L1210 leukemia cells, *Cancer Res.*, 45, 2050, 1985.
192. **Aigner-Held, R. and Daves, G. D.,** Polyamine metabolites and conjugates in man and higher animals: a review of the literature, *Physiol. Chem. Phys.*, 12, 389, 1980.
193. **Seiler, N.,** Amide-bond-forming reactions of polyamines, in *Polyamines in Biology and Medicine*, Morris, D. R. and Marton, L. J., Eds., Marcel Dekker, Basel, 1981, 127.
194. **Wallace, H. M., McGowan, S. H., and Keir, H. M.,** Polyamine metabolism in mammalian cells in culture, *Biochem. Soc. Trans.*, 13, 329, 1985.
195. **Seiler, N., Bolkenius, F. N., and Knödgen, B.,** Acetylation of spermidine in polyamine catabolism, *Biochim. Biophys. Acta*, 633, 181, 1980.
196. **Bey, P., Bolkenius, F. N., Seiler, N., and Casara, P.,** *N*-2,3-Butadienyl-1,4-butanediamine derivatives: potent irreversible inactivators of mammalian polyamine oxidase, *J. Med. Chem.*, 28, 1, 1985.
197. **Bolkenius, F. N., Bey, P., and Seiler, N.,** Specific inhibition of polyamine oxidase in vivo is a method for the elucidation of its physiological role, *Biochem. Biophys. Acta*, 838, 69, 1985.
198. **Seiler, N. and Bolkenius, F. N.,** Polyamine reutilization and turnover in brain, *Neurochem. Res.*, 10, 529, 1985.
199. **Seiler, N., Bolkenius, F. N., and Knodgen, B.,** The influence of catabolic reactions on polyamine excretion, *Biochem. J.*, 225, 219, 1985.
200. **Allen, J. C. and Smith, C. J.,** Chalones: a reappraisal, *Biochem. Soc. Trans.*, 7, 584, 1979.
201. **Patt, L. M. and Houck, J. C.,** The incredible shrinking chalone, *FEBS Lett.*, 120, 163, 1980.
202. **Allen, J. C., Smith, C. J., Curry, M. C., and Gaugas, J. M.,** Identification of a thymic inhibitor ('chalone') of lymphocyte transformation as a spermine complex, *Nature*, 267, 623, 1977.
203. **Scalabrino, G. and Ferioli, M. E.,** Polyamines in mammalian tumors. I, *Adv. Cancer Res.*, 35, 151, 1981.
204. **Jänne, J., Pösö, H., and Raina, A.,** Polyamines in rapid growth and cancer, *Biochim. Biophys. Acta*, 473, 241, 1978.
205. **Adachi, O., Yamada, H., and Ogata, K.,** Purification and properties of putrescine oxidase of *Micrococcus rubens, Agric. Biol. Chem.*, 30, 1202, 1966.
206. **Yamada, H.,** Putrescine oxidase *(Micrococcus rubens), Meth. Enzymol.*, 17B, 726, 1971.
207. **Suzuki, Y. and Yanagisawa, H.,** Purification and properties of maize polyamine oxidase — a flavoprotein, *Plant Cell Physiol.*, 21, 1085, 1980.
208. **Smith, I. A.,** Polyamine oxidase from barley and oats, *Phytochemistry*, 15, 633, 1976.

Chapter 14

5′-DEOXY-5′-METHYLTHIOADENOSINE: NOVEL METABOLIC AND PHYSIOLOGICAL ASPECTS

Fulvio Della Ragione, Maria Cartenì-Farina, and Vincenzo Zappia

TABLE OF CONTENTS

I. INTRODUCTION

Crystals of 5′-deoxy-5′-methylthioadenosine (MTA) were first isolated in 1912 by Mandel and Dunham,[1] although the correct chemical structure of this natural thioether was elucidated later on by Suzuki and co-workers.[2]

The nucleoside is ubiquitously distributed in prokaryotes and eukaryotes and represents one of the main products of *S*-adenosylmethionine (AdoMet) metabolism. Its biogenesis from methionine was demonstrated by Schlenk in 1952,[3] even before the discovery of its direct precursor, AdoMet, by Cantoni.[4]

For several years MTA has been regarded just as a minor by-product of polyamines biosynthesis and only in the last decade the observation of its antiproliferative effect, as well as the elucidation of its role in purine salvage and thiomethyl group recycling, led to an upsurge of interest of biochemists and pharmacologists towards this thioether.[5-11] Moreover, the recently observed absence of MTA phosphorylase in several malignant cells[12,13] has further attracted the attention on the possible correlation(s) between MTA metabolism and malignancy.

In this respect it sounds less provocative today the hypothesis formulated in 1973 by Williams-Ashman[14] that "in certain cells the primary function of the spermidine and spermine synthase reactions would be to produce MTA rather than polyamines."

The purpose of this chapter is to discuss some of the recent achievements on MTA metabolism, with particular emphasis towards the molecular properties of the enzymes involved in the control of the cellular content of this molecule.

II. BIOSYNTHETIC PATHWAYS

MTA is derived from AdoMet by several enzymatic reactions, involving in all instances a nucleophilic attack to the activated methylenic γ-carbon atom adjacent to the electron withdrawing trivalent sulfur (Figure 1).

In the synthesis of polyamines the aminopropyl group of decarboxylated AdoMet is transferred to a nucleophilic primary amino group of putrescine or spermidine with concomitant release of MTA.[15,6] This well-known pathway represents quantitatively the major route of MTA synthesis in mammals; indeed, up to 97% of its formation is prevented by methylglyoxal-bis-guanylhydrazone (MGBG), a specific inhibitor of polyamine biosynthesis.[17] Further details on this metabolic pathway and on the physiological role(s) of polyamines are discussed elsewhere in this book (see Volume II, Chapter 3).

Although the other enzymatic reactions leading to the formation of MTA are, at least in mammals, quantitatively less important, they will be discussed in view of their biological implications. Among them, two postsynthetic AdoMet-dependent modifications of tRNAs, namely the aminocarboxylpropylation of uridine and guanosine,[18-20] deserve particular attention for their regulatory significance in protein biosynthesis.[21] In the biogenesis of 3-(3-amino-3-carboxypropyl)uridine, the 2-aminobutyric group of the sulfonium compound is transferred to N-3 of uridine with concomitant release of MTA.[18] The modified nucleoside has been first isolated in *Escherichia coli* $tRNA^{Arg}$[18] and then detected in several tRNAs ($tRNA^{Ile}$, $tRNA^{Met}$, and $tRNA^{Val}$) from bacteria and eukaryotes;[18,20] the enzyme responsible for such a modification has been partially purified and characterized from rat liver.[20] The other AdoMet-dependent base modification leads to the formation of the so-called nucleoside Y. This compound has been isolated in tRNA of *Saccharomyces cerevisiae*[19] and its biogenesis has been partially elucitated: the 2-aminobutyric group of the sulfonium compound is transferred to guanosine yielding MTA and an unusual imidazo-guanosine ring.[19]

The biological significance of these reactions is probably related to the control of protein biosynthesis through modifications of the three-dimensional structure of the polynucleotide.[21]

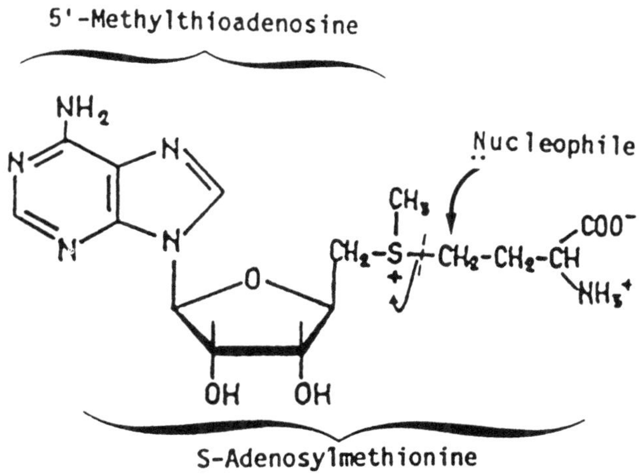

FIGURE 1. Catalytic mechanism responsible for the cleavage of *S*-adenosylmethionine.

On the other hand the possible role of such tRNAs modifications in other cellular process(es), such as the synthesis of the so-called "alarmones",[22-24] should also be considered.

MTA is also produced in eukaryotes during the biosynthesis of diphthamide (2-[3-carboxyamido-3-(trimethylammonio)propyl]histidine), a posttranslational modified amino-acid identified in the elongation factor 2 (EF-2) (Figure 2).[25,26] Detailed labeling studies on the diphthamide biogenesis in *S. cerevisiae* have indeed demonstrated that AdoMet, in analogy with the above-described tRNA modifications, acts as a donor of a 3-carboxy-3-aminopropyl group.[25,26] The catalytic mechanism, reponsible for the addition of the backbone of methionine to histidine, implies a nucleophilic attack on the γ-methylene of the sulfonium compound by an imidazolium-2-carbanion. The biological role of diphthamide, which represents the site of NAD-dependent ADP-ribosylation of EF-2,[27] is probably related to the regulation of the protein synthesis machinery.

In plant tissues MTA is formed during the synthesis of ethylene, a phytohormone involved in plant growth and development;[28] the reaction is catalyzed by the enzyme 1-aminocyclopropane carboxylate synthetase (Figure 3).[29] The first step of catalysis involves the formation of a Schiff base between the pyridoxal phosphate bound to the enzyme and AdoMet. An intramolecular nucleophilic attack by the N-C double bond on the γ-methylene of methionine is then responsible for the production of MTA with concomitant release of a cyclopropane ring.[29]

Another route of MTA biosynthesis, which also involves an initial formation of a Schiff base between the sulfonium compound and pyridoxal phosphate, has been described in *E. coli*.[30] In such a reaction AdoMet plays the unusual role of amine group donor to 7-keto-8-amino pelargonic acid, an intermediate of biotin biosynthesis. The keto product, *S*-adenosyl-2-oxo-4-methylthiobutyric acid, in turn undergoes a rapid nonenzymatic breakdown to MTA and 2-oxo-3-butenoic acid.[30]

In this context it is noteworthy that a nonenzymatic reaction between AdoMet and pyridoxal phosphate has also been described.[31] This process is greatly enhanced by the addition of divalent cations, such as Cu^{++}. The formation of a ternary complex between AdoMet, Cu^{++}, and pyridoxal phosphate results in a significant decarboxylation of the sulfonium compound[31] with a concomitant breakdown of the molecule into MTA.[31a] However, although no definitive data are available in the literature on the in vivo occurrence of complex(es) between AdoMet and pyridoxal phosphate, it has been suggested that in yeast, which ac-

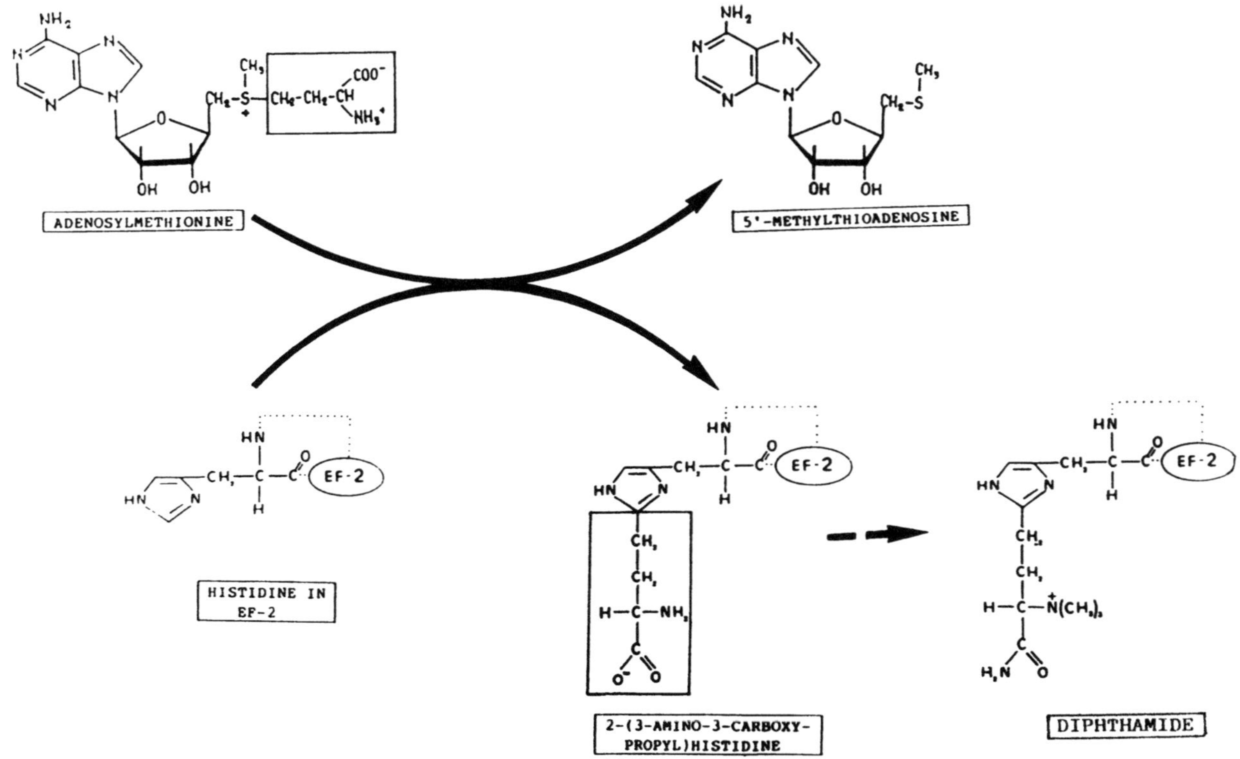

FIGURE 2. Biosynthesis of diphthamide, (2-[3-carboxyamide-3-(trimethylammonio)propyl]histidine).

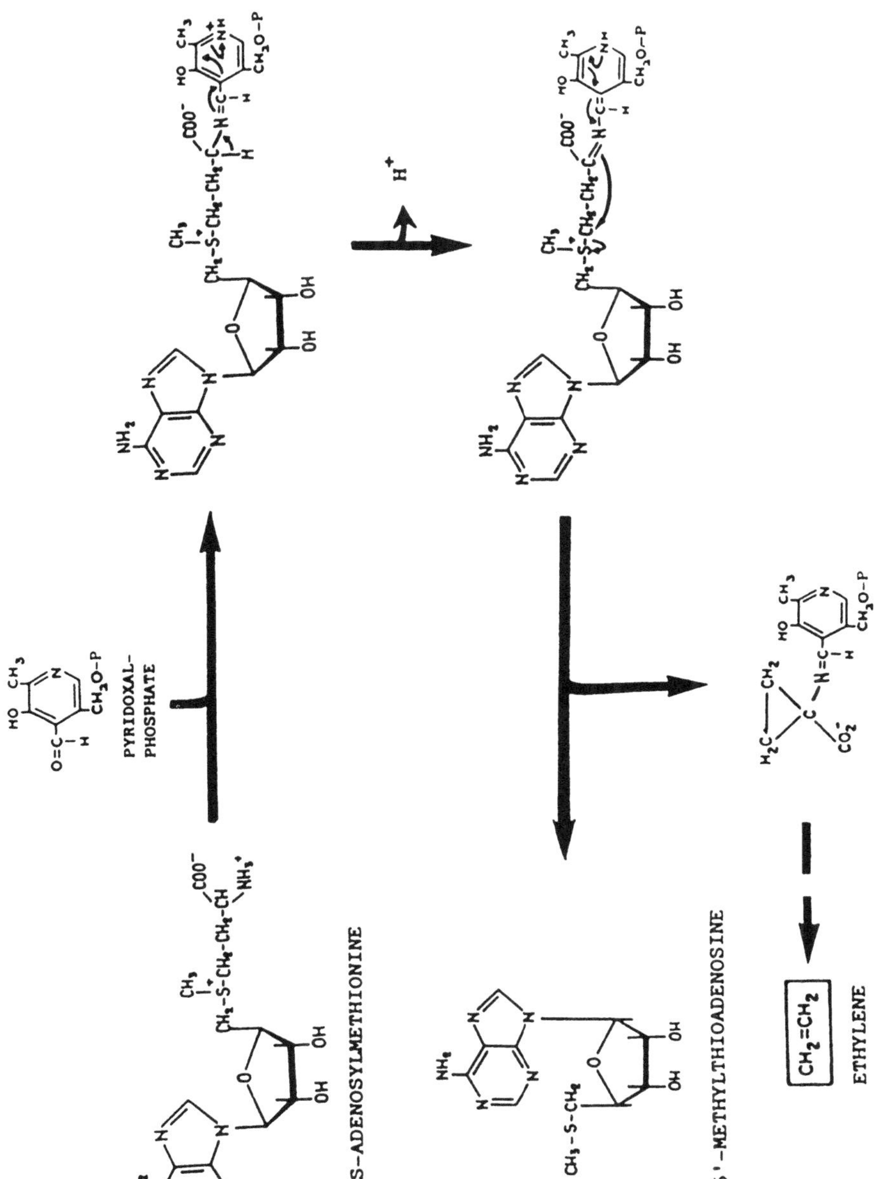

FIGURE 3. Biosynthesis of ethylene in plant tissues.

cumulates AdoMet, the sulfonium compound interferes with some pyridoxal phosphate-enzymes.[32]

Finally, MTA is also formed by direct enzymatic and nonenzymatic cleavage of AdoMet, the carboxylic group of the methionine moiety being responsible for the intramolecular attack on the γ-C, followed by the breakdown of the molecule into MTA and homoserine.[33] The enzyme responsible for this reaction, i.e., AdoMet hydrolase, has been purified to homogeneity in *E. coli* B infected by phage T3.[34,35] Such an enzyme activity, absent in uninfected cells, is expressed immediately after the phage infection, thus suggesting a close relationship between the cleavage of AdoMet and the resistance to host restriction.[34,35] Enzymatic activities cleaving AdoMet into MTA and homoserine have also been described in other bacteria,[36] yeasts,[36] and mammalian tissues,[37,38] although recent studies in rat liver extracts have seriously questioned the occurrence of this enzyme in mammalian tissues.[39]

Although a significant nonenzymatic cleavage of AdoMet into MTA is observable in vitro at physiological temperature and pH values, it is difficult to evaluate the relevance of this reaction in vivo. However, the substantial inhibition (97%) of MTA production following the administration of MGBG[17] can be taken as an indirect proof that this process contributes only negligibly to the synthesis of the thioether and that AdoMet presumably is protected in vivo against its nonenzymatic degradation into MTA. A specific interaction with macromolecule(s) blocking the amino acid side chain and the sulfonium pole of AdoMet at mutually inaccessible positions may be responsible for such a stabilization. Further studies would be useful to substantiate this view.

III. MTA CATABOLISM AND DISPOSITION

In spite of the occurrence of multiple biosynthetic pathways, the intracellular concentration of the thioether is remarkably low when compared with that of spermidine, spermine, and AdoMet.[40-44] Several investigations, utilizing reversed-phase high performance liquid chromatography,[40,41] demonstrated that the cellular content of the molecule ranges between 0.5 and 7 nmol/g of wet tissue. Similar results have been reported by Pegg and Coward[45] employing a more sensitive and specific radioimmunoassay. An efficacious MTA catabolism associated with an active excretion of the nucleoside into extracellular fluids are indeed responsible for the low cellular level of the molecule (Figure 4).

In bacteria, plants, and protozoa MTA is hydrolytically cleaved by specific enzymes into 5-methylthioribose (MTR) and adenine.[33,46-48,50-53] Such an activity was first identified by Shapiro and Mather[33] in cell-free extracts of *A. aerogenes* and then confirmed by Walker and Duerre[46] employing *E. coli B*, *A. aerogenes* and *S. thyphimurium* as enzymatic sources. The enzyme from *E. coli* has been partially characterized by Ferro et al.[47] and very recently purified to homogeneity by our group.[48]

It is noteworthy in this respect that the bacterial MTA nucleosidases so far investigated are also responsible for the cleavage of *S*-adenosylhomocysteine (AdoHcy) into adenine and ribosylhomocysteine.[46-48] This observation led to the hypothesis that in prokaryotes a single enzyme is responsible for the catabolism of the two thioethers.[46,47] However, the occurrence of a specific AdoHcy hydrolase, recently demonstrated in some bacterial species,[49] suggests that alternative mechanism(s) may be involved in microbial MTA catabolism. In this connection the presence of MTA phosphorylase in archaebacteria is of remarkable interest.[50]

MTA nucleosidase activity has also been described in several plants.[51,52] The enzyme has been purified to homogeneity and characterized in detail only from *Lupineus luteus* seeds by Guranowsky et al.;[51] it is specific for MTA and inactive towards AdoHcy.

Few data are available from literature on MTA catabolism in protozoa. The occurrence of a nucleosidase specific towards the thioether has been reported only in the protozoan *Ochromonas malhamensis*.[53] Although the properties of this enzyme have not yet been

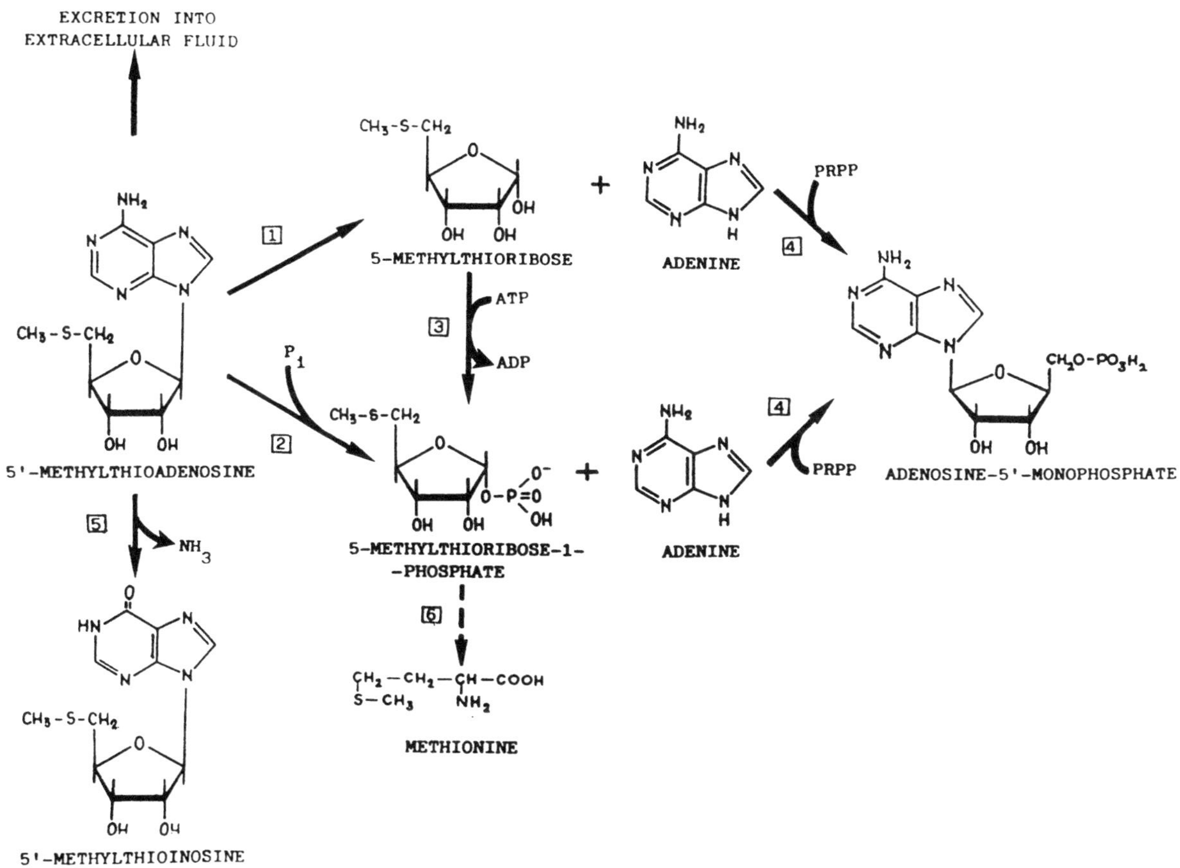

FIGURE 4. Catabolic pathways of MTA. (1) MTA nucleosidase, (2) MTA phosphorylase, (3) MTR kinase, (4) Adenine phosphoribosyl transferase, (5) nonspecific adenosine deaminase, and (6) methionine-recycling pathway.

elucitated, its presence is of remarkable physiological and pharmacological interest. Indeed, the adenine salvage pathway plays a key role in the survival of protozoa where the *de novo* synthesis of purines is not operative.[54,55]

In mammals the glycosidic bond of MTA is cleaved through a well-known phosphorolytic mechanism into adenine and 5-methylthioribose-1-phosphate (MTR-1-P).[56-62] A similar enzymatic activity has also been detected in the insect *Drosophila melanogaster*.[63]

The mammalian enzyme was first described by Pegg and Williams-Ashman in rat ventral prostate[56] and then partially purified from several tissues and cells.[57-61] Recently it has been purified more than 29,000-fold to homogeneity from human placenta in our laboratory.[62] The tissue distribution of this enzyme has been poorly investigated.[59] Among rat tissues the lowest activity has been detected in heart and kidney and the highest in lung and liver.[59] Unfortunately, these data are expressed in terms of enzyme units per total organ, so that it is hard to calculate the actual amount of MTA phosphorylase in the analyzed tissues.

Few data are also available in the literature on the regulation of the enzyme activity. Ferro et al.[64] reported that MTA phosphorylase activity significantly increases in human lymphocytes stimulated by phytohemoagglutinin; an enhancement of 12% over the nonstimulated cells is observable during the first 12 hr of growth reaching an 81 and 108% increase after 24 and 48 hr, respectively. It has also been reported that gonadectomy results in a significant decrease of the enzyme activity in ventral prostrate and in the uterus.[65] Treatment with testosterone propionate of 17-β-estradiol partially restores the enzyme activity in male and female rats, respectively.[65]

It has recently been demonstrated that a sixfold increase of MTA phosphorylase activity during the exponential growth phase of Chinese hamster ovary cells.[66] Moreover, a significant cell cycle dependence of the enzyme has been observed in synchronized HeLa cells;[66] the specific activity is higher during the late G1 phase suggesting that the increase in the enzyme activity is probably related to the synthesis of DNA.

In conclusion, MTA phosphorylase appears to be a finely regulated enzyme, although more extensive and detailed studies are needed to elucidate the molecular mechanism(s) involved in its regulation.

Before discussing the metabolic fate of adenine and MTR-1-P, other two possible ways of MTA removal must be mentioned. A nonspecific adenosine deaminase from *Aspergillus oryzae*, catalyzes the deamination of MTA to its corresponding inosine derivative.[67] However, the other adenosine deaminases so far investigated are more specific and therefore unable to attack the adenosylthioether at a significant rate.[68-70]

The excretion of MTA into extracellular fluids must be envisioned as a further mechanism for its disposition. It has indeed been demonstrated that malignant cell lines lacking MTA phosphorylase excrete the thioether at significant rate into the extracellular medium.[17] Probably this mechanism is not confined to MTA phosphorylase-negative cells since significant amounts of MTA sulfoxide are also detectable in urine of healthy volunteers.[71,72]

Both the products of the phosphorolytic cleavage of MTA are actively metabolized. Particularly two catabolic pathways have been proposed for MTR-1-P: the recycling to methionine and its dephosphorylation to MTR.[56,73-75] However, while the latter reaction has not been confirmed, being probably an artifact,[56] a large number of evidences demonstrate the conversion of the phosphorylated sugar into methionine.[73-75] So far only two steps of this unique metabolic route, where all the carbon atoms of MTR-1-P except C-1 are retained in the sulfur amino acid, have been identified[74,75] (Figure 5). The first involves the conversion of MTR-1-P to methylthioribulose-1-phosphate[75] while the transamination of 2-keto-4-S-methylthiobutyrate to methionine represents the final step.[74]

The same pathway is operative in bacteria where MTR is first phosphorylated to MTR-1-P through the action of a specific ATP-dependent kinase (Figure 4).[76,77]

The other product of MTA catabolism, adenine, is in turn recycled to AMP through the

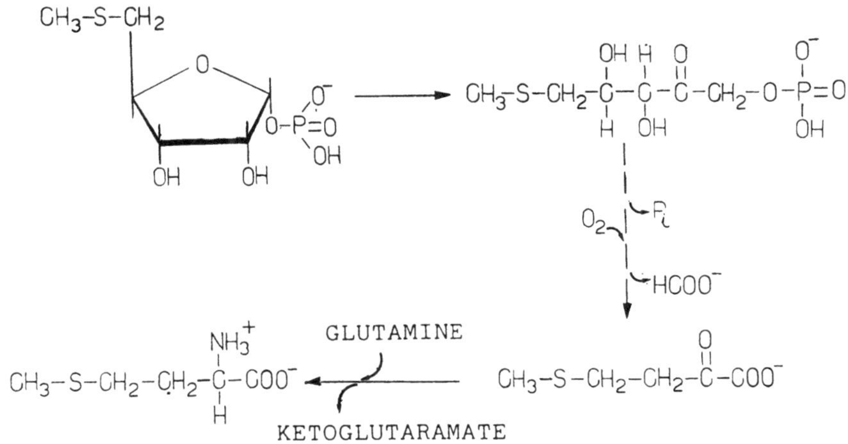

FIGURE 5. Methionine-recycling pathway from 5-methylthioribose-1-phosphate.

action of the widely distributed adenine phosphoribosyl transferase. It must be underlined that the enzymatic breakdown of MTA represents, at least in mammals, the only source of intracellular free adenine.

IV. HUMAN PLACENTAL MTA PHOSPHORYLASE: MOLECULAR AND KINETIC PROPERTIES

In spite of the large number of studies on MTA phosphorylase,[56-66] the enzyme has been purified to homogeneity and characterized only recently.[62] The procedure employs, as essential steps, covalent chromatography on organomercurial-agarose and anion exchange on Mono Q (Table 1). The specific activity of the homogeneous enzyme is 10.2 μmol of MTA cleaved per minute per milligram of protein which represents a 29,700-fold purification over the supernatant at 15,000 g.

As determined by gel filtration chromatography,[62] the molecular weight (Mr) of the enzyme is 98,000. This value is in agreement with previously reported data[58,60] with the only exception of rat lung enzyme.[59] Polyacrylamide gel electrophoresis under denaturing conditions indicates that the enzyme is composed by three apparently identical subunits (Mr 32,500). It is interesting to note that purine nucleoside phosphorylase (PNP) shares the same trimeric molecular structure,[78] thus suggesting the possible occurrence of a common ancestral gene.

The enzyme does not require for its activity any specific ion or cofactor while reducing agents, such as dithiothreitol, mercaptoethanol, or reduced glutathione, are necessary for maximal activity. Moreover, MTA phosphorylase is effectively inhibited by sulfhydryl (SH)-blocking agents, thus suggesting the involvement of SH groups in the substrate binding or in the catalytic mechanism.

The equilibrium constant of the reaction, calculated by incubating the enzyme with different amounts of MTA and phosphate, is 1.39×10^{-2} at 37°C. The reversibility of the reaction has also been directly demonstrated by incubating adenine and MTR-1-P in the presence of the purified enzyme (Figure 6).

The kinetic constants for the phosphorolytic and synthetic reaction are reported in Table 2. The Km for MTA, calculated at saturating level of phosphate, is 5 μ*M*, in agreement with the value reported for the enzyme from other sources.[57,58,60] The affinity for phosphate is somewhat lower (Km 320 μ*M*) and in agreement with that reported for the enzyme from rat liver.[60]

Table 1
PURIFICATION OF HUMAN PLACENTA MTA PHOSPHORYLASE

Step	Protein (mg)	Specific activity (units/mg)	Purification (fold)	Yield (%)
Supernatant at 15,000 g	29,000	0.00035	1	100
Ammonium sulfate (55 — 75%)	16,000	0.00061	1.8	97
Acetone (40 — 60%)	3,045	0.0031	8.6	93
DEAE Sephace	725	0.077	220	55
Hydroxyapatite	10.38	0.43	1,354	44
Agarose organomercurial	0.5	5.7	14,800	25
Mono Q	0.2	10.2	29,700	23

Note: One unit is the amount of enzyme which cleaves 1 μmol of MTA/min at 37°C.

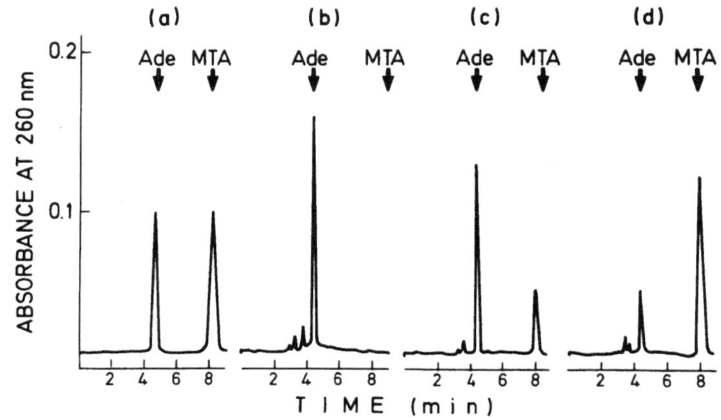

FIGURE 6. Effect of enzyme concentration on MTA phosphorylase activity in the direction of MTA synthesis. (a) Separation of adenine (Ade) and MTA by reversed-phase HPLC chromatography, (b through d) chromatographic analysis of assay mixtures containing (b) 0 units, (c) 0.001 units, and (d) 0.002 units of phosphate-free MTA phosphorylase.

Table 2
KINETIC CONSTANTS OF HUMAN PLACENTA MTA PHOSPHORYLASE

Substrate	K_m (μM)
5′-Deoxy-5′-methylthioadenosine	5
Orthophosphate	320
Adenine	23
5-Methylthioribose-1-phosphate	8

However, in spite of the equilibrium constant favoring the synthesis of MTA and the high affinity of the enzyme for MTR-1-P and adenine, it is unlikely that the synthetic reaction is operative in vivo, mainly for the rapid enzymatic removal of adenine and MTR-1-P. On the other hand, it is possible that the exposure of cells to adenine analogues or antimetabolites blocking adenine recycling may result in the reversal of the phosphorolytic reaction, leading

Table 3
PURIFICATION OF *ESCHERICHIA COLI* 5′-METHYLTHIOADENOSINE NUCLEOSIDASE

Step	Protein (mg)	Specific activity[a] (units/mg)	Purification (fold)	Yield (%)
Supernatant at 15,000 g	12,000	0.038	1	100
$(NH_4)_2SO_4$ (40 — 60%)	8,300	0.045	1.2	81
DEAE-Sephadex A-50	500	0.7	15.5	75
Hydroxyapatite	70	4.2	110	64
Sephacryl S-200	15	15.3	400	50
S-Formycinylhomocysteine Sepharose	0.5	373	9,810	40

[a] One unit is the amount of the enzyme which cleaves 1 μmol of MTA/min at 37°C.

Table 4
MOLECULAR AND CATALYTIC PROPERTIES OF *ESCHERICHIA COLI* MTA NUCLEOSIDASE

Specific activity[a]	373 units/mg of protein
Molecular weight	26,500
Turnover number	9.32×10^3
Stokes radius	2.1 nm
Protein structure	Monomeric
Kinetic constant	Km (MTA) 0.43 μ*M*

[a] One unit is the amount of enzyme which catalyzes the cleavage of 1 μmol of MTA/min at 37°C.

in turn to the formation of sulfur nucleoside derivatives of modified purines. This consideration should be taken into account in investigating the cytostatic mechanism of some adenine analogues, such as 2-fluoroadenine and 2-chloroadenine.

V. MTA NUCLEOSIDASE FROM *ESCHERICHIA COLI*: PHYSICOCHEMICAL PROPERTIES AND MECHANISM OF CATALYSIS

Escherichia coli MTA nucleosidase has been purified to homogeneity in our laboratory (about 10,000-fold, 40% yield) by using a procedure involving conventional methodologies followed by affinity chromatography on *S*-formycinylhomocysteine-Sepharose (Table 3).[48]

Several physical and kinetic properties of the homogeneous enzyme are reported in Table 4. The nucleosidase has a molecular weight of 26,500 and is composed by a single polypeptide chain. The calculated Mr value is slightly lower than that previously reported by Ferro[47] (31,000) employing a 220-fold purified enzyme.

The enzyme is highly effective in cleaving both MTA and AdoHcy; the specific activities are in fact 373 μmol of MTA and 156 μmol of AdoHcy cleaved per minute per milligram of protein, which is significantly higher than that of homogeneous mammalian AdoHcy hydrolase[79-83] and homogeneous human placental MTA phosphorylase.[62] *E. coli* MTA nucleosidase has a remarkably high affinity for MTA (Km 0.43 μ*M*) as well as for AdoHcy (Km 4.3 μ*M*). The Vmax/Km ratio, which is a useful probe to evaluate the relative efficiency towards the two substrates, is 867 for MTA and 36 for AdoHcy, thus suggesting that the

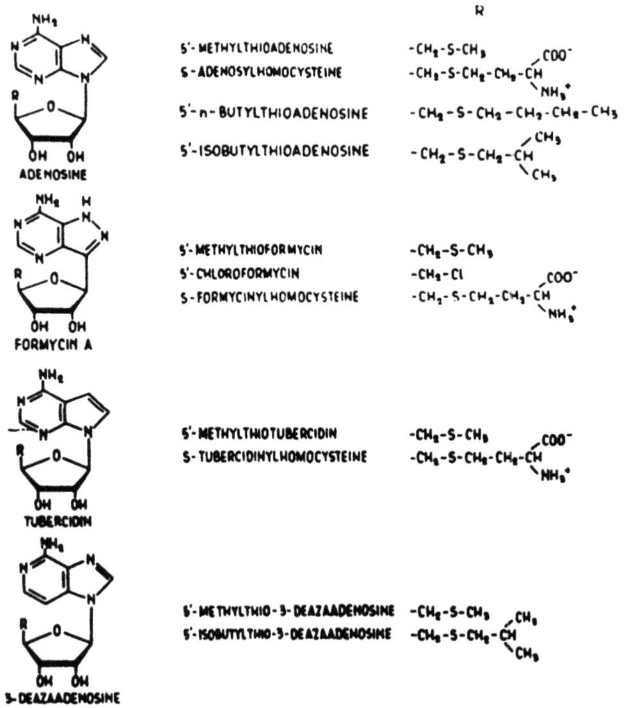

Substrate	Relative activity (%)
5'-Methylthioadenosine	100
S-Adenosylhomocysteine	42
5'-n-Butylthioadenosine	40
5'-Isobutylthioadenosine	54
5'-Chloroformycin	0.01
5'-Methylthioformycin	0.01
S-Formycinylhomocysteine	0.01
5'-Methylthiotubercidin	0.01
S-Tubercidinylhomocysteine	0.01
5'-Methylthio-3deazaadenosine	0.01
5'-Isobutylthio-3deazaadenosine	0.01

FIGURE 7. Substrate specificity of *Escherichia coli* MTA nucleosidase.

nucleosidase is more effective towards MTA. However, the lack of information about the cellular contents of AdoHcy and MTA in bacteria prevents conclusive inferences.

MTA nucleosidase is highly specific with respect to substrate requirements.[48] Indeed, among the analogues studied (Figure 7), only 5'-isobutylthioadenosine and 5'-*n*-butylthioadenosine are actively cleaved. These data are in good agreement with the results reported by Ferro et al.[47] on a number of 5'-modified MTA analogues. Conversely, the replacement of N-7 or N-3 by a methynic carbon as well as the substitution of the imidazole moiety by a pyrazole ring results in a complete loss of activity (Figure 7). In conclusion the structural integrity of adenine moiety is required for the catalytic mechanism while the chemical nature of 5'-substituent is not critical in this regard. The analogues resistant to the enzymatic cleavage have also been tested as inhibitors (see Table 5). The tubercidinyl analogues exert a remarkable competitive inhibition thus suggesting that N-7 is not involved in the recognition

Table 5
INHIBITION OF *ESCHERICHIA COLI* MTA NUCLEOSIDASE

Compound	I_{50} (μ*M*)
5′-Methylthioformycin	0.06
5′-Chloroformycin	0.4
S-Formycinylhomocysteine	0.02
5′-Methylthiotubercidin	7.7
S-Tubercidinylhomocysteine	3.2
S-8-Azaadenosylhomocysteine	1.8
5′-Isobutylthioinosine	132
5′-Methylthioinosine	>500
5′-Methylthio-3-deazaadenosine	79
5′-Isobutylthio-3-deazaadenosine	13
Adenine	216
Hypoxanthine	>500
Guanine	>500
AMP	>500
5-Methylthioribose	>500
Sinefungin	24
S-Inosylhomocysteine	>500

Note: The I_{50} values (concentrations giving 50% inhibition) were calculated with a concentration of 5′-methylthioadenosine of 0.7 μ*M*.

process while it is critical in the catalytic mechanism. This result also indicates that the protonation at N-7 atom of the substrate is an essential step in the catalysis (Figure 8), in analogy with the mechanism postulated for acidic hydrolysis of adenosine.

The powerful inhibition exerted by the formycinyl derivatives can be explained by the occurrence in their purine moiety of a protonated N-7 at physiological pH thus resembling a likely intermediate of reaction or the transition state (Figure 8). On the basis of these results the 5′-modified formycinyl analogues can be envisioned as potential antibacterial agents. Indeed, this class of compounds may increase the cellular content of AdoHcy and MTA in bacteria, thus indirectly inhibiting methylation reactions and polyamine biosynthesis. Further studies would be useful to verify the therapeutic efficacy of such compounds.

VI. PHYSIOLOGICAL AND PUTATIVE PHARMACOLOGICAL ROLES OF MTA

The occurrence of an efficacious cleavage of MTA under biological conditions characterized by an increase of polyamine biosynthesis (e.g., accelerated growth) is of great importance in the overall balance of the purine nucleotide pool. It must be underlined that the phosphorolytic breakdown of the thioether represents the only nondietary source of free adenine. On the other hand, no data are available from the literature on the quantitative relevance of methionine formation from MTR-1-P compared to that of the other well-established routes of methionine recycling, namely the methylation of homocysteine catalyzed by 5-methyltetrahydrofolate homocysteine methyltransferase[84] and betaine homocysteine methyltransferase.[85]

Mudd and Poole[86] reported that in humans the daily biosynthesis of creatine (calculated by the excretion of creatine and creatinine) accounts for 89% of the net flow of methionine through AdoMet. Part of this methionine comes from the diet and part (about 50%) from endogenous homocysteine. On the other hand less than 5% of AdoMet is transformed to

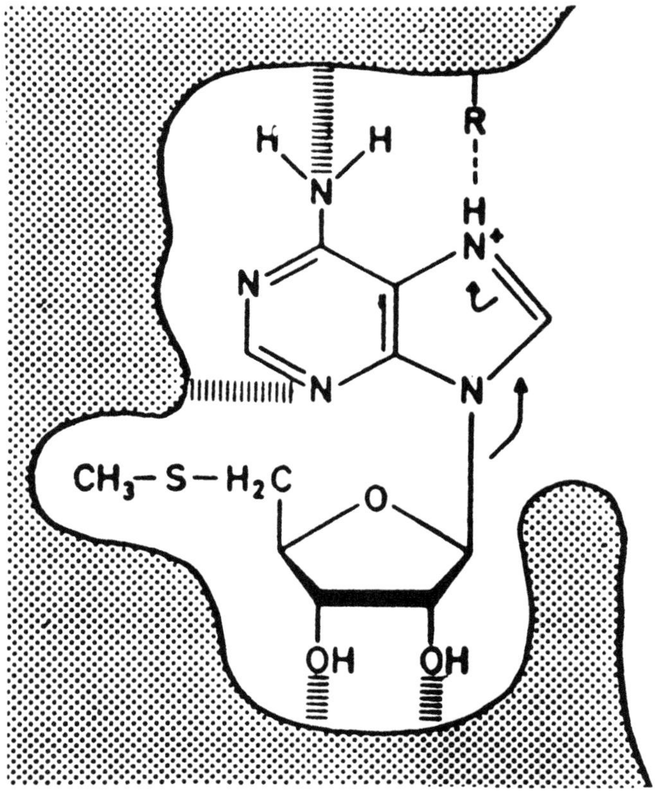

FIGURE 8. Hypothetical catalytic mechanism of *Escherichia coli* MTA nucleosidase.

MTR-1-P (via MTA), as judged by the excretion of polyamines.[86] The amount of methionine originating from MTR-1-P should therefore be not higher than 10% of that formed from homocysteine.

However, since creatine is formed almost exclusively in the liver,[87] the above reported balance reflects the metabolism of a single specialized organ. It is conceivable that in other tissues or cells the amount of MTR-1-P recycled to methionine could be comparable to that of homocysteine. In this regard it is worthy to mention that Iizasa and Carson[88] showed that in human CEM lymphoblasts the ratio of MTA formation vs. homocysteine synthesis fluctuates during the cell growth, being higher than 1 during mid-G1 phase and falling to 0.4 in the exponential growth phase.

Besides its role in purine and amino acid metabolism, a number of regulatory roles have been attributed to the intact thioether on the basis of its in vitro effects on several enzymatic systems (see Table 6). Among them, the inhibition of mammalian spermidine and spermine synthases is the most important.[89,90] Pajula and Raina,[89] employing an homogeneous preparation of bovine brain spermine synthase, reported a competitive inhibition of MTA with respect to decarboxylated AdoMet; the calculated Ki value (0.3 μM) is significantly lower than the Km for the sulfonium compound (Km = 0.6 μM).[89] Spermidine synthase is also inhibited by the thioether but with a minor efficacy (I_{50} = 50 μM).[90] An additional interesting effect exerted by MTA is the suicide-like inhibition of AdoHcy hydrolase,[91,92] the Ki values ranging between 37 and 117 μM and the K_2 from 0.1 to 0.6 min^{-1}.[92]

Whether these inhibitory effects of MTA are operative in vivo as well is a matter of investigation. However, considering the low cellular content of the thioether (<5 μM), it

Table 6
INHIBITORY EFFECT OF 5′-DEOXY-5′-METHYLTHIOADENOSINE

Enzyme	Inhibition (μ*M*)	Ref.
Spermine synthase (bovine brain)	0.3 (Ki)	89
Spermidine synthase (rat ventral prostate)	50 (42% inhibition)	90
AdoHcy hydrolase (human erythrocytes, rat	36 (Ki)	91
liver)	117 (Ki)	92
cAMP phosphodiesterase (mice spleen)	225 (Ki)	111
Adenosine kinase (human placenta)	120 (Ki)	112
Protein methylase I (rat liver)	35 (Ki)	113
Protein methylase II (bovine brain)	41 IKi)	114
Histamine *N*-methyltransferase (guinea pig brain)	450 (20% inhibition)	115
N-Acetylserotonin *O*-methyltransferase (pineal gland)	500 (29% inhibition)	115
tRNA(Guanine-7)-methyltransferase (*Salmonella thyphimurium*)	458 (Ki)	116
DNA modification methylase (*Escherichia coli*)	19 (Ki)	117
DNA Restriction endonuclease (*Escherichia coli*)	10 (Ki)	117

is conceivable that only spermine synthase is regulated in vivo, while any other effect reported in vitro appears unlikely (see Table 6).

In the context of the physiological role(s) of MTA, it has been proposed that the thioether may regulate cell differentiation; the DMSO-induced differentiation of murine erythroleukemic cells is indeed inhibited by μmolar amounts of the nucleoside.[93] On the other hand, Fitchen and co-workers[94] have recently reported that MTA does not affect the differentiation of human granulocytes; hence, the role of the nucleoside in the maturative process is still a matter of investigation.

In recent years a large number of studies (Table 7 and references therein) have indicated that MTA, at concentrations exceeding the celllular content, inhibits cell proliferation. The thioether levels capable of eliciting a 50% inhibition of cell growth vary from 3 to 500 μ*M* and are generally lower for MTA phosphorylase-negative cells. The structural integrity of the molecule is a requirement for the cytostatic effect that cannot be ascribed to its metabolites since adenine and/or MTR do not significantly affect the rate of cell growth.[68,70]

The cellular targets as well a the molecular mechanism(s) responsible for the cytostatic effect of MTA have not been so far identified. In this regard it is noteworthy that Carson and co-workers[95] selected mutant clones of a lymphoblastoid cell line 10- to 30-fold more resistant to the antiproliferative effect of MTA than parental cells. The major biochemical differences between normal and mutant cells are the higher level of AdoMet and AdoMet synthase activity occurring in the MTA-resistant clones; this metabolic response could be envisioned as a sort of mechanism of cell protection. Such an observation, together with the data reported in Table 6, leads to hypothesize that the cytostatic effect of MTA is probably mediated by a concerted inhibition of AdoMet-dependent reaction(s), namely (1) propylamino transfer reactions, (2) transmethylation reactions, and (3) aminocarboxylpropylations of EF-2 and tRNAs.

Polyamine depletion, caused by inhibition of spermidine and spermine synthase (1), has been proposed by several authors as a major mechanism of MTA antiproliferative action.[89,90] Nevertheless, the observation that the addition of polyamines does not reverse the cytostatic effect of the thioether[96] casts serious doubts upon this mechanism. On the other hand, in

Table 7
ANTIPROLIFERATIVE EFFECT OF 5′-DEOXY-5′-METHYLTHIOADENOSINE

Biological system	LD_{50} value (μM)	Ref.
Human lymphocytes	500	118
BW5147 Murine lymphoma cells	100	119
Rat T lymphocytes	50	120
SV40-transformed 3T3 mouse fibroblasts	100	121
Friend erythroleukemic cells	180	122
Human colon carcinoma DLD-1	340	123
SV49-Transformed lymphoma cells	50	124
R 1.1 murine T lymphoma cell line: clone H[a]	3	125
L1210 leukemia cells[a]	90	126
L5178Y leukemia cells[a]	33	126

[a] MTA phosphorylase-deficient cell line.

spite of the evidences on the inhibition of MTA on various transmethylases in vitro (see Table 6), no convincing proof is so far available from the literature that the inhibition is operative in vivo (2).

The third hypothesis, namely the inhibition of aminocarboxylpropylation of histidine in EF-2[25,26] and uridine in tRNAs[20] (3) appears particularly promising in view of the role of these reactions in the regulation of protein synthesis. In addition, this mechanism is indirectly supported by the observation that several MTA analogues exert a significant inhibition on protein biosynthesis in BHK cells.[97]

Besides the mentioned antiproliferative effects the in vivo administration of the molecule causes a number of additional pharmacological actions: it depresses body temperature[98] and blood pressure,[99] causes relaxation of intestinal strips in rabbits, and contraction of isolated guinea pig uterus.[99] Recently a significant antiinflammatory and analgesic activity of MTA has also been reported; the molecule exhibits a dose-related inhibitory effect on carrageenin-induced edema in rats and phenylquinone-induced writhings in mice.[100] The mechanism of the antiinflammatory effect is probably related to the inhibition of the metabolism of arachidonic acid,[100] although an adenosine-like mechanism[101] cannot be ruled out.

VII. TRANSPORT OF MTA

By contrast with the parent compounds AdoMet, decarboxylated AdoMet, and AdoHcy, MTA easily enters biological membranes.[102] Indeed, as mentioned above, cells lacking MTA phosphorylase excrete large amounts of such a thiother in the culture medium. This observation is in agreement with the data on isolated and perfused rat liver,[103] yeast,[104] and protozoa[105] indicating that the thioether is actively taken up by eukaryotic cells.

A detailed kinetic analysis of MTA transport and metabolism, carried out in human red cells, demonstrated that the thioether transport occurs via a facilitated-diffusion mechanism.[102] MTA uptake is the result of a tandem action of a transport step of high capacity (Vmax = 604 ± 51 pmol/10^6 cells per minute) and low affinity (Km = 3270 ± 321 μM) followed by a metabolic step of low capacity (Vmax = 6.6pmol/10^6 cells per minute) and high affinity (Km = 30 μM).[102]

These data suggest that at physiological level of the thioether (below 5 μM) the initial velocity values of transport represent the limiting step of MTA influx thus explaining the small amounts of the intact adenosyl nucleoside detectable intracellularly. Conversely, when

Table 8
EFFECT OF 5′-METHYLTHIOADENOSINE ANALOGUES AND RELATED COMPOUNDS ON THE UPTAKE OF 5′-METHYLTHIOADENOSINE BY HUMAN ERYTHROCYTES

Compound	Concentration of compound (μ*M*)	Methylthioadenosine concentration (μ*M*)	Inhibition (%)
Nucleosides			
Adenosine	1000	10	4
Inosine	1000	10	2
Guanosine	1000	10	4
Xanthosine	1000	10	2
7-Deazaadenosine	1000	10	3
Bases			
Adenine	250	500	21
	1000	500	58
	2000	500	91
	20	10	35
7-Deazaadenine	1000	10	42
4-Aminopyrazolo-3,4-pyrimidine	1000	10	31
Guanine	1000	10	53
Various			
AMP	1000	10	—
5-Methylthioribose	1000	10	—
S-Adenosyl-L-homocysteine	1000	10	—

the cells are exposed to cytostatic concentration of the thioether (above 200 μ*M*), the initial velocity of transport largely exceeds that of cleavage by MTA phosphorylase and the cellular pool of free nucleoside approaches the concentration of MTA in extracellular fluids.

The specificity of the erythrocytic MTA carrier has also been investigated (see Table 8); adenine, among the bases assayed, is the most potent inhibitor while a variety of nucleosides do not interfere with MTA uptake. Probably the thioether and adenine share a common carrier while the involvement of the adenosine transport system in MTA uptake must be ruled out.

VIII. MTA PHOSPHORYLASE DEFICIENCY AND MALIGNANCY

MTA phosphorylase-deficient malignant cell lines requiring for their growth an exogenous source of "methylthio" groups were first described by Toohey in 1978.[106,107] Later on, Kamatani et al.[13] reported a deficiency in MTA phosphorylase activity in 7 out of 31 (23%) human malignant cell lines deriving from leukemia, melanoma, and breast cancer.

More recently, two cases of MTA phosphorylase deficiency were described in human leukemic cells in vivo[108] thus permitting to exclude artifacts due to in vitro culture conditions.

The molecular mechanisms responsible for the absence of this enzymatic activity are yet unclear. However, recent cytogenetic studies have shown some cases of lymphoblastic leukemia with lymphomatous features associated with abnormalities of the short arm of chromosome 9 and MTA phosphorylase deficiency.[109] Genetic research based on the expression of the enzyme in somatic cell hybrids demonstrated that the structural gene for MTA phosphorylase is localized on the chromosome 9.[110] Therefore, it is possible to explain the enzyme deficiency on the basis of genetic abnormalities. However, other molecular mechanism(s), including a chromosomal deletion involving a regulator gene for MTA phosphorylase or a generalized aberration occurring in the transformed cells, which may alter the expression and/or the regulation of MTA phosphorylase, cannot be ruled out.

Moreover, the deficiency of MTA phosphorylase has been demonstrated only in malignant cells while all the normal tissues or cells so far investigated show detectable levels of the enzyme activity. In this respect the enzyme deficiency should be envisioned as a very promising tumor marker.

In addition, the absence of an enzyme involved in the purine salvage pathway is of large therapeutic interest. It has indeed been reported that the transformed cells lacking MTA phosphorylase were in culture selectively killed by inhibiting *de novo* purine synthesis in the presence of exogenous MTA.[13] In such a therapeutic approach the thioether supplied solely the phosphorylase-positive cells with adenine thus permitting their effective protection against the cytostatic drugs. This scheme of therapy should also permit the administration of higher dose of the selected antimetabolite(s) without a concomitant increase of general toxicity.

Although this pharmacological scheme is of remarkable interest it should be considered that few studies on MTA pharmacokinetics have been so far carried out and that no data are available on the MTA metabolism in human serum.

IX. CONCLUDING REMARKS

Some of the most promising observations on MTA metabolism in eukaryotes, discussed in the present chapter, concern the occurrence and regulation of MTA phosphorylase activity in normal and malignant cells.

The striking changes in the phosphorylase activity, occurring during the logarithmic growth phase, indicate that the enzyme plays a key role in cell proliferation. If these changes are due to synthesis of new protein, decrease in the rate of enzyme degradation or release of the phosphorylase from a cryptic form is still a matter of investigation. However, development of immunological techniques should allow to clarify the fine regulatory mechanisms of this enzyme. In the same context, detailed studies on the kinetic mechanism and substrate specificity should also be useful in the design of MTA analogues endowed with pharmacological activity.

Apart from the interest in the regulatory aspects, two additional lines of investigation are particularly rich of clinical interest. The absence of the phosphorylase solely in malignant cell lines may be indeed of value from both a diagnostic and a pharmacological point of view. From a diagnostic standpoint, the availability of rapid and specific screening methods (i.e., immunohistochemical techniques) should permit the evaluation of the incidence of enzyme-deficiency among the malignant cells as well as its correlation to specific tumors.[110] On the other hand, taking into account the relevance of purine recycling in cell growth and differentiation, an aberration in the solely adenine-producing pathway could be exploited for a selective antineoplastic therapy.

The other possible development in the area of clinical investigation would be the design of MTA analogues endowed with antiprotozoal effects. As it is well known, protozoa do not possess *de novo* synthesis of purine and depend for their survival on specialized and effective purine recycling pathways. In this light the occurrence of well-defined enzymatic differences between mammalian and protozoal MTA cleaving enzymes is of obvious relevance in the design of efficacious drugs characterized by low toxicity.

ACKNOWLEDGMENTS

We wish to express our appreciation to Drs. Vincenzo Gragnaniello, Antonella Ravel, and Giuliana Clemente for their assistance in the reported experiments. This work was supported by Grant 84.00852.44 from the Italian National Research Council, Project "Oncologia".

REFERENCES

1. **Mandel, J. A. and Dunham, E. K.,** *J. Biol. Chem.*, 11, 85, 1912.
2. **Suzuki, U., Odake, S., and Mori, T.,** *Biochem. Z.*, 154, 278, 1924.
3. **Smith, R. L. and Schlenk, F.,** *Arch. Biochem. Biophys.*, 38, 159, 1952.
4. **Cantoni, G. L.,** *J. Biol. Chem.*, 204, 403, 1953.
5. **Williams-Ashman, H. G., Corti, A., and Coppoc, G. L.,** in *The Biochemistry of Adenosylmethionine,* Salvatore, F., Borek, E., Zappia, V., Williams-Ashman, H. G., and Schlenk, F. Eds., Columbia University Press, New York, 1977, 493.
6. **Williams-Ashman, H. G. and Canellakis, Z. N.,** *Perspect. Biol. Med.*, 22, 421, 1979.
7. **Ferro, A. J.,** in *Transmethylation,* Usdin, E., Borchardt, R. T., and Creveling, C. R., Eds., Elsevier/North-Holland, New York, 1979, 117.
8. **Zappia, V., Cartenì-Farina, M., Cacciapuoti, G., Oliva, A., and Gambacorta, A.,** in *Natural Sulfur Compounds,* Cavallini, D., Gaull, G. E. and Zappia, V., Eds., Plenum Press, New York, 1980, 138.
9. **Parks, R. E., Jr., Stoeckler, J. D., Cambor, C., Savarese, T. M., Crabtree, G. W., and Chu. S.-H.,** in *Molecular Actions and Targets for Cancer Chemotherapeutic Agents,* Sartorelli, A. C., Laso, J. S., and Berting, J. R., Eds., Academic Press, New York, 1981, 229.
10. **Williams-Ashman, H. G., Seidenfeld, J., and Galletti, P.,** *Biochem. Pharmacol.*, 31, 277, 1982.
11. **Schlenk, F.,** *Adv. Enzymol.*, 54, 195, 1983.
12. **Kamatani, N. and Carson, D. A.,** *Cancer Res.*, 40, 4178, 1980.
13. **Kamatani, N., Nelson-Rees, N. A., and Carton, D. A.,** *Proc. Natl. Acad. Sci. U.S.A.*, 78, 1219, 1981.
14. **Williams-Ashman, H. G., Coppoc, G., Schenone, G. L., and Weber, G.,** in *Polyamines in Normal and Neoplastic Growth,* Russell, D. H., Ed., Raven Press, New York, 1973, 181.
15. **Tabor, H. and Tabor, C. W.,** *Annu. Rev. Biochem.*, 45, 285, 1976.
16. **Williams-Ashman, H. G. and Pegg, A. E.,** in *Polyamines in Biology and Medicine,* Morris, D. R. and Marton, L. J., Eds., Marcel Dekker, New York, 1981, 43.
17. **Kamatani, N. and Carson, D. A.,** *Biochim. Biophys. Acta,* 675, 344, 1981.
18. **Ohashi, Z., Maeda, M., McCloskey, J. A., and Nishimura, S. H.,** *Biochemistry,* 13, 2620, 1979.
19. **Munch, H. J. and Thiebe, R.,** *FEBS Lett.*, 51, 257, 1975.
20. **Nishimura, S.,** in *The Biochemistry of Adenosylmethionine,* Salvatore, F., Borek, E., Zappia, V., Williams-Ashman, H. G., and Schlenk, F., Eds., Columbia University Press, New York, 1977, 510.
21. **Kersten, H.,** in *Progress in Nucleic Acid Research and Molecular Biology,* Moldove, K. and Cohn, W. E., Eds., Academic Press, New York, 31, 53, 1984.
22. **Neinemann-Dorsch, C. H., Hedl, A., Grummt, I., Albert, W., Ferminant, F. J., Friss, R. R., Pierron, G., Holl, W., and Grummt, F.,** *Eur. J. Biochem.*, 83, 179, 1983.
23. **Surowj, C. S. and Berger, N. A.,** *J. Biol. Chem.*, 258, 579, 1983.
24. **Jakubowski, H.,** *Proc. Natl. Acad. Sci. U.S.A.*, 83, 2378, 1986.
25. **Van Ness, B. G., Howard, J. B., and Bodley, W.,** *J. Biol. Chem.*, 255, 10717, 1980.
26. **Van Ness, B. G., Howard, J. B., and Bodley, W.,** *J. Biol. Chem.*, 255, 10710, 1980.
27. **Honjo, T., Nishizuka, Y., and Hayaishi, O.,** *J. Biol. Chem.*, 243, 3553, 1968.
28. **Abeles, F. B.,** *Ethylene in Plant Biology,* Academic Press, New York, 1973.
29. **Adams, D. O. and Jang, S. F.,** *Proc. Natl. Acad. Sci. U.S.A.*, 76, 174, 1979.
30. **Stoner, G. L. and Eisenberg, M. A.,** *J. Biol. Chem.*, 250, 4029, 1975.
31. **Zappia, V., Cartenì-Farina, M., and Galletti, P.,** in *The Biochemistry of Adenosylmethionine,* Salvatore, F., Borek, E., Zappia, V., Williams-Ashman, H. G., and Schlenk, F., Eds., Columbia University Press, New York, 1977, 473.

31a. **Della Ragione, F.,** unpublished results.

32. **Trewyn, R. W., Nakamura, K. D., O'Connor, M. L., and Parks, L. W.,** *Biochim. Biophys. Acta,* 327, 336, 1973.
33. **Shapiro, S. K. and Mather, A. N.,** *J. Biol. Chem.*, 233, 631, 1958.
34. **Spoerel, N., and Herlich, P.,** *Eur. J. Biochem.*, 95, 227, 1979.
35. **Spoerel, N., Herlich, P., and Bickle, T. A.,** *Nature,* 278, 30, 1979.
36. **Mudd, S. H.,** *J. Biol. Chem.*, 239, 87, 1959.
37. **Swiatek, K. R., Simon, L. N., and Chao, K. L.,** *Biochemistry,* 12, 4670, 1973.
38. **Wilson, J., Corti, A., Hawkins, H., Williams-Ashman, H. G., and Pegg, A. E.,** *Biochem. J.*, 180, 515, 1979.
39. **Eloranta, T. O. and Kajander, E. O.,** *Biochem. J.*, 229, 137, 1984.
40. **Della Ragione, F., Cartenì-Farina, M., Porcelli, M., Cacciapuoti, G., and Zappia, V.,** *J. Chromatogr.*, 226, 243, 1981.
41. **Seidenfeld, J., Wilson, J., and Williams-Ashman, H. G.,** *Biochem. Biophys. Res. Commun.*, 95, 1861, 1980.

42. **Hibasami, H., Hoffman, J., and Pegg, A. E.,** *J. Biol. Chem.*, 255, 6675, 1980.
43. **Pegg, A. E., Lockwood, D. H., and Williams-Ashman, H. G.,** *Biochem. J.*, 117, 17, 1970.
44. **Danzin, C., Jung, H. J., Claverie, N., Glove, J., Sjoderma, A., and Koch-Weser, J.,** *Biochem. J.*, 160, 507, 1979.
45. **Pegg, A. E., and Coward, J. K.,** in *Advances in Polyamine Research,* Vol. 3, Caldarera, C. M., Zappia, V., and Bachrach, U., Eds., Raven Press, New York, 1981, 153.
46. **Walker, R. D. and Duerre, J. A.,** *Can. J. Biochem.*, 53, 312, 1975.
47. **Ferro, A. J., Barrett, A., and Shapiro, S. K.,** *Biochim. Biophys. Acta,* 438, 487, 1976.
48. **Della Ragione, F., Porcelli, M., Cartenì-Farina, M., Zappia, V., and Pegg, A. E.,** *Biochem. J.*, 232, 335, 1985.
49. **Shimizu, S., Shiozaki, S., Ohshiro, T., and Yamada, H.,** *Eur. J. Biochem.*, 141, 385, 1984.
50. **Cartenì-Farina, M., Oliva, A., Romeo, G., Napolitano, G., De Rosa, M., Gambacorta, A., and Zappia, V.,** *Eur. J. Biochem.*, 101, 317, 1979.
51. **Guranowski, A. B., Chiang, P. K., and Cantoni, G. L.,** *Eur. J. Biochem.*, 114, 243, 1981.
52. **Baxter, C. and Coscia, C. J.,** *Biochem. Biophys. Res. Commun.*, 59, 147, 1973.
53. **Sugimoto, Y., Toraya, T., and Fukui, S.,** *Arch. Microbiol.*, 108, 175, 1976.
54. **Sherman, I. W.,** *Microbiol. Rev.*, 43, 453, 1979.
55. **Webster, H. K. and Nhaun, M.,** *J. Clin. Invest.*, 70, 461, 1982.
56. **Pegg, A. E. and Williams-Ashman, H. G.,** *Biochem. J.*, 115, 241, 1969.
57. **Zappia, V., Oliva, A., Cacciapuoti, G., Galletti, P., Mignucci, G., and Cartenì-Farina, M.,** *Biochem. J.*, 175, 1043, 1978.
58. **Cacciapuoti, G., Oliva, A., and Zappia, V.,** *Int. J. Biochem.*, 9, 35, 1978.
59. **Garbers, D. L.,** *Biochim. Biophys. Acta,* 523, 82, 1978.
60. **Ferro, A. J., Wobel, N. C., and Nicolette, J. A.,** *Biochim. Biophys. Acta,* 570, 65, 1979.
61. **Riscoe, M. K. and Ferro, A. J.,** *J. Biol. Chem.*, 259, 5465, 1984.
62. **Della Ragione, F., Cartenì-Farina, M., Gragnaniello, V., Schettino, M. I., and Zappia, V.,** *J. Biol. Chem.*, 261, 2324, 1986.
63. **Shugart, L., Tancer, M., and Moore, J.,** *Int. J. Biochem.*, 10, 901, 1979.
64. **Ferro, A. J., Vandenbark, A. A., and Marchitto, K.,** *Biochim. Biophys. Acta,* 588, 294, 1979.
65. **Nicolette, J. A., Wrobel, N. C., and Ferro, A. J.,** *Biochim. Biophys. Acta,* 627, 190, 1980.
66. **Sunkara, P. S., Chang, C. C., and Lachman, P. I.,** *Biochem. Biophys. Res. Commun.*, 127, 596, 1985.
67. **Zappia, V., Galletti, P., Cartenì-Farina, M., and Servillo, L.,** *Anal. Biochem.*, 58, 130, 1974.
68. **Savarese, T. M., Crabtree, G. W., and Parks, R. E., Jr.,** *Biochem. Pharmacol.*, 28, 2227, 1979.
69. **Pegg, A. E. and Williams-Ashman, H. G.,** *Arch. Biochem. Biophys.*, 137, 156, 1970.
70. **Savarese, T. M., Crabtree, G. W., and Parks, R. E., Jr.,** *Biochem. Pharmacol.*, 30, 189, 1981.
71. **Mills, J. S., Mills, G. C., and McAdoo, D. J.,** *Nucleosides Nucleotides,* 2, 465, 1983.
72. **Mills, G. C., Schmalstieg, F. C., and Goldblum, R. M.,** *Biochem. Med.*, 34, 37, 1985.
73. **Backlund, P. S., Jr. and Smith, R. A.,** *J. Biol. Chem.*, 256, 1533, 1981.
74. **Backlund, P. S., Jr., Chang, C. P., and Smith, R. A.,** *J. Biol. Chem.*, 257, 4196, 1982.
75. **Trackman, P. C. and Abeles, R. H.,** *J. Biol. Chem.*, 258, 6717, 1983.
76. **Ferro, A. J., Barrett, A., and Shapiro, S. K.,** *J. Biol. Chem.*, 253, 60, 1978.
77. **Ferro, A. J. and Marchitto, K. S.,** in *Methods in Enzymology,* Tabor, H. and Tabor, C. W., Eds., Academic Press, New York, 94, 361, 1983.
78. **Osborne, W. R. A.,** *J. Biol. Chem.*, 255, 7089, 1980.
79. **Richards, H. H., Chiang, P. K., and Cantoni, G. L.,** *J. Biol. Chem.*, 253, 4476, 1978.
80. **Palmer, J. L. and Abeles, R. H.,** *J. Biol. Chem.*, 254, 1217, 1979.
81. **Ueland, P. M. and Saebo, J.,** *Biochemistry,* 18, 4130, 1979.
82. **Fujoka, M. and Tokata, Y.,** *J. Biol. Chem.*, 256, 1631, 1981.
83. **Kajander, E. O. and Raina, A. M.,** *Biochem. J.*, 193, 503, 1981.
84. **Finkelstein, J. D., Kyle, W. E., and Harus, B. J.,** *Arch. Biochem. Biophys.*, 146, 84, 1972.
85. **Finkelstein, J. D. and Mudd, S. H.,** *J. Biol. Chem.*, 242, 873, 1967.
86. **Mudd, S. H. and Poole, J. R.,** *Metabolism,* 24, 721, 1975.
87. **Im, Y. S., Chiang, P. K., and Cantoni, G. L.,** *J. Biol. Chem.*, 254, 11047, 1979.
88. **Iizasa, T. and Carson, D. A.,** *Biochim. Biophys. Acta,* 844, 280, 1985.
89. **Pajula, R. L. and Raina, A.,** *FEBS Lett.*, 99, 343, 1979.
90. **Hibasami, H., Borchardt, R. T., Cheng, S. Y., Coward J. K., and Pegg, A. E.,** *Biochem. J.*, 187, 419, 1980.
91. **Ferro, A. J., Vandenbark, A. A., and MacDonald, M. R.,** *Biochem. Biophys. Res. Commun.*, 100, 523, 1981.
92. **Della Ragione, F. and Pegg, A. E.,** *Biochem. J.*, 210, 429, 1983.
93. **Di Fiore, P. P., Grieco, M., Pinto, A., Attadia, V., Porcelli, M., Cacciapuoti, G., and Cartenì-Farina, M.,** *Cancer Res.*, 44, 4096, 1984.

94. **Fitchen, J. H., Riscoe, M. K., and Ferro, A. J.,** in *Biological Methylation and Drug Design,* Borchardt, R. T., Creveling, C. R., and Ueland, P. M., Eds., Humana Press, Clifton, N.J., 1986, 301.
95. **Kubota, M., Kajander, E. O., and Carson, D. A.,** *Cancer Res.,* 45, 3567, 1985.
96. **Pegg, A. E., Borchardt, R. T., and Coward, J. K.,** *Biochem. J.,* 194, 79, 1981.
97. **Raina, A., Tuomi, K., and Pajula, R. L.,** *Biochem. J.,* 204, 697, 1982.
98. **Kuhn, R. and Henkel, K.,** *Hoppe-Seyler's Z. Physiol. Chem.,* 269, 41, 1941.
99. **Ewing, P. L. and Schlenk, F.,** *J. Pharmacol. Exp. Ther.,* 79, 164, 1943.
100. **Stramentinoli, G. and Gualano, M.,** in *Biomedical Studies of Natural Polyamines,* Caldarera, C. M., Clo, C., and Guarneri, C., Eds., CLUEB, Bologna, Italy, 1986, 97.
101. **Snyder, S. H., Bruns, R. F., Daly, J. N., and Innis, R. B.,** *Fed. Proc. Fed. Am. Soc. Exp. Biol.,* 40, 142, 1981.
102. **Cartenì-Farina, M., Della Ragione, F., Cacciapuoti, G., Porcelli, M., and Zappia, V.,** *Biochim. Biophys. Acta,* 727, 221, 1983.
103. **Zappia, V., Galletti, P., Porcelli, M., Ruggiero, G., and Andreana, A.,** *FEBS Lett.,* 90, 331, 1978.
104. **Di Carlo, F., Schultz, A. S., and McManus, D. K.,** *J. Biol. Chem.,* 189, 151, 1951.
105. **Sugimoto, Y., Toraya, T., and Fukui, S.,** *Arch. Microbiol.,* 108, 175, 1976.
106. **Toohey, J. I.,** *Biochem. Biophys. Res. Commun.,* 83, 27, 1978.
107. **Toohey, J. I.,** *Biochem. Biophys. Res. Commun.,* 78, 1273, 1977.
108. **Kamatani, N., Yu, A. L., and Carson, D. A.,** *Blood,* 60, 1387, 1982.
109. **Carrera, C. J., Eddy, R. L., Showsand, T. B., and Carson, D. A.,** *Proc. Natl. Acad. Sci. U.S.A.,* 81, 2665, 1984.
110. **Chilcote, R. H., Brown, E., and Rowley, J. D.,** *N. Engl. J. Med.,* 313, 286, 1985.
111. **Wolberg, G., Zimmerman, P. T., Schmitges, C. J., Duncan, G. S., and Deeprose, R. D.,** *Biochem. Pharmacol.,* 31, 2201, 1980.
112. **Fox, I. H., Palella, T. D., Thompson, D., and Herring, C.,** *Arch. Biochem. Biophys.,* 215, 302, 1982.
113. **Casellas, P. and Jeanteur, P.,** *Biochim. Biophys. Acta,* 519, 255, 1978.
114. **Oliva, A., Galletti, P., Zappia, V., Paik, W. K., and Kim, S.,** *Eur. J. Biochem.,* 104, 595, 1980.
115. **Zappia, V., Zydek-Cwick, C. R., and Schlenk, F.,** *J. Biol. Chem.,* 249, 4499, 1969.
116. **Salvatore, F., Traboni, C., Colonna, A., Ciliberto, G., Paolella, G., and Cimino, F.,** in *Natural Sulfur Compounds,* Cavallini, D., Gaull, G. E., and Zappia, V., Eds., Plenum Press, New York, 1980, 25.
117. **Linn, S., Eskin, B., Lautenberger, J., Lackey, D., and Kimball, M.,** in *The Biochemistry of Adenosylmethionine,* Salvatore, F., Borek, E., Zappia, V., Williams-Ashman, H. G., and Schlenk, F., Eds., Columbia University Press, New York, 1977, 521.
118. **White, M. W., Vandenbark, A. A., Barney, C. L., and Ferro, A. J.,** *Biochem. Pharmacol.,* 31, 503, 1982.
119. **Wolford, R. W., MacDonald, M. R., Zehfus, B., Rogers, T. J., and Ferro, A. J.,** *Cancer Res.,* 41, 3035, 1981.
120. **Christa, L., Thuillier, L., and Perignon, J.-L.,** *Biochem. Biophys. Res. Commun.,* 113, 425, 1983.
121. **Pegg, A. E., Borchardt, R. T., and Coward, J. K.,** *Biochem. J.,* 194, 79, 1981.
122. **Cartenì-Farina, M., Porcelli, M., Cacciapuoti, G., Zappia, V., Grieco, M., and Di Fiore, P. P.,** *Advances in Polyamine Research,* Vol. 4, Bachrach, U., Kaye, A., and Chayen, R., Eds., Raven Press, New York, 1983, 779.
123. **Ghoda, L. Y., Savarese, T. M., Dexter, D. L., Parks, R. E., Jr., Trackman, P. C., and Abeles, R. H.,** *J. Biol. Chem.,* 259, 6715, 1984.
124. **Riscoe, M. K., Tower, P. A., and Ferro, A. J.,** *Biochem. Pharmacol.,* 33, 3639, 1984.
125. **Kubota, M., Kamatani, N., and Carson, D. A.,** *J. Biol. Chem.,* 258, 7288, 1983.
126. **Savarese, T. M.,** The Role of 5′-Deocy-5′-Methylthioadenosine Phosphorylase in the Metabolism and Action of 5′-Deoxyadenosine, 5′-Deoxy-5′-Methylthioadenosine and their Analogs in Murine Tumor Cells, Ph.D. thesis, Brown University, Providence, R.I., 1980.

Section D.
Ornithine and S-Adenosylmethionine Decarboxylases

Chapter 15

THE PHYSIOLOGICAL SIGNIFICANCE OF ORNITHINE DECARBOXYLASE

Diane Haddock Russell

TABLE OF CONTENTS

I. INTRODUCTION

Ornithine decarboxylase catalyzes the decarboxylation of ornithine, an amino acid not incorporated into proteins, to form the initial diamine in the polyamine biosynthetic pathway, diaminobutane (also referred to as putrescine, a glaring misnomer). The polyamines in turn are the organic cations of all mammalian tissues, and have specific as well general effects on many key reactions in the regulation of macromolecular synthesis, growth, and differentiation. A schematic of the polyamines pathway is illustrated in Figure 1.

A critical step in growth and differentiation is the ability of signals such as partial hepatectomy, hormones, and nutrients to rapidly elevate the activity of ornithine decarboxylase in a peak-function manner.[1-7]

Characteristically, ornithine decarboxylase has been shown to increase in receptor-containing target tissues within 1 to 2 hr of the administration of a polypeptide hormone.[8] The activity in response to a single bolus of a hormone reaches a maximal excursion rather predictably between 3 to 6 hr depending on the tissue and the hormone. Figure 2 illustrates the time course of ornithine decarboxylase activity in rat liver in response to growth hormone administration in both adult and weanling rats.[8] The response to the same dose of growth hormone in the adult is diminished, suggesting decreased receptor-mediated activity in the liver of adult rats. However, the time course of the ornithine decarboxylase response is identical in weanling and adult rats with a peak activity detectable at 4 hr and a rapid decline toward baseline. The basal activity of ornithine decarboxylase activity in rat liver is low to nondetectable and increases in ornithine decarboxylase activity in this organ in response to the administration of polypeptide trophic hormones are generally dependent on new protein synthesis as well as new DNA-dependent RNA synthesis. However, this issue becomes complex as a function of time after an inductive event or a series of such events, and will be discussed individually for hormones if appropriate studies have been conducted. The ornithine decarboxylase activity increment serves as an early biochemical marker of elevated protein and RNA synthesis in the specific target organ, and as such, is a valuable tool to probe the action of compounds which alter the growth response.

This review will deal with the evidence that ornithine decarboxylase activity is a rapidly turning over enzyme which reflects receptor-mediated activity for hormones with different membrane or intracellular modes of action. These include polypeptide and amine trophic hormones whose major second messenger is cyclic AMP and the activation of cyclic AMP-dependent protein kinases, steroid hormones, and hormones and growth factors that are internalized and may exert at least part of their action through the activation of protein kinase C. We also will summarize the evidence from the use of phorbol esters that ornithine decarboxylase may be a required event in the process of tumor promotion. It will not deal with the molecular biology of the ornithine decarboxylase gene, nor inhibitors and antizymes of ornithine decarboxylase, since these subjects will be addressed in other reviews in this volume.

II. EXTREMELY RAPID TURNOVER OF ORNITHINE DECARBOXYLASE

Ornithine decarboxylase has the shortest half-life, 10 to 15 min, of any reported eukaryotic enzyme.[9] Figure 3 shows the time course of the decline of ornithine decarboxylase activity in 4 hr regenerating rat liver ($t_{1/2}$ = 11.5 min), 24 hr regenerating rat liver ($t_{1/2}$ = 11.5 min), and normal rat liver ($t_{1/2}$ = 9.75 min) after the administration of 50 mg/kg of cycloheximide intraperitoneally. Table 1 demonstrates the relative half-life of some other mammalian enzymes. Studies of enzyme turnover rates based solely on the measurement of enzyme activity must be interpreted with caution, since drugs such as cycloheximide may block enzyme degradation as well as enzyme synthesis, and may produce effects unrelated

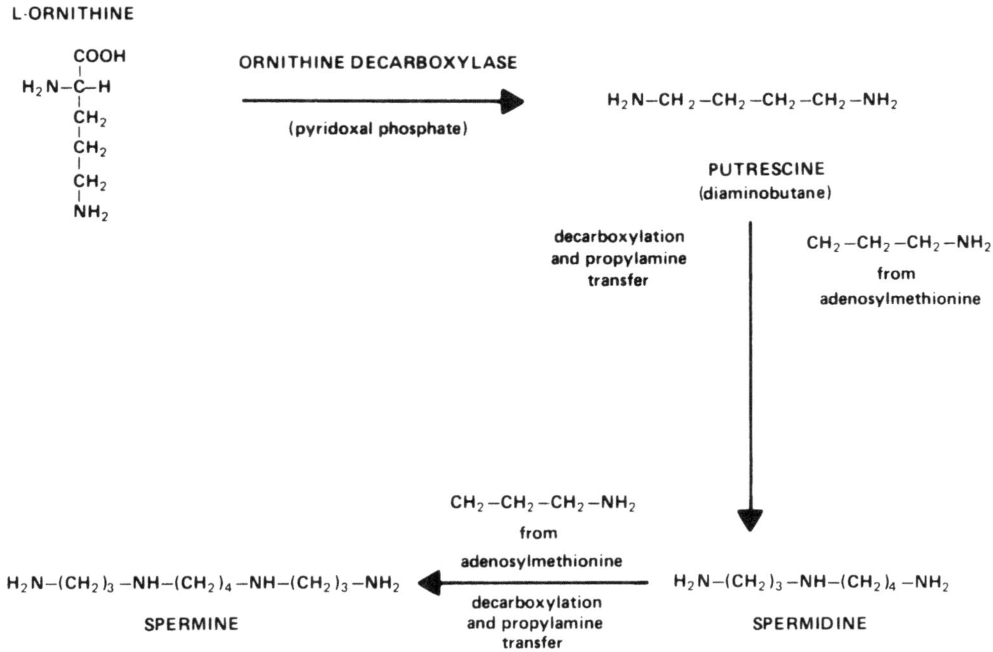

FIGURE 1. Schematic of polyamine biosynthesis in mammals. (From Russell, D. H. and Durie, B. G. M., *Polyamines as Biochemical Markers of Normal and Malignant Growth,* Raven Press, New York, 1978. With permission.)

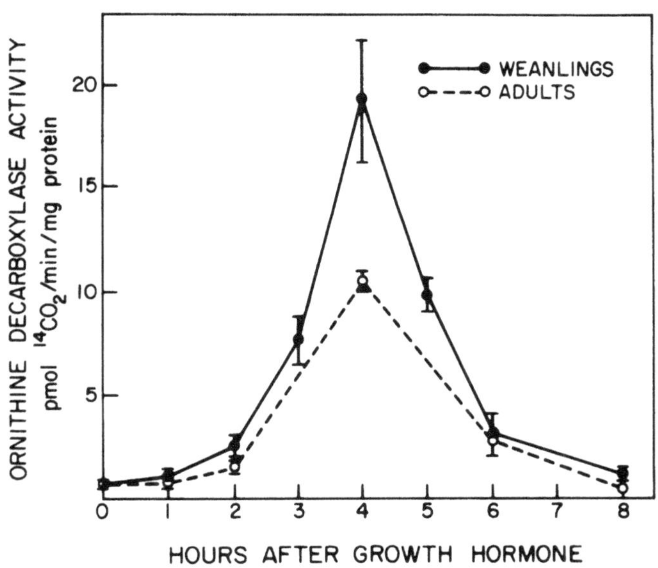

FIGURE 2. Time-course of ornithine decarboxylase activity enhancement after growth hormone administration in adult and weanling rats. All rats received 5-mg/kg doses of growth hormone and were killed at varying time intervals. Each point represents the mean ! SEM for five rats. (From Russell, D. H., Snyder, S. H., and Medina, V. J., *Endocrinology,* 86, 1414, 1970. With permission.)

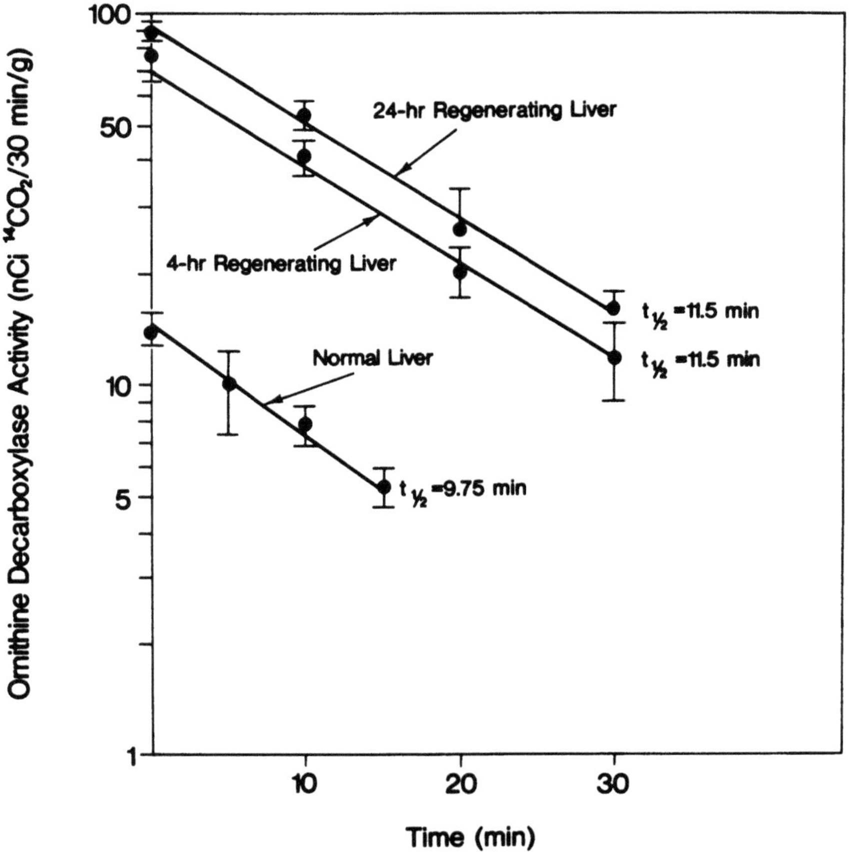

FIGURE 3. Time-course of the decline of ornithine decarboxylase activity in 4 hr regenerating rat liver, 24 hr regenerating rat liver, and normal rat liver after the administration of cycloheximide, 50 mg/kg, i.p. The lines were plotted by the least squares method. Each point represents the mean ! SEM for five animals. (From Russell, D. H. and Snyder, S. H., *Mol. Pharmacol.*, 5, 253, 1969. With permission.)

Table 1
HALF-LIVES OF SOME MAMMALIAN ENZYMES

Enzyme	$t_{1/2}$
Ornithine decarboxylase	11 min
δ-Aminolevulinic acid synthetase[10]	67—72 min
Tyrosine transaminase[11]	1.5 hr
Histidine decarboxylase[12]	1.8 hr
Ribonucleotide reductase[13] (conversion of CMP to dCMP)	2.0 hr
Tryptophan pyrrolase[14,15]	2.5 hr
Threonine dehydratase[16]	3.0 hr
Serine dehydratase[17]	5.2 hr
Catalase[18]	1—1.5 days
Glutamic-pyruvic transaminase[19]	3.5 days
Arginase[20]	4—5 days

From Russell, D. H. and Snyder, S. H., *Mol. Pharmacol.*, 5, 253, 1969.

to protein synthesis. Many authors have demonstrated the rapid half-life of ornithine decarboxylase by measuring the amount of immunoprecipitable protein[21,22] and, more recently, by determining the binding of [^{3}H]-difluoromethylornithine, a suicide substrate for the enzyme.[23-25] Russell and co-workers[8] also calculated the natural decline of ornithine decarboxylase after growth hormone induction from data similar to that presented in Figure 2.[8] Weanling rats received growth hormone and were killed at five intervals, between 5 1/2 and 6 hr after growth hormone administration. The half-life for the fall of enzyme activity in this time interval was about 24 min. The rationale for estimating enzyme turnover by measuring the decline of induced enzyme activity is that this decline occurs when all the inducer has been processed so that the time for enzymatic activity to fall to basal levels reflects the degradation or turnover rate. The longer half-life for ornithine decarboxylase that we found in such experiments may be due to a small amount of new enzyme synthesis resulting in a slower return to basal level than after the complete cessation of new enzyme synthesis produced by cycloheximide. The synthesis of most mammalian enzymes is a linear function of time whereas enzyme degradation is an exponential function of time. Therefore, rates of change of enzyme levels from one state to another are determined solely by the degradation rate of the enzyme.[19,26,27] The very high degradative rate of ornithine decarboxylase suggests that its activity changes rapidly in response to stimuli for altered synthesis, both in terms of its rapid accumulation and rapid decline. Since ornithine decarboxylase is the rate-limiting enzyme in the biosynthesis of the polyamines, this suggests that polyamine synthesis is a finely modulated process, a concept which has received considerable confirmation over the last 2 decades. This characteristic is also responsible for the uniqueness of ornithine decarboxylase activity as a biochemical marker of altered macromolecular synthesis.

III. PHYSIOLOGICAL REGULATION OF ORNITHINE DECARBOXYLASE BY ANTERIOR PITUITARY POLYPEPTIDE HORMONES AND AMINE TROPHIC HORMONES

Table 2 demonstrates the generality of ornithine decarboxylase induction by anterior pituitary polypeptide trophic hormones and amine trophic hormones in their specific target tissues. These hormones are coupled to ornithine decarboxylase induction, at least in part, by the elevation of the second messenger, cyclic AMP, and the activation of cyclic AMP-dependent protein kinases.[6,39,83] The majority of these studies also indicate that new protein and ribosomal RNA synthesis are part of the mechanism of the elevation of ornithine decarboxylase activity rather than posttranslational modification. An exception to the involvement of cyclic AMP as the major second messenger in the enhancement of ornithine decarboxylase activity by these hormones may be the case of growth hormone. An activation of cyclic AMP-dependent protein kinases without a detectable increase in the cyclic AMP concentration has been demonstrated in weanling rat liver in response to growth hormone, followed rapidly by ornithine decarboxylase induction.[39] This same activation of the cyclic AMP-dependent protein kinases in response to growth hormone is difficult to demonstrate in older rats and, therefore, there is the possibility that all the effects of growth hormone are not mediated through a cyclic AMP and cyclic AMP-dependent kinase activation pathway. Because recent studies have implicated protein kinase C activation as part of the pathway for ornithine decarboxylase induction in response to prolactin, a member of the growth hormone family, it is possible that protein kinase C also may be involved in transmembrane signaling in response to growth hormone.[84,85] The effects of certain amine trophic hormones, such as epinephrine and norepinephrine, on such tissues as the heart may also involve a cyclic AMP-mediated mechanism, as well as biochemical events involved with the endocytosis of the β-receptor, and perhaps membrane phosphoinositide turnover and protein kinase C activation.

Table 2
GENERALITY OF ORNITHINE DECARBOXYLASE INDUCTION BY ANTERIOR PITUITARY POLYPEPTIDE TROPHIC HORMONES AND AMINE TROPHIC HORMONES

Hormone	Target tissue(s)	Ref.
Growth hormone	Liver, adrenal, heart, kidney, thymus, developing brain	8, 28—47
ACTH	Adrenal	34, 43, 48, 49
LH, FSH, LHRH	Ovary, testis, CHO cells	50—62
TSH	Thyroid	63—69
β-Endorphin	Kidney	70
α-MSH	Melanoma cells	71
Catecholamines	Heart	72—76
	Parotid	77—80
	Granulosa cells	81
	Glioma cells	82

Note: ACTH = adrenocorticotropin, LH = leuteinizing hormone, FSH = follicle-stimulating hormone, LHRH = leuteinizing hormone releasing factor, and TSH = thyrotropin.

IV. CYCLIC AMP AND CYCLIC AMP-DEPENDENT PROTEIN KINASES CONSTITUTE ONE PHYSIOLOGICAL ROUTE FOR ORNITHINE DECARBOXYLASE INDUCTION

There is now substantial evidence that a major route of induction of ornithine decarboxylase is via a cyclic AMP-mediated series of events. In a variety of growth systems investigated, many of which are summarized in Table 2, ornithine decarboxylase induction occurs after the rise in cyclic AMP and the activation of cyclic AMP-dependent protein kinases. In lymphocytes which contain small concentrations of type I and type II cyclic AMP-dependent protein kinase, the ability to induce ornithine decarboxylase in response to mitogenic stimuli is dependent on an early specific activation of only type I protein kinase.[86] Table 3 summarizes the initial studies demonstrating ornithine decarboxylase induction in response to cyclic AMP analogues and/or phosphodiesterase inhibitors, and Table 4 references the studies in which cyclic AMP-dependent protein kinases have been shown to be activated prior to the induction of ornithine decarboxylase.

Since much of the literature which deals with the role of cyclic AMP in the growth processes suggest that cyclic AMP is inhibitory to growth,[123-125] it is important to discuss how the cyclic AMP mediation of ornithine decarboxylase activity might relate to the growth process. Cyclic AMP probably drives hypertrophy or G_1 of cell cycle and not hyperplasia even though these processes may be coupled. Furthermore, one of the clear-cut roles of cyclic AMP as a second messenger is related to the demonstration by Robison and co-workers[126,127] that cyclic AMP could rapidly alter metabolism by phosphorylation of enzymes regulating glycogen synthesis in glycogenolysis. Thus, the generation of an increased energy supply within a cell provides the necessary requisites for the cell to increase RNA and protein synthesis, again events which initiate in G_1 of cell cycle. In fact, when cyclic AMP is not allowed to fall naturally at the G_1/S border in CHO cells, there is an alteration in the pattern of expression of the cyclic AMP-dependent protein kinases with a decrease in type II kinase and a severalfold increase of type I kinase that results in an arrest of the cells in G_1 prior

Table 3
ORNITHINE DECARBOXYLASE INDUCTION IN RESPONSE TO CYCLIC AMP ANALOGUES AND/ OR PHOSPHODIESTERASE INHIBITORS

Tissue	Ref.
Adrenal cortex	48, 49, 87
Adrenal medulla	87, 88
Kidney	87
Liver	21, 89—94
Mammary tissue	95, 96
Quail oviduct	97
Testis	53
Ventral prostate	98—100
Adrenocortical tumor cell clones	101
Astrocytoma	102
BHK	103, 104
CHO cells	105—108
Chondrocytes	109
Fibroblasts	110
Glioma and neuroblastoma	82
L-cell	111
Macrophage J774.1	112
Reuber H35 hepatoma	113

Table 4
CYCLIC-AMP-DEPENDENT PROTEIN KINASE ACTIVATION PRECEDES ORNITHINE DECARBOXYLASE INDUCTION

Stimulus	Tissue	Ref.
Chlorea toxin	Thyroid	68
Cold exposure reserpine carbamylcholine	Adrenal medulla	88, 114
Compensatory regeneration	Liver, adrenal	115
PHA, Con A	Lymphocytes	86, 116
3-Methylcholanthrene, phenobarbital	Liver	117
Biliverdin	Liver	118
PGE_1	Neuroblastoma cells	119
Glucagon	Liver	120
G_1 progression		121, 122

to DNA synthesis.[122] For a more detailed discussion of the role of cyclic AMP and cyclic AMP-dependent protein kinases in the regulation of ornithine decarboxylase, see References 123 to 125 and 128.

Since ornithine decarboxylase genetic activity is regulated by cyclic AMP, at least one gene in this multigene family may contain a cyclic-AMP-responsive promoter sequence. Recent studies suggest that the responsiveness of eukaryotic genes to cyclic AMP involves a highly conserved "core" promoter sequence, a palindrome 5′-TGACGTCA-3′ that is present in a number of cyclic AMP-response genes.[129] Since the hypothesis of Kuo and Greengard[130] that all the biochemical and physiological effects of cyclic AMP are mediated by regulation of the activity of cyclic AMP-dependent protein kinase, the specificity of the

Table 5
ALTERATION OF ORNITHINE DECARBOXYLASE ACTIVITY BY ANDROGENS AND ESTROGENS

Organ(s)	Hormone(s)	Magnitude of ODC increase	Ref.
Ventral prostate of castrated rats	Testosterone propionate	3-fold	131
Oviduct of immature chicks	Estradiol-17β, diethylstilbestrol	55-fold	132
Uterus of castrated or hypophysectomized rats	Estradiol-17β	10- to 66-fold	132
Uterus of immature rats	Estradiol-17β	14- to 25-fold	133
Liver of adrenalectomized rats	Testosterone	2-fold	30
Uterus of castrated rats	Estradiol-17β	2-fold	134
Uterus of immature rats	Estradiol-17β	100-fold	135
Uterus of adrenalectomized and ovariectomized rats	Estradiol-17β	12-fold	135
Liver of male *Xenopus laevis*	Estradiol	44-fold	136
MNR1 male mouse kidney, intact or castrated	Testosterone propionate	100- to 20,000-fold	137
Baby hamster kidney cells (BHK-21)	Diethylstilbestrol > estradiol-17β > testosterone	50—100%	138
L929 mouse thigh muscle cells	Diethylstilbestrol	50—100%	138
F344/CV male rat kidney after castration	Estradiol-17β > diethylstilbestrol > estriol > estrone	11-fold	139
Male Syrian gold hamster kidney after castraction	Estradiol-17β	300%	140
Nylar male rat kidney	Testosterone propionate	2000-fold	141, 142
MNR1 male mouse kidney, intact or castrated	Testosterone propionate	100- to 20,000-fold	143
Ventral prostate of aged AXC rats	Testosterone propionate	3-fold	144
Nelson Collins Swiss male mouse kidney after castration	5-α-Dihydrotestosterone	600-fold	145
Achyla ambisexualis mycelium	Antheridiol	5-fold	146
Female DBA/2J mouse kidney	Testosterone propionate	1000-fold	147
Female BALB/c mouse kidney	Testosterone propionate	400-fold	148

response would be produced by the nature and localization of the endogenous protein substrate(s) phosphorylated in the responsive cell type as well as by the types of cyclic AMP-dependent protein kinase present and activated by the nucleotide. This conserved core may be aligned with specific protein substrates for phosphorylation which would confer the cyclic AMP-responsiveness pattern.

V. TROPHIC ASPECTS OF STEROID HORMONES EXPRESSED THROUGH THE INDUCTION OF ORNITHINE DECARBOXYLASE

Steroid hormones have trophic effects during the development of steroid-specific tissues and also after the administration of steroids to steroid-depleted animals. Steroid hormones are thought to exert most of their effects by interaction with specific high affinity receptors in target organs. Some evidence now suggests that this occurs in the nucleus to directly alter genetic transcription. Elevated macromolecular synthesis in response to steroid hormones has been coupled in a dose-dependent manner to an early, significant elevation of ornithine decarboxylase in a target organ. Table 5 summarizes the alterations of ornithine decarboxylase activity in response to androgens and estrogens. Pegg and Williams-Ashman[131] were the first to report an elevation in ornithine decarboxylase activity in response to a steroid hormone, testosterone. They found a relatively high basal activity of ornithine decarboxylase in the rat ventral prostate which declined markedly after castration. Within 6 hr after the injection of testosterone propionate to castrated rats, ornithine decarboxylase activity was increased significantly and it was nearly threefold above castrated controls within 24 hr. A daily injection of the androgen restored the ornithine decarboxylase activity to the level of intact controls. Uterine ornithine decarboxylase activity also can be regulated by estrogens. Russell and Taylor[134] demonstrated a significant elevation of ornithine decarboxylase activity in the uterus of ovariectomized female rats in response to estradiol-17β administration. A 14- to 25-fold increase in the specific activity of ornithine decarboxylase in the uterus occurred in response to estradiol administration to immature rats.[133] If estrogen is administered to immature chicks, a 30- to 50-fold increase in ornithine decarboxylase activity occurred in oviduct homogenates within a 4-hr period. Progesterone also has been implicated in the regulation of ornithine decarboxylase activity during the process of decidualization of the endometrium[149] and in the regulation of uterine ornithine decarboxylase activity during the early postimplantation events in the pregnant hamster.[150]

There is no direct evidence, at this time, that calcium fluxes and alterations in the intracellular concentration of cyclic AMP are involved in the response of target tissues to steroid hormones, although there is considerable evidence that the intracellular concentration of proteins required for the action of second messengers such as type I and type II cyclic AMP-independent protein kinases, may be regulated by steroids. For instance, in rat ventral prostate and in another steroid-sensitive tissue, the levator anus muscle, castration results in a rapid decrease in the specific concentration of type I cyclic AMP-dependent protein kinase in both tissues.[151] The amount of type I cyclic AMP-dependent protein kinase present in the ventral prostate could be restored to normal by daily injections of dihydrotestosterone. Therefore, it is likely that steroid hormone action may prove necessary for maintenance of discrete pools of proteins involved in the regulation of growth and differentiated function.

The regulation of ornithine decarboxylase by glucocorticoids is much less straightforward, with glucocorticoids having either a positive or negative effect on ornithine decarboxylase activity (Table 6). As a generality, hormones such as hydrocortisone and dexamethasone elevate ornithine decarboxylase in liver and kidney and tend to inhibit ornithine decarboxylase in other tissues such as thymus and spleen. Complex interactions are known to occur between glucocorticoids, thyroid hormones, and prolactin.[159-161] In fact, β-adrenoreceptor number and β-adrenergic response in the heart is altered by the serum prolactin level.[159] Therefore, the interaction of many hormones to affect serum levels, receptor number, and receptor responsitivity of tissues to a specific hormone now appears to be a common feature in the physiological regulation of function, growth, and development.

VI. INTERNALIZED POLYPEPTIDE HORMONES AND GROWTH FACTORS

Ornithine decarboxylase activity is elevated in target tissues by hormones and growth factors that do not appear to involve cyclic AMP as a second messenger (Table 7). The single clarifying factor for these substances appears to be their internalization into target tissues. Although portions of their mechanism of action are known particularly for insulin, there is not a common second messenger that has been elucidated. In the case of prolactin, it appears that in both mammary gland and Nb 2 mode lymphoma cells, inositol lipid metabolism and protein kinase C activation are involved. The breakdown of inositol lipids with the generation of inositol triphosphate as a calcium mobilizing second messenger has gained considerable support.[197-199] Interestingly, tumor-promoting phorbol esters induce 100- to 1000-fold elevation of ornithine decarboxylase activity in carcinogen-initiated skin, and these esters drive cell proliferation and/or cell differentiation in a variety of different cell types.[200,201] Active phorbol esters are now thought to be analogues of diacylglycerol, an

Table 6
ALTERATION OF ORNITHINE DECARBOXYLASE ACTIVITY BY GLUCOCORTICOIDS

Organ(s)	Hormone(s)	Magnitude of ODC change	Ref.
Rat liver after adrenalectomy	Hydrocortisone	60-fold	30
Female NMR1 mouse kidney, intact and after adrenalectomy	Cortisone acetate	2-fold	152
Female rat liver	Cortisone	6-fold	153
	Dexamethasone	2-fold	153
Female rat thymus	Cortisone	31%	153
	Dexamethasone	0% of control	153
Female rat kidney	Dexamethasone	165% of control	153
	Dexamethasone	30-fold	153
Female rat spleen	Dexamethasone	32% of control	153
Mouse placenta	Dexamethasone	50% of control	154
Bone matrix-induction of ODC in thoracic region of rats	Hydrocortisone	70% of control	155
	Corticosterone	63% of control	155
	Prednisolone	55% of control	155
	Triamcinolone	42% of control	155
	Dexamethasone	17% of control	155
	Betamethasone	15% of control	155
	Flucinolone	3% of control	155
Baby hamster kidney cells (BHK-21)	Progesterone	50—100%	138
L929 mouse thigh muscle cells	Progesterone	50—100%	138
S49 lymphoma cells	Dexamethasone	40% of control	156
Female rat kidney	Dexamethasone	43-fold	157
Liver of adrenalectomized rat	Corticosterone	3-fold	158
	Dexamethasone	2-fold	158
Brain of adrenalectomized rat	Dexamethasone		
Hippocampus and cortex		3-fold	158
Cerebellum		8-fold	158
Brain of adrenalectomized rat	Corticosterone		
Hippocampus and cortex		3-fold	158
Cerebellum		6-fold	158

endogenously generated compound, which activates protein kinase C.[202] Although protein kinase C requires calcium for activation, the presence of diacylglycerol or the phorbol ester TPA dramatically increases the affinity of the enzyme for calcium to less than 10^{-7} *M* allowing the activation of protein kinase C to occur without the need for any detectable mobilization of calcium. Although protein kinase C is not known to be a second messenger for insulin, phorbol esters modulate insulin receptor phosphorylation and insulin action in hepatoma cells, cause similar increases in phosphofructokinase activity, and both elevate the phosphorylation of S6 ribosomal protein in a number of cell lines.[203-205] In addition to these relationships between insulin action and protein kinase C, epidermal growth factor (EGF) receptor is a substrate for protein kinase C. Phosphorylation of the EGF receptor occurs on serine and threonine residues and results in the inhibition of EGF-stimulated tyrosine phosphorylation.[206] Other investigators have demonstrated a positive relationship between protein kinase C activity and EGF receptors in a number of human mammary tumor cell lines.[207] In Nb 2 node lymphoma cells that are prolactin dependent for mitogenesis, the addition of the active tumor promoter, TPA, results in about 25% of the amount of proliferation elicited by an optimal concentration of prolactin.[85] In these cells, the addition of cyclic AMP analogue is inhibitory to both ornithine decarboxylase induction and proliferation.

Table 7
EFFECT OF HORMONES AND GROWTH FACTORS KNOWN TO BE INTERNALIZED ON ORNITHINE DECARBOXYLASE ACTIVITY

Organ(s) or cell line	Hormone	Magnitude of ODC Increase	Ref.
Rat liver after adrenalectomy	Insulin	26-fold	30
Mouse L-cells	Insulin	1.5-fold	30
Mouse mammary gland organ culture	Insulin	2-fold	95
Baby hamster kidney cells	Insulin	3-fold	103
HTC cells	Insulin	2- to 10-fold	162
Chick embryo fibroblasts	Insulin	4- to 5-fold	163
Baby hamster kidney cells	Insulin	4-fold	164
Rat small intestine (starved)	Insulin	6- to 7-fold	165
Diabetic rat liver	Insulin	3-fold	166
Diabetic rat liver, muscle, heart, thymus	Insulin	2- to 9-fold	167
Chick embryo brain cells in culture	Insulin	2.7-fold	168
Rat brain	Insulin	4-fold	169
	Insulin	8-fold	170
Chick pelvic cartridge	Insulin	1.8-fold	171
Liver of pancreatectomized rat	Insulin	2-fold	172
Rat liver after adrenalectomy	T_4	9-fold	30
Tadpole liver	T_4	25-fold	173
Rat heart	T_3	2-fold	174
Rat liver	T_3	6-fold	175
2-day-old rat heart	T_3	45%	174
Embryonic chick pelvic cartridge	T_3	2-fold	171
Thyroidectomized male rat liver	T_3	7-fold	176
Rat liver, kidney, adrenal, spleen, thymus, heart	Prolactin	2- to 60-fold	177 178 179
Rat liver, kidney, adrenal, thymus, and spleen in hypophysectomized rats	Prolactin	5- to 950-fold	180 181
Primary cultures of DMBA-induced rat mammary tumors	Prolactin	2-fold	182
Nb 2 node lymphoma cells	Prolactin	220-fold	183 84
Chick embryo epidermis	EGF	50-fold	184
Neonatal mouse testes, kidney, liver	EGF	2- to 18-fold	185
Mouse mammary gland organ culture	EGF	2- to 3-fold	95
Mouse skin	EGF	2-fold	186
Human skin fibroblasts	EGF	6-fold	186
Mouse digestive tract	EGF	2- to 5-fold	187
Porcine granulosa cells (serum-free)	EGF	133% of control	188
	EGF + MSA	2.5-fold	188
Rat adrenal pheochromocytoma (PC12)	EGF	30-fold	189
Human hepatoma cell line PLC/PRF 15	EGF	2- to 6-fold	190
Rat adrenal pheochromocytoma (PC12) cells	NGF	13- to 30-fold	191
	NGF	55-fold	192
Rat brain	NGF	15- to 19-fold	193
Rat brain after adrenalectomy	NGF	6-fold	193
Rat brain	NGF	15- to 19-fold	194
Rat brain, liver, kidney, adrenal	NGF	4- to 25-fold	195
Rat superior cervical ganglia	NGF	4-fold	169
Rat adrenal pheochromocytoma (PC12) cells	NGF	30-fold	189
	NGF	64-fold	189

Table 7 (continued)
EFFECT OF HORMONES AND GROWTH FACTORS KNOWN TO BE INTERNALIZED ON ORNITHINE DECARBOXYLASE ACTIVITY

Organ(s) or cell line	Hormone	Magnitude of ODC Increase	Ref.
Chick embryo fibroblasts	MSA	7-fold	196
Porcine granulosa cells	MSA	2-fold	188
Embryonic chick pelvic cartridge	MSA	1.8-fold	171

Note: MSA = multiplication stimulating activity and NGF = nerve growth factor.

To date, protein kinase C activation is coupled to ornithine decarboxylase activity and cellular proliferation in a variety of tissues and seems likely to play a role as a "mitogenic" kinase. There are many unanswered questions, and this area of research is moving very rapidly. It is also unknown how protein kinase C might alter genetic transcription processes. However, phosphorylation of topoisomerase II by protein kinase C has been reported, and results in the activation of topoisomerase suggesting that this enzyme may be a target for the regulation of nuclear events by protein kinase C.[208]

Transglutaminase may be important in the internalization process for at least some of these hormones and growth factors (see Reference 209 for a more-detailed discussion). The translocation of physiologically active hormones to important regulatory organelles in the cell, such as the golgi and the nucleus, may be important to the long-term action of these hormones on ornithine decarboxylase activity and on RNA and DNA synthesis.[210] Hormones and pyhysiologically active molecules that may be taken up by this mechanism include insulin, prolactin, growth hormone, epidermal growth factor, nerve growth factor, triiodothyronine, low density lipoprotein, and α_2-macroglobulin.[211] That catecholamines may also be internalized is suggested by the rapid down-regulation of β-receptors in erythrocytes in response to catecholamine administration, and the ability to inhibit this down-regulation process by the addition of competitive inhibitors of transglutaminase such as dansyl cadaverine and methylamine.[212]

Further studies of transmembrane signaling and nuclear events that occur in response to these hormones and growth factors will likely lead to important information related to our understanding of normal and neoplastic growth processes. These events are likely also to be involved in differentiation, since, in certain systems, TPA administration results in differentiation, not proliferation. It is also not known, at this time, whether the internalization process plays any physiological role other than as a degradative pathway.

VII. ORNITHINE DECARBOXYLASE INDUCTION RELATED TO EXTENT OF TUMOR PROMOTION BY TUMOR-PROMOTING PHORBOL ESTERS AND RELATED COMPOUNDS.

Ornithine decarboxylase induction has been shown to be a consistent event in response to tumor promoters such as TPA, a diterpine compound (Table 8). O'Brien and co-workers[214] first linked TPA-stimulated ornithine decarboxylase activity in mouse skin to tumor promotion. In mouse skin, TPA application directly to skin resulted in a 230-fold elevation of ornithine decarboxylase within 4 hr.[213] Induction was dependent on protein synthesis but not RNA synthesis since it was blocked by cycloheximide pretreatment but not affected by actinomycin D pretreatment. The half-life of ornithine decarboxylase after TPA induction in skin was 17 min, similar to that first reported by Russell and Snyder[9] for ornithine

Table 8
ORNITHINE DECARBOXYLASE INDUCTION BY TUMOR PROMOTERS

Organ or cell line	Extent of ODC elevation	Ref.
TPA application to mouse skin	200- to 400-fold	213 214 215 216
TPA addition to mouse epidermal cells	10-fold	217
Hexa-analogue	10-fold	217
Phorbol dibenzoate	10-fold	217
Phorbol dibutyrate	3-fold	217
TPA addition to hamster embryo cells, 3T3 mouse cells, and HE68BP cells	7- to 10-fold	218 219
Mouse skin after HHPA (10-*O*-hexadecanoyl-16-hydroxyphorbol-13-acetate) and 4-deoxyHHPA application	60- to 110-fold	220
TPA application to rat skin	4-fold	221
TPA administration on rat brain, liver, and lung and mouse liver	8- to 250-fold	222 223
Rat liver at i.p. injection of TPA	5-fold	224
Teleocidin B application to mouse skin	300-fold	225
TPA addition to bladder cancer cell cultures	39-fold	226
Rat liver after topical TPA	30-fold	227
TPA on proliferating basal cells of mouse epidermis	10- to 25-fold	228
Mouse skin explants in culture in response to TPA	40-fold	229
TPA addition to Reuber H35 hepatoma cells	10- to 33-fold	230
TPA addition to Chinese hamster ovary cells	13-fold	231
TPA administration to mouse embryo fibroblasts	2-fold	232
Swiss 3T3 cells after TPA administration	7-fold	233
Mouse liver after TPA administration	9- to 30-fold	234
Multiple applications of TPA to mouse skin	6000-fold after fifth application	235

decarboxylase in regenerating rat liver. Cordycepin (3′-deoxyadenosine), a compound that partially blocks mRNA synthesis, partially blocked the elevation in ornithine decarboxylase activity. The increase was not dependent on elevations of either cyclic AMP or cyclic GMP.[216] Both phospholipase and lipoxygenase activities have been implicated as intermediates in the ability of TPA to induce ornithine decarboxylase in mouse skin.[236-239] Pretreatment with colchicine or other microtubule-disrupting agents[240] also inhibited the ability of TPA to induce ornithine decarboxylase in mouse epidermis, suggesting that an intact microtubular structure is required for TPA-stimulated induction of ornithine decarboxylase. In mouse skin explants maintained in serum-free Eagle's HeLa cell medium, chelation of extracellular calcium by EGTA prevented the ornithine decarboxylase induction by TPA.[229] Addition of calcium to the medium, but not magnesium, restored the induction. Trifluoperazine, a calmodulin inhibitor, also was able to inhibit ornithine decarboxylase induction in the presence of TPA.

Further evidence that ornithine decarboxylase induction was relevant to tumor promotion was suggested by studies of a series of phorbol esters, with different abilities to promote tumor formation. Their relative tumor-promoting capacity was directly related to the magnitude of the ornithine decarboxylase induction.[241] Other weaker tumor promoters, such as Tween 60, anthraline, and iodoacetic acid, also induced ornithine decarboxylase activity but at doses 100- to 1000-fold greater than TPA.

Multiple applications of TPA to mouse skin resulted in ornithine decarboxylase induction after each application, with similar kinetic parameters. However, after eight applications, the induction reached a level 600-fold above controls.[241] Boutwell et al.[243] hypothesized that ornithine decarboxylase induction was specifically associated with tumor promotion. They maintained that the demonstrated induction of ornithine decarboxylase activity by carcinogens, such as DMBA application to mouse skin, was related, not to their initiation properties, but to their promoter capacity. Promoters stimulate growth-related events such as increased protein, RNA, and DNA synthesis and decrease terminal cell differentiation events. A long-term study of dietary phenobarbital as a promoter of diethylnitrosamine-induced hepatic carcinogenesis did not demonstrate elevation of ornithine decarboxylase activity monitored at 1, 11, and 20 weeks after a 500-ppm phenobarbital diet.[242] However, evidence that dietary phenobarbital significantly increased the number of neoplastic nodules or hepatic cellular carcinomas was not presented either. It is possible that the timing of monitoring ornithine decarboxylase activity related to tumorlike phenotypic expression was premature. Olson and Russell[241] found that dietary 2-AAF after diethylnitrosamine initiation of liver carcinogenesis also did not result in an early elevation of ornithine decarboxylase activity. Other studies which do not support the correlation of ornithine decarboxylase induction to tumor promotion are difficult to interpret.[244,245]

The correlation of the potency of phorbol esters in promoting tumor formation, as opposed to their ability to induce ornithine decarboxylase and to stimulated DNA synthesis, has been shown to be dependent upon the dose of phorbol ester.[217] For instance, the addition of 0.1 μg/mℓ TPA to epidermal cells resulted in the same extent of induction of ornithine decarboxylase as a tenfold higher amount of TPA, 1.0 μg/mℓ. At the lower TPA concentration, there was a consistent relationship between the ornithine decarboxylase induction and subsequent DNA synthesis and tumor promotion. However, at high concentrations, the relationship was less clear, and certain phorbol esters and hyperplastic agents without promoting ability resulted in small increases of ornithine decarboxylase activity. Boutwell[246,247] suggested that tumor promotion itself might consist of more than one stage. Since that time, Slaga and co-workers[248-251] have provided evidence that tumor promotion can be divided into at least two stages, designated stage I and stage II. They suggested that promotion is reversible in stage I but later becomes irreversible. Limited applications of TPA act as stage I promoters, whereas mezerein, an analogue of TPA, is a weak stage I promoter but an effective stage II agent when given after stage I promotion. The characteristic, apparently common to stage I promoters, is the ability to stimulate altered function and proliferation of keratinocytes.[252-254] This selective ability to stimulate a particular type of cell may be characteristic of stage I promotion. However, stage II promotion now appears related to polyamine accumulation, particularly that of putrescine, since stage II promoters such as mezerein can be blocked by the administration of a suicide inhibitor of ornithine decarboxylase, α-difluoromethylornithine (DFMO). This results in an inhibition of ornithine decarboxylase activity, decreased putrescine accumulation, and marked reduction in papilloma size and number, followed by a dramatic reduction in tumor incidence. It is important to note that putrescine concentrations return to normal within 24 hr after a single application of TPA, and that putrescine significantly enhanced tumor yield when given with TPA whereas putrescine alone did not affect epidermal ornithine decarboxylase or result in papilloma formation. Therefore, ornithine decarboxylase activity marks both stage I and stage II promoter events. Blocking of the stage-I-specific ornithine decarboxylase activity blocks tumor promotion and limits tumor expression whereas blocking ornithine decarboxylase in stage II selectively can be overcome by the addition of putrescine. Further, the tumor promoter potency of TPA can be markedly enhanced by the simultaneous administration of putrescine. That both ornithine decarboxylase expression and the capacity for marked putrescine accumulation may be characteristic of the tumor phenotype expression now has been carefully demonstrated both in response to oncogenic viral transformation and after tumor development in response to sequential initiation and promotion events.

There is no doubt that tumor promoters, particularly TPA, are the most powerful inducers of ornithine decarboxylase reported to date. Although carcinogens elevate ornithine decarboxylase, this elevation has been attributed to their capacity as promoters and not to their initiation properties. The ability to block ornithine decarboxylase induction of initiated mouse skin in response to tumor promoters such as TPA is highly correlated with inhibition of tumor formation. This is discussed in more detail in relation to the effect of vitamin A analogues on ornithine decarboxylase activity in a recent review.[209] Retinoid analogues have been widely studied as blockers of TPA induction of ornithine decarboxylase activity as well as for their growth inhibitory properties.

In the case of the tumor-promoter studies referenced above, TPA is likely to exert, at least part of its action, to induce ornithine decarboxylase by the direct activation of protein kinase C. Tumor promoters and growth factors were first shown to act synergistically in mitogen-activated bovine lymphocytes.[255] We have already discussed the similarity between the structure of diacylglycerol and TPA, both presumably capable of activating protein kinase C. At least in the case of prolactin stimulation of mitogenesis in the Nb 2 node lymphoma cells, a calcium signal pathway coupled to the generation of IP_3, protein kinase C activation and the subsequent increase in ornithine decarboxylase has been demonstrated, the magnitude of the events related to the subsequent proliferation of these cells.[84,85] However, in the prolactin-stimulated Nb 2 cells, the extent of ornithine decarboxylase excursion in G_1 of cell cycle can be uncoupled to some degree by certain inhibitors from the extent of proliferation, suggestive of the importance of modulatory events at the G_1/S border.[85]

VIII. CONCLUDING STATEMENTS

Although this review certainly has not covered all of the studies implicating ornithine decarboxylase as an important event marking the endogenous or exogenous regulation of growth and development, it does underline the importance of ornithine decarboxylase activity as an index of macromolecular synthesis. Studies of embryological and fetal development have not been discussed in spite of the fact that these studies corroborate the conclusions of this review. Studies utilizing the suicide substrate inhibitor of ornithine decarboxylase, difluoromethylornithine, will be discussed in another review in this volume, and again, suggest the importance of ornithine decarboxylase and polyamine biosynthesis in the regulation of both normal and neoplastic growth. Monitoring ornithine decarboxylase activity can be important in elucidating the effects of biological response modifiers on particular target tissues, and also has proven efficacious in defining the mechanisms of action of toxic chemicals. Further studies of the relationship of ornithine decarboxylase expression to protooncogene expression and to the expression of cell division cycle genes may assist in further definition of key regulatory points in cell differentiation versus cell proliferation. Most of the scientists working on various aspects of ornithine decarboxylase expression and polyamine biosynthesis would still agree that the physiological importance of these compounds, and their biosynthetic pathways, merit a greater research effort as well as a more serious evaluation of their significance in textbooks and college courses related to physiology, pharmacology, biochemistry, and molecular biology.

REFERENCES

1. **Russell, D. H. and Snyder, S. H.,** Amine synthesis in rapidly growing tissues: ornithine decarboxylase activity in regenerating rat liver, chick embryo and various tumors, *Proc. Natl. Acad. Sci. U.S.A.*, 60, 1420, 1968.
2. **Jänne, J. and Raina, A.,** Stimulation of spermidine synthesis in the regenerating rat liver: relation to increased ornithine decarboxylase activity, *Acta Chem. Scand.*, 22, 1349, 1968.
3. **Cohen, S. S.,** *Introduction to the Polyamines,* Prentice-Hall, Englewood Cliffs, N.J., 1971.
4. **Bachrach, U.,** *Function of Naturally Occurring Polyamines,* Academic Press, New York, 1973.
5. **Russell, D. H.,** Polyamines on growth — normal and neoplastic, in *Polyamines in Normal and Neoplastic Growth,* Russell, D. H., Ed., Raven Press, New York, 1973, 1.
6. **Russell, D. H. and Durie, B. G. M.,** *Polyamines as Markers of Normal and Malignant Growth,* Raven Press, New York, 1978.
7. **Russell, D. H.,** Putrescine and spermidine biosynthesis in the development of normal and anucleolate mutants of *Xenopus laevis, Proc. Natl. Acad. Sci. U.S.A.*, 68, 523, 1971.
8. **Russell, D. H., Snyder, S. H., and Medina, V. J.,** Growth hormone induction of ornithine decarboxylase in rat liver, *Endocrinology,* 86, 1414, 1970.
9. **Russell, D. H. and Snyder, S. H.,** Amine synthesis in regenerating rat liver: extremely rapid turnover of ornithine decarboxylase, *Mol. Pharmacol.*, 5, 253, 1969.
10. **Marver, H. S., Collins, A., Tschudy, D. P., and Recheigl, M., Jr.,** α-Aminolevulinic acid synthetase. II. Induction in rat liver, *J. Biol. Chem.*, 241, 4323, 1966.
11. **Kennedy, F. T.,** Turnover of rat liver tyrosine transaminase stabilization after inhibition of protein synthesis, *Science,* 156, 525, 1967.
12. **Snyder, S. H. and Epps, L.,** Regulation of histidine decarboxylase in rat stomach by gastrin: the effect of inhibitors of protein synthesis, *Mol. Pharmacol.*, 4, 187, 1968.
13. **Turner, M. K., Abrams, R., and Lieberman, I.,** Levels of ribonucleotide reductase activity during the division cycle of the L cell, *J. Biol. Chem.*, 243, 3725, 1968.
14. **Goldstein, L., Stella, E. J., and Knox, W. E.,** The effect of hydrocortisone on tyrosine-α-ketoglutarate transaminase and tryptophan pyrrolase activities in the isolated, perfused rat liver, *J. Biol. Chem.*, 237, 1723, 1965.
15. **Schimke, R. T., Sweeney, E. W., and Berlin, C. M.,** The roles of synthesis and degradation in the control of rat liver tryptophan pyrrolase, *J. Biol. Chem.*, 240, 322, 1965.
16. **Pitot, H. C. and Peraino, C.,** Studies on induction and repression of enzymes in rat liver. I. Induction of threonine dehydrase and ornithine-delta-transaminase by oral intubation of casein hydrolysate, *J. Biol. Chem.*, 239, 1783, 1964.
17. **Jost, J.-P., Khairallah, E. A., and Pitot, H. C.,** Studies on the induction and repression of enzymes in rat liver. V. Regulation of the rate of synthesis and degradation of serine dehydratase by dietary amino acids and glucose, *J. Biol. Chem.*, 243, 3057, 1968.
18. **Price, V. E., Sterling, W. R., Tarantola, V. A., Hartley, R. W., Jr., and Recheigl, M., Jr.,** The kinetics of catalase synthesis and destruction in vivo, *J. Biol. Chem.*, 237, 3468, 1962.
19. **Segal, H. L. and Kim, Y. S.,** Environmental control of enzyme synthesis and degradation, *J. Cell. Comp. Physiol.*, 66 (Suppl. 1), 11, 1965.
20. **Schimke, R. T.,** The importance of both synthesis and degradation in the control of arginase levels in rat liver, *J. Biol. Chem.*, 239, 3808, 1964.
21. **Canellakis, Z. N. and Theoharides, T. C.,** Stimulation of ornithine decarboxylase synthesis and its control by polyamines in regenerating rat liver and cultured rat hepatoma cells, *J. Biol. Chem.*, 251, 4436, 1976.
22. **Hölttä, E.,** Immunochemical demonstration of increased accumulation of ornithine decarboxylase in rat liver after partial hepatectomy and growth hormone induction, *Biochim. Biophys. Acta,* 45, 301, 1975.
23. **Pösö, H. and Pegg, A. E.,** Measurement of the amount of ornithine decarboxylase in *Saccharomyces cerevisiae* and *Saccharomyces uvarum* by using α-[5-^{14}C]-difluoromethylornithine, *Biochim. Biophys. Acta,* 747, 209, 1983.
24. **Persson, L., Oredsson, S. M., Anehus, S., and Heby, O.,** Ornithine decarboxylase inhibitors increase the cellular content of the enzyme: implications for translational regulation, *Biochem. Biophys. Res. Commun.*, 131, 239, 1985.
25. **Pegg, A. E., Seeley, J., and Zagon, I. S.,** Autoradiographic identification of ornithine decarboxylase in mouse kidney by means of α-[5-^{14}C]-difluoromethylornithine, *Science,* 217, 68, 1982.
26. **Schimke, R. T., Ganschow, R., Doyle, D., and Arias, I. M.,** Regulation of protein turnover in mammalian tissues, *Fed. Proc. Fed. Am. Soc. Exp. Biol.*, 27, 1223, 1968.
27. **Goldstein, A., Aronow, L., and Kalman, S. M.,** *Principles of Drug Addiction,* Harper and Row, New York, 1968.
28. **Jänne, J. and Raina, A.,** On the stimulation of ornithine decarboxylase and RNA polymerase activity in rat liver after treatment with growth hormone, *Biochim. Biophys. Acta,* 174, 769, 1969.

29. **Russell, D. H. and Snyder, S. H.,** Amine synthesis in regenerating rat liver: effect of hypophysectomy and growth hormone on ornithine decarboxylase, *Endocrinology,* 84, 223, 1969.
30. **Panko, W. B. and Kenney, F. T.,** Hormonal stimulation of hepatic ornithine decarboxylase, *Biochem. Biophys. Res. Commun.,* 43, 346, 1971.
31. **Richman, R. A., Underwood, L. E., VanWyk, J. J., and Voina, S. J.,** Synergistic effect of cortisol and growth hormone on hepatic ornithine decarboxylase activity, *Proc. Soc. Exp. Biol. Med.,* 138, 880, 1971.
32. **Sogani, R. K., Matsushita, S., Mueller, J. F., and Raben, M. S.,** Stimulation of ornithine decarboxylase activity in rat tissues by growth hormones and by serum growth factor from rats infested with spargana of *Spirometra mansonoides, Biochim. Biophys. Acta,* 279, 377, 1972.
33. **Brandt, J. T., Pierce, D. A., and Fausto, N.,** Ornithine decarboxylase activity and polyamine synthesis during kidney hypertrophy, *Biochim. Biophys. Acta,* 279, 184, 1972.
34. **Levine, J. J., Nicholson, W. E., Liddle, G. W., and Orth, D. H.,** Stimulation of adrenal ornithine decarboxylase by adrenocorticotropin and growth hormone, *Endocrinology,* 92, 1089, 1973.
35. **Suzuki, F., Inoue, H., and Takeda, Y.,** Relation between growth hormone and thioamide derivatives in induction of ornithine decarboxylase in rat liver, *J. Biochem.,* 74, 661, 1973.
36. **Roger, L. J., Schanberg, S. M., and Fellows, R. E.,** Growth and lactogenic hormone stimulation of ornithine decarboxylase in neonatal rat brain, *Endocrinology,* 95, 904, 1974.
37. **Hölttä, E.,** Immunochemical demonstration of increased accumulation of ornithine decarboxylase in rat liver after partial hepatectomy and growth hormone induction, *Biochim. Biophys. Acta,* 399, 420, 1975.
38. **Feller, D. D., Neville, E. D., and Ellis, S.,** In vivo response of ornithine decarboxylase activity to growth hormone as demonstrated by oxidation of L-ornithine-1-^{14}C in hypophysectomized rats, *Physiol. Chem. Phys.,* 9, 55, 1977.
39. **Byus, C. V., Haddox, M. K., and Russell, D. H.,** Activation of cyclic AMP-dependent protein kinase(s) by growth hormone in the liver and adrenal gland of the rat, *J. Cyclic Nucleotide Res.,* 4, 45, 1978.
40. **Ikeno, T. and Guroff, G.,** The effect of vasopressin on the activity of ornithine decarboxylase in rat brain and liver, *J. Neurochem.,* 33, 973, 1979.
41. **Dogterom, A. A. and Robles, B. R.,** Stimulation of ornithine decarboxylase activity in *Lymnaea stagnalis* after a single injection with molluscan growth hormone, *Gen. Comp. Endocrinol.,* 40, 238, 1980.
42. **Hurley, T. W., Kuhn, C. M., Schanberg, S. M., and Handwerger, S.,** Differential effects of placental lactogen, growth hormone and prolactin in rat liver ornithine decarboxylase activity in the perinatal period, *Life Sci.,* 27, 2269, 1980.
43. **Levine, J. H., Leaming, A. B., and Gross, A. J.,** Location of renal ornithine decarboxylase activity, *Life Sci.,* 27, 1653, 1980.
44. **Klingensmith, M. R., Freifeld, A. G., Pegg, A. E., and Jefferson, L. S.,** Effects of growth hormone and glucagon on ornithine decarboxylase activity and adenosine 3′,5′-monophosphate levels in isolated rat hepatocytes, *Endocrinology,* 106, 125, 1980.
45. **Snow, E. C., Feldbush, T. L., and Oaks, J. A.,** The effect of growth hormone and insulin upon MLC responses and the generation of cytotoxic lymphocytes, *J. Immunol.,* 126, 161, 1981.
46. **Gringnolo, A., Kuhn, C. M., and Schanberg, S. M.,** Maturation of growth hormone stimulation of kidney ornithine decarboxylase in the rat, *Life Sci.,* 30, 383, 1982.
47. **Bartolome, J., Gringnolo, A., Bartolome, M., Trepanier, P., Lerea, L., Weigel, S., Whitmore, W., Michalopoulos, G., Kavlock, R., and Slotkin, T.,** Postnatal methyl mercury exposure effects on ontogeny of renal and hepatic ornithine decarboxylase responses to trophic stimuli, *Toxicol. Appl. Pharmacol.,* 80, 147, 1985.
48. **Richman, R., Dobbins, C., Voina, S., Underwood, L., Mahaffee, D., Gitelman, H. J., VanWyk, J., and Ney, R. L.,** Regulation of adrenal ornithine decarboxylase by adrenocorticotropic hormone and cyclic AMP, *J. Clin. Invest.,* 52, 2007, 1973.
49. **Levine, J. H., Nicholson, W. E., Peytremann, A., and Orth, D. H.,** The mechanism of ACTH stimulation of adrenal ornithine decarboxylase activity, *Endocrinology,* 97, 136, 1975.
50. **Kobayashi, Y., Kupelian, J., and Maudsley, D. V.,** Ornithine decarboxylase stimulation in rat ovary by luteinizing hormone, *Science,* 172, 379, 1971.
51. **Kaye, A. M., Icekson, I., Lamprecht, S. A., Gruss, R., Tsafriri, A., and Lindner, H. R.,** Stimulation of ornithine decarboxylase activity by luteinizing hormone in immature and adult rat ovaries, *Biochemistry,* 12, 3072, 1973.
52. **Maudsley, D. V. and Kobayashi, Y.,** Induction of ornithine decarboxylase in rat ovary after administration of luteinizing hormone or human chorionic gonadotrophin, *Biochem. Pharmacol.,* 26, 2697, 1974.
53. **Reddy, P. R. K. and Villee, C. A.,** Stimulation of ornithine decarboxylase activity by gonadotropic hormones and cyclic AMP in the testis of immature rats, *Biochem. Biophys. Res. Commun.,* 65, 1350, 1975.
54. **Johnson, D. C. and Sashida, T.,** Temporal changes in ovarian ornithine decarboxylase and cyclic AMP in immature rats stimulated by exogenous or endogenous gonadotrophins, *J. Endocrinol.,* 73, 463, 1977.

55. **Osterman, J. and Hammond, J. M.,** FSH and LH stimulation of ornithine decarboxylase activity: studies with porcine granulosa cells *in vitro, Endocrinology,* 101, 1335, 1977.
56. **Nureddin, A.,** Ovarian ornithine decarboxylase regulation in the immature, the pubescent, and the pseudopregnant rat, *Proc. Natl. Acad. Sci. U.S.A.,* 75, 2530, 1978.
57. **Osterman, J., Demers, L. M., and Hammond, J. M.,** Gonadotropin stimulation of porcine ovarian ornithine decarboxylase *in vitro*: the role of 3′,5′-adenosine monophosphate, *Endocrinology,* 103, 1718, 1978.
58. **Veldhuis, J. D., Demers, L. M., and Hammond, J. M.,** Regulation of ornithine decarboxylase in isolated granulosa cells *in vitro* by constituents of follicular fluid, *Endocrinology,* 105, 1143, 1979.
59. **Costa, M., Meloni, M., and Jones, M. K.,** Regulation of ornithine decarboxylase activity by amino acids, cyclic AMP and luteinizing hormone in cultured mammalian cells, *Biochim. Biophys. Acta,* 608, 1143, 1980.
60. **Francis, F. L., Triche, T. J., Brown, T. J., Brown, H. C., and Bercu, B. B.,** *In vitro* gonadotropin stimulation of bovine Sertoli cell ornithine decarboxylase activity, *J. Androl.,* 2, 312, 1981.
61. **White, S. S. and Ojeda, S. R.,** Changes in ovarian luteinizing hormone and follicle-stimulating hormone receptor content and in gonadotropin-induced ornithine decarboxylase activity during prepubertal and pubertal development of the female rat, *Endocrinology,* 109, 152, 1981.
62. **Madhubala, R. and Reddy, P. R. K.,** Stimulation of ornithine decarboxylase activity by luteinizing hormone releasing hormone in the testis of immature rat, *Biochem. Biophys. Res. Commun.,* 109, 269, 1982.
63. **Matsuzaki, S. and Suzuki, M.,** Thyroid function and polyamines. I. Rapid fluctuation of thyroid ornithine decarboxylase activity in response to change in circulating thyrotropin level in the rat, *Endocrinol. Jpn.,* 21, 529, 1974.
64. **Richman, R., Park, S., Akbar, M., Yu, S., and Burke, G.,** Regulation of thyroid ornithine decarboxylase (ODC) by thyrotropin. I. The rat, *Endocrinology,* 96, 1403, 1975.
65. **Scheinman, S. J., Burrow, G. N., Theoharides, T. C., and Canellakis, Z. N.,** Stimulation of ornithine decarboxylase synthesis in the rat thyroid, *Life Sci.,* 21, 1143, 1977.
66. **Combest, W., Chiasson, R. B., Klandorf, H., Hedge, G. A., and Russell, D. H.,** Effect of thyroid state on ornithine decarboxylase activity of the adenohypophysis of the rat and chicken, *Gen. Comp. Endocrinol.,* 35, 146, 1978.
67. **Matsuzaki, S., Kakegawa, T., Suzuki, M., and Hamana, K.,** Thyroid function and polyamines. III. Changes in ornithine decarboxylase activity and polyamine contents in the rat thyroid during hyperplasia and involution, *Endocrinol. Jpn.,* 25, 129, 1978.
68. **Friedman, Y., Lang, M., Levasseur, S., and Burke, G.,** Demonstration of a tonic regulatory thyrotropin effect on thyroid function, *Endocrinology,* 104, 467, 1979.
69. **Combest, W. L. and Russell, D. H.,** Alteration in cyclic AMP-dependent protein kinase and polyamine biosynthetic enzymes during hypertrophy and hyperplasia of the thyroid in the rat, *Mol. Pharmacol.,* 23, 641, 1983.
70. **Haddox, M. K. and Russell, D. H.,** β-Endorphin is a kidney trophic hormone, *Life Sci.,* 25, 615, 1979.
71. **Scott, K. F. F., Meyskens, F. L., Jr., and Russell, D. H.,** Retinoids increase transglutaminase activity and inhibit ornithine decarboxylase activity in Chinese hamster ovary cells and in melanoma cells stimulated to differentiate, *Proc. Natl. Acad. Sci. U.S.A.,* 79, 4093, 1982.
72. **Warnica, J. W., Antony, P., Gibson, K., and Harris, P.,** The effect of isoprenaline and propranolol on rat myocardial ornithine decarboxylase, *Cardiovasc. Res.,* 9, 793, 1975.
73. **Bareis, D. L. and Slotkin, T. A.,** Responses of heart ornithine decarboxylase and adrenal catecholamines to methadone and sympathetic stimulants in developing and adult rats, *J. Pharmacol. Exp. Ther.,* 205, 164, 1978.
74. **Fuller, R. W. and Hemrick, S. K.,** Studies of noradrenergic influence on cardiac ornithine decarboxylase in rats, *J. Mol. Cell. Cardiol.,* 10, 1031, 1978.
75. **Haddox, M. K., Womble, J. R., Larson, D. F., Roeske, W. R., and Russell, D. H.,** Isoproterenol stimulation of ornithine decarboxylase blocked by propranolol during ontogeny of the murine heart, *Mol. Pharmacol.,* 20, 382, 1981.
76. **Copeland, J. G., Larson, D. F., Roeske, W. R., Russell, D. H., and Womble, J. R.,** β_2-Adrenoceptors regulate induction of myocardial ornithine decarboxylase in mice *in vivo, Br. J. Pharmacol.,* 75, 479, 1982.
77. **Inoue, H., Tanioka, H., Shiba, K., Asada, A., Kato, Y., and Takeda, Y.,** Effect of isoproterenol on polyamine metabolism in mouse salivary glands, *J. Biochem.,* 75, 679, 1974.
78. **Inoue, H., Kato, Y., Takigawa, M., Adachi, K., and Takeda, Y.,** Effect of DL-α-hydrazino-aminovaleric acid, an inhibitor of ornithine decarboxylase, on polyamine metabolism in isoproterenol-stimulated mouse parotid glands, *J. Biochem.,* 77, 879, 1975.
79. **Inoue, H., Asada, A., Kato, Y., and Takeda, Y.,** Interconversion of aliphatic polyamines in isoproterenol-stimulated mouse parotid glands, *J. Biochem.,* 84, 719, 1978.

80. **Tsang, B. K., Rixon, R. H., and Whitfield, J. F.,** A possible role for cyclic AMP in the initiation of DNA synthesis by isoproterenol-activated parotid gland cells, *J. Cell. Physiol.*, 102, 19, 1980.
81. **Veldhuis, J. D., Harrison, R. S. and Hammond, J. M.,** β_2-Adrenergic stimulation of ornithine decarboxylase activity in porcine granulosa cells in vitro, *Biochim. Biophys. Acta*, 627, 123, 1980.
82. **Bachrach, U.,** Cyclic AMP-mediated induction of ornithine decarboxylase of glioma and neuroblastoma cells, *Proc. Natl. Acad. Sci. U.S.A.*, 72, 3087, 1975.
83. **Russell, D. H.,** Type I cyclic AMP-dependent protein kinase as a positive effector of growth, in, *Advances in Cyclic Nucleotide Research*, Vol. 9, George, W. J. and Ignarro, L. J., Eds., Raven Press, New York, 1978, 493.
84. **Buckley, A. R., Montgomery, D. W., Kibler, R., Putnam, C. W., Zukoski, C. F., Gout, P. W., Beer, C. T., and Russell, D. H.,** Prolactin stimulation of ornithine decarboxylase and mitogenesis in Nb 2 node lymphoma cells: the role of protein kinase C and calcium mobilization, *Immunopharmacology*, 12, 37, 1986.
85. **Russell, D. H., Buckley, A. R., Montgomery, D. W., Larson, N. A., Gout, P. W., Beer, C. T., Putnam, C. W., Zukoski, C. F., and Kibler, R.,** Prolactin-dependent mitogenesis in Nb 2 node lymphoma cells: effects of immunosuppressive cyclopeptides, *J. Immunol.*, 1380, 276, 1987.
86. **Byus, C. V., Klimpel, G. R., Lucas, D. O., and Russell, D. H.,** Implications of type I or type II cyclic AMP-dependent protein kinase as independent positive or negative effectors of lymphocyte mitogenesis, *Nature (London)*, 268, 63, 1977.
87. **Byus, C. V. and Russell, D. H.,** Effects of methylxanthine derivatives on cyclic AMP levels and ornithine decarboxylase activity of rat tissues, *Life Sci.*, 15, 1991, 1974.
88. **Russell, D. H. and Byus, C. V.,** Ornithine decarboxylase: transynaptic induction in the medulla, in *Advances in Biochemical Psychopharmacology*, Vol. 15, Costa, E., Giacobini, E., and Paoletti, R., Eds., Raven Press, New York, 1976, 445.
89. **Beck, W. T., Bellantone, R. A., and Canellakis, E. W.,** In vivo stimulation of rat liver ornithine decarboxylase activity by dibutyryl cyclic adenosine 3',5'-monophosphate, theophylline and dexamethasone, *Biochem. Biophys. Res. Commun.*, 48, 1649, 1972.
90. **Beck, W. T. and Canellakis, Z. N.,** The in vivo chemical stimulation of hepatic ornithine decarboxylase activity: modifications of activity at the transcriptional and post-transcriptional levels of protein synthesis, in *Polyamines in Normal and Neoplastic Growth*, Russell, D. H., Ed., Raven Press, New York, 1973, 261.
91. **Hölttä, E. and Raina, A.,** Stimulation of ornithine decarboxylase and nuclear RNA polymerase activity in rat liver by glucagon and dibutyryl cyclic AMP, *Acta Endocrinol.*, 73, 794, 1973.
92. **Eloranta, T. and Raina, A.,** Involvement of hypophysis in the stimulation of liver ornithine decarboxylase by dibutyryl cyclic AMP, *FEBS Lett.*, 55, 22, 1975.
93. **Manen, C. A. and Russell, D. H.,** Relationship of ornithine decarboxylase to RNA polymerase I activity, *Life Sci.*, 17, 1769, 1975.
94. **Costa, M., Costa, E. R., Manen, C. A., Sipes, I. G., and Russell, D. H.,** Cyclic AMP-dependent protein kinase and ornithine decarboxylase involvement in the induction of cytochrome P-450 and hepatic hypertrophy, *Mol. Pharmacol.*, 12, 871, 1976.
95. **Aisbitt, R. P. G. and Barry, J. M.,** Stimulation by insulin of ornithine decarboxylase activity in cultured mammary tissue, *Biochim. Biophys. Acta*, 320, 610, 1973.
96. **Oka, T. and Perry, J. W.,** Studies on regulatory factors of ornithine decarboxylase activity during development of mouse mammary epithelium in vitro, *J. Biol. Chem.*, 251, 1738, 1976.
97. **Preslock, J. P. and Hampton, J. K.,** Ornithine decarboxylase stimulation by 3',5'-cyclic AMP in oviduct of *Coturnix* quail, *Am. J. Physiol.*, 225, 903, 1973.
98. **Tsang, B. K. and Singhal, R. L.,** Polyamine and cAMP metabolism in rat ventral prostate (VP) following administration of cyproterone acetate (CA), *Pharmacologist*, 18, 164, 1976.
99. **Fuller, D. J. M., Gerner, E. W., and Russell, D. H.,** Polyamine biosynthesis and accumulation during the G_1 to S phase transition, *J. Cell. Physiol.*, 93, 81, 1977.
100. **Russell, D. H. and Haddox, M. K.,** Cyclic AMP-mediated induction of ornithine decarboxylase in normal and neoplastic growth, in *Advances in Enzyme Regulation*, Vol. 17, Weber, G., Ed., Pergamon Press, Oxford, 1979, 61.
101. **Kudlow, J. E., Rae, P. A., Schimmer, B. P., and Burrow, G. N.,** Hormonal regulation of ornithine decarboxylase in adrenal cell clones by cAMP and cAMP dependent protein kinase, *Clin. Res.*, 26, 844A, 1978.
102. **Gibbs, J. B., Hsu, C. Y., Terasaki, W. L., and Brooker, G.,** Calcium and microtubule dependence for increased ornithine decarboxylase activity stimulated by β-adrenergic agonists, dibutyryl cyclic AMP, or serum in a rat astrocytoma cell line, *Proc. Natl. Acad. Sci. U.S.A.*, 77, 995, 1980.
103. **Hogan, B., Shields, R., and Curtis, D.,** Effect of cyclic nucleotides on the induction of ornithine decarboxylase in BHK cells by serum and insulin, *Cell*, 2, 229, 1974.

104. **Hibasami, H., Tanaka, M., Nagai, J., and Ikeda, T.,** Changes in ornithine decarboxylase activity and cyclic adenosine-3′,5′-monophosphate concentrations during the cell cycle by synchronized BHK cells, *Aust. J. Exp. Biol. Med. Sci.*, 55, 379, 1977.
105. **Costa, M. and Nye, J. S.,** Calcium, asparagine and cAMP are required for ornithine decarboxylase activation in intact Chinese hamster ovary cells, *Biochem. Biophys. Res. Commun.*, 85, 1156, 1978.
106. **Costa, M.,** Cyclic AMP-dependent regulation of ornithine decarboxylase activity in Chinese hamster ovary cells maintained with a salts/glucose medium, *J. Cyclic Nucleotide Res.*, 5, 375, 1978.
107. **Costa, M.,** Superinduction of ornithine decarboxylase by actinomycin D and cordycepin, *Biochem. Biophys. Res. Commun.*, 81, 832, 1978.
108. **Costa, M.,** The regulation of ornithine decarboxylase activity in intact normal and transformed cells maintained with a minimal salts/glucose medium, *Life Sci.*, 25, 2113 1979.
109. **Takigawa, M., Takano, T., and Suzuki, F.,** Effects of parathyroid hormone and cyclic AMP analogues on the activity of ornithine decarboxylase and expression of the differentiated phenotype of chondrocytes in culture, *J. Cell, Physiol.*, 106, 259, 1981.
110. **Boucek, R. J., Weiss, J., and Lembach, K. J.,** Cell-type differences in the serum and cyclic AMP-dependent regulation of ornithine decarboxylase activity in two cultured fibroblast types, *Biochem. J.*, 204, 357, 1982.
111. **Yamasaki, Y. and Ichihara, A.,** Induction of ornithine decarboxylase in cultured mouse L cells. I. Effects of cellular growth, hormones, and actinomycin D, *J. Biochem.*, 80, 557, 1976.
112. **Nichols, W. K. and Prosser, F. H.,** Induction of ornithine decarboxylase in macrophages by bacterial lipopolysaccharides (LPS) and mycobacterial cell wall material, *Life Sci.*, 27, 913, 1980.
113. **Byus, C. V., Wicks, W. D., and Russell, D. H.,** Induction of ornithine decarboxylase in Reuber H35 rat hepatoma cells, *J. Cyclic Nucleotide Res.*, 2, 241, 1976.
114. **Byus, C. V. and Russell, D. H.,** Possible regulation of ornithine decarboxylase activity in the adrenal medulla of the rat by a cAMP-dependent mechanism, *Biochem. Pharmacol.*, 25, 1595, 1976.
115. **Byus, C. V., Hedge, G. A., and Russell, D. H.,** The involvement of cyclic AMP-dependent protein kinase(s) in the induction of ornithine decarboxylase in the regenerating rat liver and in the adrenal gland after unilateral adrenalectomy, *Biochim. Biophys. Acta*, 498, 39, 1977.
116. **Klimpel G. R., Byus, C. V., Russell, D. H., and Lucas, D. O.,** Cyclic AMP-dependent protein kinase activation and the induction of ornithine decarboxylase during lymphocyte mitogenesis, *J. Immunol.*, 123, 817, 1979.
117. **Byus, C. V., Costa, M., Sipes, I. G., Brodie, B. B., and Russell, D. H.,** Activation of cyclic AMP-dependent protein kinase and induction of ornithine decarboxylase as early events in the induction of ornithine decarboxylase as early events in the induction of mixed-function oxygenases, *Proc. Natl. Acad. Sci. U.S.A.*, 73, 1241, 1976.
118. **Matsui, I., Otani, S., and Morisawa, S.,** Induction of ornithine decarboxylase by biliverdin in rat liver, *Life Sci.*, 24, 2231, 1979.
119. **Bachrach, U., Benalal, D., and Reches, A.,** Morphine inhibits cyclic AMP-dependent protein kinase and ornithine decarboxylase activities in neuroblastoma X glioma hybrid cells, *Life Sci.*, 25, 1879, 1979.
120. **Sudilovsky, O.,** In vivo regulation of hepatic protein kinase by adenosine-3′,5′-monophosphate mediated glucagon stimulation, *Biochem. Biophys. Res. Commun.*, 58, 85, 1975.
121. **Costa, M., Gerner, E. W., and Russell, D. H.,** G_1 specific increases in cyclic AMP and protein kinase activity, *Biochim. Biophys. Acta*, 425, 246, 1976.
122. **Haddox, M. K., Magun, B. E., and Russell, D. H.,** Differential expression of type I and type II cyclic AMP-dependent protein kinase during cell cycle and cyclic AMP-induced growth arrest, *Proc. Natl. Acad. Sci. U.S.A.*, 77, 3445, 1980.
123. **Anderson, W. B. and Pastan, I.,** Altered adenylate cyclase activity: its role in growth regulation and malignant transformation of fibroblasts, *Adv. Cyclic Nucleotide Res.*, 5, 681, 1975.
124. **Sheppard, J. R.,** Restoration of contact-inhibited growth to transformed cells by dibutyryl adenosine 3′,5′:cyclic monophosphate, *Proc. Natl. Acad. Sci. U.S.A.*, 68, 1316, 1971.
125. **Sheppard, J. R.,** Difference in the cyclic adenosine 3′,5′:monophosphate levels in normal and transformed cells, *Nature (London)*, 236, 14, 1972.
126. **Robison, G. A., Butcher, R. W., and Sutherland, E. W.,** Cyclic AMP, *Annu. Rev. Biochem.*, 37, 149, 1968.
127. **Robison, G. A., Butcher, R. W., and Sutherland, E. W.,** *Cyclic AMP*, Academic Press, New York, 1971.
128. **Haddox, M. K. and Russell, D. H.,** Cyclic AMP-dependent protein kinase implicated in the transcriptional induction of ornithine decarboxylase, in *Cold Springs Harbor Conferences on Cell Proliferation-Protein Phosphorylation*, Book B, Vol. 8, Rosen, O. M. and Krebs, E. G., Eds., Cold Spring Harbor Laboratory, Cold Spring Harbor, New York, 1981, 1013.

129. **Montminy, M. R., Sevarino, K. A., and Goodman, R. H.,** A conserved promoter sequence mediates induction of cAMP-responsive genes, The Endocrine Soc. Program and Abstracts, 68th Annual Mtg., 1986, 50.
130. **Kuo, J. F. and Greengard, P.,** Cyclic nucleotide-dependent protein kinases. IV. Widespread occurrence of adenosine 3′,5′-monophosphate-dependent protein kinase in various tissues and phylla of the animal kingdom, *Proc. Natl. Acad. Sci. U.S.A.*, 64, 1349, 1969.
131. **Pegg, A. E. and Williams-Ashman, H. G.,** Rapid effects of testosterone on prostatic polyamine synthesizing enzyme systems, *Biochem. J.*, 109, 32, 1968.
132. **Cohen, S., O'Malley, B. W., and Stastny, M.,** Estrogenic induction of ornithine decarboxylase in vivo and in vitro, *Science*, 170, 336, 1970.
133. **Kaye, A. M., Icekson, I., and Lindner, H. R.,** Stimulation by estrogens or ornithine and *S*-adenosylmethionine decarboxylases in the immature rat uterus, *Biochim. Biophys. Acta*, 252, 150, 1971.
134. **Russell, D. H. and Taylor, R. L.,** Polyamine synthesis and accumulation in the castrated rat uterus after estradiol-17β stimulation, *Endocrinology*, 88, 1397, 1971.
135. **Bulger, W. H. and Kupfer, D.,** Studies on the induction of rat uterine ornithine decarboxylase by DDT analogs. I. Comparison with estradiol-17β activity, *Pestic. Biochem. Physiol.*, 8, 253, 1978.
136. **Capuco, A. V. and Tseng, M. T.,** Ornithine decarboxylase activity and incorporation of ^{3}H-thymidine into DNA in estrogen-treated male *Xenopus laevis, IRCS Med. Sci.*, 7, 547, 1979.
137. **Persson, L. and Rosengren, E.,** Inhibition of ornithine decarboxylase activity in mouse kidney by administration of diamines, *Acta Chem. Scand. B*, 33, 537, 1979.
138. **Lin, Y. C., Loring, J. M., and Villee, C. A.,** Diethylstilbestrol stimulates ornithine decarboxylase in kidney cells in culture, *Biochem. Biophys. Res. Commun.*, 95, 1393, 1980.
139. **Nawata, H., Yamamoto, R. S., and Poirier, L. A.,** Ornithine decarboxylase induction and polyamine levels in the kidney of estradiol-treated castrated male rats, *Life Sci.*, 26, 689, 1980.
140. **Nawata, H., Yamamoto, R. S., and Poirier, L. A.,** Elevated levels of ornithine decarboxylase and polyamines in the kidneys of estradiol-treated male hamsters, *Carcinogenesis*, 2, 1207, 1981.
141. **Pass, K. A., Bintz, J. E., and Postulka, J. J.,** Ontogenic development of the renal ornithine decarboxylase response to testosterone, *Experientia*, 37, 1119, 1981.
142. **Pass, K. A., Bintz, J. E., and Postulka, J. J.,** Effects of testosterone on renal ornithine decarboxylase and kidney function, *Enzyme*, 27, 108, 1982.
143. **Persson, L.,** Decarboxylation of ornithine and lysine by ornithine decarboxylase from kidneys of testosterone treated mice, *Acta Chem. Scand. B*, 35, 451, 1981.
144. **Shain, S. A. and Moss, A. L.,** Aging in the AXC rat: differential effects of chronic testosterone treatment on restoration of diminished prostate L-ornithine decarboxylase and *S*-adenosyl-L-methionine decarboxylase activities, *Endocrinology*, 109, 1184, 1981.
145. **Pajunen, A. E. I., Isomaa, V. V., Jänne, O. A., and Bardin, C. W.,** Androgenic regulation of ornithine decarboxylase activity in mouse kidney and its relationship to changes in cytosol and nuclear androgen receptor concentrations, *J. Biol. Chem.*, 257, 8190, 1982.
146. **Wright, J. M., Gulliver, W. P., Michalski, C. J., and Boyle, S. M.,** Ornithine decarboxylase activity and polyamine content during zoospore germination and hormone-induced sexual differentiation of *Achyla ambisexualis, J. Gen. Microbiol.*, 128, 1509, 1982.
147. **Bullock, L. P.,** Androgen and progestin stimulation of ornithine decarboxylase activity in the mouse kidney, *Endocrinology*, 112, 1903, 1983.
148. **Seely, J. E. and Pegg, A. E.,** Changes in mouse kidney ornithine decarboxylase activity are brought about by changes in the amount of enzyme protein as measured by radioimmunoassay, *J. Biol. Chem.*, 258, 2496, 1983.
149. **Barkai, U., Shaham, Y., and Kraicer, P. F.,** Inhibition of induction of deciduoma by acute administration of antibody to progesterone, *Life Sci.*, 23, 679, 1978.
150. **Luzzani, F., Colombo, G., and Galliani, G.,** Evidence for a role of progesterone in the control of uterine ornithine decarboxylase in the pregnant hamster, *Life Sci.*, 31, 1553, 1982.
151. **Fuller, D. J. M., Byus, C. V., and Russell, D. H.,** Specific regulation by steroid hormones of the amount of type I cyclic AMP-dependent protein kinase holoenzyme, *Proc. Natl. Acad. Sci. U.S.A.*, 75, 223, 1978.
152. **Andersson, A. C., Henningsson, S., and Rosengren, E.,** Activities of kidney decarboxylases of histidine and ornithine in adrenalectomized mice substituted with cortisone, *Experientia*, 31, 1101, 1975.
153. **Richards, J. F.,** Ornithine decarboxylase activity in lymphoid tissues of rats: effects of glucocorticoids, *Life Sci.*, 23, 1619, 1978.
154. **Tye, L. M., Richards, J. F., and Burton, A. F.,** Corticosteroids, ornithine decarboxylase activity and the incorporation of leucine, uridine, and thymidine into mouse placenta, *Pediatr. Res.*, 12, 1155, 1978.
155. **Rath, N. C. and Reddi, A. H.,** Evaluation of anti-inflammatory drugs based upon the inhibition of matrix-induced ornithine decarboxylase activity during connective tissue proliferation, *Proc. Soc. Exp. Biol. Med.*, 162, 320, 1979.

156. **Russell, D. H., Haddox, M. K., and Gehring, U.,** Effects of dexamethasone and dibutyryl cyclic AMP on polyamine synthesizing enzymes in mouse lymphoma cells, *J. Cell. Physiol.*, 106, 375, 1981.
157. **Richards, J. F., Lit, K., Fuca, R., and Bourgeault, C.,** Multiple species of ornithine decarboxylase in rat tissues: effects of dexamethasone, *Biochem. Biophys. Res. Commun.*, 99, 1461, 1981.
158. **Cousin, M. A., Lando, D., and Moguilewsky, M.,** Ornithine decarboxylase induction by glucocorticoids in brain and liver of adrenalectomized rats, *J. Neurochem.*, 38, 1296, 1982.
159. **Katovich, M. J., Baker, S. P., and Nelson, C.,** Effects of elevated prolactin and its normalization on thyroid hormone, cardiac β-adrenoreceptor number and β-adrenergic responsiveness, *Life Sci.*, 34, 889, 1984.
160. **Gelato, M., Marshall, S., Boudreau, M., Brunt, J., Campbell, G. A., and Meites, J.,** Effects of thyroid and ovaries on prolactin binding activity in rat liver, *Endocrinology*, 96, 1292, 1975.
161. **Dave, J. R., Wotorsch, R. J., and Kalimi, M. Y.,** Endocrine-mediated parallel changes in hepatic glucocorticoid and prolactin, *Biochim. Biophys. Acta*, 845, 276, 1985.
162. **Hogan, B. L. M., McIlhinney, A., and Murden, S.,** Effect of growth conditions on the activity of ornithine decarboxylase in cultured hepatoma cells. II. Effect of serum and insulin, *J. Cell. Physiol.*, 83, 353, 1974.
163. **Haselbacher, G. K. and Humbel, R. E.,** Stimulation of ornithine decarboxylase activity in chick fibroblasts by non-suppressible insulin-like activity (NSILA), insulin and serum, *J. Cell. Physiol.*, 88, 239, 1976.
164. **Hibasami, H., Tanaka, M., Nagai, J., and Ikeda, T.,** Changes in ornithine decarboxylase activity and polyamine concentrations in growing BHK cells, *Mie Med. J. Tsu. Jpn.*, 25, 223, 1976.
165. **Maudsley, D. V., Leif, J., and Kobayashi, Y.,** Ornithine decarboxylase in rat small intestine: stimulation with food or insulin, *Am. J. Physiol.*, 231, 1557, 1976.
166. **Brosnan, M. E., Roebothan, B. V., and Hall, D. E.,** Polyamine and amino acid content, and activity of polyamine-synthesized decarboxylases, in liver of streptozotocin-induced diabetic and insulin-treated diabetic, rats, *Biochem. J.*, 190, 395, 1980.
167. **Conover, C. A., Rozovski, S. J., Belur, E. R., Aoki, T. T., and Ruderman, N. B.,** Ornithine decarboxylase activity in insulin-deficient states, *Biochem. J.*, 192, 725, 1980.
168. **Parker, K. and Vernadakis, A.,** Stimulation of ornithine decarboxylase activity in neural cell culture: potential role of insulin, *J. Neurochem.*, 35, 155, 1980.
169. **Hendry, I. A. and Bonyhady, R.,** Retrogradely transported nerve growth factor increases ornithine decarboxylase activity in rat superior cervical ganglia, *Brain Res.*, 200, 39, 1980.
170. **Roger, L. J. and Fellows, R. E.,** Stimulation of ornithine decarboxylase activity by insulin in developing rat brain, *Endocrinology*, 106, 619, 1980.
171. **Burch, W. M. and Lebovitz, H. E.,** Hormonal activation of ornithine decarboxylase in embryonic chick pelvic cartilage, *Am. J. Physiol.*, 241, E454, 1981.
172. **Murakami, Y., Noguchi, T., and Hayashi, S.-I.,** Role of pancreatic hormones in dietary induction of ornithine decarboxylase of rat liver, *J. Biochem.*, 90, 141, 1981.
173. **Fischer, M. S. and Cohen, P. O.,** Regulation of the activity of ornithine decarboxylase in tadpole hepatic tissues, *Dev. Biol.*, 36, 357, 1974.
174. **Lau, C. and Slotkin, T. A.,** Regulation of rat heart ornithine decarboxylase: change in affinity for ornithine evoked by neuronal, hormonal, and ontogenetic stimuli, *Mol. Pharmacol.*, 16, 504, 1979.
175. **Combest, W. L. and Russell, D. H.,** Separate mechanisms of induction for ornithine decarboxylase by triiodothyronine and aminophylline, *Life Sci.*, 27, 651, 1980.
176. **Palacios-Romero, R. and Mowbray, J.,** The rapid transient stimulation of both cytoplasmic and mitochondrial ornithine decarboxylase in the liver of thyroidectomized rats by 3,5,3′-triodo-L-thyronine, *Biosci. Rep.*, 1, 71, 1981.
177. **Richards, J. F.,** Ornithine decarboxylase activity in tissues of prolactin-treated rats, *Biochem. Biophys. Res. Commun.*, 63, 292, 1975.
178. **Russell, D. H., Larson, D. F., Cardon, S. B., and Copeland, J. G.,** Cyclosporine inhibits prolactin induction of ornithine decarboxylase in rat tissues, *Mol. Cell. Endocrinol.*, 35, 159, 1984.
179. **Buckley, A. R., Putnam, C. W., Montgomery, D. W. and Russell, D. H.,** Prolactin administration stimulates rat hepatic DNA synthesis, in press.
180. **Thomson, M. J. and Richards, J. F.,** Ornithine decarboxylase and thymidine kinase activity in tissues of prolactin-treated rats: effect of hypophysectomy, *Life Sci.*, 22, 337, 1978.
181. **Russell, D. H. and Larson, D. F.,** Prolactin-induced polyamine biosynthesis in spleen and thymus: specific inhibition by cyclosporine, *Immunopharmacology*, 9, 165, 1985.
182. **Frazier, R. P. and Costlow, M. E.,** Prolactin stimulation of ornithine decarboxylase activity in cultured rat mammary tumors, *Exp. Cell Res.*, 138, 39, 1982.
183. **Richards, J. F., Beer, C. T., Bourgeault, C., Chen, K., and Gout, P. W.,** Biochemical response of lymphoma cells to mitogenic stimulation by prolactin, *Mol. Cell. Endocrinol.*, 26, 41, 1982.
184. **Stastny, M. and Cohen, S.,** Epidermal growth factor. IV. The induction of ornithine decarboxylase, *Biochim. Biophys. Acta*, 204, 578, 1970.

185. **Stastny, M. and Cohen, S.**, The stimulation of ornithine decarboxylase activity in testes of the neonatal mouse, *Biochim. Biophys. Acta*, 261, 177, 1972.
186. **DiPasquale, A., White, D., and McGuire, J.**, Epidermal growth factor stimulates putrescine transport and orithine decarboxylase activity in cultivated human fibroblasts, *Exp. Cell. Res.*, 116, 317, 1978.
187. **Feldman, E. J., Aures, D., and Gorssman, M. I.**, Epidermal growth factor stimulates ornithine decarboxylase activity in the digestive tract of mouse, *Proc. Soc. Exp. Biol. Med.*, 159, 400, 1978.
188. **Veldhuis, J. D. and Hammond, J. M.**, Multiplication stimulating activity regulates ornithine decarboxylase in isolated porcine granulosa cells *in vitro*, *Endocr. Res. Commun.*, 6, 299, 1979.
189. **Guroff, G., Dickens, G., and End, D.**, The induction of ornithine decarboxylase by nerve growth factor and epidermal growth factor in PC12 cells, *J. Neurochem.*, 37, 342, 1981.
190. **Moriarity, D. M., DiSorbo, D. M., Litwack, G., and Savaga, C. R., Jr.**, Epidermal growth factor stimulation of ornithine decarboxylase activity in a human hepatoma cell line, *Proc. Natl. Acad. Sci. U.S.A.*, 78, 2752, 1981.
191. **Greene, L. A. and McGuire, J. C.**, Induction of ornithine decarboxylase by nerve growth factor dissociated from effects on survival and neurite outgrowth, *Nature*, 276, 191, 1978.
192. **Hatanaka, H., Otten, U., and Hotenen, H.**, Nerve growth factor-mediated selective induction of ornithine decarboxylase in rat pheochromocytoma; a cyclic AMP-independent process, *FEBS Lett.*, 92, 313, 1978.
193. **Ikeno, T., MacDonnell, P. C., Nagaiah, K., and Guroff, G.**, The permissive effect of cortical steroids on the induction of brain ornithine decarboxylase by nerve growth factor, *Biochem. Biophys. Res. Commun.*, 82, 957, 1978.
194. **Lewis, M. E., Lakshmanan, J., Nagaiah, K., MacDonnell, P., and Guroff, G.**, Nerve growth factor increases activity of ornithine decarboxylase in rat brain, *Proc. Natl. Acad. Sci. U.S.A.*, 75, 1021, 1978.
195. **Nagaiah, K., Ikeno, T., Lakshmanan, J., MacDonnell, P., and Guroff, G.**, Intraventricular administration of nerve growth factors induces ornithine decarboxylase in peripheral tissues of the rat, *Proc. Natl. Acad. Sci. U.S.A.*, 75, 2512, 1978.
196. **Smith, G. L. and Stange, A. W.**, Polyamines and the stimulation of DNA synthesis in chicken embryo fibroblasts by purified multiplication-stimulating activity, *Life Sci.*, 23, 1871, 1978.
197. **Berridge, M. J. and Irvine, R. F.**, Inositol trisphosphate, a novel second messenger in cellular signal transduction, *Nature*, 312, 315, 1984.
198. **Nishizuka, Y.**, The role of protein kinase C in seltzer for signal transduction and turnover promotion, *Nature (London)*, 308, 693, 1984.
199. **Hoken, L. E.**, Receptors in phosphoinositide — generated second messengers, *Annu. Rev. Biochem.*, 54, 205, 1985.
200. **Rovera, G., O'Brien, T. G., and Diamond, L.**, Induction of differentiation in human promyelocytic leukemia cells by tumor promoters, *Science*, 204, 868, 1979.
201. **Cohen, R., Pacifici, M., Rubinstein, N., Biehl, J., and Holtzer, H.**, Effect of a tumor promoter on myogenesis, *Nature (London)*, 266, 538, 1977.
202. **Nishizuka, Y.**, Perspectives on the role of protein kinase C in stimulus-response coupling, *J. Natl. Cancer Inst.*, 76, 363, 1986.
203. **Takayama, S., White, M. F., Lauris, V., and Kahn, C. R.**, Phorbol esters modulate insulin receptor phosphorylation and insulin action in cultured hepatoma cells, *Proc. Natl. Acad. Sci. U.S.A.*, 81, 7797, 1984.
204. **Bosca, L., Rousseau, G. G., and Hue, L.**, Phorbol 12-myristate 13-acetate and insulin increase the concentration of fructose 2,6-bisphosphate and stimulate glycolysis in chicken embryo fibroblasts, *Proc. Natl. Acad. Sci. U.S.A.*, 82, 6440, 1985.
205. **Trevillyan, J. M., Kulkarni, R. K., and Byus, C. V.**, Tumor-promoting phorbol esters stimulate the phosphorylation of ribosomal protein S6 in quiescent Reuber H35 hepatoma cells, *J. Biol. Chem.*, 259, 897, 1984.
206. **Beguinot, L., Hanover, J. A., Ito, S., Richert, N. D., Willingham, M. C., and Pastan, I.**, Phorbol esters induce transient internalization without degradation of unoccupied epidermal growth factor receptors, *Proc. Natl. Acad. Sci. U.S.A.*, 82, 2774, 1985.
207. **Fabbro, D., Kung, W., Roos, W., Regazzi, R., and Eppenberger, U.**, Epidermal growth factor binding and protein kinase C activities in human breast cancer cell lines: possible quantitative relationship, *Cancer Res.*, 46, 2720, 1986.
208. **Sahyoun, N., Wolf, M., Besterman, J., Hsieh, T.-S., Sander, M., LeVine, H., III, Chang, K.-J., and Cuatrecasas, P.**, Protein kinase C phosphorylates topoisomerase II: topoisomerase activation and its possible role in phorbol ester-induced differentiation of HL-60 cells, *Proc. Natl. Acad. Sci. U.S.A.*, 83, 1603, 1986.
209. **Russell, D. H.**, Ornithine decarboxylase: a key regulatory enzyme in normal and neoplastic growth, in *Drug Metab. Rev.*, 16, 1, 1985.
210. **Goldfine, I. D.**, Insulin receptors and the site of action of insulin, *Life Sci.*, 23, 2639, 1978.

211. **Davies, P. J. A., Davies, D. R., Levitzki, A., Maxfield, F. R., Milhaud, P., Willingham, M. C., and Pastan, I. H.**, Transglutaminase is essential in receptor-mediated endocytosis of α^2-macroglobulin and polypeptide hormones, *Nature*, 283, 162, 1980.
212. **Chuang, D. M.**, Inhibitors of transglutaminase prevent agonist-mediated internalization of β-adrenergic receptors, *J. Biol. Chem.*, 256, 8291, 1981.
213. **O'Brien, T. G., Simsiman, R. C., and Boutwell, R. K.**, Induction of the polyamine-biosynthetic enzymes in mouse epidermis by tumor-promoting agents, *Cancer Res.*, 35, 1662, 1975.
214. **O'Brien, T. G., Simsiman, R. C., and Boutwell, R. K.**, Induction of the polyamine-biosynthetic enzymes in mouse epidermis and their specificity for tumor promotion, *Cancer Res.*, 35, 2426, 1975.
215. **O'Brien, T. G.**, The induction of ornithine decarboxylase as an early, possibly obligatory, event in mouse skin carcinogenesis, *Cancer Res.*, 36, 2644, 1976.
216. **Mufson, R. A., Astrup, E. G., Simsiman, R. C., and Boutwell, R. K.**, Dissociation of increases in levels of 3′:5′-cyclic AMP and 3′:5′-cyclic GMP from induction of ornithine decarboxylase by the tumor promoter 12-*O*-tetradecanoylphorbol-13-acetate in mouse epidermis in vivo, *Proc. Natl. Acad. Sci. U.S.A.*, 74, 657, 1977.
217. **Yuspa, S. H., Lichti, U., Ben., T., Patterson, E., Hennings, H., Slaga, T. J., Colburn, N., and Kelsey, W.**, Phorbol esters stimulate DNA synthesis and ornithine decarboxylase activity in mouse epidermal cell cultures, *Nature*, 262, 402, 1976.
218. **O'Brien, T. G., Lewis, M. A., and Diamond, L.**, Ornithine decarboxylase activity and DNA synthesis after treatment of cells in culture with 12-*O*-tetradecanoylphorbol-13-acetate, *Cancer Res.*, 39, 4477, 1979.
219. **O'Brien, T. G., Saladik, D., and Diamond, L.**, Regulation of polyamine biosynthesis in normal and transformed hamster cells in culture, *Biochim. Biophys. Acta*, 632, 270, 1980.
220. **Fujiki, H., Mori, M., Sugimura, T., Hirota, M., Ohigashi, H., and Koshimizu, K.**, Relationship between ornithine decarboxylase-inducing activity and configuration at C-4 in phorbol ester derivatives, *J. Cancer Res. Clin. Oncol.*, 98, 9, 1980.
221. **Lesiewicz, J., Morrison, D. M., and Goldsmith, L. A.**, Ornithine decarboxylase in rat skin. II. Differential response to hair plucking and a tumor promoter, *J. Invest. Dermatol.*, 75, 411, 1980.
222. **Weiner, R. A. and Byus, C. V.**, Induction of ornithine decarboxylase by 12-*O*-tetradecanoylphorbol-13-acetate in rat tissues, *Biochem. Biophys. Res. Commun.*, 97, 1575, 1980.
223. **Byus, C. V. and Weiner, R. A.**, Tumor promoting phorbol-ester derivatives increase ornithine decarboxylase activity and polyamine biosynthesis in the liver of the rat and mouse, *Carcinogenesis*, 3, 751, 1982.
224. **Bisschop, A., van Rooijen, L. A. A., Derks, J. H. G. M., and van Wyk, R.**, Induction of rat hepatic ornithine decarboxylase by the tumor promoters of 12-*O*-tetradecanoylphorbol-13-acetate and phenobarbital *in vivo*; effect of retinyl-acetate, *Carcinogenesis*, 2, 1283, 1981.
225. **Fujiki, H., Mori, M., Nakayasu, M., Terada, M., Sugimura, T., and Moore, R. E.**, Indole alkaloids: dihydroteleocidin B, teleocidin, and lyngbyatoxin A as members of a new class of tumor promoters, *Proc. Natl. Acad. Sci. U.S.A.*, 78, 3872, 1981.
226. **Izumi, K., Hirao, Y., Hopp, L., and Oyasu, R.**, *In vitro* induction of ornithine decarboxylase in urinary bladder carcinoma cells, *Cancer Res.*, 41, 405, 1981.
227. **Kishore, G. S. and Boutwell, R. K.**, Induction of mouse hepatic ornithine decarboxylase by skin application of 12-*O*-tetradecanoylphorbol-13-acetate, *Experientia*, 37, 179, 1981.
228. **Lichti, U., Patterson, E., Hennings, H., and Yuspa, S. H.**, The tumor promoter 12-*O*-tetradecanoylphorbol-13-acetate induces ornithine decarboxylase in proliferating basal cells but not in differentiating cells from mouse epidermis, *J. Cell. Physiol.*, 107, 261, 1981.
229. **Verma, A. K. and Boutwell, R. K.**, Intracellular calcium and skin tumor promotion: calcium regulation of the induction of epidermal ornithine decarboxylase activity by the tumor promoter 12-*O*-tetradecanoylphorbol-13-acetate, *Biochem. Biophys. Res. Commun.*, 101, 375, 1981.
230. **Wu, K., Wang, D., and Feinman, R. D.**, Inhibition of proteases by α_2-macroglobulin: the role of lysyl amino groups of trypsin in covalent complex formation, *J. Biol. Chem.*, 20, 10409, 1981.
231. **Lichti, U. and Gottesman, M. M.**, Genetic evidence that a phorbol ester tumor promoter stimulates ornithine decarboxylae activity by a pathway that is independent of cyclic AMP-dependent protein kinases in CHO cells, *J. Cell. Physiol.*, 113, 433, 1982.
232. **Lillehaug, J. R. and Djurhuus, R.**, Effect of diethylstilbestrol on the transformable mouse embryo fibroblast C3H/10T1/2C18 cells. Tumor promotion, cell growth, DNA synthesis and ornithine decarboxylase, *Carcinogenesis*, 3, 797, 1982.
233. **Lockyer, J. M. and Magun, B. E.**, The effects of serum and phorbol ester tumor promoter on ornithine decarboxylase activity in Swiss 3T3 cells, *Cancer Lett.*, 18, 215, 1983.
234. **Raunio, H. and Pelkonen, O.**, Effect of polycyclic aromatic compounds and phorbol esters on ornithine decarboxylase and aryl hydrocarbon hydroxylase activities in mouse liver, *Cancer Res.*, 43, 782, 1983.
235. **Takigawa, M., Simsiman, R. C., and Boutwell, R. K.**, The difference between the effects of single and double applications of 12-*O*-tetradecanoylphorbol-13-acetate, a potent tumor promoter, on polyamine metabolism and nucleic acid synthesis in mouse epidermis, *Carcinogenesis*, 4, 5, 1983.

236. **Verma, A. K., Rice, J. M., and Boutwell, R. K.,** Prostaglandins and skin tumor promotion: inhibition of tumor promoter-induced ornithine decarboxylase activity in epidermis by inhibitors of prostaglandin synthesis, *Biochim. Biophys. Res. Commun.*, 79, 1160, 1977.
237. **Verma, A. K., Ashendel, C. L., and Boutwell, R. K.,** Inhibition by prostaglandin synthesis inhibitors of the induction of epidermal ornithine decarboxylase activity, the accumulation of prostaglandins, and tumor promotion caused by 12-*O*-tetradecanoylphorbol-13-acetate, *Proc. Natl. Acad. Sci. U.S.A.*, 77, 5312, 1980.
238. **Yuspa, S. H., Lichti, U., and Ben, T.,** Local anesthetics inhibit induction of ornithine decarboxylase by the tumor promoter 12-*O*-tetradecanoylphorbol-13-acetate, *Proc. Natl. Acad. Sci. U.S.A.*, 77, 5312, 1980.
239. **Nakadate, T., Yamamoto, S., Ishii, M., and Kato, R.,** Inhibition of 12-*O*-tetradecanoylphorbol-13-acetate-induced epidermal ornithine decarboxylase activity by phospholipase A_2 inhibitors and lipoxygenase inhibitors, *Cancer Res.*, 42, 2841, 1982.
240. **O'Brien, T. G., Simsiman, R. C., and Boutwell, R. K.,** The effect of colchicine on the induction of ornithine decarboxylase by 12-*O*-tetradecanoylphorbol-13-acetate, *Cancer Res.*, 36, 3766, 1976.
241. **Olson, J. W. and Russell, D. H.,** Prolonged ornithine decarboxylase induction in regenerating carcinogen-treated liver, *Cancer Res.*, 40, 4373, 1980.
242. **Farwell, D. C., Nolan, C. E., and Herbst, E. J.,** Liver ornithine decarboxylase during phenobarbital promotion of nitrosamine carcinogenesis, *Cancer Lett.*, 5, 139, 1978.
243. **Boutwell, R. K., O'Brien, T. G., Verma, A. K., Weekes, R. G., De Young, L. M., Ashendel, C. L., and Astrup, E. G.,** in *Naturally Occurring Carcinogens — Mutagens and Modulators of Carcinogenesis,* Miller, E. C., Miller, J. A., Hirono, I., Sugimura, T., and Takayama, S., Eds., Japan Scientific Society Press, Tokyo, 1979, 287.
244. **Marks, F., Bertsch, S., and Furstenberger, G.,** Ornithine decarboxylase activity, cell proliferation, and tumor promotion in mouse epidermis *in vivo, Cancer Res.*, 39, 4183, 1979.
245. **Savage, R. E., Jr., Westrich, C., Guion, C., and Pereira, M. A.,** Chloroform induction of ornithine decarboxylase activity in rats, *Environ. Health Perspect.*, 46, 157, 1982.
246. **Boutwell, R. K.,** Some biological aspects of skin carcinogenesis, *Prog. Exp. Tumor Res.*, 4, 207, 1950.
247. **Boutwell, R. K.,** The function and mechanism of promoters of carcinogenesis, *Crit. Rev. Toxicol.*, 2, 419, 1974.
248. **Klein-Szanto, A. J. P., Major, S. K., and Slaga, T. J.,** Induction of dark keratinocytes by 12-*O*-tetradecanoylphorbol-13-acetate and mezerein as an indicator of tumor-promoting efficiency, *Carcinogenesis,* 1, 399, 1980.
249. **Slaga, T. J., Fischer, S. M., Nelson, K., and Gleason, G. L.,** Studies on the mechanism of skin tumor promotion: evidence for several stages in promotion, *Proc. Natl. Acad. Sci. U.S.A.*, 77, 3659, 1980.
250. **Slaga, T. J., Klein-Szanto, A. J. P., Fischer, S. M., Weeks, C. E., Nelton, K., and Major, S.,** Studies on mechanism of action of anti-tumor-promoting agents: their specificity in two-stage promotion, *Proc. Natl. Acad. Sci. U.S.A.*, 77, 2251, 1980.
251. **Weeks, C. E., Herrmann, A. L., Nelson, F. R., and Slaga, T. J.,** α-Difluoromethylornithine, an irreversible inhibitor of ornithine decarboxylase, inhibits tumor promoter-induced polyamine accumulation and carcinogenesis in mouse skin, *Proc. Natl. Acad. Sci. U.S.A.*, 79, 6028, 1982.
252. **Raick, A. N.,** Ultrastructural, histological, and biochemical alterations produced by 12-*O*-tetradecanoylphorbol-13-acetate on mouse epidermis and their relevance to skin tumor promotion, *Cancer Res.*, 33, 269, 1973.
253. **Raick, A. N.,** Cell proliferation and promoting action in skin carcinogenesis, *Cancer Res.*, 34, 920, 1974.
254. **Raick, A. N.,** Cell differentiation and tumor-promoting action in skin carcinogenesis, *Cancer Res.*, 34, 2915, 1974.
255. **Mastro, A. M. and Mueller, G. C.,** Synergistic action of phorbol esters in mitogen-activated bovine lymphocytes, *Exp. Cell. Res.*, 88, 40, 1974.

Chapter 16

MOLECULAR GENETICS OF MAMMALIAN ORNITHINE DECARBOXYLASE

Chaim Kahana

TABLE OF CONTENTS

I. INTRODUCTION

Ornithine decarboxylase (ODC; L-ornithine carboxylyase EC. 4.1.1.17) is the first and a key enzyme in the biosynthesis of polyamines in mammalian cells.[1] This pyridoxal phosphate-dependent enzyme which decarboxylates ornithine to form putrescine is characterized by an extremely short half-life of 10 to 30 min,[2-5] and by rapid and dramatic changes in its level in response to various trophic stimuli.[6-9] Recent studies[2-4,10] using specific antibodies and difluoromethylornithine (DFMO), a suicide inhibitor of ODC,[11] have indicated that changes in *de novo* synthesis and stability of ODC may contribute to alteration in its activity during androgen stimulated mouse kidney hyperthrophy. The combination of rapid turnover and a rapid and dramatic response to a variety of growth stimuli strongly suggest an important role for ODC activity and the generated polyamines during cellular growth and differentiation, and indicates that these changes may be brought about by various mechanisms. Since the availability of antibody reagent and a molecular DNA clone are of a major value for the analysis of gene expression, I will prepare highly specific anti-ODC antiserum and isolate ODC cDNA and use them to explore potential genetic mechanisms that might control ODC activity in mammalian cells.

II. PREPARATION OF ODC AND ANTI-ODC ANTISERUM[16]

ODC was purified to homogeneity from the kidneys of androgen-treated female BALB/c mice using a combination of published procedures.[12,13] In a typical preparation, cytosal was isolated for 200 kidneys and the material precipitating between 30 and 50% $(NH_4)_2SO_4$ was purified by sequential chromatography on DEAE-cellulose, heparin sepharose, ultrogel-34, and finally on an affinity column composed of pyridoxamine phosphate coupled to affigel-10 beads as described by Boucek and Lembach.[14] ODC was eluted from this last column with pyridoxal phosphate. As shown in Figure 1A, 50 to 100 μg of homogenous enzyme were obtained. Of the purified enzyme, 50 μg were subcutaneously injected in Freund's complete adjuvant into New Zealand rabbits at 0, 3, and 6 weeks. Serum was collected 10 days after the injection and when analyzed by the Ouchterlony method showed a single precipitation line (Figure 1B). The antibody protein-A Sepharose complex efficiently eliminated the activity of the soluble enzyme. Currently, ODC is isolated from overproducer cells (described in the next section) yielding miligram amounts of pure enzyme, thus utilizing only DEAE-cellulose and pyridoxamine phosphate-affigel-10 columns during purification.

III. SELECTION OF MOUSE MYELOMA CELLS OVERPRODUCING ODC[15]

Even after strong stimulations of ODC activity in various tissues and cells, the enzyme still represents only a minor fraction of the cellular proteins and efforts were made to obtain an enriched source for ODC protein and mRNA. To do so several cell lines were tested for their ability to give rise to a DFMO resistant variant. Of the cell lines tested, a mouse myeloma cell line emerged as rapidly acquiring resistancy to increasing concentrations of DFMO. Its selection was stopped at 30 m*M* DFMO, 60 times the initial inhibitory concentration. Similar DFMO-resistant murine lymphoma lines[17,18] and hamster line[19] have been reported. The myeloma variant described here, designated 653-1, massively overproduces ODC protein as shown by immunoprecipitation of cellular proteins labeled in vivo with $[^{35}S]$-methionine (Figure 2A), and ODC mRNA as shown by translation in vitro of RNA isolated from the parental and the DFMO resistant cells (Figure 2B).

IV. ISOLATION OF ODC cDNA[15,16]

Since the above-mentioned translation in vitro of RNA isolated from the parental and

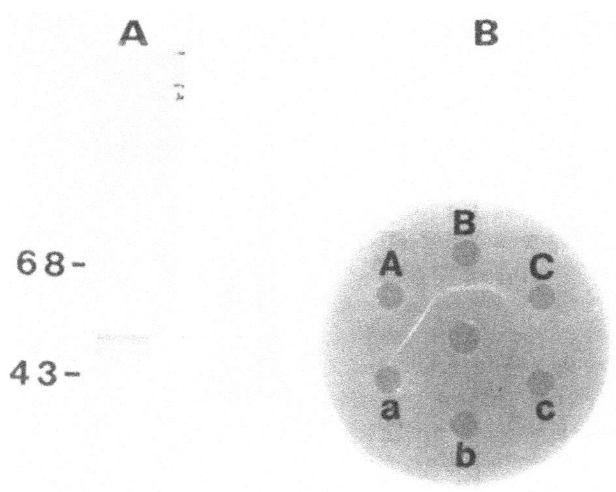

FIGURE 1. Production of anti-ODC antiserum. ODC was isolated from kidneys of female BALB/c mice after stimulation with testosterone, as described in the test. (A). Of the material eluted from the pyridoxamine phosphate affigel-10 column, 5 μg was fractionated by electrophoresis in 10% SDS polyacrylamide gel. The molecular sizes (in kilobases) of bovine serum albumin and ovalbumin are indicated. (B) Ouchterlony analysis of antiserum. Of highly purified ODC, 5 μg was placed in the central well of the Ouchterlony plate, 10 μℓ of preimmune serum of three different rabbits were placed in wells A, B, and C and 10 μℓ of immune serum from the same rabbits in wells A, B, and C.

653-1 cells strongly suggested that ODC RNA is overproduced in 653-1 cells, poly(A)$^+$ RNA from these cells to prepare cDNA library in pBR322 was used. This library was screened in duplicate by differential hybridization to [^{32}P] cDNA probes representing mRNA isolated from the parental and DFMO-resistant 653-1 cells (Figure 3). About 5% of the screened colonies showed marked preferential hybridization to the probe representing the 653-1 cells. The 1.1 kilobase (kb) insert of the largest clone (f2 in Figure 3) was used to demonstrate that all the preferentially hybridizing clones share nucleotide sequences. Hybridization selection followed by translation in vitro demonstrated that the selected RNA directed the synthesis of 53 kilodaltons (kDa) protein recognized by anti-ODC antiserum confirming that clone f2 encodes ODC (Figure 4). Other ODC clones were reported using mouse kidney mRNA[20,21] and RNA from DFMO-resistant murine lymphoma cells[22] as a source of the RNA used for cDNA synthesis. All these clones share nucleotide sequence with the exception of one;[21] this clone is of interest since despite a lack of nucleic acid homology it was able to select mRNA which directed the synthesis in vitro, of a 53-kDa protein recognized by anti-ODC antiserum. Clone f2 was next used as a hybridization probe in characterizing DNA and RNA from 653-1 and the parental cells. As shown in Figure 5B, DNA from BALB/c mouse myeloma cells yielded several restriction fragments hybridizing to the ODC probe, suggesting that the mouse genome may contain a family of ODC related genes. Of the hybridizing fragments, only one of 6.7 kg hybridized to a greater degree in the 653-1 DNA demonstrating that amplification of a ODC gene was involved in the acquisition of DFMO resistancy by the 653-1 cells. Hybridization analysis of RNA

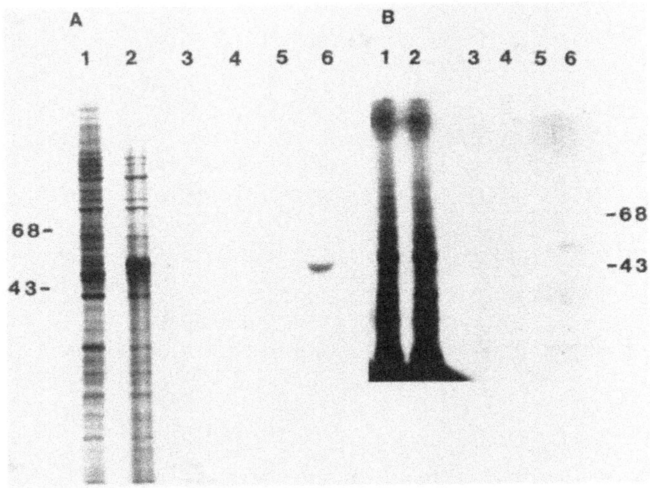

FIGURE 2. Overproduction of ODC protein and mRNA in 653-1 cells. (A) Exponentially growing parental (sensitive) and 653-1 DFMO-resistant cells were pulse-labeled for 20 min with [^{35}S]-methionine. Cytosolic aliquotes containing equal radioactivity were analyzed by acrylamide gel electrophoresis either directly (lanes 1 and 2) or after immunoprecipitation (lanes 3 to 6). Lanes 1, parental; 2, 653-1; 3 and 4, parental and 653-1, respectively, with preimmune serum; 4 and 6, parental and 653-1 with immune serum. (B) Total cellular RNA was isolated from exponentially growing parental and 653-1 cells. The RNA was translated in vitro in reticulocyte lysate. Lanes 1 to 6 are as in (A).

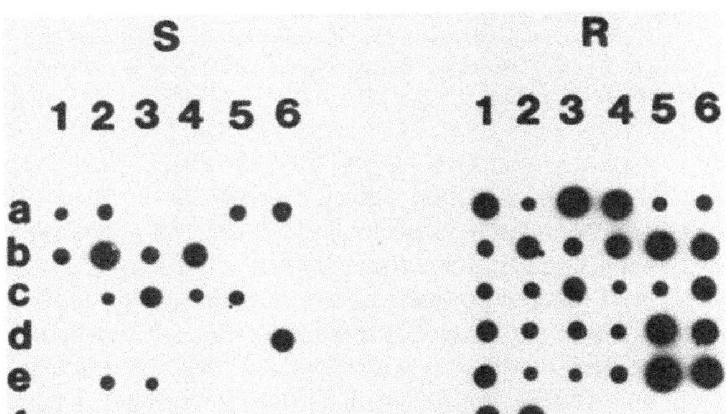

FIGURE 3. Screening of cDNA library. A selected group of colonies that showed preferential hybridization to the cDNA probe representing 653-1 cells or equal hybridization to both 653-1 and parental cell probes were grown in duplicates on nitrocellulose filters. The colonies were lyzed and hybridized to ^{32}P-cDNA probes synthesized from poly(A)$^+$ RNA which was isolated from the parental (S) or 653-1 (R) cells.

isolated from the parental and 653-1 cells demonstrated massive overproduction of ODC RNA in 653-1 cells (Figure 5A). Comparison of 653-1 cells DNA and RNA analysis demonstrates that there appears to be a much greater increase in ODC mRNA in 653-1 cells than expected from the degree of gene amplification, raising the possibility that the amplified genes are more active than the normal ones, either because another change occurred in them

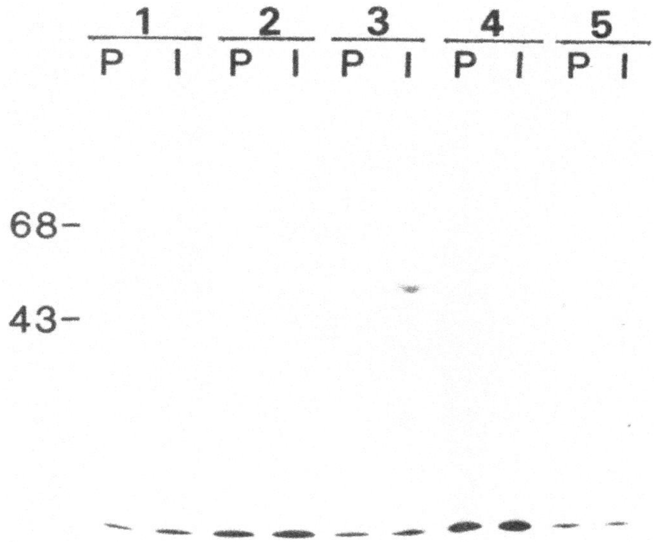

FIGURE 4. Hybridization-selection analysis of cDNA clone f2. Recombinant plasmid DNA was bound to nitrocellulose and hybridized with 1 mg of total polysomal RNA isolated from the 653-1 cells. Complementary mRNA was eluted and characterized by in vitro translation and immunoprecipitation with preimmune serum (P) or immune antiserum (I). Lane 1, translation of total RNA; Lane 2, RNA that did not hybridize to clone f2; Lane 3, RNA that hybridizes to clone f2; Lane 4, RNA that did not hybridize to clone b2 (see Figure 3); and Lane 5, RNA that hybridizes to clone b2. The molecular weight in kilobases of bovine serum albumin and ovalbumine are indicated.

during amplification or because the extra copies escaped regulation by a transcriptional regulator of this gene which is present in limited amounts in the cells. Hybridization analysis of RNA also demonstrated the existence of two species of ODC RNA — a predominant one of about 2.4 kb and a minor one of about 2.6 kb, demonstrating that the 1.1 kb insert of clone f2 represents only part of the ODC mRNA sequences. To obtain a complete ODC cDNA clone a second library was prepared using 653-1 mRNA for cDNA synthesis and utilizing RNase H and DNA polymerase for second strand synthesis.[15,23] With the insert of clone f2 as a hybridization probe, several positive colonies were detected that contained plasmids with about 2.4 kb of inserted sequences corresponding to the estimated length of the major species of ODC RNA. One of these clones, designated pmODC-1, was used in the next study. Its restriction map is shown in Figure 6 and its nucleotide sequence as determined by sequencing both strands by the dideoxy chain termination method[24,25] is shown in Figure 7. It consists of a noncoding 5′ segment of 737 nucleotides, a coding segment of 1383 nucleotide encoding a protein of 461 amino acids whose calculated molecular weight is 51,105, and 342-nucleotide 3′ noncoding region. Sequence information of two other ODC clones has been also reported,[26,27] and good agreement in parallel regions was noted. The sequence of clone pODC 74 reported by Hickok et al.,[27] demonstrated that extra sequences at the 3′ nontranslated region are unique to the minor 2.6 kb species of ODC RNA.

V. PREDICTED AMINO ACID SEQUENCE OF ODC[16]

The amino acid sequence of ODC as deduced from the nucleotide sequence of the cDNA is presented in Figure 7. Comparison of this sequence with the other reported ODC sequences

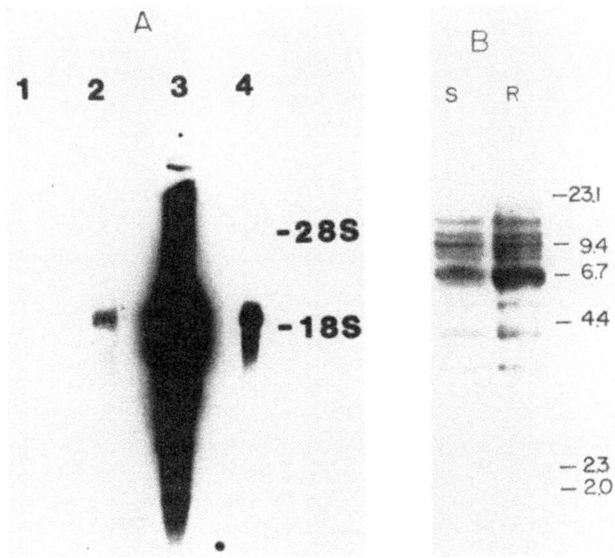

FIGURE 5. Analysis of 653-1 RNA and DNA. (A) Total cellular RNA from the parental myeloma or 653-1 cells was fractionated by electrophoresis in formaldehyde/agarose gel, transferred to a nitrocellulose filter, and hybridized with nick-translated DNA of clone f2. Lanes 1 and 2, 10 and 20 μg, respectively, of parental cells RNA; Lane 3, 10 μg of 653-1 RNA; and Lane 4, short exposure of Lane 3. The positions of 28S and 18S ribosomal RNA are indicated. (B) Total genomic cellular DNA was isolated from the parental myeloma (S) and from 653-1 (R) cells. Portions of 10 μg each were digested with the restriction endonuclease EcoRI, fractionated in 1% agarose gel, transferred to nitrocellulose, and hybridized to nick-translated ODC DNA. The sizes of HindIII fragments of λ phage DNA are indicated (in Kb).

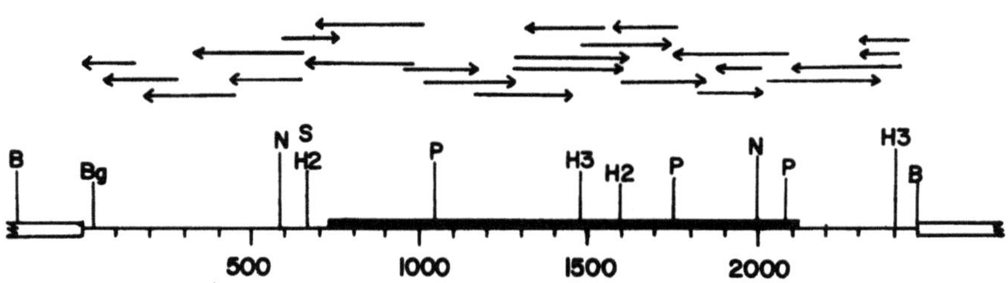

FIGURE 6. Diagram of the mouse ODC cDNA clone pmODC-1. The numbers refer to the nucleotide position along the insert. The thick line indicates the coding region and the open boxes indicate pBR322 sequences surrounding the G-C homopolymer tails added during cloning. The sequencing strategy is indicated by the arrows. The restriction endonuclease sites shown above the line are as follows: B, BamHI; Bg, BglII; H3, HindIII; H 2, HincII; N, NcoI; p, PstI; S, SalI.

revealed a perfect agreement with the amino acids sequence deduced from the nucleotide sequence of ODC mRNA from murine lymphoma cells[26] and only a few differences were noted when compared to the sequence of the ODC clone which was derived from mouse kidney RNA perhaps reflecting differences between kidney and plasmacytoma ODC. As seen in Figure 7, of the 461 amino acids there are 12 cysteine residues, 62 acidic and 57

```
-737  TTCCTCTGTCTCTTCCGGGGTTTTTTGCTTATTAAAGATCTTTTCAGCTTCTCTCACTAAATCTCCTAAGGTCTTGCTCTTTAAATCTTTTAACCGCTCTAACTTTCGCCCAATATCCG

-618  GGCAGACTGCCAGATGAATGACATAGACACATTGGTTTCTTGCCCTGGGTCCTCAGGGTCATAAGGAGTGTACCTGCGATAGGCCTCCTTGAGTCTCTCTAAAAAGGCTGAGGGAGACT

-499  CATTAGGTCCCTGGGTTATCCCTGTTACCTAGGCCAAATTGGTGGCTTCTGCCCGCGTTTTGGAGACCCGCTAAGAGCAACTGGCGATAGAGGACTAGGTGGTTCCTACCTTCTGTAGT

-380  GGTGTAATCTCAATCGGGGCGCTCAAGGGGAAAAGCAGCATTGACTTCATTAGGCAACTGAGTGGGGCGTCCATCATTGCCCCGGACTGCCTTTCTAGCCTCTAGGAGCACCCGCTGCT

-261  TTTCTTCTCCGGTCAGCAGGGTCCCCAACAACTGCTGACAGTCGTCCCAGGTGGGCTGGTGGGTGATGAGGACGGACTGACGGGCGCTCCTCGGGGTTTGGCGCGGCGCCTCCATGGGT

-142  CAGGCCAGCCGGGCCACCGTGCTGTGAGTGTTTCCACCACTCCAAGAAGGCAGCATTCAGAGTTCTTGGCTAAGTCGACCTTGTGAGGAGCTGGTGATAATTTGATTCCATCTCCAGGT

 -23  TCCCTGTAAGCACATCGAGAACC ATG AGC AGC TTT ACT AAG GAC GAG TTT GAC TGC CAC ATC CTT GAT GAA GGC TTT ACT GCT AAG GAC ATT CTG
                              MET Ser Ser Phe Thr Lys Asp Glu Phe Asp Cys His Ile Leu Asp Glu Gly Phe Thr Ala Lys Asp Ile Leu  24

  73  GAC CAA AAA ATC AAT GAA GTC TCT TCC TCT GAC GAT AAG GAT GCG TTC TAT GTT GCG GAC CTC GGA GAC ATT CTA AAG AAG CAT CTG AGG
      Asp Gln Lys Ile Asn Glu Val Ser Ser Ser Asp Asp Lys Asp Ala Phe Tyr Val Ala Asp Leu Gly Asp Ile Leu Lys Lys His Leu Arg  54

 163  TGG CTA AAA GCT CTT CCC CGC GTC ACT CCC TTT TAC GCA GTC AAG TGT AAC GAT AGC AGA GCC ATA GTG AGC ACC CTA GCT GCC ATT GGG
      Trp Leu Lys Ala Leu Pro Arg Val Thr Pro Phe Tyr Ala Val Lys Cys Asn Asp Ser Arg Ala Ile Val Ser Thr Leu Ala Ala Ile Gly  84

 253  ACA GGA TTT GAC TGT GCA AGC AAG ACT GAA ATA CAG TTG GTG CAG GGG CTT GGG GTG CCT GCA GAG AGG GTT ATC TAT GCA AAT CCT TGT
      Thr Gly Phe Asp Cys Ala Ser Lys Thr Glu Ile Gln Leu Val Gln Gly Leu Gly Val Pro Ala Glu Arg Val Ile Tyr Ala Asn Pro Cys 114

 343  AAG CAA GTC TCT CAA ATC AAG TAT GCT GCC AGT AAC GGA GTC CAG ATG ATG ACT TTT GAC AGT GAA ATT GAA TTG ATG AAA GTC GCC AGA
      Lys Gln Val Ser Gln Ile Lys Tyr Ala Ala Ser Asn Gly Val Gln MET MET Thr Phe Asp Ser Glu Ile Glu Leu MET Lys Val Ala Arg 144

 433  GCA CAT CCA AAG GCA AAG TTG GTT CTA CGG ATT GCC ACT GAT GAT TCC AAA GCT GTC TGT CGC CTC AGT GTT AAG TTT GGT GCC ACA CTC
      Ala His Pro Lys Ala Lys Leu Val Leu Arg Ile Ala Thr Asp Asp Ser Lys Ala Val Cys Arg Leu Ser Val Lys Phe Gly Ala Thr Leu 174

 523  AAA ACC AGC AGG CTT CTC TTG GAA CGG GCA AAA GAG CTA AAT ATT GAC GTC ATT GGT GTG AGC TTC CAT GTG GGC AGT GGA TGT ACT GAT
      Lys Thr Ser Arg Leu Leu Leu Glu Arg Ala Lys Glu Leu Asn Ile Asp Val Ile Gly Val Ser Phe His Val Gly Ser Gly Cys Thr Asp 204

 613  CCT GAT ACC TTC GTT CAG GCA GTG TCG GAT GCC CGC TGT GTG TTT GAC ATG GCA ACA GAA GTT GGT TTC AGC ATG CAT CTG CTT GAT ATT
      Pro Asp Thr Phe Val Gln Ala Val Ser Asp Ala Arg Cys Val Phe Asp MET Ala Thr Glu Val Gly Phe Ser MET His Leu Leu Asp Ile 234

 703  GGT GGT GGC TTT CCT GGA TCT GAA GAT ACA AAG CTT AAA TTT GAA GAG ATC ACC AGT GTA ATC AAC CCA GCT CTG GAC AAG TAC TTC CCA
      Gly Gly Gly Phe Pro Gly Ser Glu Asp Thr Lys Leu Lys Phe Glu Glu Ile Thr Ser Val Ile Asn Pro Ala Leu Asp Lys Tyr Phe Pro 264

 793  TCA GAC TCT GGA GTG AGA ATC ATA GCT GAG CCA GGC AGA TAC TAT GTC GCA TCA GCT TTC ACG CTT GCA GTC AAC ATC ATT GCC AAA AAA
      Ser Asp Ser Gly Val Arg Ile Ile Ala Glu Pro Gly Arg Tyr Tyr Val Ala Ser Ala Phe Thr Leu Ala Val Asn Ile Ile Ala Lys Lys 294

 883  ACC GTG TGG AAG GAG CAG CCC GGC TCT GAC GAT GAA GAT GAG TCA AAT GAA CAA ACC TTC ATG TAT TAT GTG AAT GAT GGA GTA TAT GGA
      Thr Val Trp Lys Glu Gln Pro Gly Ser Asp Asp Glu Asp Glu Ser Asn Glu Gln Thr Phe MET Tyr Tyr Val Asn Asp Gly Val Tyr Gly 324

 973  TCA TTT AAC TGC ATT CTT TAT GAT CAT GCC CAT GTG AAG GCC CTG CTG CAG AAG AGA CCC AAG CCA GAC GAG AAG TAT TAC TCA TCC AGC
      Ser Phe Asn Cys Ile Leu Tyr Asp His Ala His Val Lys Ala Leu Leu Gln Lys Arg Pro Lys Pro Asp Glu Lys Tyr Tyr Ser Ser Ser 354

1063  ATC TGG GGA CCA ACA TGT GAT GGC CTT GAT CGG ATC GTG GAG CGC TGT AAC CTG CCT GAA ATG CAT GTG GGT GAT TGG ATG CTG TTT GAG
      Ile Trp Gly Pro Thr Cys Asp Gly Leu Asp Arg Ile Val Glu Arg Cys Asn Leu Pro Glu MET His Val Gly Asp Trp MET Leu Phe Glu 384

1153  AAC ATG GGT GCA TAC ACC GTT GCT GCT GCT TCT ACT TTC AAT GGG TTC CAG AGG CCA AAC ATC TAC TAT GTA ATG TCA CGG CCA ATG TGG
      Asn MET Gly Ala Tyr Thr Val Ala Ala Ala Ser Thr Phe Asn Gly Phe Gln Arg Pro Asn Ile Tyr Tyr Val MET Ser Arg Pro MET Trp 414

1243  CAA CTC ATG AAA CAG ATC CAG AGC CAT GGC TTC CCG CCG GAG GTG GAG GAG CAG GAT GAT GGC ACG CTG CCC ATG TCT TGT GCC CAG GAG
      Gln Leu MET Lys Gln Ile Gln Ser His Gly Phe Pro Pro Glu Val Glu Glu Gln Asp Asp Gly Thr Leu Pro MET Ser Cys Ala Gln Glu 444

1333  AGC GGG ATG GAC CGT CAC CCT GCA GCC TGT GCT TCT GCT AGG ATC AAT GTG TAG ATGCCATTCTTGTAGCTCTTGCCTGCAAGTTTAGCTTGAATTAAGGC
      Ser Gly MET Asp Arg His Pro Ala Ala Cys Ala Ser Ala Arg Ile Asn Val  *

1434  ATTTGGGGGGACCATTTAACTTACTGCTAGTTTGGGATGTCTTTGTGAGAGTAGGGTTGGCACCAATGCAGTATGGAAGGCTAGGAGATGGGGGGTCACACTTACTGTGTTCCTATGGA

1553  AACTTTGAATATTTGTATTACATGGATTTTTATTCACTTTTCAGACATTGATACTAACGTGTGCCCCTCAGCTGCTGAGCAAGCGTTTGTAGCTTGTACATTGGCAGAATGGGCCAGAA

1672  GCTTATTTTGTGACCCATTGTGAAAATAAAATATCTTTAAATAAAAAAAAAAAAAAA  1728
```

FIGURE 7. Nucleotide sequence of mouse ODC cDNA clone pmODC-1. The sequence of the 2465-nucleotide long cDNA insert of the mouse ODC clone pmODC-1 is presented, as is the amino acid sequence translated from the long open reading frame. The numbers on the left refer to the nucleotide number starting with the adenosine of the initiating ATG as number 1. The numbers on the right refer to the amino acid number starting with the initiating methionine.

basic amino acids, and there are two contiguous Lys-Lys or Lys-Arg residues which could be important for proteolytic cleavage.[28] Of interest is the serine residue located at position 303 which is followed by acidic amino acids conforming a site similar to the one reported for casein kinase II phosphorylation sites.[29] Preliminary experiments demonstrated that [^{32}P]-labeled ODC containing alkali labile phosphate can be immunoprecipitated with anti-ODC antiserum from extracts of 653-1 cells labeled in vivo with [^{32}P]-orthophosphate (Figure 8), and, as revealed by acid hydrolysis and high voltage paper electrophoresis, the phosphate resides exclusively in phosphoserine residues. In this respect it is important to note that Meggio et al.[30] reported recently that rat heart ODC can be phosphorylated in vitro by casein kinase II. The observation of ODC phosphorylation in vivo might suggest a role for this posttranslational modification in controlling ODC in mammalian cells. Such a role, as well as the kinase that phosphorylates ODC in cells and the exact location of the phosphorylation have not yet been demonstrated.

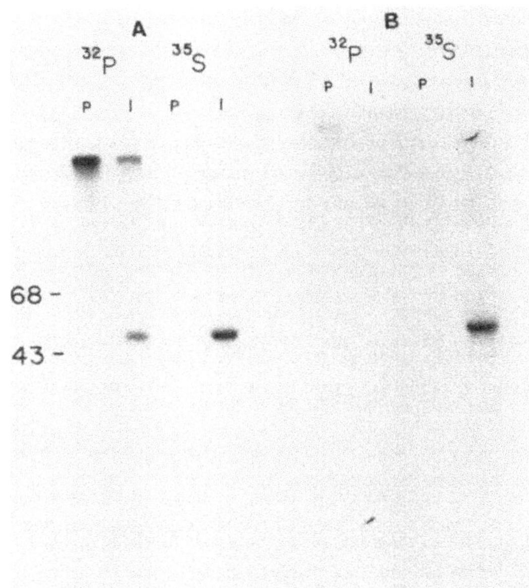

FIGURE 8. Phosphorylation of ODC in 653-1 cells. (A) Exponentially growing 653-1 cells were labeled with [^{32}P]-orthophosphate or with [^{35}S]-methionine. The cellular cytosolic fraction was subjected to immunoprecipitation with preimmune (P) or with immune (I) serum, and the immunoprecipitated material fractionated by SDS-polyacrylamide electrophoresis. (B) The gel presented in (A) was treated with 1*N* KOH at 50°C for 60 min. The molecular sizes of bovine serum albumin and ovalbumin are presented (in kilodaltons).

VI. REGULATION OF ODC RNA DURING CELLULAR GROWTH[15]

Having a specific ODC probe it was determined whether the increase in ODC activity measured during various growth stimulations is accompanied by changes in the level of ODC mRNA. Two experimental systems were used for this purpose: androgen-stimulated mouse kidney hypertrophy during which a striking increase in ODC protein and enzyme activity were observed[2-4,31] and BALB/c 3T3 mouse fibroblasts stimulated to grow by serum or purified growth factors.[32] As shown in Figure 9, in each case the growth stimulation was paralleled by a marked and rapid increase in the level of ODC mRNA. Hybridization of [^{32}P]-labeled nuclear runoff transcripts synthesized in vitro in nuclei which were isolated at various stages after mitogenic stimulation of quiescent 3T3 cells to ODC DNA revealed that the observed increase in ODC RNA results from increased transcriptional activity of an ODC gene.[32a] Hopefully, the isolation of an active ODC genomic clone will enable the characterization of the molecular mechanisms leading to this mitogen-mediated activation of ODC gene.

VII. TRANSLATIONAL REGULATION OF ODC mRNA[33]

Attention has been drawn to the possibility that ODC mRNA is translationally regulated by the following findings: (1) ODC RNA isolated from testosterum-stimulated mouse kidneys in which ODC activity was strongly stimulated, and ODC mRNA isolated from 653-1 cells

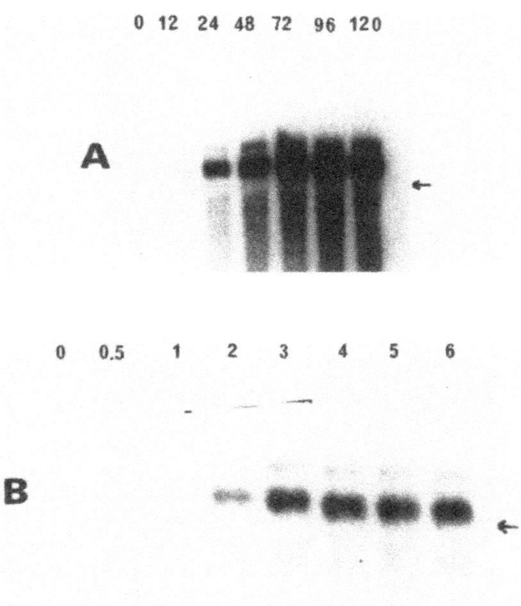

FIGURE 9. Effect of androgen and serum on the level of ODC RNA. (A) Female, 10-week old BALB/c mice were injected subcutaneously with 10 mg of testosterone propionate in sesame oil. Kidneys were removed at various times thereafter, RNA was isolated, and 20-μg portions were fractionated by electrophoresis, transferred to nitrocellulose, and hybridized to nick-translated ODC cDNA. Each sample represents RNA prepared from six individual mice. Times at which RNA was isolated are indicated. (B) Confluent cultures of BALB/c 3T3 cells were maintained in medium containing 0.5% fetal calf serum for 3 days and then stimulated by replacing the medium with one containing 20% fetal calf serum. At specific times, cells were harvested and the RNA fractionated and analyzed as in (A). Times at which RNA was isolated are indicated. The position of 18S ribosomal RNA is indicated by an arrow.

in which it is massively overproduced are very poorly translated in vitro (compare Figures 2A and B); (2) the sequence of clone pmODC-1 demonstrated that ODC mRNA has a long 5′ noncoding leader containing four initiator codons,[16] each of which is followed by an in-frame termination codon all located 5′ to the initiator ATG codon, a situation similar to the one observed in the case of the translationally regulated yeast mRNA encoding a regulator of amino acids biosynthesis;[34,35] and (3) translational regulation of ODC was also suggested from studies of the enzyme during liver regeneration[37-41] and the growth of cells in culture,[41-45] in which it was shown that polyamines cause a rapid fall in ODC activity similar to the one caused by cyclohexmide, but had no effect on the rate of enzymatic activity inactivation in vivo nor on its activity in vitro.[40] To test this possibility more directly, polyamine-mediated translational regulation of ODC mRNA was analyzed in the 653-1 cells which massively overproduce ODC mRNA and protein enabling measurement of changes in ODC RNA and in the rate of ODC synthesis by short pulsing with [^{35}S]-methionine in response to putative regulators. The growth of 653-1 cells in the absence of DFMO (but not in its presence) is highly sensitive to added ornithine, presumably due to large amounts of polyamines accumulating in the cells. In exploring the basis of ornithine toxicity, a rapid

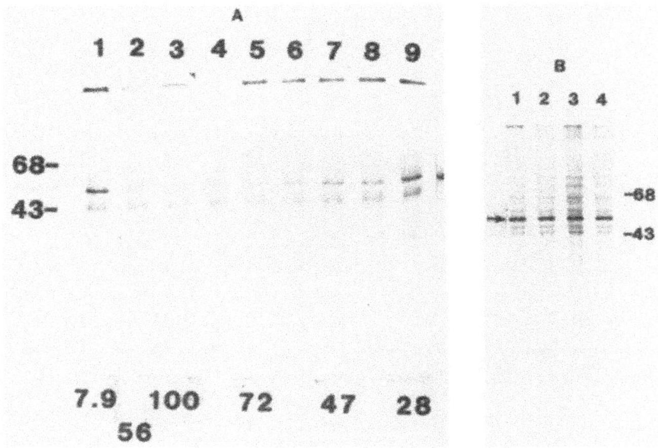

FIGURE 10. Effect of ornithine on ODC synthesis. (A) 653-1 cells growing in the absence of DFMO were exposed to 5 m*M* L-ornithine for 0, 30, and 60 min (lanes 1 to 3). After 60 min the medium was replaced with ornithine-free medium and the cells harvested at 30 min intervals thereafter (lanes 4 to 9). At 5 min prior to harvesting the cells were exposed to [^{35}S]-methionine and equal amounts of radioactive total cellular proteins were analyzed by gel electrophoresis. The numbers at the bottom of the lanes refer to the cellular content of putrescine (nanomoles per 10^6 cells) of the samples indicated. (B) 653-1 cells were grown and analyzed as in (A) but DFMO was present in the growth medium. Lanes 1 to 4 are 0, 15, 30, and 45 min, respectively. The molecular sizes of bovine serum albumin and ovalbumine are indicated (in kilodaltons).

and selective decrease in ODC synthesis (Figure 10A) was noticed. As seen in the figure, this effect is totally reversed by the removal of ornithine from the growth medium; in contrast, in the presence of DFMO, the addition of ornithine does not effect the synthesis of ODC (Figure 10B) suggesting that the effect of ornithine was mediated by its intracellular conversion to putrescine. Measurement of the intracellular level of polyamines following the addition of ornithine to the growth medium (in the absence of DFMO) demonstrated a prompt accumulation of putrescine, as expected from the high ODC activity in 653-1 cells (Figure 10A). In contrast to ornithine, the addition of putrescine to 653-1 cells dramatically prevented the synthesis of ODC regardless of whether DFMO was present or absent in the growth medium (Figure 11). In addition to putrescine, spermidine, spermine, cadavarine, and diaminopopane also strongly and specifically prevented the synthesis of ODC, and these polyamines had no or a very slight effect on the overall protein synthesis. In contrast, pactamycin (an inhibitor of initiation of protein synthesis) and cycloheximide (an inhibitor of elongation of polypeptide chains) at levels of inhibition varying from 20 to 80% showed no differential effect on ODC synthesis. Since during the inhibition of ODC synthesis, no changes in ODC RNA or in ODC stability were observed it was concluded that ODC mRNA was subjected to translational regulation by polyamines. In this respect, Kanamoto and co-workers[46] recently reported that the rapid decay in ODC activity in primary cultured hepatocytes after the addition of putrescine is partially due to the suppression of ODC synthesis.

It was recently demonstrated that in contrast to intact ODC mRNA which is very poorly translated in vitro, ODC RNA which was synthesized in vitro with no 5′ nontranslated region is very efficiently translated in vitro,[32a] strongly suggesting that sequences from the 5′ nontranslated region of the mRNA are part of the machinery regulating its translatability. Further studies should determine the specific details of this regulatory mechanism and its relative contribution in controlling ODC in cells under physiological conditions.

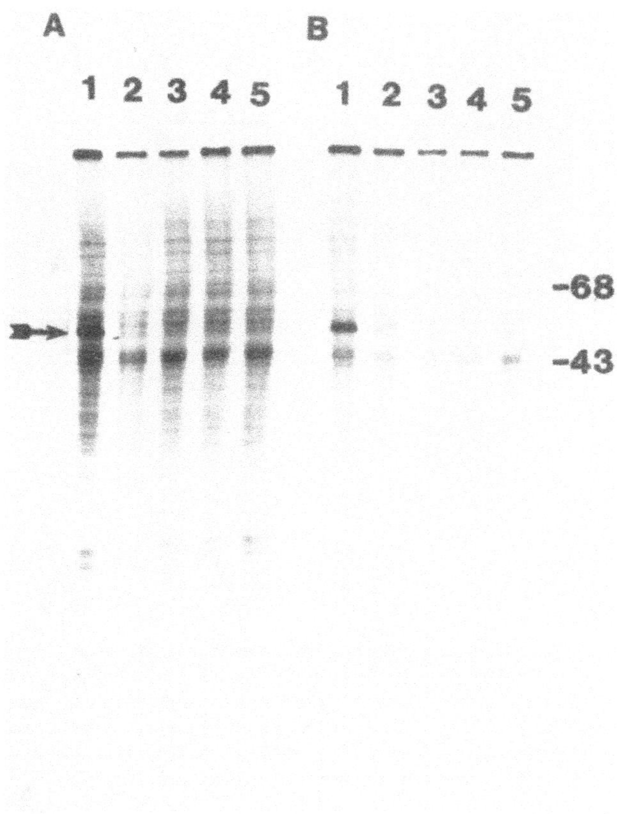

FIGURE 11. 653-1 cells were grown and labeled as in Figure 9 in the (A) presence or (B) absence of DFMO. Putrescine (30 m*M*) was added at time 0 (lane 1) and cells were harvested at 15-min intervals (lanes 2 to 5).

VIII. FAMILY OF ODC-RELATED GENES IN MAMMALIAN GENOMES[16]

Using cloned ODC cDNA as a hybridization probe, several restriction fragments resulting from digestion of mouse cellular DNA with one of a number of restriction endonucleases were shown to contain ODC sequences.[16,26] This finding, together with the observation that only one 7.6-kb fragment was amplified in the DFMO-resistant cells suggested that the mouse genome may contain a family of ODC related genes. However, because the probes used in the above-mentioned studies contained both coding and noncoding sequences it was not established that the various hybridizing fragments shared ODC coding sequences. Knowing the nucleotide sequence of ODC mRNA[16] enabled a set of cloned probes specifically representing 5′ noncoding, coding, and 3′ noncoding sequences (BglII-SalI, PstI-HindIII, and PstI-BamI, respectively — Figure 6) to be prepared. As seen in Figure 12, each of these clones when used to probe a Southern blot containing EcoRI digested mouse cellular DNA hybridized to all bands detected by the entire cDNA, strongly indicating the existence of a multimember family of ODC related genes. As shown in Figure 13, the existence of multiple ODC genes is not unique to the mouse genome, but is also observed in rat and human genomes. As mentioned previously, only the 6.7 kb fragment from the mouse genome can be pointed out as representating a functional gene since its amplification enabled the acquisition of DFMO resistancy by 653-1 cells. The relationship between the various mem-

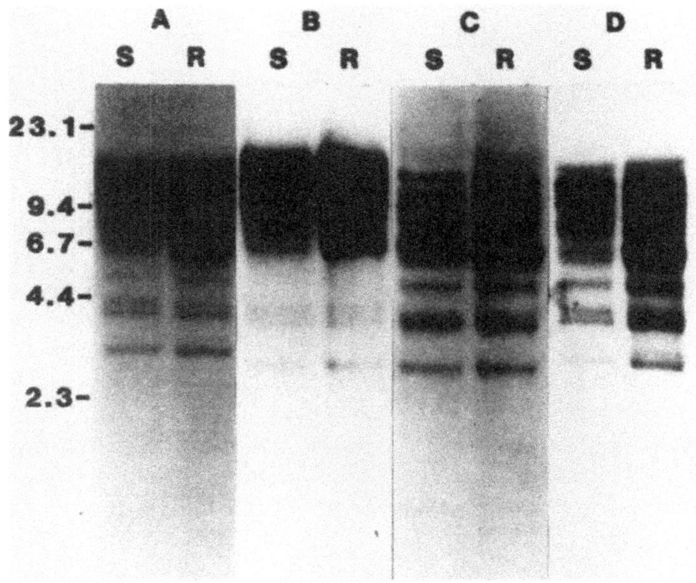

FIGURE 12. Restriction analysis of ODC sequences in mouse genomic DNA with various ODC probes. Total genomic DNA was isolated from the parental (S) or 653-1 (R) cells. Portions of 10 μg each were digested with restriction endonuclease EcoRI, fractionated in 1% agarose gel, transferred to nitrocellulose, and hybridized to ^{32}P-labeled ODC cDNA probes (A). The entire pmODC-1 sequence: (B) 5′ noncoding sequence, (C) coding sequence, and (D) 3′ noncoding sequence. The lengths in kilobase pairs of DNA markers consisting of HindIII fragments of phage DNA are indicated.

bers of the ODC gene family and the possibility that other members are also functional should still be determined.

IX. CONCLUDING REMARKS

For many years now, ODC has been in the focus of extensive enzymological studies yielding extremely valuable information concerning its characterization as an enzyme and its tremendous induction by a variety of growth stimulations. This accumulated information was also extremely valuable in enabling the isolation of a molecular cDNA clone in several laboratories. As reflected in this and other chapters of this book, tremendous progress has already been made in pointing to various genetic mechanisms as potentially controlling ODC in mammalian cells. Hopefully, rapid progress will continue in exploring the specific details of these mechanisms and in evaluating the relative contribution of each of them in controlling ODC activity in various biological systems and under various physiological conditions.

ACKNOWLEDGMENTS

The studies presented here were carried out during my stay as a postdoctoral fellow in Dr. Daniel Nathan's laboratory at the Johns Hopkins University School of Medicine. I would like to thank Mrs. Peggy Wynick for her help in preparing the manuscript.

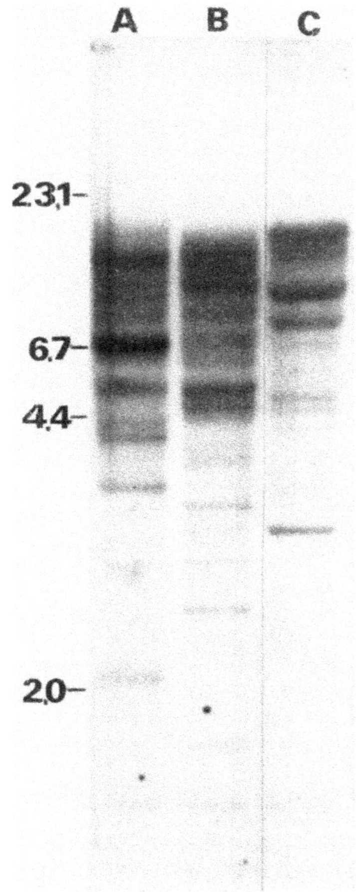

FIGURE 13. Restriction analysis of ODC sequences in genomic DNA of various mammalian species. Total genomic DNA was isolated from (A) 653-1 mouse myeloma cells, (B) rat embryo fibroblasts, and (C) human HeLa cells. Digestion and analysis was as in Figure 12. pmODC-1 was used as an hybridization probe. Molecular weight markers are as in Figure 12.

REFERENCES

1. **Pegg, A. E. and Williams-Ashman, H. G.,** Biosynthesis of putrescine, in *Polyamines in Biology and Medicine,* Morris, D. R. and Marton, L. J., Eds., Marcel Dekker, New York, 1981, chap. 1.
2. **Isomaa, V. V., Pajunen, A. E. I., Bardin, C. W., and Jänne, O. A.,** Ornithine decarboxylase in mouse kidney, *J. Biol. Chem.,* 258, 6735, 1983.
3. **Seeley, J. E. and Pegg, A. E.,** Changes in mouse kidney ornithine decarboxylase activity are brought about by changes in the amount of enzyme protein as measured by radioimmunoassay, *J. Biol. Chem.,* 258, 2496, 1983.
4. **Seeley, J. E., Pösö, H., and Pegg, A. E.,** Effect of androgenes on turnover of ornithine decarboxylase in mouse kidney, *J. Biol. Chem.,* 257, 7549, 1982.
5. **Seeley, J. E., Pösö, H., and Page, A. E.,** Measurement of the number of ornithine decarboxylase molecules in rat and mouse tissues under various physiological conditions by binding of radiolabeled α-difluoromethylornithine, *Biochem. J.,* 206, 311, 1982.

6. **Tabor, C. W. and Tabor, H.,** 1-4-Diaminobutane (putrescine), spermidine, and spermine, *Annu. Rev. Biochem.*, 45, 285, 1976.
7. **Jänne, J., Pösö, H., and Raina, A.,** Polyamines in rapid growth and cancer, *Biochim. Biophys. Acta*, 473, 241, 1978.
8. **Pegg, A. E. and McCann, P. P.,** Polyamines metabolism and function, *Am. J. Physiol.*, 243, 212, 1982.
9. **Tabor, C. N. and Tabor, H.,** Polyamines, *Annu. Rev. Biochem.*, 53, 749, 1984.
10. **Kameji, T. and Hayashi, S. I.,** Affinity labeling of purified ornithine decarboxylase by α-difluoromethylornithine, *Biochim. Biophys. Acta*, 705, 405, 1982.
11. **Metcalfe, B. B., Bey, P., Danzin, C., Jung, M. J., Casara, P., and Vevert, J. P.,** Catalytic irreversible inhibition of mammalian ornithine decarboxylase, *J. Am. Chem. Soc.*, 100, 2551, 1978.
12. **Seeley, J. E., Pösö, H., and Pegg, A. E.,** Purification of ornithine decarboxylase from kidneys of androgen-treated mice, *Biochemistry*, 21, 3394, 1982.
13. **Kameji, T., Murakami, Y., Fujita, F., and Hayashi, S. I.,** Purification and some properties of ornithine decarboxylase from rat liver, *Biochim. Biophys. Acta*, 717, 111, 1872.
14. **Boucek, R. J. and Lembach, K. J.,** Purification by affinity chromatography and preliminary characterization of ornithine decarboxylase from simian virus 40 transformed 3T3 mouse fibroblasts, *Arch. Biochem. Biophys.*, 184, 408, 1977.
15. **Kahana, C. and Nathans, D.,** Isolation of cloned cDNA encoding mammalian ornithine decarboxylase, *Proc. Natl. Acad. Sci. U.S.A.*, 81, 3645, 1984.
16. **Kahana, C. and Nathans, D.,** Nucleotide sequence of murine ornithine decarboxylase mRNA, *Proc. Natl. Acad. Sci. U.S.A.*, 82, 1673, 1985.
17. **McConlogue, L. and Coffino, P.,** Ornithine decarboxylase in difluoromethylornithine-resistant mouse lymphoma cells, *J. Biol. Chem.*, 258, 1983, 1983.
18. **McConlogue, L. and Coffino, P.,** A mouse lymphoma cell mutant whose major protein product is ornithine decarboxylase, *J. Biol. Chem.*, 258, 12083, 1983.
19. **Choi, J. H. and Scheffler, I.,** Chinese hamster ovary cells resistant to β-difluoromethylornithine are overproducers of ornithine decarboxylase, *J. Biol. Chem.*, 258, 12601, 1983.
20. **Kontula, K. K., Torkkeli, T. K., Bardin, C. W., and Jänne, O. A.,** Androgen induction of ornthinine decarboxylase mRNA in mouse kidney as studied by complementary DNA, *Proc. Natl. Acad. Sci. U.S.A.*, 81, 731, 1984.
21. **Berger, F. G., Szymanski, P., Read, E., and Watson, G.,** Androgen-regulated ornithine decarboxylase mRNAs of mouse kidney, *J. Biol. Chem.*, 258, 2496, 1984.
22. **McConlogue, L., Gupta, M., Wu, L., and Coffino, P.,** Molecular cloning and expression of the mouse ornithine decarboxylase gene, *Proc. Natl. Acad. Sci. U.S.A.*, 81, 540, 1984.
23. **Gubler, V. and Hoffman, B. J.,** A simple and very efficient method for generating cDNA libraries, *Gene*, 25, 263, 1983.
24. **Sanger, F. and Coulson, A. R.,** The use of think acrylamide gels for DNA sequencing, *FEBS Lett.*, 87, 107, 1978.
25. **Sanger, F., Coulson, A. R., Barrall, B. G., Smith, A. J. H., and Rol, B. A.,** Cloning in a single-stranded bacteriophage as an aid to rapid DNA sequencing, *J. Mol. Biol.*, 143, 161, 1980.
26. **Gupta, M. and Coffino, P.,** Mouse ornithine decarboxylase complete amino acid sequence deduced from cDNA, *J. Biol. Chem.*, 260, 2971, 1985.
27. **Hickok, N. J., Seppanen, P. J., Kontula, K. K., Jänne, P. A., Bardin, C. W., and Jänne, O. A.,** Two ornithine decarboxylase mRNA species in mouse kidney arise from size heterogeneity at their 3′ termini, *Proc. Natl. Acad. Sci. U.S.A.*, 83, 584, 1986.
28. **Steiner, D. F., Quinn, P. S., Chan, S. J., Marsh, J., and Tager, H. S.,** Processing mechanisms in the biosynthesis of proteins, *Ann. N.Y. Acad. Sci.*, 343, 1, 1980.
29. **Hathaway, G. M. and Traugh, J. A.,** Caseom lomases — multipotential protein kinases, *Curr. Top. Cell. Regul.*, 21, 101, 1982.
30. **Meggio, F., Flamini, F., Caldarera, C. M., Guarnieri, C., and Pinna, L. A.,** Phosphorylation of rat heart ornithine decarboxylase by type-2 casein kinase, *Biochem. Biophys. Res. Commun.*, 122, 997, 1984.
31. **PaJunen, A. E. I., Isomaa, V. V., Jänne, O. A., and Bardin, C. W.,** Androgenic regulation of ornithine decarboxylase activity in mouse kidney and its relationship to changes in cytosol and nuclear androgen receptor concentrations, *J. Biol. Chem.*, 257, 8190, 1982.
32. **Linzer, D. I. H. and Nathans, D.,** Growth-related changes in specific mRNAs of cultured mouse cells, *Proc. Natl. Acad. Sci. U.S.A.*, 80, 4271, 1983.

32a. **Kahana, C.,** unpublished results.

33. **Kahana, C. and Nathans, D.,** Translational regulation of mammallian ornithine decarboxylase by polyamines, *J. Biol. Chem.*, 260, 15390, 1985.
34. **Thireos, G., Driscol-Penn, M., and Greer, H.,** 5′ untranslated sequences are required for the translational control of a yeast regulatory gene, *Proc. Natl. Acad. Sci. U.S.A.*, 81, 5096, 1984.

35. **Hinnebusch, A. G.,** Evidence for translational regulation of the activator of general amino acid control in yeast, *Proc. Natl. Acad. Sci. U.S.A.*, 81, 6442, 1984.
36. **Jänne, J. and Hölttä, E.,** Regulation of ornithine decarboxylase activity by putrescine and spermidine in rat liver, *Biochem. Biophys. Res. Commun.*, 61, 449, 1947.
37. **Guha, S. K. and Jänne, J.,** Inhibition of ornithine decarboxylase *in vivo* in rat ovary, *Biochem. Biophys. Res. Commun.*, 75, 136, 1977.
38. **Pösö, H.,** Regulation of ornithine decarboxylase and *S*-adenosyl-L-methionine decarboxylase in regenerating rat liver by various amines: evidence for translational control, *Acta Chem. Scand. B*, 31, 71, 1977.
39. **Pegg, A. E., Canover, C., and Wrona, A.,** Effects of aliphatic diamines on rat liver ornithine decarboxylase activity, *Biochem. J.*, 170, 651, 1978.
40. **Heby, O. and Jänne, J.,** Polyamine antimetabolites: biochemistry, specificity and biological effects of inhibitors of polyamine synthesis, in *Polyamines in Biology and Medicine,* Morris, D. R. and Marton, L. J., Eds., Marcel Dekker, New York, 1981, chap. 13.
41. **Pett, D. M. and Ginsberg, H. S.,** Metabolism of polyamines in KB cells, *Fed. Proc. Fed. Am. Soc. Exp. Biol.*, 27, 615, 1968.
42. **Kay, J. E. and Lindsay, V. J.,** Control of ornithine decarboxylase activity in stimulated human lymphocytes by putrescine and spermidine, *Biochem. J.*, 132, 791, 1973.
43. **Clarke, J. L.,** Specific induction of ornithine decarboxylase in 3T3 mouse fibroblasts by pituitary growth factors: cell density-dependent biphasic response and alteration of half life, *Biochemistry,* 13, 4668, 1974.
44. **Clark, J. L. and Fuller, J. L.,** Regulation of ornithine decarboxylase in 3T3 cells by putrescine and spermidine: indirect evidence for translational control, *Biochemistry,* 14, 4403, 1975.
45. **Bethell, P. R. and Pegg, A. E.,** Effects of diamines on ornithine decarboxylase activity in control and virally transformed mouse fibroblasts, *Biochem. J.*, 180, 87, 1979.
46. **Kanamoto, R., Utsumoniya, K., Kameji, T., and Hayashi, S. I.,** Effects of putrescine on synthesis and degradation of ornithine decarboxylase in primary cultured hepatocytes, *Eur. J. Biochem.*, 154, 589, 1986.

Chapter 17

THE MOLECULAR BIOLOGY OF MOUSE ORNITHINE DECARBOXYLASE

Philip Coffino and Lisa McConlogue

TABLE OF CONTENTS

I. ODC OVERPRODUCING CELLS AND CLONING OF cDNA

The near simultaneous appearance in early 1984 of papers from three different laboratories reporting the cloning of ornithine decarboxylase (ODC) cDNA[1-3] signaled molecular biologists' interest in and appreciation of an enzyme whose properties have entranced biochemists, cell biologists, and physiologists for decades. My laboratory has chosen to focus on ODC because it is biologically important, appears to play a significant role in cell growth, and is subject to a fascinating variety of modes of regulation. An extensive literature demonstrates that activity of ODC can be made to undergo very large and rapid changes in response to a great variety of stimuli that include hormones, mitogens, tumor promoters, and agents that stimulate differentiation. The multiple mechanisms that appear to underlie these changes require the tools of molecular biology if they are to be adequately studied.

A major obstacle to the biochemical study of mammalian ODC has been its low absolute abundance, even in tissues with the highest enzymatic activity. Low abundance of the ODC mRNA also makes it more difficult to clone the cDNA, the logical first step in the application of molecular biology to the enzyme. We used difluoromethylornithine (DFMO), an inhibitor of ODC, to select a series of S49 lymphoma tissue culture cell variants that overproduce the enzyme; in the most floridly productive of these (termed D4.1) 15 to 20% of the newly synthesized protein is made as ODC.[4] There is a single distinctive band, of molecular mass about 53,000, produced in greater abundance in the mutants. The protein is ODC. Its identity was confirmed by multiple criteria, including specific immunoprecipitation with anti-ODC antibody.

Interestingly, no labeled polypeptides other than full-size ODC are seen to be present in increased amount in the D4.1 overproducer. This is true even if they are sought by two-dimensional electrophoresis, a far more sensitive method. This observation is relevant to an unusual aspect of ODC metabolism, its short in vivo half-life, the shortest known for any enzyme.[5] It suggests that the degradation of ODC cannot be initiated by a process that noticeably changes its electrophoretic mobility, e.g., ubiquitination, unless the modified form has only a fleeting existence. Secondly, whatever the mechanism of ODC rapid degradation, the first step that causes a significant change in mobility is rate limiting, else degradation intermediates would readily be seen in the overproducer.

Poly-A^+ RNA was isolated from the D4.1 cells and double-stranded cDNA was made and cloned.[1] A series of ODC cDNA clones were thus isolated. Using a cDNA probe obtained in that way, the amount of ODC mRNA in the mutant cells was determined to be 300-fold greater than that in wild-type cells. Sequencing the cDNA[6] revealed an open reading frame encoding a protein with 461 amino acids, with a molecular weight of 51,172. The size and isoelectric point (predicted from the amino acid composition) are consistent with the values observed for purified mouse ODC.

II. ALTERATION OF ODC mRNA LEVEL EXPLAINS SOME FORMS OF REGULATION

Availability of a cDNA probe for ODC makes it possible to determine in what cases and to what degree changes in intracellular ODC mRNA level can explain changes in enzyme activity. The results have varied with the system studied. For example,[15] after treating PC12 cultured rat pheochromocytoma cells with nerve growth factor, their ODC activity and ODC mRNA content change together, with the same time course and to the same degree. This represents a simple case: induction of activity in response to the growth factor is apparently the result solely of changes in ODC mRNA level. In contrast, when mice are treated with androgen, induction of enzymatic activity in the kidney exceeds induction of mRNA;[1] at least a portion of the increased activity results from prolongation of ODC protein half-life.[7,8] Alteration of enzyme stability thus introduces itself as a second regulatory theme.

III. NONTRANSCRIPTIONAL REGULATION OF ODC ACTIVITY

Adding putrescine to cultured cells reduces their intracellular ODC activity. Similar changes in ODC activity have been observed in the tissues of intact animals treated with polyamines. Conversely, competitive inhibitors of ODC, by reducing polyamine pools, increase enzymatic activity. Polyamines can therefore mediate a form of feedback inhibition of ODC. The mechanism by which this occurs, however, is not evident. A series of mutant S49 cell lines was used to investigate this question.[10] These cells were generated as intermediates in the multistep selection required to obtain the D4.1 overproducing variant of S49 cells described above, or were isolated in independent single-step selections. The intermediate lines examined have basal ODC activities that range from about two- to tenfold that of the wild-type cells from which they were derived. Experiments were performed to determine what happens to ODC activity and ODC mRNA levels when the cells were treated overnight with putrescine to increase their polyamines or with α-methylornithine, a reversible competitive inhibitor of the enzyme, to decrease their polyamines. Use of the inhibitor does not interfere with the assay of ODC activity if extracts first are dialyzed. Northern blot analysis of mRNA from each cell type revealed, as expected, that under basal conditions the variant cells, which have more ODC activity than do the wild-type cells, also have more ODC mRNA. In all of the cells, treatment with either putrescine or α-methylornithine had no discernible effect on the amount of ODC mRNA. These treatments did, however, have effects on ODC activity, and these effects were strikingly different in the wild-type and mutant cells. Putrescine treatment greatly depressed ODC activity in wild-type cells, but had a relatively small effect on activity in the overproducers. In contrast, treatment with α-methylornithine had very little effect in wild-type cells and greatly enhanced the activity in the mutants tested.

Two conclusions reasonably follow from this series of experiments. Polyamines exert regulatory effects on ODC activity and this change in activity is not a reflection of changes in steady-state ODC mRNA levels. The differing response to treatment of mutant and wild-type cells is most readily explained by the hypothesis that the former contain higher levels of endogenously generated polyamines than do wild-type cells and that further experimental augmentation of their polyamines is capable of exerting little additional regulatory effect. Conversely, the reduction of polyamine levels that results from use of the enzyme inhibitor causes relief of supression in the mutants, but no change in the wild-type cells, because polyamine levels in the latter are not sufficient to suppress activity.

These results showed that alteration of polyamine pools can impel a 12-fold change in ODC activity; this may constitute an underestimate of the magnitude of the effect because of limitations on the degree to which we can control those pools. Further experiments demonstrated that two components together explain the regulation of ODC activity: protein half-life and synthetic rate. To measure the half-life of ODC, cells were treated with cycloheximide to inhibit protein synthesis and the rate of decline of ODC activity was measured. These experiments showed that changes in polyamine pools can cause a two- to threefold change in the half-life of the ODC. By implication, there should be a four- to sixfold change in synthetic rate of ODC. Direct determination, by measuring with two-dimensional gels, of the rate of ^{35}S-methionine incorporation into ODC (using pulse label times that were short compared to ODC half-life) confirmed this. Using similar means, others have recently confirmed in a different cultured cell line that changes in polyamine levels alter translational efficiency of ODC mRNA.[14] Because polyamines change the synthetic rate of the enzyme without altering the steady-state level of its mRNA, it seems a plausible hypothesis that regulation of translational efficiency of ODC mRNA by polyamines could play a role in this process. Studies we have carried out to define the structure of the mRNA suggest a mechanism by which this could occur.

IV. CLONING OF MOUSE ODC GENOME DNA

To assess the genetic organization of mouse ODC genes and the nature of their alteration in D4.1 cells, high molecular weight DNA was prepared from wild-type and D4.1 cells, digested with EcoR I or BamH I and analyzed by the Southern procedure, using ODC cDNA as probe. With each restriction enzyme, multiple hybridizing bands were seen. Using EcoR I, one band of about 6.9 kilobase (kb) is present in very much higher copy number in D4.1 than in wild-type DNA. Similarly, among the BamH I fragments, one, of about 6.7 kb, is amplified in D4.1 DNA. These results implied that ODC may belong to a multigene family, one member of which is amplified in the D4.1 mutant.[1] This analysis suggested that the entire gene could be contained within a 5.7-kb EcoR I-BamH I fragment. Using DNA from the amplified cells, that fragment was cloned in the pUC8 vector. Dideoxy sequencing of that gene fragment, however, revealed sequence diversion from the 5′-most portion of the ODC cDNA sequence. This suggested, as proved to be the case, that the site of transcriptional initiation as well as one or more noncoding exons was located 5′ to the genomic fragment we had cloned. We therefore isolated an overlapping clone containing the missing exon. Sequencing relevant parts of that fragment showed it to contain all the sequence information present in the 5′-most part of the longest cDNA we have sequenced. Primer extension and S_1 protection experiments, using mRNA preparations from both D4.1 and wild-type S49 cells, as well as RNA from the kidney of androgen treated Balb/c mice, were carried out to define the site of initiation of transcription. All of these sources of RNA gave concordant results and demonstrated that the transcription start site lies 32 bases 3′ to a canonical TATAA box. Because the 5′ terminus of ODC mRNA is identical, regardless of whether it is obtained from wild-type or overproducing cultured cells or from normal mouse tissue, the cloned genomic DNA from the mutant cells must offer a faithful representation of the sequence transcribed in normal cells. Hence, we can use that sequence to infer the structure of the 5′ region of ODC mRNA, which proves to have interesting features.

V. STRUCTURE OF THE 5′ UNTRANSLATED LEADER OF ODC mRNA

The codon in ODC mRNA used for initiation of translation is proceeded by 312 bases of 5′ untranslated leader. The sequence is structurally unusual. Between positions −298 and −196, 85 of 103 bases Gs or Cs [numbering is with respect to the A of the AUG initiation codon]. A computer algorithm (rna-dp, H. Martinez, Biomathematics Computation Laboratory, UCSF) designed to predict RNA secondary structure by free energy minimization (Tinoco energy rules) was applied to the segment −306 to −115. The resulting structure has a predicted free energy of −145 kcal, of which −75 kcal resides in a single stem-loop structure between positions −297 and −196. The predictive value of calculations of this kind are sufficiently feeble that one should put little confidence in the correctness of this particular secondary structure. On the other hand, it is highly likely that this RNA can assume *some* very stable secondary structure and that there exist a series of alternate secondary structures closely related in energy.

It is our hypothesis that some of these are more readily translated than others and that polyamines can bind to the RNA and thus change the equilibrium among alternative forms. That polyamines can bind to RNA with high affinity and can stabilize a specific secondary structure has been demonstrated for tRNA[11] and for MS2 single-stranded RNA virus.[12] Much interest has been elicited by observations that secondary structure of mRNA leader sequences can greatly interfere with translation (e.g., Reference 9), but there is little understanding of (or even plausible speculation about) the biological role of such structure. There is considerable evidence that in eukaryotes translation is initiated by binding of the 40S subunit of ribosomes at the capped 5′ end of mRNA, and that the message is then scanned in the 3′ direction until an initiation codon is encountered (Reference 9 and references within). Secondary structure interposed between the cap site and initiation codon could therefore plausiby affect translational efficiency. Kozak has found that hairpins in the vicinity of the initiation codon do not interfere with translation so long as their predicted free energy is less than -30 kcal, but that more stable hairpins can do so (personal communication). Recall that the predicted free energy of the major stem-loop structure of the ODC mRNA is -75 kcal. It is of course possible that polyamines regulate translational efficiency of ODC mRNA, not by binding to the leader sequence, but in some less direct fashion. The hypothesis, however, is simple and can be readily tested.

VI. PROSPECTS

In our view three significant problems must be addressed if we are to understand how ODC is regulated. First, how do multiple classes of effectors including hormones, mitogens, and tumor promoters act to regulate ODC mRNA levels? Second, how do polyamines affect the translation of ODC mRNA? Third, what makes ODC such an unstable protein and how is its stability regulated? The work described here should facilitate obtaining answers to each of these questions. It makes available molecular probes for assessment of ODC mRNA, genomic DNA for determination of cis-acting sequences and the *trans*-acting regulatory cellular factors that interact with these sequences, and mutant cells in which the enzyme is made in sufficient abundance that its fate and mechanism of immolation may be conveniently investigated.

CAUTION

The manuscript for this paper was submitted May 12, 1986, and has not been subsequently revised. Therefore, the information provided is not current, and some of the expressed opinions and interpretations are contrary to those now held by the authors.

ACKNOWLEDGMENTS

This work was supported by grants to Philip Coffino from the NIH (CA29048), NSF (PCM78-07382), and the American Cancer Society (NP-477B).

REFERENCES

1. **McConlogue, L., Gupta, M., Wu, L., and Coffino, P.,** Molecular cloning and expression of the mouse ornithine decarboxylase gene, *Proc. Natl. Acad. Sci. U.S.A.*, 81, 540, 1984.
2. **Kontula, K. K., Torkkeli, T. K., Bardin, C. W., and Jänne, O. A.,** Androgen induction of ornithine decarboxylase mRNA in mouse kidney as studied by complementary DNA, *Proc. Natl. Acad. Sci. U.S.A.*, 81, 731, 1984.
3. **Kahana, C. and Nathans, D.,** Isolation of cloned cDNA encoding mammalian ornithine decarboxylase, *Proc. Natl. Acad. Sci. U.S.A.*, 81, 3645, 1984.
4. **McConlogue, L. and Coffino, P.,** A mouse lymphoma cell mutant whose major protein product is ornithine decarboxylase, *J. Biol. Chem.*, 258, 12083, 1983.

5. **Russell, D. H. and Snyder, S. H.,** Amine synthesis in regenerating rat liver: extremely rapid turnover of ornithine decarboxylase, *Mol. Pharmacol.*, 5, 253, 1969.
6. **Gupta, M. and Coffino, P.,** Mouse ornithine decarboxylase: complete amino acide sequence deduced from cDNA, *J. Biol. Chem.*, 260, 2941, 1985.
7. **Isomaa, V. V., Pajunen, A. E. I., Barden, C. W., and Jänne, O. A.,** Ornithine decarboxylase in mouse kidney, *J. Biol. Chem.*, 258, 6735, 1983.
8. **Seely, J. E. and Pegg, A. E.,** Changes in mouse kidney ornithine decarboxylase activity are brought about by changes in the amount of enzyme protein as measured by radioimmunoassay, *J. Biol. Chem.*, 258, 2496, 1983.
9. **Kozak, M.,** Point mutations define a sequence flanking the AUG initiator codon that modulates translation by eukaryotic ribosomes, *Cell*, 44, 283, 1986.
10. **McConlogue, L., Dana, S. L., and Coffino, P.,** Multiple mechanisms are responsible for altered expression of ornithine decarboxylase in overproducing variant cells, *Mol. Cell. Biol.*, 6, 2865, 1986.
11. **Quigley, G. J., Teeter, M. M., and Rich, A.,** Structural analysis of spermine and magnesium ion binding to yeast phenylalanine transfer RNA, *Proc. Natl. Acad. Sci. U.S.A.*, 75, 64, 1978.
12. **Jacobson, A. B., Kumar, H., and Zuker, M.,** Effect of spermidine on the conformation of bacteriophage MS2 RNA electron microscopy and computer modeling, *J. Mol. Biol.*, 181, 517, 1985.
13. **Spena, A., Krause, E., and Dobberstein, B.,** Translation efficiency of zein mRNA is reduced by hybrid formation between the 5'- and 3'-untranslated region, *EMBO J.*, 4, 2153, 1985.
14. **Kahana, C. and Nathans, D.,** Translational regulation of mammalian ornithine decarboxylase by polyamines, *J. Biol. Chem.*, 260, 15390, 1985.
15. **Feinstein, S. C., Dana, S. L., McConlogue, L., Shooter, E. M., and Coffino, P.,** Nerve growth factor rapidly induces ornithine decarboxylase mRNA in PC12 rat pheochromocytoma cells, *Proc. Natl. Acad. Sci. U.S.A.*, 82, 5761, 1985.

Chapter 18

INHIBITORS OF ORNITHINE AND *S*-ADENOSYLMETHIONINE DECARBOXYLASES

Anthony E. Pegg

TABLE OF CONTENTS

I. INTRODUCTION

The development of inhibitors of enzymes involved in polyamine biosynthesis has been an area in which major advances have been made in the past few years. The availability of potent and specific inhibitors of ornithine decarboxylase (ODC) has provided powerful research tools to investigate the function of polyamines. Although studies with inhibitors of *S*-adenosylmethionine (AdoMet) decarboxylase are not yet so far advanced, it is likely that equivalent compounds to block the action of this enzyme will soon be more widely available. This chapter gives a brief account of the compounds which are presently in use to study the function of polyamines and to investigate the importance of these biosynthetic decarboxylases as therapeutic targets. It is in no way comprehensive and more detailed coverage of this work including reference to pharmacological use of these agents can be found in numerous more detailed recent reviews[1-8] as well as in some of the other chapters of this book.

II. ORNITHINE DECARBOXYLASE

A. Available Inhibitors and Their Mechanism of Action

1. Early Studies

Detailed summaries of the extensive early work on the development of ODC inhibitors have been provided.[9-12] These inhibitors were mainly structural analogues of ornithine or putrescine and these studies provided valuable information on the requirements for binding to the active site even though, in many cases, the inhibitors were not sufficiently active in the cell to be useful agents for modifying cellular polyamine levels. Two compounds, which were of some importance in this regard, are α-methylornithine and α-hydrazinoornithine (more correctly D,L-α-hydrazino-δ-aminovaleric acid). D,L-α-methylornithine is a reasonably specific, competitive inhibitor having a K_i of about 40 μM (virtually all of the inhibitory activity resides in the L-isomer which has K_i of 19 μM). It can be used to block the action of ODC in some cultured cells where the cellular ornithine content is low but is not sufficiently active for general use. D,L-α-Hydrazino-δ-aminovaleric acid is a more potent inhibitor with K_i values in the range of 1 μM having been reported but it reacts with pyridoxal phosphate and thus may lack specificity and biological stability. Both of these inhibitors were found to stabilize ODC in the cell and the increase in ODC protein brought about by them reduces their effectiveness. These considerations led to the attempts to develop more active or irreversible inhibitors.

Coward and Pegg[13] synthesized *N*-(5′-phosphopyridoxyl)-ornithine in the hope that it would act as a "multisubstrate adduct" inhibitor of ODC. It was indeed a good inhibitor being competitive with pyridoxal phosphate and having a K_i of 0.6 μM but it was very poorly taken up by cells (probably because of the negatively charged phosphate group) and had no effect on putrescine production in cells treated with it.[13]

Some amines having activity as ODC inhibitors have been described and are covered in the reviews given above. More recently, 1-aminooxy-3-aminopropane has been reported to be a very potent competitive inhibitor of ODC.[11,12,14] This compound is not specific since it also affects AdoMet decarboxylase and spermidine synthase, but it may be of value in perturbing cellular polyamine content.

2. α-Difluoromethylornithine

By far the most widely used inhibitor of ODC is the enzyme-activated, irreversible inhibitor α-difluoromethylornithine (DFMO). Extensive studies have now been carried out with DFMO and related compounds and are covered at length in the reviews listed above.[1-12] The rationale behind the synthesis of DFMO was that it would serve as a substrate for ODC. It was thought that after enzymatic decarboxylation a reactive intermediate would be generated which would inactivate the enzyme by forming a covalent bond with a residue at the active site.[15] This mechanism is supported by the kinetics of inactivation of the enzyme[12] and by experiments with radioactive DFMO labeled with ^{14}C at the 5 position and at the 1 position.[16-20] The partition ratio of decarboxylation to inactivation is about 3.3.[20] The exact nature of the linkage between DFMO and ODC has not yet been determined but preliminary studies suggest that it binds to the lysine at residue 298 of the rat kidney enzyme.[7]

The binding of DFMO to ODC is sufficiently specific that the reaction of labeled DFMO with cell extracts can be used to estimate the number of molecules of active ODC protein present.[17,19,21,22] Such titrations have shown that ODC is a very minor component of the total soluble protein amounting to less than one part in a million of the soluble protein in most cells.[2,7,17,19] The reaction of ODC protein with [^{3}H]-DFMO of high specific activity has also been used to generate a labeled ligand for radioimmunoassay[23] and for the screening of hybridoma cells producing monoclonal antibodies to this protein.[24] The reaction with DFMO can also be used to identify the protein corresponding to the subunit of ODC in complex mixtures either by labeling with radioactive drug[16,28,25-28] or by studying the change in mobility on isoelectric focusing gels.[27-29]

DFMO has been used to study the turnover of ODC.[19] Measurement of the fate of the ODC protein labeled by administration of [^{14}C]-DFMO to mice in vivo confirmed directly the rapid turnover of this protein and indicated that androgen administration increases its half-life. No labeled peptide fragments were detected in these experiments indicating that the rate-limiting step in the degradation is likely to be the initial cleavage and that further degradation is even more rapid.[19]

The specific covalent binding of labeled DFMO to ODC also permits it to be used to study the localization of the enzyme by autoradiography.[30-32] The majority of the ODC in the mouse kidney was present in the cells of the proximal tubules with considerably lower amounts in the distal tubules and in the collecting tubules.[30,32] Most ODC appears to be cytosolic but a significant fraction is present in the nucleus in both mouse kidney cells[32] and in the cells of a polychaete, *Ophryotrocha labronica*.[31]

3. Other Irreversible Inhibitors

After the success achieved with DFMO, a large number of compounds designed to act as irreversible inhibitors according to the same principle were synthesized.[3,11,12] Some of these inhibitors are also derivatives of the substrate ornithine whereas others are related to the product putrescine, but all are thought to act by binding to the active site and undergoing an enzyme-catalyzed reaction which generates an active species. Some of these inhibitors may have advantages over DFMO. DFMO is relatively poorly taken up by cells[33,34] and is rapidly excreted.[35] Also, the K_i of DFMO (39 μM) is quite high and its rate of inactivation of ODC is rather slow (the $t_{1/2}$ at saturating concentrations of DFMO is 3.1 min).[15] The uptake problem may be solved by the use of putrescine analogues such as (2R,5R)-δ-methylacetylenicputrescine (MAP) or by the use of methyl esters of ornithine derivatives which are rapidly cleaved to form the amino acid within the cell.[3,8,11,12,34] Some of the newer ODC inhibitors also have lower K_i values, e.g., 2-fluoromethyl-(E)-dehydroornithine (2.7 μM) and MAP (3 μM) and somewhat faster inactivation rates. At saturating levels of MAP, the $t_{1/2}$ of ODC is only 1.7 min.[3,11,12]

B. Effect on Cellular Polyamines

Treatment of cultured cells with ODC inhibitors such as DFMO leads to a decline of putrescine and spermidine to almost undetectable levels. Spermine levels (on a per cell basis) are depleted only slightly. The maintenance of cellular spermine is due to a number of factors. These include the continued synthesis of a small amount of putrescine which is completely converted to spermine because of the excess decarboxylated AdoMet, the virtual cessation of cell growth, and the lack of substantial degradation of spermine.[2,7] Effects of DFMO on tissues of treated animals are similar but the extent of depletion is usually less. It has been found that MAP does bring about spermine depletion in HTC cells[11,36] showing that this more potent inhibitor of ODC does have some advantages.

C. Effect on Cell Growth

Although there are important exceptions (see Reference 7) it has been found in most cases that DFMO brings about cytostatic rather than cytotoxic effects towards mammalian cells. The reduction in cell growth rate can be completely overcome by addition of putrescine, spermidine, or spermine.[2,6,7,10,37-39] It should be noted that spermine reverses the growth inhibitory effects of DFMO even though it itself is not depleted in DFMO-treated cells.

The mechanism by which polyamine depletion by DFMO leads to the cessation of cell growth is not clear. The most likely possibility is that the effect is due to a reduction in protein synthesis which may lead to a deficiency of certain key proteins needed for cell division 2,4,40. Although polyamine content may influence the synthesis of all proteins it is clear that some proteins are much more sensitive to polyamine levels than others (see Reference 41 and references therein).

The absence of cytotoxic effects of DFMO in many cell types may be linked to the lack of effect on spermine. Treatments with a combination of DFMO and other inhibitors or with MAP or with other amines which bring about spermine depletion are accompanied by a progressive loss of viability.[6,7,38,39,42] The loss of cellular viability in cells depleted of both spermine and spermidine may relate to a need for polyamines to maintain normal chromatin structure.[6,7,43]

D. Effect on Decarboxylated *S*-Adenosylmethionine

A very large increase (several hundredfold) in the content of decarboxylated AdoMet occurs in cells treated with DFMO.[2,6,7,37,44,45] This rise is due to the absence of putrescine and spermidine to serve as substrates for the aminopropyltransferases and to the increased amount of AdoMet decarboxylase protein (see below). The increase in AdoMet decarboxylase protein is more than sufficient to compensate for the absence of putrescine-mediated activation of this enzyme and it produces decarboxylated AdoMet at an increased rate at a time when there is no use for it as a precursor of spermidine and spermine. A portion of the decarboxylated AdoMet is converted to an acetyl derivative.[46,47] The significance of the accumulation of these two nucleosides in the effects of DFMO on cell growth and differentiation is not yet known. In HTC and SV-3T3 cells it appears that the effects on cell growth are mediated by the depletion of spermidine.[37,42,45] However, the possibility that the large increases in the content of decarboxylated AdoMet and its acetyl derivative might influence growth and differentiation in some cell types perhaps by influencing methylation or histone acetylation should not be overlooked.

E. Effect on AdoMet Decarboxylase

Studies with DFMO have shown that the amount of AdoMet decarboxylase activity increases substantially in cells depleted of spermidine.[37,44,48-51] These increases result from a rise in the total AdoMet decarboxylase protein as measured by radioimmunoassay.[50,51] This rise is accomplished by the combination of a decreased rate of degradation[44,48,49,51] and an increased rate of synthesis of the protein.[51,52] The increase in synthesis results from an increase in the mRNA content[51,52] and from an increased translation of the mRNA since the translation is inhibited by polyamines.[41] The eight- to tenfold rise in mRNA for AdoMet decarboxylase which is brought about in rat prostate by treatment with DFMO[52] has proved to be very useful in providing enriched mRNA preparations which have been used to obtain cDNA probes for the rat enzyme.[53]

F. Development of Resistant Cells

Prolonged exposure of cultured cells to DFMO with stepwise increases in the DFMO concentration leads to the development of resistance to the antiproliferative effects of this drug. This is mediated via the overproduction of the ODC protein.[29,54-58] In most cases this overproduction is due to a large increase in the amount of ODC mRNA due to the amplification of the ODC gene. However, other cases in which the DFMO-resistant cells produce higher levels of ODC mRNA than can be explained by the degree of gene amplification or where the increase in ODC protein synthesis is due to an increased translation efficiency are also known.[59,60]

Another established alternative route to resistance involves an increased stability of the enzyme. Several cell lines have been described in which this change has occurred. The increase in half-life produces a higher amount of the enzyme protein at the steady state.[61-63]

III. *S*-ADENOSYLMETHIONINE DECARBOXYLASE

A. Available Inhibitors and Their Specificity

1. *Methylglyoxal bis(Guanylhydrazone) and Other Nonnucleoside Inhibitors*

Early studies of inhibitors of AdoMet decarboxylase were reviewed by Porter et al.[64] and by Pegg.[65] Much interest was stirred by the discovery that the antileukemic agent, methylglyoxal bis(guanylhydrazone) [MGBG], was a potent inhibitor of this enzyme.[66] MGBG is a competitive inhibitor of the putrescine-activated mammalian AdoMet decarboxylase having a K_i of less than 1 μM.[66-68] It is a less potent inhibitor of the Mg^{2+}-activated AdoMet decarboxylase from *Escherichia coli* or the AdoMet decarboxylases from the plant, *Brassica pekinensis,* or the slime mould, *Physarum polycephalum,* which do not require putrescine or metal cations. However, these enzymes are still sensitive to the drug.[44,69,70]

A number of analogues of MGBG such as dimethylglyoxal bis(guanylhydrazone) and ethylglyoxal bis(guanylhydrazone) [EGBG] are even more potent than MGBG itself whereas others, such as glyoxal bis(guanylhydrazone) and di-*N'''*-methylglyoxal bis(guanylhydrazone), are less active.[65,68,69] Of particular interest is 1,1′-[(methylethanediylidene)dinitrilo]bis(3-aminoguanidine) [MBAG], a derivative which has additional amino groups at each end and is an irreversible inactivator of AdoMet decarboxylase.[44,67,71] This inactivation is probably due to the reaction of the hydrazine moiety of MBAG with the pyruvate prosthetic group of the enzyme.

The inhibition of AdoMet decarboxylase activity brought about by MGBG in vivo may be quite transient because the drug brings about a large increase in the amount of the enzyme protein. This increase appears to be produced by means of the stabilization of the enzyme against breakdown and the half-life of AdoMet decarboxylase is increased by more than tenfold.[44,51,52,67,72] MBAG also leads to an increase in the amount of AdoMet decarboxylase protein as measured by radioimmunoassay but this protein was inactive since MBAG is an irreversible inhibitor.[73] This is a significant advantage of MBAG but unfortunately MBAG is too toxic and is not sufficiently stable in the cell to be used in long-term experiments.

A number of other nonnucleoside substances have been described as inhibitors of AdoMet decarboxylase. Some of these, such as Berenil and Pentamidine,[74,75] resemble MGBG in some ways. Others include guanethidine,[76] phenothiazines such as chlorpromazine and imipramine,[77] and various norharmane derivatives.[78] It is not yet clear whether the pharmacological actions of these drugs are in any way related to their interference with the action of AdoMet decarboxylase and their specificity has not yet been established.

Khomutov et al.[14] have reported that 1-aminooxy-3-aminopropane inhibits mammalian AdoMet decarboxylase and this compound also acts as an inhibitor of other enzymes in the polyamine biosynthetic pathway including ODC and spermidine synthase. It may therefore have some utility to deplete polyamines by inhibiting several biosynthetic steps. A similar idea has been advanced by Hibasami et al.[79] who synthesized a compound related to MGBG termed methylglyoxal bis(butylamidinohydrazone) which inhibited ODC, AdoMet decarboxylase, and spermidine synthase.

All AdoMet decarboxylases which have so far been characterized have a pyruvate prosthetic group.[4,44] Pankaskie and Abdel-Monem[80] attempted to produce inhibitors of AdoMet decarboxylase by the synthesis of various pyruvate derivatives. These attempts were not successful probably because the pyruvate prosthetic group is covalently attached to the enzyme protein.

2. Nucleoside Inhibitors

Another idea which unfortunately was not successful was the synthesis of the α-difluoromethyl derivative of AdoMet. This was inactive as an inhibitor of AdoMet decarboxylase.[81] It is not clear whether this is due to its inability to serve as a substrate for the enzyme or to the absence of an appropriate residue for reaction of an activated species at the active site.

Potential nucleoside inhibitors of AdoMet decarboxylase were investigated by several groups.[69,81-84] Some factors needed for binding of the substrate to the enzyme were revealed by these studies but no very potent inhibitors were obtained. The most active compound, apart from the product decarboxylated AdoMet, was *S*-methyl-5′-methylthioadenosine which had a K_i of 2 μM[81] or 6 μM.[69] However, this nucleoside is a considerably more potent inhibitor of spermine synthase.[84,85] It was essential for inhibitory activity towards AdoMet decarboxylase that compounds contained adenine or a close derivative as the base[69] but the sulfonium center in AdoMet could be replaced with a nitrogen.[81,82]

Much better progress has been made in the design of nucleoside inhibitors using a rational approach for the production of compounds which would bind to the active site and form a covalent bond with the pyruvate prosthetic group. Such compounds would therefore be irreversible inactivators of the enzyme. *S*-(5′-deoxy-5′-adenosyl)-methylthioethylhydroxylamine was synthesized and found to inhibit AdoMet decarboxylase.[86-88] The synthesis of 5′-[3-aminooxypropyl)amino]-5′-deoxyadenosine as an inhibitor of AdoMet decarboxylase was reported.[89] At present there is no published information on the biological stability of these compounds or of their ability to affect polyamine synthesis within the cell. A number of potential irreversible inhibitors including 5′-[(3-aminooxypropyl)methylamino]-5′-deoxyadenosine [MAOPA], 5′-[(3-aminooxyethyl)methylamino]-5′-deoxyadenosine [MAOEA], and 5′-[(3-hydrazinopropyl)methylamino]-5′-deoxyadenosine [MHZPA] were synthesized by Secrist and colleagues. When tested in vitro these compounds were confirmed as extremely potent irreversible inactivators of mammalian AdoMet decarboxylase.[90]

B. Effect on Cellular Polyamines and Cell Growth

The biochemistry and pharmacology of MGBG have been reviewed.[6,64,67,68] Many studies have shown that MGBG can bring about changes in cellular polyamine content when administered to intact animals or to cells in culture. These changes include a large increase in putrescine and a decline in spermidine and spermine. They are therefore consistent with the action of MGBG at the AdoMet decarboxylase step. Findings that MGBG reduced decarboxylated AdoMet content in rat tissues and in cultured cells are also in agreement with this inhibition. However, MGBG is not very specific an inhibitor and the changes in polyamines brought about by it may also be influenced by its acting as an inhibitor of diamine oxidase[64,67,68,91] and as an inducer of spermidine/spermine-N^1-acetyltransferase.[68,92,93]

Although it is known that the administration of spermidine can prevent many of the effects of MGBG including its ability to inhibit cell growth this cannot be used as evidence that the changes in polyamine content produced by MGBG are responsible for the growth inhibition. MGBG is actively transported into mammalian cells[6,64,67,68] and can reach very high intracellular concentrations. At these concentrations, it may have many effects on cellular metabolism including the production of severe mitochondrial damage that may not be directly related to the depletion of polyamine levels.[64,68] Its transport into the cell uses the same carrier system as polyamines and the prevention of its action by spermidine may be due to competition for this transport.

The inhibitors synthesized by Secrist and colleagues appear to give much more specific effects towards polyamine synthesis than MGBG. Both MAOEA and MHZPA blocked the synthesis of spermidine and spermine in L1210 cells in culture and had a similar growth inhibitory effect to DFMO in these cells.[90] Their ability to inhibit the rate of growth was overcome if exogenous spermidine was provided.[90] The growth inhibitory action of MHZPA was manifest even in a DFMO-resistant L1210 cell line.[57]

C. Effect of AdoMet Decarboxylase Inhibitors on ODC Activity

MGBG leads to an increase in the activity of ODC in rat tissues and in cells in culture (see References 68 and 92 and citations therein). This increase is partly due to an increase in the half-life of the enzyme[92] but it appears likely that its synthesis is also enhanced. In view of the lack of specificity of MGBG it would be premature to conclude from this data that the increase is due to the decline in polyamines but this may well be the case since, as described below, other AdoMet decarboxylase inhibitors also increase ODC content. It is well documented that ODC is regulated by the cellular polyamine content and the amount of ODC protein is inversely related to polyamine levels.[2,4,6,7,27,41,60,94-100] This regulation is not accompanied by changes in the mRNA level and appears to be due to polyamines decreasing the half-life of the enzyme.[94,95,98] and inhibiting the translation of the mRNA.[41,100]

MHZPA and MAOEA also bring about an increase (about 15-fold) in the amount of ODC protein (measured both by radioimmunoassay and by measurement of activity).[101] This increase can be prevented by the addition of spermidine and appears to be regulated by a threefold increase in the half-life of the enzyme and a substantially increased synthetic rate. This increase in synthesis was not accompanied by a change in the mRNA and appears therefore to be brought about by an increased translational efficiency of existing mRNA.[101]

IV. CONCLUSIONS

ODC inhibitors are now well established as research tools and as useful pharmacological agents.[1-12] Although there are still some problems in using these inhibitors to bring about substantial polyamine depletion in all systems, their availability has led to an explosion of knowledge and interest in the role of polyamines and this pathway as a target for the design of therapeutic agents.

Although the field of AdoMet decarboxylase inhibition is not so well advanced at present, the irreversible nucleoside inhibitors produced by several groups have considerable promise if the problems associated with their cellular uptake and biological stability can be overcome. Suitable compounds are now available as research tools to investigate the function and importance of AdoMet decarboxylase in cultured cells.

The tight regulation of cellular polyamine levels by means of changes in the biosynthetic pathway and the close interrelation of ODC and AdoMet decarboxylase are illustrated by the results with the inhibitors. When inhibitors of either enzyme are given alone, the polyamine depletion which they produce leads to a large increase in the activity of the other enzyme. Such increases complicate the interpretation of some experiments with these inhibitors, and the simultaneous administration of both ODC and AdoMet decarboxylase inhibitors may have some advantages both therapeutically and in the design of experiments to investigate the importance of polyamine biosynthesis and cell growth and differentiation.

In this context, it is worth mentioning another possible means of influencing cellular polyamine levels by changing the activities of these decarboxylases. This approach uses the fact that both ODC and AdoMet decarboxylase activities are reduced in the presence of high levels of polyamines as described above. Porter and colleagues[6,102,103] have identified compounds such as N^1,N^8-bis(ethyl)spermidine and N^1,N^{12}-bis(ethyl)spermine which are able to bring about a decline in the amounts of ODC and AdoMet decarboxylase by this repression but cannot themselves substitute for the normal polyamines in growth. Therefore, they block the ability of the cell to make polyamines by means of ODC and AdoMet decarboxylase and may have applications similar to those of the direct inhibitors.

ACKNOWLEDGMENTS

Research on polyamines in the author's laboratory was supported by grants CA-18138, CA-37606, and GM-26290.

REFERENCES

1. **Heby, O.,** Role of polyamines in the control of cell proliferation and differentiation, *Differentiation,* 19, 1, 1981.
2. **Pegg, A. E. and McCann, P. P.,** Polyamine metabolism and function, *Am. J. Physiol.,* 243, C212, 1982.
3. **Sjoerdsma, A. and Schechter, P. J.,** Chemotherapeutic implications of polyamine biosynthesis inhibition, *Clin. Pharm. Ther.,* 35, 287, 1984.
4. **Tabor, C. W. and Tabor, H.,** Polyamines, *Annu. Rev. Biochem.,* 53, 749, 1984.
5. **Williamson, J. D. and Tyms, A. S.,** Polyamine inhibitors as antimicrobial agents?, *Med. Microbiol.,* 4, 239, 1984.
6. **Porter, C. W. and Sufrin, J. R.,** Interference with polyamine biosynthesis and/or function by analogs of polyamines or methionine as a potential anticancer chemotherapeutic strategy, *Anticancer Res.,* 6, 525, 1986.
7. **Pegg, A. E.,** Recent advances in the biochemistry of polyamines in eukaryotes, *Biochem. J.,* 234, 249, 1986.
8. **McCann, P. P., Pegg, A. E., and Sjoerdsma, A., Eds.,** *Inhibition of Polyamine Metabolism: Biological Significance and Basis for New Therapies,* Academic Press, San Diego, 1987, 1.
9. **Stevens, E. and Stevens, E.,** Inhibitors of the biosynthesis of putrescine, spermidine and spermine, in *Polyamines in Biomedical Research,* Gaugas, J. M., Ed., John Wiley & Sons, New York, 1980, 167.
10. **Mamont, P. S., Bey, P., and Koch-Weser, J.,** Biochemical consequences of drug-induced polyamine deficiency in mammalian cells, in *Polyamines in Biomedical Research,* Gaugas, J. M., Eds., John Wiley & Sons, New York, 1980, 147.
11. **Danzin, C. and Mamont, P.,** Polyamine inhibition *in vivo* and in organ growth and repair, in *Biological Significance and Basis for New Therapies,* McCann, P. P., Pegg, A. E., and Sjoerdsma, A., Eds., Academic Press, San Diego, in press.
12. **Bey, P., Danzin, C., and Jung, M.,** Inhibition of basic amino acid decarboxylases involved in polyamine biosynthesis, in *Biological Significance and Basis for New Therapies,* McCann, P. P., Pegg, A. E., and Sjoerdsma, A., Eds., Academic Press, San Diego, in press.
13. **Coward, J. K. and Pegg, A. E.,** Inhibition of polyamine biosynthesis, *Adv. Enzyme Regul.,* 26, 107, 1987.
14. **Khomutov, R. M., Hyvonen, T., Karvonen, E., Kauppinen, L., Paalanen, T., Paulin, L., Eloranta, T., Pajula, R.-L., Andersson, L. C., and Pösö, H.,** 1-Aminooxy-3-aminopropane, a new and potent inhibitor of polyamine biosynthesis that inhibits ornithine decarboxylase, adenosylmethionine decarboxylase and spermidine synthase, *Biochem. Biophys. Res. Commun.,* 130, 596, 1985.
15. **Metcalf, B. W., Bey, P., Danzin, C., Jung, M. J., Casara, P., and Vevert, J. P.,** Catalytic irreversible inhibition of mammalian ornithine decarboxylase (E.C. 4.1.1.17) by substrate and product analogues, *J. Am. Chem. Soc.,* 100, 2551, 1978.
16. **Pritchard, M. L., Seely, J. E., Pösö, H., Jefferson, L. S., and Pegg, A. E.,** Binding of radioactive α-difluoromethylornithine to rat liver ornithine decarboxylase, *Biochem. Biophys. Res. Commun.,* 100, 1597, 1981.

17. **Seely, J. E., Pösö, H., and Pegg, A. E.,** Measurement of the number of ornithine decarboxylase molecules in rat and mouse tissues under various physiological conditions by binding of radiolabelled α-difluoromethylornithine, *Biochem. J.*, 206, 311, 1982.
18. **Seely, J. E., Pösö, H., and Pegg, A. E.,** Purification of ornithine decarboxylase from kidneys of androgen-treated mice, *Biochemistry*, 21, 3394, 1982.
19. **Seely, J. E., Pösö, H., and Pegg, A. E.,** Effect of androgens on turnover of ornithine decarboxylase in mouse kidney, *J. Biol. Chem.*, 257, 7549, 1982.
20. **Pegg, A. E., McGovern, K. A., and Wiest, L.,** Decarboxylation of α-difluoromethylornithine by ornithine decarboxylase, *Biochem. J.*, 241, 305, 1987.
21. **Erwin, B. G., Seely, J. E., and Pegg, A. E.,** Mechanism of stimulation of ornithine decarboxylase activity in transformed mouse fibroblasts, *Biochemistry*, 22, 3027, 1983.
22. **Kyriakidis, D. A., Flamigni, F., Pawlak, J. W., and Canellakis, E. S.,** Mode of interaction of ornithine decarboxylase with antizyme and α-difluoromethylornithine, *Biochem. Pharmacol.*, 33, 1575, 1984.
23. **Seely, J. E. and Pegg, A. E.,** Changes in mouse kidney ornithine decarboxylase activity are brought about by changes in the amount of enzyme protein as measured by radioimmunoassay, *J. Biol. Chem.*, 258, 2496, 1983.
24. **Pegg, A. E., Seely, J. E., Persson, Lo., Herlyn, M., Ponsell, K., and O'Brien, T. G.,** Studies of mammalian ornithine decarboxylase using a monoclonal antibody, *Biochem., J.*, 217, 123, 1984.
25. **Isomaa, V. V., Pajunen, A. E. I., Bardin, C. W., and Jänne, O. A.,** Ornithine decarboxylase in mouse kidney, *J. Biol. Chem.*, 258, 6735, 1983.
26. **Kameji, T. and Hayashi, S.-I.,** Affinity labeling of purified ornithine decarboxylase by α-difluoromethylornithine, *Biochim. Biophys. Acta*, 705, 405, 1982.
27. **Persson, L., Seely, J. E., and Pegg, A. E.,** Investigation of structure and rate of synthesis of ornithine decarboxylase protein in mouse kidney, *Biochemistry*, 23, 3777, 1984.
28. **Seely, J. E., Persson, L., Sertich, G. J., and Pegg, A. E.,** Comparison of ornithine decarboxylase from rat liver, rat hepatoma and mouse kidney, *Biochem. J.*, 226, 577, 1985.
29. **McConlogue, L. and Coffino, P.,** Ornithine decarboxylase in difluoromethylornithine-resistant mouse lymphoma cells, *J. Biol. Chem.*, 258, 8384, 1983.
30. **Pegg, A. E., Seeley, J., and Zagon, I. S.,** Autoradiographic identification of ornithine decarboxylase in mouse kidney by means of α-[5-^{14}C]difluoromethylornithine, *Science*, 217, 68, 1982.
31. **Emanuelsson, H. and Heby, O.,** Nuclear ornithine decarboxylase. Electron microscope autoradiographic identification of the active enzyme using α-[5-^{3}H]difluoromethylornithine, *Eur. J. Cell. Biol.*, 30, 288, 1983.
32. **Zagon, I. S., McLaughlin, P. J., Seely, J. E., Hoeksema, G. W., and Pegg, A. E.,** Autoradiographic localization of ornithine decarboxylase in mouse kidney by use of radiolabeled α-difluoromethylornithine, *Cell Tissue Res.*, 235, 371, 1984.
33. **Erwin, B. G. and Pegg, A. E.,** Uptake of α-difluoromethylornithine by mouse fibroblasts, *Biochem. Pharmacol.*, 31, 2820, 1982.
34. **Bitonti, A. J., Bacchi, C. J., McCann, P. P., and Sjoerdsma, A.,** Uptake of α-difluoromethylornithine by *Trypanosome brucei brucei*, *Biochem. Pharmacol.*, 35, 351, 1986.
35. **Grove, J., Fozard, J. R., and Mamont, P. S.,** Assay of α-difluoromethylornithine in body fluids and tissues by automatic amino-acid analysis, *J. Chromatogr.*, 223, 409, 1981.
36. **Mamont, P. S., Siat, M., Joder-Ohlenbusch, A.-M., Bernhardt, A., and Casara, P.,** Effects of (2R,5R)-6-heptyne-2,5-diamine, a potent inhibitor of L-ornithine decarboxylase, on rat hepatoma cells cultured *in vitro*, *Eur. J. Biochem.*, 142, 457, 1984.
37. **Pegg, A. E.,** The role of polyamine depletion and accumulation of decarboxylated *S*-adenosylmethionine in the inhibition of growth of SV-3T3 cells treated with α-difluoromethylornithine, *Biochem. J.*, 224, 29, 1984.
38. **Gerner, E. W. and Mamont, P. S.,** Restoration of the polyamine contents in rat hepatoma tissue-culture cells after inhibition of polyamine biosynthesis, *Eur. J. Biochem.*, 156, 31, 1986.
39. **Casero, R. A., Bergeron, R. J., and Porter, C. W.,** Treatment with α-difluoromethylornithine plus a spermidine analog leads to spermine depletion and growth inhibition in cultured L1210 leukemia cells, *J. Cell. Physiol.*, 121, 476, 1984.
40. **Rudkin, B. B., Mamont, P. S., and Seiler, N.,** Decreased protein-synthetic activity is an early consequence of spermidine depletion in rat hepatoma tissue-culture cells, *Biochem. J.*, 217, 731, 1984.
41. **Kameji, T. and Pegg, A. E.,** Inhibition of translation of mRNAs for ornithine decarboxylase and *S*-adenosylmethionine decarboxylase by polyamines, *J. Biol. Chem.*, 262, 2427, 1987.
42. **McGovern, K. A., Clark, R. S., and Pegg, A. E.,** Effect of 1,3,6-triaminohexane and 1,4,7-triaminoheptane on growth and polyamine metabolism in SV-3T3 cells treated with 2-difluoromethylornithine, *J. Cell. Physiol.*, 127, 311, 1986.
43. **Knuutila, S. and Pohjanpelto, P.,** Polyamine starvation causes parallel increase in nuclear and chromosomal aberrations in a polyamine-dependent strain of CHO, *Exp. Cell Res.*, 145, 222, 1983.

44. **Pegg, A. E.,** *S*-Adenosylmethionine decarboxylase: a brief review, *Cell Biochem. Function,* 2, 11, 1984.
45. **Mamont, P. S., Danzin, C., Wagner, J., Siat, M., Joder-Ohlenbusch, A.-M., and Claverie, N.,** Accumulation of decarboxylated *S*-adenosyl-L-methionine in mammalian cells as a consequence of the inhibition of putrescine biosynthesis, *Eur. J. Biochem.,* 123, 499, 1982.
46. **Wagner, J., Hirth, Y., Piriou, F., Zakett, D., Claverie, N., and Danzin, C.,** *N*-Acetyl decarboxylated *S*-adenosylmethionine, a new metabolite of decarboxylated *S*-adenosylmethionine: isolation and characterization, *Biochem. Biophys. Res. Commun.,* 133, 546, 1985.
47. **Pegg, A. E., Wechter, R. S., Clark, R. S., Wiest, L., and Erwin, B. G.,** Acetylation of decarboxylated *S*-adenosylmethionine by mammalian cells, *Biochemistry,* 25, 379, 1986.
48. **Mamont, P. S., Joder-Ohlenbusch, A.-M., Nussli, M., and Grove, J.,** Indirect evidence for a strict negative control of *S*-adenosyl-L-methionine decarboxylase by spermidine in rat hepatoma cells, *Biochem. J.,* 196, 411, 1981.
49. **Alhonen Hongisto, L.,** Regulation of *S*-adenosylmethionine decarboxylase by polyamines in Ehrlich ascites-carcinoma cells grown in culture, *Biochem. J.,* 190, 747, 1980.
50. **Shirahata, A., Christman, K. L., and Pegg, A. E.,** Quantitation of *S*-adenosylmethionine decarboxylase protein, *Biochemistry,* 24, 4417, 1985.
51. **Shirahata, A. and Pegg, A. E.,** Regulation of *S*-adenosylmethionine decarboxylase activity in rat liver and prostate, *J. Biol. Chem.,* 260, 9583, 1985.
52. **Shirahata, A. and Pegg, A. E.,** Increased content of mRNA for a precursor of *S*-adenosylmethionine decarboxylase in rat prostate after treatment with 2-difluoromethylornithine, *J. Biol. Chem.,* 261, 13833, 1986.
53. **Pajunen, A. and Pegg, A. E.,** Cloning of rat prostate cDNA for *S*-adenosylmethionine decarboxylose, *Fed. Proc., Fed. Am. Soc. Exp. Biol.,* 46, 2046, 1987.
54. **Choi, J. H. and Scheffler, I. E.,** Chinese hamster ovary cells resistant to α-difluoromethylornithine are overproducers of ornithine decarboxylase, *J. Biol. Chem.,* 258, 12601, 1983.
55. **Kahana, C. and Nathans, D.,** Nucleotide sequence of murine ornithine decarboxylase mRNA, *Proc. Natl. Acad. Sci. U.S.A.,* 82, 1673, 1985.
56. **Pohjanpelto, P., Hölttä, E., Jänne, O. A., Knuutila, S., and Alitalo, K.,** Amplifiction of ornithine decarboxylase gene in response to polyamine deprivation in chinese hamster ovary cells, *J. Biol. Chem.,* 260, 8532, 1985.
57. **Alhonen-Hongisto, L., Kallio, A., Sinervirta, R., Seppänen, P., Kontula, K. K., Jänne, O. A., and Jänne, J.,** Difluoromethylornithine-induced amplification of ornithine decarboxylase genes in Ehrlich ascites carcinoma cells, *Biochem. Biophys. Res. Commun.,* 126, 734, 1985.
58. **Pegg, A. E., Madhubala, R., and Secrist, J. A.,** Properties of a DFMO-resistant L1210 cell line, *Cancer Res.,* 48, 2678, 1988.
59. **Alhonen-Hongisto, L., Sinervirta, R., Jänne, O. A., and Jänne, J.,** Gene expression of ornithine decarboxylase in L1210 leukaemia cells exposed to DL-2-difluoromethylornithine in the presence of cadaverine, *Biochem. J.,* 232, 605, 1985.
60. **McConlogue, L., Dana, S. L., and Coffino, P.,** Multiple mechanisms are responsible for altered expression of ornithine decarboxylase in overproducing variant cells, *Mol. Cell. Biol.,* 6, 2865, 1986.
61. **Mamont, P. S., Duchesne, M.-C., Grove, J., and Tardif, C.,** Initial characterization of a HTC cell variant partially resistant to the anti-proliferative effect of ornithine decarboxylase inhibitors, *Exp. Cell Res.,* 115, 387, 1978.
62. **Pritchard, M. L., Pegg, A. E., and Jefferson, L. S.,** Ornithine decarboxylase from hepatoma cells and a variant cell line in which the enzyme is more stable, *J. Biol. Chem.,* 257, 5892, 1982.
63. **Pösö, H., Karvonen, E., Suomalainen, H., and Andersson, L. C.,** A human neuroblastoma cell line with an altered ornithine decarboxylase, *J. Biol. Chem.,* 259, 12307, 1984.
64. **Porter, C. W., Dave, C., and Mihich, E.,** Inhibition of *S*-adenosylmethionine decarboxylase as an approach in cancer therapeutics, in *Polyamines in Biology and Medicine,* Morris, D. R. and Marton, L. S., Eds., Marcel Dekker, New York, 1981, 407.
65. **Pegg, A. E.,** Inhibitors of *S*-adenosylmethionine decarboxylase, *Methods Enzymol.,* 94, 239, 1983.
66. **Williams-Ashman, H. G. and Schenone, A.,** Methyl glyoxal bis(guanylhydrazone) as a potent inhibitor of mammalian and yeast *S*-adenosylmethionine decarboxylases, *Biochem. Biophys. Res. Commun.,* 46, 288, 1972.
67. **Williams-Ashman, H. G. and Pegg, A. E.,** Aminopropyl group transfers, in *Polyamines in Biology and Medicine,* Morris, D. R. and Marton, L. J., Eds., Marcel Dekker, New York, 1981, 43.
68. **Williams-Ashman, H. G. and Seidenfeld, J.,** Aspects of the biochemical pharmacology of methyl glyoxal bis(guanylhydrazone), *Biochem. Pharmacol.,* 35, 1217, 1986.
69. **Pegg, A. E. and Jacobs, G.,** Comparison of inhibitors of *S*-adenosylmethionine decarboxylase from different species, *Biochem. J.,* 213, 495, 1983.
70. **Yamanoha, B. and Cohen, S. S.,** *S*-Adenosylmethionine decarboxylase and spermine synthase from Chinese cabbage, *Plant Physiol.,* 78, 784,1985.

71. **Pegg, A. E. and Conover, C. C.,** Irreversible inhibition of mammalian and yeast *S*-adenosylmethionine decarboxylase by 1,1′-(methylethanediylidenedinitrilo)-bis(3-aminoguanidine), *Biochem. Biophys. Res. Commun.*, 69, 766, 1976.
72. **Pegg, A. E.,** Investigation of the turnover of rat liver *S*-adenosylmethionine decarboxylase using a specific antibody, *J. Biol. Chem.*, 254, 3249, 1979.
73. **Pegg, A. E.,** Inhibition of mammalian *S*-adenosylmethionine decarboxylase activity by 1.1′-[methylethanediylidene)dinitrilo]bis-(3-aminoguanidine), *J. Biol. Chem.*, 253, 539, 1978.
74. **Bitonti, A. J., Dumont, J. A., and McCann, P. P.,** Characterization of *Trypanosoma brucei brucei S*-adenosyl-L-methionine decarboxylase and its inhibition by Berenil, pentamide and methylglyoxal bis(guanylhydrazone), *Biochem. J.*, 237, 685, 1986.
75. **Karvonen, E., Kauppinen, L., Partanen, T., and Pösö, H.,** Irreversible inhibition of putrescine-stimulated *S*-adenosyl-L-methionine decarboxylase by Berenil and Pentamide, *Biochem. J.*, 231, 165, 1985.
76. **Johnson, E. M. and Taylor, A. S.,** Inhibition of *S*-adenosyl methionine decarboxylase by guanethidine, *Biochem. Pharmacol.*, 29, 113, 1979.
77. **Hietala, O. A., Lapinjoki, S. P., and Pajunen, A. E. I.,** Phenothiazines and structurally related compounds as inhibitors of adenosylmethionine decarboxylase: effects on the purified enzyme, *Biochem. Int.*, 8, 245, 1984.
78. **Lapinjoki, S. P. and Gynther, J.,** Contribution of π-electrons to the inhibition of *S*-adenosyl-L-methionine decarboxylase by aromatic tricyclic compounds, *Biochem. Int.*, 12, 847, 1986.
79. **Hibasami, H., Tsukada, T., Maekawa, S., and Nakashima, K.,** Methylglyoxal bis(butylamidinohydrazone), a new inhibitor of polyamine biosynthesis that simultaneously inhibits ornithine decarboxylase, adenosylmethionine decarboxylase and spermidine synthase, *Biochem. Pharmacol.*, 35, 2982, 1986.
80. **Pankaskie, M. and Abdel-Monem, M. M.,** Inhibitors of polyamine biosynthesis. VII Evaluation of pyruvate derivatives as inhibitors of *S*-adenosyl-L-methionine decarboxylase, *J. Pharm. Sci.*, 69, 1000, 1980.
81. **Kolb, M., Danzin, C., Barth, J., and Claverie, N.,** Synthesis and biochemical properties of chemically stable produce analogues of the reaction catalyzed by *S*-adenosyl-L-methionine decarboxylase, *J. Med. Chem.*, 25, 550, 1982.
82. **Pankaskie, M. and Abdel-Monem, M. M.,** Inhibitors of polyamine biosynthesis. VIII. Irreversible inhibition of mammalian *S*-adenosyl-L-methionine decarboxylase by substrate analogues, *J. Med. Chem.*, 23, 121, 1980.
83. **Yamonoha, B. and Samejima, K.,** Inhibition of *S*-adenosylmethionine decarboxylase from rat liver by synthetic decarboxylated *S*-adenosylmethionine and its analogs, *Chem. Pharm. Bull.*, 28, 2232, 1980.
84. **Pegg, A. E., Coward, J. K., Talekar, R. R., and Secrist, J. A.,** Effects of certain 5′-substituted adenosines on polyamine synthesis: selective inhibitors of spermine synthase, *Biochemistry*, 25, 4091, 1986.
85. **Pegg, A. E. and Coward, J. K.,** Growth of mammalian cells in the absence of the accumulation of spermine, *Biochem. Biophys. Res. Commun.*, 133, 82, 1985.
86. **Khomutov, R. M., Zavalova, L. L., Siirku, V. I., Artamonova, E. Y., and Khomutov, A. R.,** Effective inhibitor of *S*-adenosylmethionine decarboxylase, *Bioorg. Khim.*, 9, 130, 1983.
87. **Artamonova, E. Y., Zavalova, L. L., Khomutov, R. M., and Khomutov, A. R.,** Irreversible inhibition of *S*-adenosylmethionine decarboxylase by hydroxylamine-containing analogues of decarboxylated *S*-adenosylmethionine, *Bioorg. Khim.*, 12, 206, 1986.
88. **Paulin, L.,** The effects of 1-aminooxy-3-aminopropane and *S*-(5′-deoxy-5′-adenosyl)methylthioethylhydroxylamine on ornithine decarboxylase and *S*-adenosyl-L-methionine decarboxylase from *Escherichia coli*, *FEBS Lett.*, 202, 323, 1986.
89. **Kolb, M. and Barth, J.,** Synthesis of 5′-[(3-aminooxypropyl)-amino]-5′-deoxyadenosine, *Leibigs. Ann. Chem.*, 1036, 1985.
90. **Pegg, A. E., Jones, D. B., and Secrist, J. A.,** Effect of inhibitors of *S*-adenoxylmethionine decarboxylase on polyamine content and growth of L1210 cells, *Biochemistry*, 27, 1408, 1988.
91. **Kallio, A. and Jänne, J.,** Role of diamine oxidase during the treatment of tumour-bearing mice with combinations of polyamine anti-metabolites, *Biochem. J.*, 212, 895, 1983.
92. **Karvonen, E. and Pösö, H.,** Stabilization of ornithine decarboxylae and N^1-spermidine acetyltransferase in rat liver by methylglyoxal bis(guanylhidrazone), *Biochim. Biophys. Acta*, 791, 239, 1984.
93. **Pegg, A. E., Erwin, B. G., and Persson, L.,** Induction of spermidine/spermine N^1-acetyltransferase by methylglyoxal bis(guanylhydrazone), *Biochim. Biophys. Acta*, 842, 111, 1985.
94. **Seely, J. E. and Pegg, A. E.,** Effect of 1,3-diaminopropane on ornithine decarboxylase enzyme protein in thioacetamide-treated rat liver, *Biochem. J.*, 216, 701, 1983.
95. **Kanamoto, R., Utsunomiya, K., Kameji, T., and Hayashi, S.-I.,** Effects of putrescine on synthesis and degradation of ornithine decarboxylase in primary cultured hepatocytes, *Eur. J. Biochem.*, 154, 539, 1986.
96. **Dircks, L., Grens, A., Slezynger, T., and Scheffler, I. E.,** Posttranscriptional regulation of ornithine decarboxylase activity, *J. Cell. Phys.*, 126, 371, 1986.

97. **Persson, L., Holm, I., and Heby, O.,** Translational regulation of ornithine decarboxylase by polyamines, *FEBS Lett.*, 205, 175, 1986.
98. **Hölttä, E. and Pohjanpelto, P.,** Control of ornithine decarboxylase in Chinese hamster ovary cells by polyamines, *J. Biol. Chem.*, 261, 9502, 1986.
99. **Sertich, G. J., Glass, J. R., Fuller, D. J. M., and Gerner, E. W.,** Altered polyamine metabolism in Chinese hamster cells growing in a defined medium, *J. Cell. Phys.*, 127, 114, 1986.
100. **Kahana, C. and Nathans, D.,** Translational regulation of mammalian ornithine decarboxylase by polyamines, *J. Biol. Chem.*, 260, 15390, 1985.
101. **Pegg, A. E. and Secrist, J. A.,** Increase in ornithine decarboxylase in response to administration of inhibitors of *S*-adenosylmethionine decarboxylase, in preparation.
102. **Porter, C. W., Cavanaugh, P. F., Stolowich, N., Ganis, B., Kelly, E., and Bergeron, R. J.,** Biological properties of N^4- and N^8-spermidine derivatives in cultured L1210 leukemia cells, *Cancer Res.*, 45, 2050, 1985.
103. **Porter, C. W., McManis, J., Casero, R. A., and Bergeron, R. J.,** The relative abilities of bis(ethyl) derivatives of putrescine, spermidine, and spermine to regulate polyamine biosynthesis and inhibit cell growth, *Cancer Res.*, 47, 2821, 1987.

Chapter 19

THE ROLE OF ANTIZYME IN THE REGULATION OF ORNITHINE DECARBOXYLASE ACTIVITY IN EUKARYOTIC CELLS

Evangelos S. Canellakis and Shin-ichi Hayashi

TABLE OF CONTENTS

I. INTRODUCTION

We shall consider the control of ornithine decarboxylase (ODC) activity in eukaryotic cells especially as it may be affected by the antizyme to ornithine decarboxylase.[1,2] An earlier evaluation has been presented recently by Canellakis et al.[3]

II. IDENTIFICATION OF ANTIZYME AND ITS RELATIONSHIP TO ODC

By exposing H-35 hepatoma cells to increasing concentrations of putrescine, Fong et al.[1] found that the activity of ODC in the extracts of the H-35 cells progressively decreased until it reached undetectable levels. They also noted that putrescine had induced a protein inhibitor of ODC whose induction was dependent upon ongoing protein synthesis. This study was extended by Heller et al.[2] using neuroblastoma N-18 and other cells. The phenomenon was found to be general. The protein that was synthesized upon addition of putrescine or other polyamines was identified as a specific, noncompetitive, salt-dissociable inhibitor of ODC. The protein exhibited various metabolic similarities to ODC, among which was the relationship of its T1/2 to that of ODC. As indicated in Table 1, although the T1/2 of the inhibitor of ODC varies with the cells tested, its T1/2 was found to be comparable to that of ODC of the same tissue. For these reasons, and because its synthesis was induced by the end product of the enzyme it inhibited, thus representing a novel mechanism of control of enzyme activity, this protein inhibitor was named antizyme to ODC.

In addition to the cells mentioned above, antizyme to ODC has been induced in a variety of eukaryotic cells including fibroblasts, nerve, and hepatic cells[4] as well as HeLa[3] and hepatoma cells,[6,7] thyroid slices,[8] chicken liver,[9] rat skin,[10] and rat mammary glands,[11] in tissue secretions such as milk[12] and in plant cells.[13,14] In most of these cases, the antizyme was induced by exposing the cells to a variety of diamines or polyamines, both natural and unnatural. The maximal activity of antizyme that can be induced in rat liver by the administration of putrescine is approximately equivalent to the maximal amount of ornithine decarboxylase that can be induced following administration of dexamethasone.[15,16] There is also an excellent correspondence between the conditions that permit the induction of ODC and those that permit the induction of antizyme, as well as between those conditions that contribute to the stabilization or elimination of ODC or antizyme, after the latter have been induced (Table 2).

As noted, the above results are obtained with cells maintained in a salts/glucose solution where cell division does not occur. Where experiments have been performed with animals or with cells maintained in growth media, similar results have been obtained. The T1/2 of ODC activity[7,24-26] and of immunologically reacting material to anti-ODC antibody where this has been measured[27] is shorter in the presence of putrescine alone than in the presence of the same concentration of putrescine plus a protein synthesis inhibitor (cycloheximide or puromycin). The results presented in Table 2 bring to the forefront two very interesting points:

1. The induction of ODC or of antizyme by their respective inducers, requires the continued presence of the inducer; following the removal of their respective inducers, a rapid fall in the activity of ODC or of antizyme occurs.
2. Both ODC and antizyme are stabilized by their respective inducers, even in the presence of cycloheximide.

These conclusions are in agreement with the observation of Rinehart and Canellakis[28] that the induction of ODC occurs only when the intracellular concentration of the inducer attains a critical threshold concentration and that ODC activity falls very rapidly, if the inducer

Table 1
HALF-LIVES OF ODC AND ANTIZYME

Source	Half-life (min) ODC	Half-life (min) Antizyme
Rat liver	15—24	24
H-35	15	18
L-1210	29	38
Neuroblastoma, N18	68	66

From Heller, J. S., Fong, W. F., and Canellakis, E. S., *Proc. Natl. Acad. Sci. U.S.A.*, 73, 1858, 1976.

Table 2
COMPARISON OF CONDITIONS THAT LEAD TO THE INDUCTION, STABILIZATION, AND DESTABILIZATION OF ODC AND ANTIZYME OF CELLS MAINTAINED IN A SALTS-GLUCOSE MEDIUM[1,17,18]

Condition	ODC activity	Antizyme activity
Effect on the Induction of Activity		
Add inducer	Induction[a]	Induction[b]
Inducer + actD	Induction minimally affected	Induction minimally affected
Inducer + CHX	No induction	No induction
Effect on Previously Induced Activity		
Remove inducer	Rapid fall; short T1/2	Rapid fall; short T1/2
Remove inducer, add cycloheximide	Rapid fall; short T1/2	Rapid fall; short T1/2
Retain inducer, add cycloheximide	Stabilization, long T1/2	Stabilization, long T1/2

[a] Inducers for ODC activity can be asparagine, a-amino isobutyric acid, *N*-methyl α-aminoisobutyric acid,[18-20] or fresh growth medium.[18-22]
[b] Inducer for antizyme can be putrescine, spermidine, spermine, or various diamines,[4,23] actD = actinomycin D and CHX = cycloheximide.

falls below this critical concentration. They also conform with the observation of Solano et al.[29] that the active form of ODC is stabilized in salt solutions by ornithine.

To the extent that antizyme is required for the more rapid elimination of ODC (see discussion below), the inhibition of protein synthesis by cycloheximide should prevent the synthesis of the antizyme necessary to eliminate ODC, resulting in its stabilization in the presence of cycloheximide. Conversely, since ODC is required for the rapid elimination of antizyme[17] inhibition of the synthesis of ODC by cycloheximide should prevent the elimination of antizyme, resulting in its stabilization in the presence of cycloheximide.

In the subsequent sections, we shall also explore what other modes of regulation of ODC activity may exist.

III. PHYSICAL PROPERTIES OF ANTIZYME

The rat liver antizyme of ornithine decarboxylase has been purified 600,000-fold to homogeneity by Kitani and Fujisawa.[16] This feat was only made possible following the comparably extensive purification of rat liver ornithine decarboxylase by Kameji and collaborators[30] and by Kitani and Fujisawa.[31] The use of affinity columns of highly purified ornithine decarboxylase was a major tool in the purification procedure, resulting in a 5500-fold purification of rat liver antizyme in a single passage. The rat liver antizyme, molecular weight MW = 22,000, and isoelectric point = 6.8 exhibits a high affinity for rat liver ornithine decarboxylase.

Kitani and Fujisawa[32] have determined the equilibrium constant of the reaction between ODC and antizyme to be about 1.4×10^{11} M^{-1} in 50 m*M* phosphate buffer, pH 7.0 at 0°C. Sodium chloride had a minimal effect on the binding constant between ODC and antizyme; however, the velocity of their interaction was greatly increased by higher sodium chloride concentrations and by lower temperatures. Low salt concentrations also promoted the association of ODC subunits while lowering the K_m for ornithine.[33] Solano et al.[29] demonstrated that L-ornithine and ionic strength have opposing effects on ODC; the salt favors enzyme dissociation into monomeric, enzymatically inactive forms while ornithine prevents the dissociation of the enzyme and favors a catalytically active dimer. Examination of these results indicates that in the absence of ornithine and in an isotonic salt medium, about 50% of the ODC exists in the monomeric, inactive form. From a series of measurements, on the extent and rate of interaction of the ODC-antizyme complex with DFMO in 50 m*M* salt, Kyriakidis et al.[34] find that the ODC-antizyme complex interacts with DFMO and concluded that native ODC binds with two molecules of antizyme. Kitani and Fujisawa[35] find that when ODC is incubated in the presence of a high molar excess of antizyme and in 10 m*M* salt and 20% ethylene glycol to minimize the dissociation of the ODC-antizyme complex, 13% of the ODC present in the ODC-antizyme complex interacts with DFMO. They suggest that the interaction with DFMO may occur through the dissociation of the ODC-antizyme complex, which would be greater in 50 m*M* salt and greater still in the isotonic environment (150 m*M* salt) of the cell. It would be interesting to follow up this observation with ODC of higher specific activity and with rate studies to understand the relationship between the active site of the enzyme and the site of antizyme interaction with ODC. Nevertheless, regardless of the underlying molecular mechanism by which this interaction occurs, it can be concluded that under physiological conditions, DFMO does interact with the ODC-antizyme complex.

Flamigni et al.[36] have noted that ODC is inhibited by basic proteins in proportion to their content of arginine, poly-L-arginine being the most effective. This finding is very interesting when we consider the association of ODC with nuclei, especially since ODC bound to basic proteins can be reactivated by nucleic acids.[37]

IV. THE CELLULAR DISTRIBUTION OF ANTIZYME

We have suggested[19,38] that inactive or cryptic forms of ODC and of antizyme exist in subcellular structures and that these can act as reservoirs from which their active forms can be released rapidly upon demand. An inactive, or cryptic form of antizyme was shown to be a normal component of uninduced rat liver and of H-35 cells.[39] In addition to its presence in cell nuclei, antizyme was associated in a bound and inactive form with the postribosomal ribonucleoprotein particles. These particles are obtained following prolonged centrifugation at $100,000 \times g$ of the ribosome-free deoxycholate treated microsome fraction. Whereas, high salt concentrations (0.3 *M* NaCl) were required to extract antizyme from these ribonucleoprotein particles, antizyme of a high specific activity could be extracted with only micromolar concentrations of putrescine or the other natural polyamines. It was proposed that this inactive or cryptic form of antizyme in these ribonucleoprotein particles is bound to its mRNA and prevents its translation. It was also suggested that antizyme through an allosteric reaction with putrescine could be released from its mRNA thus liberating the mRNA in a form that permits translation of antizyme. Such a mechanism would provide for the rapid synthesis of antizyme without resorting to transcription from DNA. Support for such a possibility is that synthesis of antizyme can occur in the presence of actinomycin D.[2] Within this context, we should keep in mind that under certain conditions, the synthesis of ODC can also proceed in the presence of actinomycin D.[46]

McCann et al.[7] while studying the kinetics of antizyme induction in HMO_A cells also noted that a latent form of antizyme could be released from the cell membranes by extraction with low concentrations of putrescine. Both the membrane-bound form of antizyme[7] and that bound to the postribosomal ribonucleoprotein particles[38] constitute a small percent of the total inducible antizyme.

Murakami et al.[41] examined the distribution of antizyme in cultures of rabbit kidney RK13 cells. They found that antizyme exists not only in the cytosol but also within the particulate fractions of these cells from which it can be extracted with 1 *M* NaCl. They also showed that treatment of these cells with putrescine increases the antizyme:ODC ratio and accelerates the rate of disappearance of ODC.

V. THE CELLULAR DISTRIBUTION OF ODC

Considerable evidence exists that particle-bound active, as well as inactive, forms of ODC exist. Greenfield and colleagues[42] have studied the distribution of antigenically reactive ODC among different cell lines. They found that whereas in RAW 264 cells (murine macrophage-like cells) antigenically reactive ODC could only be observed in the cytoplasm, in the JB6 epidermal cells, both the cytoplasm and the nucleoplasm reacted to their anti-ODC antibody (although no reaction was noted in the nucleolus). Anehus et al.[43] have used a combination of immunocytochemical and autoradiographic methods using ^{3}H-DFMO, to study the distribution of ODC in the DF 2 variant of the ODC overproducer CHO cells developed by Choi and Scheffler.[44] The immunocytochemical studies indicated that the larger portion of the ODC is in the cytoplasm rather than in the nucleus, while the autoradiographic method showed that the distribution of ODC was similar in the cytoplasm and nucleus. This very interesting dual approach merits repeating with other cell lines and experimental conditions. Autoradiographic studies of female germ cells of a polychaete, indicated that ODC is present in high concentrations in the cytoplasm as well as in the nucleolus and in the nucleoplasm[45,46] while immunocytochemical studies of proximal tubule cells of mouse renal cortex suggest that ODC is predominantly located in the cytoplasm.[47]

The autoradiographic and the immunocytochemical data need not represent only the active forms of ODC. As discussed in Section II, Kyriakidis et al.[34] as well as Kitani and Fujisawa[35] found that the ODC-antizyme complex reacts with ^{3}H-DFMO to give radioactive ODC while Seely and Pegg[27,48] showed that the ODC-antizyme complex reacts with the antibody to ODC. The immunochemical data may also include other inactive ODC forms, since under the isotonic cellular conditions a large portion of ODC may well exist in an enzymatically inactive monomeric form[29] which does not react with ^{3}H-DFMO. Should ODC be bound to basic histone-like nuclear proteins,[36] such forms of ODC may also respond differently to these modes of assay.

Evidence for the association of ODC in nuclei of eukaryotic cells has been provided for rat liver,[49,50] for chick embryo,[51] for *Physarum polycephalum*,[52,53] and for *Tetrahymena pyriformis*.[54] In rat heart, Flamigni et al.[55] localized a high specific activity ODC in the cytosolic fraction and in the nucleoli.

Following partial hepatectomy or administration of dexamethasone to rats, Bartholeyns[56] found that the total ODC activity available in a homogenate was less than could be accounted for by the sum of activities in the various fractions. Following such modes of induction of ODC, an increase of cytosolic enzyme only, and not of nuclear enzyme of rat liver, occurred.

Furthermore, whereas both the ODC derived from nuclear and the cytosolic fractions are responsive to inhibition by DFMO when tested in vitro, administration of DFMO to rats inhibits only the cytosolic enzyme and not the nuclear enzyme. This finding may be related to the different mode of interaction of the monomeric and the dimeric forms of ODC to DFMO.[33] Perhaps some of the information that is available from the distribution of antizyme and ODC in plants may prove to be applicable to the above studies. Panagiotidis et al.[57,58] found that germinating barley seeds contain both active ODC as well as active antizyme bound to their chromatin. Of these, ODC is not extractable by salts or by detergents while the antizyme is extractable by 2 *M* NaCl.

Indications for the existence of inactive or cryptic forms of ODC can be elicited from various experiments in which rapid changes in ODC activity have been studied. A rapid (1 to 2 min) transient increase of ODC in rabbit heart following perfusion with catecholamines was noted by Casti et al.[59] The experiments of Koenig et al.[60] show similarly rapid transient increases in ODC activity in mouse kidney slices in response to androgens, while recently, Muscari et al.[61] have demonstrated very rapid increases in ODC activity in rabbit heart and thoracic aorta following perfusion with adrenaline. In reviewing the experimental data, it appears that the increase in ODC activity cannot account for the increase in putrescine, and definitely not for the increase in all three polyamines which is a constant feature of the Koenig experiments, nor does it appear likely that protein synthesis can be involved. The simplest explanation that would account for these results is that there exist inactive forms of ODC which are activated under these conditions. However, is it possible that pools of bound polyamines exist which can be rapidly liberated?

VI. THE PHYSIOLOGICAL CORRESPONDENCE BETWEEN ANTIZYME AND ODC

In the early experiments[2,3] free antizyme could only be determined following exposure of cells to high concentrations of putrescine (10 m*M*). The existence of the ODC-antizyme complex in vivo when cells were treated with low concentrations of polyamines was demonstrated in HTC cells by McCann et al.[6] These investigators exposed these cells to 10^{-5} *M* putrescine for prolonged periods of time to eliminate most of the ODC activity. Following treatment of these inactive extracts with KCl, they were able to isolate chromatographically, free ODC and free antizyme.

Using starved cell cultures that contain minimal levels of ornithine decarboxylase to complex the antizyme that is induced,[18] Heller and Canellakis[17] demonstrated that free antizyme could be induced with concentrations of putrescine as low as 10^{-6} to 10^{-7} *M*. However, if the cells had been preinduced to high levels of ODC, 2×10^{-2} *M* putrescine was necessary to induce enough antizyme to inactivate the endogenous ODC and produce excess free antizyme. The opposite result was obtained if the cells had been preinduced to high levels of antizyme; 2×10^{-2} *M* asparagine was required to induce enough ODC to inactivate the endogenous antizyme, and produce excess free ODC. By varying the ODC content of cells, the concentration of putrescine required to induce measurable amounts of free antizyme was found to vary with the level of preinduced ODC. Conversely, by varying the antizyme content of cells, the concentration of asparagine required to induce measurable amounts of free ODC varied with the level of preinduced antizyme. These results demonstrated that the delay in the appearance of either ODC or antizyme, was proportional to the cell content of antizyme or ODC, respectively. They also provided an explanation for the high concentrations of putrescine that were required to induce antizyme in growing cells which had high endogenous levels of ODC.[2,3]

It now appears that asparagine and putrescine added to such contact inhibited cultures precipitate both general and specific effects. At the general level asparagine causes increased chromosomal decondensation,[62] indicating a facilitation of gene transcription, while at the specific level it induces the transcription of both the c-myc[62] and the ODC genes,[62,63] as evidenced by the increase in their respective mRNAs. These effects are probably associated with the changes in Na^+ uptake that occur with both asparagine and putrescine.[1,20,21] Asparagine is among those amino acids that are transported by the sodium-dependent System A and ASC amino acid transport systems and it, as well as other such amino acids, also stimulates the uptake of putrescine.[64] With respect to the effects of asparagine on ODC protein, the results of Kanamoto et al.[65] indicate that asparagine increases the rate of synthesis of ODC and decreases its rate of disappearance (to be published), while putrescine enhances the rate of disappearance of ODC and partially suppresses its rate of synthesis. Physiological changes of ODC and antizyme in rat brain have been noted by Hietala[66] and Laitinen et al.[67,68] and their collaborators. Their detailed studies indicate that newborn mouse brain, exhibits about 70-fold higher ODC activity than adult mouse brain; a progressive decrease in ODC activity occurs during the first 3 weeks after birth, which is counterbalanced by a progressive increase in antizyme activity as evidenced by a several hundredfold difference between the immunoreactive ODC and the catalytically active enzyme. Laitinen[67] has shown that although the overall brain antizyme content does not appear to change greatly, there are significant changes in different antizyme pools, and that these correlate well with the changes in putrescine concentration. The evidence suggests that antizyme contributes to the posttranslational control of ODC activity.

The kinetics of appearance of antizyme within the cell cycle by Linden et al.[69] indicate that whereas ODC activity exhibited two maxima, in late-G1/early-S and in the late-S/G2 phases of the cell cycle, a marked decrease in ODC activity was noted in the intervening mid-S phase. On the other hand, a marked increase in antizyme was noted during this intervening mid-S phase, suggesting that antizyme is involved in the regulation of ODC activity during the cell cycle.

VII. THE METABOLIC FATE OF ODC AND OF ANTIZYME

From a comparison of the rate of decay of ODC activity and of immunoreactive ODC protein in rat liver, following the administration of 1,3-diaminopropane with or without cycloheximide to rats, Seely and Pegg[27,48] suggested that antizyme may contribute to the more rapid degradation of ODC. Additional support for such a view became available with the development of a general method for the quantitation of the ODC-antizyme complex in cell extracts by Murakami et al.[24,25] and by Mitchell.[70]

In a detailed investigation, Murakami et al.[24,25] show that DFMO-inactivated ODC exchanges with the native ODC that exists in the ODC-antizyme complex. This exchange results in the liberation of the native ODC from the complex. By assaying enzymatically the liberated ODC, the amount of ODC-antizyme complex can be quantitated. The development of this method has made it possible to determine the cellular amount of the complex under various conditions with and without exogenous polyamines and to conclude[24,25] that antizyme most likely contributes to the degradation of ODC. The evidence can be summarized as follows:

1. There was good correlation between the antizyme:ODC ratio and the ODC degradation rate in HTC cells under physiological conditions
2. The variant HMO_A cells which have a stabilized ODC, accumulated large amounts of the ODC-antizyme complex
3. Upon addition of putrescine, the cellular level of the ODC-antizyme complex increased temporarily, but then decreased in both the HTC and HMO_A cells, but the amount of total ODC (sum of free and complexed ODC) decreased more rapidly than it did upon addition of cycloheximide

4. Preincubation of HTC cells with putrescine, which increased the antizyme:ODC ratio, shortened the half-life of ODC activity; however, the half-life of ODC activity was not affected if protein synthesis was inhibited by adding cycloheximide at the time of pretreatment with putrescine.

In order to explain the slower decay rate of antizyme than ODC, these authors proposed that antizyme recycles during the process of ODC decay in a mechanism which bears some similarity to the ubiquitin-mediated proteolytic system of reticulocytes[71,72] as had been previously suggested by Seely and Pegg.[27]

Several lines of indirect evidence also support the key role played by antizyme in the degradation of ODC.

1. In addition to inhibiting the synthesis of ODC,[26,25] polyamines accelerate its decay as evidenced by the more rapid decrease of immunoreactive ODC protein in the presence of polyamines than in the presence of cycloheximide (Table 2 and References 7, 27, and 73). This finding has been reinforced by the observation that polyamines accelerate the decay of prelabeled ODC protein.[26,65,74] On the other hand, inhibition of ODC by ornithine analogues (other than DFMO), results in its stabilization.[75-80]
2. The polyamine-induced acceleration of ODC decay is blocked by cycloheximide (Table 2 and References 7, 27, and 73) but not by actinomycin D (Table 2 and References 73 and 81) which does not inhibit antizyme synthesis.[1,8,73]
3. The increase in the half-life of ODC activity that occurs upon ODC induction[22,65,71,82-84] could be a consequence of a fall in the antizyme:ODC ratio.

As discussed above in Sections IV to VI, we should also consider the alternative possibility that the antizyme is not directly responsible for the degradation of ODC but facilitates the sequestration of ODC into an inactive form (either particle or membrane bound), from which ODC can be subsequently reactivated or eventually degraded. This possibility is supported by the fact that particulate-bound forms of antizyme[7,39,41] and ODC have been detected. Furthermore, the rapid increases in ODC activity reported by Koenig et al.[60] and by Muscari et al.[61] could be due to the release of ODC from particle-bound forms.

VIII. CONDITIONS THAT MAY NOT LEAD TO THE REGULATION OF ODC ACTIVITY BY ANTIZYME

Several reports have appeared suggesting that antizyme does not participate in the regulation of ODC activity. In many cases, the absence of antizyme in an extract was determined by the inability of the extract to inhibit ODC. Careful examination of the data indicates that in some of these cases[85-87] the cell extract in which ODC had been inhibited following various manipulations was not completely devoid of ODC activity. By definition, such extracts could not contain any *free antizyme*. Of the various negative experiments that were presented, Heller et al.[4] reexamined the 3T3 cells[85] and found that large quantities of antizyme could be detected in the 3T3 cells if the cells were exposed to putrescine until all the ODC activity had disappeared.

From studies on the effects of TPA and retinoids on ODC induction in epidermal and other cells,[88,89] it was concluded that antizyme did not participate in the down regulation of ODC. Since retinoids[88-91] do not inhibit ODC activity as extensively as putrescine, it would have been difficult to detect any free antizyme in such cell extracts. With the availability of the newer assays for the ODC-antizyme complex[24,25] it may now be worthwhile to reevaluate whether these, and other experimental conditions, do in fact control ODC activity through a different posttranslational mechanism.

The results obtained from various laboratories with kidney cells present an interesting insight into the problem. Persson and Rosengren[92] did not detect any antizyme in the soluble extracts of kidneys of testosterone treated mice. Seely and Pegg[27] followed up on this observation and found no evidence for posttranslational control of ODC by antizyme or for the presence of antizyme in this tissue. Laitinen et al.[68] suggested that this may be a unique situation since putrescine does not accumulate in the kidneys of mice[93] and therefore is not available in high enough concentrations to induce antizyme. Murakami et al.[41] investigated this problem by exposing rabbit kidney RK13 cells to high putrescine concentrations and examining the distribution of antizyme in cultures of these cells. They found that antizyme exists in the cytosol of untreated RK13 cells as the ODC-antizyme complex, while it exists in free form in both the cytosol and the particulate fractions of putrescine-treated cells. They also showed that treatment of these cells with putrescine increases their antizyme:ODC ratio and accelerates the rate of disappearance of ODC.

The question whether the induction of ODC activity in mouse kidney following testosterone treatment is regulated by antizyme remains to be resolved. Nevertheless, these results emphasize the difficulty in determining the absence of antizyme with a great degree of certainty and the necessity to determine the role of latent or cryptic forms of ODC and antizyme bound to subcellular structures.

In a *spe2-4* mutant of *Saccharomyces cerevisiae* with four to six times higher ODC activity than normal cells,[94] no evidence for an inhibitor of ODC was found even when these cells were grown in the presence of 10^{-4} *M* spermidine plus 10^{-4} *M* spermine. Although these concentrations of polyamines inhibited ODC activity in the *S. cerevisiae* mutant by as much as 90%, no loss of ODC protein was noted. It was suggested that a different type of posttranslational modification of ODC may exist in these cells.

IX. DIFFERENT FORMS OF ODC AND ANTIZYME

Two ODC mRNA species (95 to 98) are apparent in hybridization experiments with ODC-specific cDNA probes in RNA fractions of various cells. Mitchell and collaborators[99,100] have isolated two distinct forms of ODC from extracts of HTC cells (a tumor cell line derived from rat liver), which differ in their charged states. In HTC cells, spermidine and spermine but not putrescine appear to stimulate the conversion of the metabolically more stable and more acidic Form II (ODC II) into the metabolically less stable and less acidic Form I (ODC I). This conversion of one form to the other may consist of a dephosphorylation reaction. Inhibitors of spermidine and spermine synthesis increase the amount of ODC II, resulting in a great degree of stabilization of ODC activity. ODC I and ODC II have been shown to interact equally well with antizyme in vitro. This property has its in vivo counterpart. Polyamines added to cells in culture, caused a rapid loss of both forms of ODC, probably through their interactions with antizyme, whereas inhibition of protein synthesis with cycloheximide resulted in a more rapid loss of the less stable ODC I.

Studies by Bishop et al.[101] indicate that a noncompetitive protein inhibitor of ODC, MW = approximately 50,000 can be elicited in thymus and upon administration of dexamethasone to rats. These authors do not indicate whether this particular protein can also be induced by polyamines. However, Hu and Brosnan[102] find that diaminopropanol induces an approximately 50,000 MW noncompetitive protein inhibitor of ODC in mammary glands of lactating rats.[102] These authors find this 50,000-MW protein to be similar to rat liver antizyme; the ODC present in complexes of ODC with this inhibitor equilibrate with the DFMO-inactivated ODC, yielding active ODC, a property that is similar to that of rat liver antizyme (see Section VI). In relation to the above, it should be noted that two forms of antizyme have also been reported for chicken liver.[9] Recently, Hayashi et al.[103] have shown that their monoclonal antibodies to rat liver antizyme do not cross react with the brain antizyme.[67,68]

These results strongly suggest that perhaps antizyme is tissue specific. Germinating barley seeds also appear to contain two forms of antizyme, a cytosolic antizyme MW = 16,000 whereas the chromatin bound antizyme is much smaller, MW = 9000.[14,15]

X. ALTERNATE MECHANISMS OF REGULATION OF ODC ACTIVITY

A. Direct Modification of Activity

Kitani and Fujisawa[104] noted that the activity of rat liver ODC can be increased or decreased approximately twofold by individual phospholipids. It is unclear to what extent this effect has a physiological function, such as explaining the rapid alterations in ODC activity[59-61] referred to in Section V or whether it is the response to the detergent-like action of phospholipids. Some detergents have a stabilizing effect on ODC and have been used during its purification.

As mentioned earlier we should also note that ODC is inhibited by basic proteins in proportion to their arginine content.[36]

B. Effects on ODC Activity Mediated Through the Cell Membrane

Work from this laboratory[17,19] indicated that concentrations of exogenous spermine, as low as 1×10^{-6} *M*, added to L1210 cells caused a precipitous fall in the induction of ODC activity. Approximately 50-fold higher concentrations of spermine, up to 5×10^{-5} *M* did not cause any further fall in ODC activity. Yet, when this spermine concentration was doubled to 1×10^{-4} *M*, ODC activity decreased precipitously to zero levels and free antizyme was subsequently induced. It was considered that this differential effect between the low spermine concentrations, that were several orders of magnitude lower than their intracellular concentration, and the high spermine concentrations could be due to different mechanisms of inhibition. It was proposed that the effects of spermine at low concentrations could be mediated through membrane receptors. Electron microscopic evidence for the existence of polyamine receptors on cell membranes has been provided in a detailed series of experiments by Moulinoux and collaborators[105] in their studies of the interaction of polyamines and antipolyamine antibodies with cell membranes.

McCann et al.[7] found different kinetics of inhibition of the ODC activity of HMO_A cells by low polyamine concentrations compared to those obtained with high concentrations of polyamines. They too came to the conclusion that at these low concentrations, polyamines may inhibit ODC activity through an alternate mechanism that did not involve the induction of antizyme.

The possibility that ODC activity may be regulated through the membrane has had additional support from two other sources. Chen et al.[106] have shown that very low concentrations of local anesthetics as well as of ionophores, both of which are presumed to exert their effects through interaction with the cell membrane, inhibit ODC activity. Earlier studies[107] had also shown that cytoskeleton-disrupting agents also prevent the induction of ODC. The cumulative evidence suggests that this particular area of research needs further investigation. The use of compounds whose biological effects have been shown to be mediated through membrane receptors including retinoids or androgens, in association with antipolyamine antibodies, may be especially useful in this regard.

C. Regulation at the Level of ODC mRNA

Kahana and Nathans[108] studied the effect of putrescine on the translation of ODC in a mutant mouse myeloma cell line 653-1 that overproduces both ODC protein and ODC mRNA as a result of gene amplification. They concluded that the major effect of polyamines is to inhibit the translation of ODC mRNA. This conclusion was based on their finding that 3×10^{-2} *M* putrescine decreased the rate of synthesis of ornithine decarboxylase, but did not affect either the cellular content of preformed ODC mRNA nor the stability of prelabeled ODC protein. In these very interesting experiments the stability of the prelabeled ODC protein was determined in cells exposed to fresh medium containing cycloheximide, in the presence and absence of polyamines. In Section II we indicated that ODC is stabilized when cells are exposed to cycloheximide in the presence of an inducer of ODC, such as fresh medium (Table 2); furthermore, the presence of cycloheximide would not permit the synthesis of antizyme which has been shown to participate in the elimination of ODC. We should also keep in mind that fresh serum induces the transcription of a number of genes[109] while the inhibition of protein synthesis promotes the accumulation of various mRNAs.[110] The authors consider that it is difficult to reach a definitive conclusion without in vitro translation or nuclear runoff experiments.

Dircks et al.,[111] who performed a similar series of experiments with an ODC overproducer CHO variant exposed to fresh medium, also suggested that polyamines may inhibit the translation of ODC mRNA, or, alternatively, that other modes of ODC regulation exist. Most recently, Pegg and associates[112] provided evidence that spermidine inhibits the translation of ODC from total mRNA in a reticulocyte system.

The information to be derived on the mechanism of regulation of ODC activity at the mRNA level should prove to be very informative. However, working with crude in vitro systems, we should also consider the possibility that they contain antizyme mRNA. If so, it is possible that the polyamines activate the translation of antizyme, and that the antizyme inhibits the translation of the ODC message. If the difference in the molecular weights of active ODC and antizyme are also mirrored in their mRNAs, the rate of synthesis of antizyme would be fivefold greater than that of ODC.

D. Regulation Through Covalent Modification of ODC

Russell[113] has reviewed the evidence that transamidation of glutamine residues of ODC with putrescine by transglutaminase inhibits ODC activity, leading to the complete loss of activity. The role of phosphorylation of ODC on the activity of ODC has also been the subject of investigation.[99,100,114,115] As the evidence accumulates in these two interesting directions, it will be possible to arrive at more definitive statements.

XI. OVERVIEW OF ODC REGULATION BY ANTIZYME

Despite the large amount of information that is presently available on the regulation of polyamine biosynthesis, we shall have to obtain more information in the following areas:

1. The extent to which the direct degradation[24,27,48] of ODC and/or antizyme are involved in the regulation of ODC activity or whether following their interaction they are first converted to latent or cryptic forms,[19,38] as well as the interesting possibility that the levels of ODC are regulated by the recycling of antizyme.[116]
2. The physiological role of antiantizyme discovered by Fujita and collaborators[117] which binds strongly to the antizyme of an ODC-antizyme complex and liberates free ODC.
3. The function of the ribonucleoprotein-bound[39] and membrane-bound[7] forms of antizyme. The role of the cellular compartmentation of ODC and of antizyme in the regulation of polyamine biosynthesis.
4. The possible regulation of ODC activity by extracellular stimuli acting upon membrane receptors.
5. The extent to which antizyme may participate directly in the control of the transcription of the ODC gene or the translation of the ODC mRNA, and conversely, whether ODC can participate in the control of the transcription of the antizyme gene or of the translation of the antizyme mRNA.

REFERENCES

1. **Fong, W. F., Heller, J. S., and Canellakis, E. S.,** The appearance of an ornithine decarboxylase inhibitory protein upon the addition of putrescine to cell cultures, *Biochim. Biophys. Acta,* 428, 456, 1976.
2. **Heller, J. S., Fong, W. F., and Canellakis, E. S.,** Induction of a protein inhibitor to ornithine decarboxylase by the end products of its reaction, *Proc. Natl. Acad. Sci. U.S.A.,* 73, 1858, 1976.
3. **Canellakis, E. S., Kyriakidis, D. A., Rinehart, C. A., Jr., Huang, S. C., Panagiotidis, C., and Fong, W. F.,** Regulation of polyamine biosynthesis by antizyme and some recent developments relating the induction of polyamine biosynthesis to cell growth, *Biosci. Rep.,* 5, 189, 1985.
4. **Heller, J. S., Chen, K. Y., Kyriakidis, D. A., Fong, W. F., and Canellakis, E. S.,** The modulation of the induction of ornithine decarboxylase by spermine, spermidine and diamines, *J. Cell Physiol.,* 96, 225, 1978.
5. **Branca, A. A. and Herbst, E. J.,** Polyamines, ODC antizyme, and HeLa cell proliferation, *Biochem. J.,* 186, 925, 1978; **Herbst, E. J. and Branca, A. A.,** *Advances in Polyamine Research,* Vol. 3, Caldarera, C. M., Zappia, V., and Bachrach, U., Eds., Raven Press, New York, 1981, 287.
6. **McCann, P. P., Tardif, C., and Mamont, P. S.,** Regulation of ornithine decarboxylase by ODC-antizyme in HTC cells, *Biochem. Biophys. Res. Commun.,* 75, 948, 1977.
7. **McCann, P. P., Tardif, C., Hornsperger, J., and Bohlen, P.,** Two distinct mechanisms for ornithine decarboxylase regulation by polyamines in rat hepatoma cells, *J. Cell Physiol.,* 94, 183, 1979.
8. **Friedman, Y., Park, S., Levasseur, S., and Burke, G.,** Regulation of thyroid ornithine decarboxylase by the polyamines, *Biochim. Biophys. Acta,* 500, 291, 1977.
9. **Grillo, M. A., Bedino, S., and Testore, G.,** Studies on ornithine decarboxylation in chicken liver, in *Advances Polyamine Research,* Vol. 3, Caldarera, C. M., Zappia, V., and Bachrach, U., Eds., Raven Press, New York, 1981, 27.
10. **Lesiewicz, J. and Goldsmith, L. A.,** Antizyme release is an early event in ornithine decarboxylase induction by hair plucking, *J. Invest. Dermatol.,* 80, 97, 1983.
11. **Brosnan, M. E., Farrell, R., Wilansky, H., and Williamson, D. H.,** Effect of starvation and refeeding on polyamine concentrations and ornithine decarboxylase antizyme in mammary gland of lactating rats, *Biochem. J.,* 212, 149, 1983.
12. **Brosnan, M. E. and Hu, Y.-W.,** Synthesis and function of polyamines in mammary gland during lactation, in *Recent Progress in Polyamine Research,* Selmeci, L., Brosnan, M. E., and Seiler, N., Eds., Akademiai Kiado, Budapest, 1985, 169.
13. **Kyriakidis, D. A.,** Macromolecular effectors of ornithine decarboxylase in germinated barley seeds, in *Advances in Polyamine Research,* Vol. 4, Caldarera, C. M., Zappia, V., and Bachrach, U., Eds., Raven Press, New York, 1983, 427.
14. **Kyriakidis, D. A.,** Effect of plant growth hormones and polyamines on ornithine decarboxylase activity during the germination of barley *(Hordeum vulgare cultivar beca)* seeds, *Physiol. Plant.,* 57, 499, 1983.
15. **Canellakis, E. S., Heller, J. S., and Kyriakidis, D. A.,** The interaction of ornithine decarboxylase with its antizyme, in *Advances in Polyamine Research,* Vol. 3, Zappia, V., and Bachrach, U., Caldarera, C. M., Eds., Raven Press, New York, 1981, 1.
16. **Kitani, T. and Fujisawa, H.,** Purification and some properties of a protein inhibitor (antizyme) of ornithine decarboxylase from rat liver, *J. Biol. Chem.,* 259, 10036, 1984.
17. **Heller, J. S. and Canellakis, E. S.,** Cellular control of ornithine decarboxylase activity by its antizyme, *J. Cell. Physiol.,* 107, 209, 1981.
18. **Chen, K. Y. and Canellakis, E. S.,** Enzyme regulation in neuroblastoma cells in a salts/glucose medium: induction of ornithine decarboxylase by asparagine and glutamine, *Proc. Natl. Acad. Sci. U.S.A.,* 74, 3791, 1977.
19. **Canellakis, E. S., Viceps-Madore, D., Kyriakidis, D. A., and Heller, J. S.,** The regulation and function of ornithine decarboxylase and of the polyamines, *Curr. Topics Cell Regul.,* 15, 155, 1979.
20. **Viceps-Madore, P., Chen, K. Y., Tsou, H. R., and Canellakis, E. S.,** Studies on the role of protein synthesis and of sodium on the regulation of ornithine decarboxylase activity, *Biochim. Biophys. Acta,* 717, 305, 1979.
21. **Rinehart, C. A., Viceps-Madore, D., Fong, W. F., Ortiz, J. G., and Canellakis, E. S.,** The effect of transport system A and N amino acids and of nerve and epidermal growth factors on the induction of ODC activity, *J. Cell. Physiol.,* 123, 435, 1985.
22. **Hogan, B. L. M. and Murdin, S.,** Effect of growth conditions on the activity of ornithine decarboxylase in cultured hepatoma cells, *J. Cell. Physiol.,* 83, 345, 1974.
23. **Janne, J. and Höltta, E.,** Regulation of ornithine decarboxylase activity by putrescine and spermidine in rat liver, *Biochem. Biophys. Res. Commun.,* 61, 449, 1974.
24. **Murakami, Y., Fujita, K., Kameji, T., and Hayashi, S.,** Accumulation of ornithine decarboxylase-antizyme complex in HMO_A cells, *Biochem. J.,* 225, 689, 1985.

25. **Murakami,Y. and Hayashi, S.**, Role of antizyme in degradation of ornithine decarboxylase in HTC cells, *Biochem. J.*, 226, 893, 1985.
26. **Holta, E. and Pohjanpelto, P.**, Control of ornithine decarboxylase in Chinese hamster ovary cells by polyamines, *J. Biol. Chem.*, 261, 9502, 1986.
27. **Seely, J. E. and Pegg, A. E.**, Effect of 1,3-diaminopropane on ornithine decarboxylase enzyme protein in thioacetamide-treated rat liver, *Biochem. J.*, 216, 701, 1983.
28. **Rinehart, C. A., Jr. and Canellakis, E. S.**, Induction of ornithine decarboxylase activity by insulin and growth factors is mediated by amino acids, *Proc. Natl. Acad. Sci. U.S.A.*, 82, 4365, 1985.
29. **Solano, F., Penafiel, R., Solano, M. E., and Lozano, J. A.**, Equilibrium between active and inactive forms of rat liver ornithine decarboxylase mediated by L-ornithine and salts, *FEBS Lett.*, 190, 324, 1985.
30. **Kameji, T., Murakami, Y., Fujita, K., and Hayashi, S. I.**, Purification and some properties of ornithine decarboxylase from rat liver, *Biochim. Biophys. Acta,* 717, 111, 1982.
31. **Kitani, T. and Fujisawa, H.**, Purification and properties of ornithine decarboxylase from rat liver, *J. Biol. Chem.*, 258, 235, 1983.
32. **Kitani, T. and Fujisawa, H.**, The velocity of the reaction between ornithine decarboxylase and antizyme highly depends on the presence of salt and temperature, *Biochem. Int.*, 10, 435, 1985.
33. **Kitani, T. and Fujisawa, H.**, Influence of salts on the activity and the subunit structure of ODC, *Biochim. Biophys. Acta,* 784, 164, 1984.
34. **Kyriakidis, D. A., Flamigni, F., Pawlak, J. W., and Canellakis, E. S.**, Mode of interaction of ornithine decarboxylase with antizyme and aa-difluoromethylornithine, *Biochem. Pharmacol.*, 33, 1575, 1984.
35. **Kitani, T. and Fujisawa, H.**, α-Difluoromethylornithine does not bind to ornithine decarboxylase-antizyme complex, *Biochem. Biophys. Res. Commun.*, 137, 1101, 1986.
36. **Flamigni, F., Guarnieri, C., Marmiroli, S., and Caldarera, C. M.**, Inhibition of rat heart ornithine decarboxylase by basic polypeptides, *Biochem. J.*, 229, 807, 1985.
37. **Huang, S., Kyriakidis, D. A., Rinehart, C. A., and Canellakis, E. S.**, Reversal of the antizyme inhibition of ODC by nucleic acids, *Biochem. Pharmacol.*, 33, 1383, 1984.
38. **Canellakis, E. S., Kyriakidis, D. A., Heller, J. S., and Pawlak, J. W.**, The complexity of regulation of ornithine decarboxylase, *Med. Biol.*, 59, 279, 1981.
39. **Heller, J. S., Kyriakidis, D. A., Fong, W. F., and Canellakis, E. S.**, Ornithine decarboxylase antizyme is a normal component of uninduced H-35 cells and rat liver, *Eur. J. Biochem.*, 81, 545, 1977.
40. **Yamasaki, Y. and Ichihara, A.**, Induction of ornithine decarboxylase in cultured mouse L cells, *J. Biochem. Tokyo,* 80, 557, 1976.
41. **Murakami, Y., Marumo, M., and Hayashi, S.**, Existence of antizyme and ornithine decarboxylase-antizyme complex in RK13 kidney cells, *FEBS Lett.*, 199, 49, 1986.
42. **Greenfield, A. R. L., Taffet, S. M., and Haddox, M. K.**, Immunocytochemical localization of ornithine decarboxylase in cultured murine cells, *Cell Tissue Res.*, 243, 33, 1986.
43. **Anehus, S., Emanuelsson, H., Persson, L., Sundler, F., Scheffler, I. E., and Heby, O.**, Localization of ornithine decarboxylase in mutant CHO cells that overproduce the enzyme. Differences between the intracellular distribution of monospecific ornithine decarboxylase antibodies and radiolabeled α-difluoromethylornithine, *J. Cell. Biol.*, 35, 264, 1984.
44. **Choi, J. H. and Scheffler, I. E.**, Chinese hamster ovary cells resistant to α-difluoromethylornithine are overproducers of ornithine decarboxylase, *J. Biol. Chem.*, 258, 12601, 1983.
45. **Emanuelsson, H. and Heby, O.**, Ornithine decarboxylase activity in nucleolus and nucleoplasm demonstrated autoradiographically with tritium-labeled α-difluoromethylornithine, *Cell Biol. Int. Rep.*, 6, 951, 1982.
46. **Emanuelsson, H.**, Nuclear ornithine decarboxylase electron microscope auto radiographic identification of the active enzyme using alpha-5 tritium labeled difluoromethyl ornithine, *Eur. J. Cell. Biol.*, 30, 288, 1983.
47. **Persson, L., Rosengren, E., and Sundler, F.**, Localization of ornithine decarboxylase by immunocytochemistry, *Biochem. Biophys. Res. Commun.*, 104, 1196, 1982.
48. **Seely, J. E. and Pegg, A. E.**, Changes in mouse kidney ornithine decarboxylase activity are brought about by changes in the amount of enzyme protein as measured by radioimmunoassay, *J. Biol. Chem.*, 258, 2496, 1983.
49. **Murphy, B. J. and Brosman, M. E.**, Subcellular localization of ornithine decarboxylase in liver of control and growth-hormone-treated rats, *Biochem. J.*, 157, 33, 1976.
50. **Bitoni, A. J. and Couri, D.**, Drug-induced migration of cytoplasmic ornithine decarboxylase into rat liver nucleus is not related to increased RNA polymerase activity, *Biochem. Biophys. Res. Commun.*, 99, 1040, 1983.
51. **Solomon, H., Vicente, J., Snyder, S. H., Kreuz, D. S., Medina, V. I., and Russel, D. H.**, Polyamines synthesis and turnover in rapidly growing tissues, *Ann. N.Y. Acad. Sci.*, 171, 749, 1970.
52. **Kuehn, G. D., Atmar, V. J., and Daniels, G. R.**, Polyamine-dependent protein kinase from the slime mold *Physarum polycepholum, Methods Enzymol.*, 94, 144, 1983.

53. **Mitchell, J. L. A.,** Ornithine decarboxylase and the ornithine decarboxylase-modifying protein of *Physarum polycephalam, Methods Enzymol.*, 94, 140, 1983.
54. **Sklaviadis, T. K., Georgatsos, J. G., and Kyriakidis, D. A.,** Purification and properties of ornithine decarboxylase from *Tetrahymena pyriformis, Biochim. Biophys. Acta*, 831, 288, 1985.
55. **Flamigni, F., Guarnieri, C., and Caldarera, C. M.,** Heart ornithine decarboxylase from control and isoproterenol-treated rats: kinetic properties, multiple forms and subcellular distribution, *Gen. Pharmacol.*, 17, 31, 1986.
56. **Bartholeyns, J.,** Subcellular distribution of ornithine decarboxylase in rat liver and accessiblilty of the enzyme to aa-difluoromethylornithine, an irreversible inhibitor of ornithine decarboxylase, *Life Sci*, 32, 1305, 1982.
57. **Panagiotidis, C. A. and Kyriakidis, D. A.,** Purification of a non-histone protein with properties of antizyme to ornithine decarboxylase from germinated barley seeds, *Plant Growth Regul.*, 3, 247, 1985.
58. **Koromilas, A. E., Panagiotidis, C. A., and Kyriakidis, D. A.,** Regulation and function of ornithine and arginine decarboxylase in germinating barley seeds, in *Recent Progress in Polyamine Research*, Selmeci, L., Brosnan, M. E. and Seiler, N., Eds., Akademiai Kiado, Budapest, 1985, 501.
59. **Casti, A., Corti, A., Reali, N., Mezzetti, G., Orlandini, G., and Caldarera, C. M.,** Modifications of major aspects of myocardial metabolism as a response to noradrenaline. Action of the hormone on cytoplasmic processing of ribonucleic acid after reserpine treatment, *Biochem. J.*, 168, 333, 1977.
60. **Koenig, H., Goldstone, A., and Lu, C. Y.,** Polyamine regulates calcium fluxes in a rapid plasma membrane response, *Nature*, 305, 530, 1983.
61. **Muscari, C., Stefanelli, C., Speroni, E., Carati, D., and Rossoni, C.,** Rapid activation of ornithine decarboxylase in rabbit heart and thoracic aorta perfused with noradrenaline and the inhibiting effect of the calcium antagonist diltiazem, Biomedical Studies of Natural Polyamines, p. 65, 1986.
62. **Vitkauskas, G. V., Palayoor, T., Pantelias, G., Rinehart, C. A., Jr., and Canellakis, E. S.,** The induction of ornithine decarboxylase by amino acids, in *Biomedical Studies of Natural Polyamines*, Caldarera, C. M., Clo, C., and Guarnieri, C., Eds., 1986, 3.
63. **Palayoor, T., Vitkauskas, G. V., Pantelias, G., and Canellakis, E. S.,** Manuscript in preparation.
64. **Rinehart, C. A., Jr. and Chen, K. Y.,** Characterization of the polyamine transport system in mouse neuroblastoma cells, *J. Biol. Chem.*, 259, 4750, 1984.
65. **Kanamoto, R., Utsunomiya, K., Kameji, T., and Hayashi, S.,** Effects of putrescine on synthesis and degradation of ornithine decarboxylase in primary cultured hepatocytes, *Eur. J. Biochem.*, 154, 539, 1986.
66. **Hietala, O. A.,** Detection of ornithine decarboxylase antizyme in mouse brain, *J. Neurochem.*, 40, 1174, 1983.
67. **Laitinen, P. H.,** Involvement of an "Antizyme" in the inactivation of ornithine decarboxylase, *J. Neurochem.*, 45, 1303, 1985.
68. **Laitinen, P. H., Huhtinen, R. L., Hietala, O. A., and Pajunen, A. E. I.,** Ornithine decarboxylase activity in brain regulated by a specific macromolecule, the antizyme, *J. Neurochem.*, 44, 1885, 1985.
69. **Linden, M., Anehus, S., Langstrom, E., Baldetorp, B., and Heby, O.,** Cell cycle phase-dependent induction of ornithine decarboxylase-antizyme, *J. Cell Physiol.*, 125, 273, 1985.
70. **Mitchell, J. L. A., Qasba, P., and Mahan, D. W.,** Alternative charge states of mammalian ornithine decarboxylase and their polyamine-induced inactivation, *Recent Prog. Polyamine Res.*, 55, 64, 1985.
71. **Ciechanover, A., Hod, Y., and Hershko, A.,** A heat-stable polypeptide component of an ATP-dependent proteolytic system from reticulocytes, *Biochem. Biophys. Res. Commun.*, 81, 1100, 1978.
72. **Wilkinson, K. D., Urban, M. K., and Haas, A.L.,** Ubiquitin is the ATP-dependent proteolysis factor I of rabbit reticulocytes, *J. Biol. Chem.*, 255, 7529, 1980.
73. **Utsunomiya, K., Kanamoto, R., and Hayashi, S.,** Role of antizyme in regulation of ornithine decarboxylase in primary cultured hepatocytes, *Jikeikai Med. J.*, 32, 575, 1985.
74. **Glass, J. R. and Gerner, E. W.,** Polyamine-mediated turnover of ornithine decarboxylase in Chinese-hamster ovary cells, *Biochem. J.*, 236, 351, 1986.
75. **Persson, L., Oredsson, S. M., Anehus, S., and Heby, O.,** Ornithine decarboxylase inhibitors increase the cellular content of the enzyme: implications for translational regulation, *Biochem. Biophys. Res. Commun.*, 131, 239, 1985.
76. **Harik, S. I., Hollenberg, M. D., and Snyder, S. H.,** Ornithine decarboxylase turnover slowed by α-hydrazinoornithine, *Mol. Pharmacol.*, 10, 41, 1974.
77. **Inoue, H., Kato, Y., Takigawa, M., Adachi, K., and Takeda, Y.,** Effect of DL-α-hydrazino-aminovaleric acid, an inhibitor of ornithine decarboxylase, on polyamine metabolism in isoproterenol-stimulated mouse parotid glands, *J. Biochem.*, 77, 879, 1975.
78. **McCann, P. P., Tardif, C., Duchesne, M. C., and Mamont, P. S.,** Effect of α-methylornithine on ornithine decarboxylase activity of rat hepatoma cells in culture, *Biochem. Biophys. Res. Commun.*, 76, 893, 1977.
79. **Stevens, L. and McKinnon, I. M.,** The effect of 1,4-diamino-butanone on the stability ornithine decarboxylase from *Aspergillus nidulans, Biochem. J.*, 166, 635, 1977.

80. **Karvonen, E. and Poso, H.,** Stabilization of ornithine decarboxylase and N1-spermidine acetyltransferase in rat liver by methyl glyoxal bis(guanylhydrazone), *Biochim. Biophys. Acta,* 791, 239, 1984.
81. **Kay, J. E. and Lindsay, V. J.,** Control of ornithine decarboxylase activity in stimulated human lymphocytes by putrescine and spermidine, *Biochem. J.,* 132, 791, 1973.
82. **Prouty, W. F.,** Ornithine decarboxylase inactivation in HeLa cells, *J. Cell Physiol.,* 89, 65, 1976.
83. **Seely, J. E., Poso, H., and Pegg, A. E.,** Effect of androgens on turnover of ornithine decarboxylase in mouse kidneys. Studies using labeling of the enzyme by reaction with [14C]-difluoro-methylornithine, *J. Biol. Chem.,* 257, 7549, 1982.
84. **Berger, F. G., Szymansky, P., Read, E., and Watson, G.,** Androgen-regulated ornithine decarboxylase mRNAs of mouse kidney, *J. Biol. Chem.,* 259, 7941, 1984.
85. **Clark, J. L. and Fuller, J. L.,** Protein inhibitor of ornithine decarboxylase does not account for effect of putrescine on 3T3 cells, *Biochem. Biophys. Res. Commun.,* 75, 948, 1976.
86. **O'Brien, T. G., Saladik, D., and Diamond, L.,** Regulation of polyamine biosynthesis in normal and transformed hamster cells in culture, *Biochim. Biophys. Acta,* 632, 270, 1980.
87. **Nikula, P., Alhonen-Hongisto, L., and Janne, J.,** Effects of bis(guanylhydrazones) on the activity and expression of ornithine decarboxylase, *Biochem. J.,* 231, 213, 1985.
88. **Jetten, A. M. and Shirley, J. E.,** Retinoids antagonize the induction of ornithine decarboxylase activity by phorbol esters and phospholipase C in rat tracheal epithelial cells, *J. Cell. Physiol.,* 123, 386, 1985.
89. **Lopriniz, C. L. and Verma, A. K.,** Dose and schedule of oral retinoic acid and indomethacin needed to effectively inhibit phorbol ester-induced ornithine decarboxylase activity, *Carcinogenesis,* 6, 1589, 1985.
90. **Bolmer, S. D. and Wolf, G.,** Retinoids and phorbol esters alter release of fibronectin from enucleated cells, *Proc. Natl. Acad. Sci. U.S.A.,* 79, 6541, 1982.
91. **Jetten, A. M.,** Modulation of cell growth by retinoids and their possible mechanisms of action, *Fed. Proc. Fed. Am. Soc. Exp. Biol.,* 43, 134, 1984.
92. **Persson, L. and Rosengren, E.,** Inhibition of ornithine decarboxylase activity in mouse kidney by administration of diamines, *Acta Chem. Scand.,* 33, 537, 1979.
93. **Persson, L.,** Putrescine turnover in kidneys of testosterone-treated mice, *Biochim. Biophys. Acta,* 674, 204, 1981.
94. **Tyagi, A. K., Tabor, C. W., and Tabor, H.,** Ornithine decarboxylase from *Saccharomyces cerevisiae, J. Biol. Chem.,* 256, 12156, 1981.
95. **Kontula, K. K., Torkkeli, T. K., Bardin, C. W., and Jänne, O. A.,** Androgen induction of ornithine decarboxylase mRNA in mouse kidney as studied by complementary DNA, *Proc. Natl. Acad. Sci. U.S.A.,* 81, 731, 1984.
96. **Kahana, C. and Nathans, D.,** Isolation of cloned cDNA encoding mammalian ornithine decarboxylase, *Proc. Natl. Acad. Sci. U.S.A.,* 81, 3645, 1984.
97. **McConlogue, L., Gupta, M., Wu, L., and Coffino, P.,** Molecular cloning and expression of the mouse ornithine decarboxylase gene, *Proc. Natl. Acad. Sci. U.S.A.,* 81, 540, 1984.
98. **Gilmour, S. K., Avdalovic, N., Madara, T., and O'Brien, T. G.,** Induction of ornithine decarboxylase by 12-*O*-tetradecanoylphorbol 13-acetate in hamster fibroblasts, *J. Biol. Chem.,* 260, 16439, 1985.
99. **Zahner, S. L., Prahlad, K. V., and Mitchell, J. L. A.,** Regulation of ODC activity in the thymus and liver of rats by adrenal hormones, *Cytobios,* 45, 25, 1986.
100. **Mitchell, J. L. A., Qasba, P., Stofko, R. E., and Franzen, M. A.,** Ornithine decarboxylase modification and polyamine-stimulated enzyme inactivation in HTC cells, *Biochem. J.,* 228, 297, 1985.
101. **Bishop, P. B., Young, J., Peng, T., and Richards, J. F.,** An inhibitor of ornithine decarboxylase in the thymus and spleen of dexamethasone-treated rats, *Biochem. J.,* 226, 205, 1985.
102. **Hu, Y. and Brosnan, M. E.,** Properties of ornithine decarboxylase-antizyme from mammary gland of lactating rats, *Arch. Biochem. Biophys.,* 254, 637, 1987.
103. **Hayashi, S., Marumo, M., Matsufuli, S., Onoue, H., Kanamoto, R., and Murakami, Y.,** Ubiquitous presence of antizyme and its role in the regulation of ornithine decarboxylase in mammalian cells and tissues, in Abstr. Int. Conf. Polyamines in Life Sciences, Lake Yamanaka, 1986, 87.
104. **Kitani, T. and Fujisawa, H.,** The effects of phospholipids on the activity of ODC from rat liver, *FEBS Lett.,* 132, 296, 1981.
105. **Moulinoux, J.-P., Le Calve, M., Quemener, V., and Quash, G.,** In vitro studies on the entry polyamines into normal red blood cells, *Biochimie,* 66, 385, 1984.
106. **Chen, K. Y., Kyriakidis, D. A., and Canellakis, E. S.,** The inhibition of the serum-stimulated increase of ornithine decarboxylase by ionophores and local anesthetics, *Biochim. Biophys. Acta,* 716, 72, 1982.
107. **Chen, K. Y., Heller, J. S., and Canellakis, E. S.,** Studies of the regulation of ornithine decarboxylase activity by the microtubules: the effect of colchicine and vinblastine, *Biochem. Biophys. Res. Commun.,* 68, 401, 1976.
108. **Kahan, C. and Nathans, D.,** Translational regulation of mammalian ornithine decarboxylase by polyamines, *J. Biol. Chem.,* 260, 15390, 1985.

109. **Muller, T., Bravo, R., Buckhard, J., and Curran, T.,** Induction of c-fos gene and protein by growth factors precedes activation of c-myc, *Nature,* 312, 716, 1984.
110. **Thompson, C. P., Challoner, P. B., Neiman, P. E., and Groudine, M.,** Expression of the c-myb proto-oncogene during cellular proliferation, *Nature,* 319, 374, 1986.
111. **Dircks, L., Grens, A., Slezynger, T. C., and Scheffler, I. E.,** Post-transcriptional regulation of ornithine decarboxylase activity, *J. Cell Physiol.,* 126, 371, 1986.
112. **Pegg, A. E., Sertich, G. J., Kameji, T., Ewin, B. G., and Shirata, A.,** Regulation of polyamine metabolism by polyamines, in Abstr. Int. Conf. Polyamines in Life Sciences, Lake Yamanaka, 1986, 73.
113. **Russell, D. H.,** Ornithine decarboxylase may be a multifunctional protein, *Adv. Enzyme Regul.,* 21, 201, 1983.
114. **Meggio, F., Flamigni, F., Caldarera, C. M., Guarnieri, C., and Pinna, L. A.,** Phosphorylation of rat heart ornithine decarboxylase by tape-2 casein kinase, *Biochem. Biophys. Res. Commun.,* 122, 997, 1984.
115. **Seely, J. E., Stetler, D. A., Jacob, S. T., and Pegg, A. E.,** Absence of inactivation on phosphorylation of ornithine decarboxylase by nuclear protein kinase NII and of immunological cross-reactivity between RNA polymerase I and ornithine decarboxylase, *Biochem. Biophys. Res. Commun.,* 120, 219, 1984.
116. **Hayashi, S., Kameji, T., Fujita, K., Murakami, Y., Kanamoto, R., Utsunomiya, K., Matsufuji, S., Takiguchi, M., Mori, M., and Tatibana, M.,** Molecular mechanism for the regulation of hepatic ornithine decarboxylase, *Adv. Enzyme Regul.,* 23, 311, 1985.
117. **Fujita, K., Murakami, Y., and Hayashi, S.,** A macromolecular inhibitor of the antizyme to ornithine decarboxylase, *Biochem. J.,* 204, 647, 1982.

Chapter 20

S-ADENOSYLMETHIONINE DECARBOXYLASE: GENES AND EXPRESSION

Michael W. White and David R. Morris

TABLE OF CONTENTS

I. INTRODUCTION

Since it was first discovered,[1] *S*-adenosylmethionine decarboxylase (AdometDC) has been the center of many investigations due to its key role in the regulation of the polyamine pathway and its potential as a chemotherapeutic target. This latter topic will not be discussed in any detail here, since the subject will be covered in another chapter. It is our intention in this chapter to briefly review some of the studies where modulation of AdometDC activity has been observed in intact cells and tissues. We will also review our current knowledge of the structure and the regulation of AdometDC protein. Finally, we will discuss the recent data concerning AdometDC mRNA structure and the control of AdometDC mRNA levels and mRNA translation in bovine lymphocytes.

II. REGULATION OF ADOMETDC ACTIVITY

AdometDC has been assayed in countless numbers of different organisms, tissues, and cells, both prokaryotic and eukaryotic, and its activity has been observed to change under a variety of conditions. In this discussion we will limit our survey to those mammalian systems in which modulation of AdometDC activity has been observed.

A. Cell Proliferation

Increases in AdometDC activity during mitogenesis have been reported by a number of investigators. One of the best characterized systems where mitogenic-stimulation of AdometDC activity has been demonstrated is in lectin-activated lymphocytes. Phytohemagglutinin stimulation of human lymphocytes leads to an increase in AdometDC activity during the first 24 hr following activation.[2] AdometDC activity increases in a biphasic manner in bovine T-lymphocytes stimulated with concanavlin A (ConA).[3] The first elevation (10-fold) was maximal by 12 hrs followed by a further 2-fold increase resulting in an overall 20-fold change in AdometDC activity by 24 hr after ConA induction. The mechanisms controlling each step in the biphasic stimulation of AdometDC activity are different and will be discussed below. The changes in AdometDC activity precede the entry of these lymphocytes into S-phase which begins at 24 hr consistent with the role of these polyamines in DNA replication.[4] Mitogenic stimulation of AdometDC activity has also been observed in rat lymphoma cells activated with prolactin[5] and in primary explants of mouse mammary gland stimulated with insulin.[6]

B. Viral Infection

Elevated AdometDC activity has been shown to be associated with the infection and growth of viruses and other microorganisms. The infection of guinea pigs with Q fever resulted in an induction of liver AdometDC activity during the 1st day following infection.[7] A threefold increase in AdometDC activity was observed in chick embryo fibroblasts by 12 hr after infection with a temperature-sensitive Rous sarcoma virus.[8] This induction in AdometDC activity occurred only at the permissive temperature and was not due to a change in the half-life of AdometDC protein. Goldstein et al.[9] examined the changes in AdometDC activity in mouse kidney cells infected with polyoma virus and found that AdometDC activity shows a biphasic stimulation reminiscent of the behavior of AdometDC in bovine lymphocytes. The kinetics of the first increase paralleled early viral mRNA synthesis while the second elevation was coincident with viral-induced cellular DNA synthesis. The addition of actinomycin D, 2 hr after infection, blocked the initial increase in AdometDC activity suggesting that the change in AdometDC activity was due to an increase in mRNA synthesis.

C. Malignancy

It is generally agreed that polyamines and therefore their synthesis are required for cell growth, so it is not surprising that many malignant cell types have been reported to have elevated AdometDC activities. Heby et al.[10] compared the kinetics of the change in AdometDC activity and the change in growth rate of a rat brain tumor cell line and found that the peak AdometDC activity coincides with the period of maximal growth rate. In SV40-transformed 3T3 fibroblasts, levels of AdometDC activity were measured and found to be higher than the enzyme levels in similarly cultured normal 3R3 fibroblasts.[11] Several tumor promoters have been reported to cause an elevation of AdometDC activity. The phorbol ester, 12-*O*-tetradecanoyl-phorbol acetate (TPA), was first shown to increase AdometDC activity (as well as ornithine decarboxylase, ODC, activity) in mouse epidermis by O'Brien et al.[12] Several other tumor promoters were tested and had similar effects on epidermal AdometDC activity.[13] Interestingly, hyperplastic agents such as cantharidin or ethyl phenylpropriolate stimulated epidermal AdometDC activity with no effect on background ODC activity. It is not clear whether this is a tissue-specific effect or whether these effects will apply to other hyperplastic agents. From their current studies, Scalabrino and Ferioli[14] have suggested that AdometDC activity (together with the activity of ODC) may be useful as an indicator of the degree of malignancy of human tumors. In this brief survey only two major types of tumors were examined and AdometDC was not found to correlate as well as ODC to the degree of malignancy. Whether AdometDC activity will be useful as a diagnostic indicator of human malignancy is unclear; the preliminary results are not promising.

D. Hormones

The effect of hormones on their target tissues has proven to be a rich area with regards to the study of polyamine biosynthesis and changes in AdometDC activity. Thyrotropin rapidly elevates AdometDC activity in rat thyroid[15] and thyrotoxin has been reported to cause an increase in AdometDC activity in rat heart.[16] The activity of AdometDC in rat uterine tissue was found to be stimulated within 4 hr following a single injection of estradiol-17-beta or estriol.[17] Large changes in AdometDC activity have been observed in prostate tissue when rats are treated with androgens. The activity of AdometDC in ventral prostate decreased to 7% of normal after castration and was stimulated eight- to tenfold after treatment of these animals with testosterone.[18] This pattern of stimulation of AdometDC in prostate tissue is in contrast to the response in other tissues such as kidney[19] and seminal vesicle[18] where the AdometDC enzyme levels were relatively unaffected. The changes in AdometDC activity seen in ventral prostate may be associated with the secretion of spermidine and spermine which is a unique function of this tissue.[20]

E. Development and Regeneration

There are a number of examples of modulation of AdometDC activity during development. A survey of AdometDC enzyme levels in various tissues during chick embryo development demonstrated that elevated activity correlated with increases in both polyamine levels and growth rates of the respective tissues.[21] In another example, AdometDC activity peaked during the last stages of sperminogenesis in the rooster and then rapidly decreased to undetectable levels in mature spermatozoa; as expected, polyamine levels were also low in the mature cells.[22]

Several groups have shown that AdometDC activity is stimulated during the regeneration period following tissue injury and in the tissues or organs of animals recovering from a metabolic stress such as dietary starvation. The elevation of AdometDC activity in the response to injury is associated with the onset of cell proliferation which occurs during the regeneration period and has been observed in regenerating rat liver,[23] intestinal mucosa after jejunectomy,[24] and diaphram muscle after nerve section[25] to list a few examples. During the recovery of animals from stress, the induction of AdometDC activity has been observed in a variety of tissues. In rats recovering from fasting, both AdometDC and ODC activities are stimulated in liver, kidney, and intestine while only AdometDC levels were reported to increase in brain and skeletal muscle.[26] These latter increases in AdometDC activity were of marginal magnitude. There is however, a unique association between AdometDC activity and brain tissue which will be discussed below. The addition of 1,25-dihydroxyvitamin D to vitamin D-deficient chickens resulted in a rapid stimulation of AdometDC activity in intestinal mucosa by 1 hr after administration of the vitamin.[27] The induction of AdometDC activity was blocked by hydrocortisone which prevents the rise in intracellular calcium stimulated by vitamin D; however, there is no evidence at this time to suggest that these events are more than casually linked.

Based on the above studies it can be concluded that for the majority of the systems examined the onset of proliferation is followed by stimulation of the activity of the polyamine biosynthetic enzymes, most notably ODC and AdometDC, and the subsequent elevation in the level of polyamines. An intriguing exception to this general scheme is the behavior of Adomet DC in the developing brain. By following the AdometDC activity in developing brain tissue, several investigators have observed that AdometDC enzyme levels, which are low in fetal brain, increase shortly after birth and reach steady-state levels in the adult brain that are as much as tenfold above fetal levels.[28] The behavior of ODC activity follows an inverse pattern; ODC enzyme levels are high in fetal brain, begin to decrease at birth, and are undetectable in the adult brain.[29] This is clearly a unique pattern of expression of AdometDC and ODC compared to most tissues. What function the elevated level of AdometDC expression is serving in the adult brain has not been determined although it has been suggested that spermidine which is high in the adult brain may be required for regulation of neural transmission.[30]

F. Regulation by Polyamine Levels

A complete discussion of the subject of feedback regulation of AdometDC expression by cellular polyamine levels can be found in a recent review by Pegg.[31] The basic conclusions from these previous studies are that decreased levels of spermidine or an elevation in putrescine levels lead to a stimulation of AdometDC activity. It has been suggested that spermidine is the key feedback regulator since the spermidine synthase inhibitor *S*-adenosyl-1,8-diamino-octane decreases spermidine levels and in parallel causes an increase AdometDC activity with no affect on putrescine levels.[32]

III. ADOMETDC AS A PROTEIN AND ENZYME

A. Purification and Properties of the Protein

The facile purification of AdometDC has been made possible by two important discoveries using the potent competitive inhibitor, methylglyoxal bis(guanylhydrazone) (MGBG). First, it was demonstrated that administration of MGBG to cells in culture[3] and to animals[33] led to striking elevations of specific enzymatic activity in extracts because of stabilization of the enzyme in vivo (see discussion below). Intraperitoneal injection of MGBG was used in the first purifications of the enzyme from rat liver[34-36] and in subsequent preparations from other sources.[37-39] Second, an affinity matrix of MGBG-Sepharose[34] has proved to be universally useful in purifying the enzyme from all sources. Application fo these two approaches, together with more traditional purification techniques, has produced highly purified preparations from rat,[34-36,39] murine,[37] bovine,[38] yeast,[40,41] and bacterial[41,42] sources.

Purified AdometDC from a variety of mammalian sources has been found to have a subunit molecular weight of about 32,000; the basic subunit seems to associate to dimers and perhaps higher multimers.[35-39] Microbial enzymes are of somewhat different subunit structures; AdometDC from *Saccharomyces cerevisiae* is a dimer of 41,000 subunit molecular weight[40] and the *Escherichia coli* enzyme contains six 17,000-molecular weight subunits.[42]

There has been little protein chemistry done with AdometDC. The amino terminus of the protein seems to be blocked and the amino acid sequence of a few proteolytic peptides has been determined as part of the isolation of cDNA clones.[43,44]

B. Enzymatic Properties

The decarboxylation of AdoMet by the *E. coli* enzyme is carried out with retention of stereochemical configuration; conservation of stereochemistry is found with all amino acid decarboxylases that have been examined.[45] The K_m of mammalian AdometDC for its substrate is 50 to 100 μM.[36,39] Inhibition by the potent inhibitor MGBG is competitive with substrate and has a K_i of about 1 μM.[10] Mammalian AdometDC is strongly activated by putrescine (K_a = 25 μM),[39] forming the basis of a proposed "feed-forward" activation scheme in which the activity of AdometDC is regulated by the product of the first step of the pathway catalyzed by ODC.[46] The enzyme is activated by several diamines, but none is as effective as putrescine.[39] The yeast enzyme also requires putrescine for maximal activity,[47] but the activity of *E. coli* AdometDC is unaffected by diamines.[48] Instead, the bacterial enzyme has a requirement for Mg^{2+};[49] the physiological significance of this requirement has not been proven, but may relate to the regulation of general polycation balance (discussed in the context of putrescine synthesis in Reference 50).

C. Prosthetic Group

Unlike most amino acid decarboxylases, AdometDC does not require pyridoxal phosphate as a cofactor,[51] but it is inhibited by carbonyl reagents.[35,36] The reactive carbonyl group in the enzymes from *E. coli,*[42] yeast,[40] and rat[35,36] has been shown to be covalently bound pyruvate. A pyruvoyl residue, instead of pyridoxal phosphate, is found as the prosthetic group of several enzymes which act on amino acids (reviewed in Reference 52). In the case of two of these, histidine decarboxylase and proline reductase, the pyruvate is bound in amide linkage at the amino terminus; although this has not been investigated with AdometDC, this observation may relate to the blocked amino terminal group noted above.[44] Analogous to the pyridoxal phosphate-dependent enzymes, the pyruvoyl enzymes are throught to form Schiff base intermediates with their substrates.[52] Consistent with this mechanism, the complex between AdometDC and either substrate or product[42,53,54] could be reduced with sodium cyanoborohydride to yield stable compounds and inactivating the enzyme. An irreversible inhibitor of AdometDC, 1,1′-[(methylethenediylidene)dinitrilo]bis(3-aminoguanidine)], an analogue of MGBG, may also form an adduct with the active site pyruvoyl group.[55] The importance of the general occurrence of pyruvate instead of pyridoxal phosphate in this enzyme is a mystery. In the case of histidine decarboxylase, the mammalian enzyme uses pyridoxal phosphate,[56] whereas, as noted above, the enzyme from *Lactobacillus* contains pyruvate, indicating a lack of preference. In principle, one might expect that pyridoxal phosphate-dependent AdometDC should occur in nature; perhaps a better understanding of the mechanism of action and the biosynthesis of this enzyme will clarify this point.

Very little is known of the biosynthesis of the pyruvoyl group of AdometDC. In the case of the prototypic pyruvoyl enzyme, histidine decarboxylase of *Lactobacillus,* the prosthetic group is generated from a serine residue during the autocatalytic cleavage of the proenzyme.[57] It seems that we are very close to extending our knowledge in this area with regard to AdometDC. Work is progressing on the molecular cloning of cDNAs[43,44] and genomic DNAs (unpublished) coding for this protein, the sequence of which should give insight into the primary translation product. Evidence already exists that products are formed during in vitro translation of the AdometDC message, which are of higher molecular weight than the mature protein.[58,59]

D. Multiple Forms

Because of the many stimuli influencing the level of AdometDC activity in various tissues and cell types (discussed above), the existence of multiple forms seems possible. The availability of antisera prepared to the enzymes isolated from rat,[39] mouse,[37] and bovine liver[38] made possible immunological comparisons. Antibody to the mouse liver enzyme was found to fully cross react with the mammary gland enzyme.[37] No difference in immunological cross reactivity could be found between bovine liver AdometDC and the enzyme in extracts of either resting or mitogenicaly activated lymphocytes from the same species.[38] In a similar study with the rat enzyme,[39] no immunological differences could be detected between enzyme preparations from liver, prostate, and psoas muscle.

Despite the immunological similarities between AdometDC in extracts of various tissues, there is compelling evidence that multiple forms may exist. AdometDC has been purified to homogeneity from both liver and psoas muscle of the rat.[39,60] These enzymes were found to differ in physical and catalytic properties. Although the mobility of both enzymes was identical on SDS gel electrophoresis, the two proteins cleanly separated on nondenaturing gels, probably due to a difference of 0.4 pH units in isoelectric point. Significant difference in substrate binding was also seen, as reflected in differences in K_m for AdoMet, K_i for MGBG, and binding to MGBG-Sepharose columns. At this time, one cannot distinguish between altered posttranslational modifications and differences in primary amino acid sequence as the source of these two forms of the enzyme. The latter possibility is of interest because of the evidence suggesting that a single gene codes for AdometDC, at least in bovine DNA (discussed below).

IV. REGULATION OF PROTEIN LEVEL

A. Rate of Synthesis

The availability of immunological reagents specific for AdometDC has made it possible to compare changes in enzyme activity in vivo with the amount of cross reactive protein present, the rate of synthesis of the protein, and the rate of its turnover. In the few instances where activity and protein have been compared, they have been found to vary in parallel. Thus, the changes in AdometDC activity elicited by liver regeneration,[54] androgen-induced prostatic growth,[54] mitogenic activation of lymphocytes,[38] and modulation of polyamine levels (see Reference 31 for an extensive discussion of control of AdometDC by polyamines) are all accompanied by corresponding alterations in the amount of immunologically reactive protein. One therefore concludes that, in these cases, changes in the activity of AdometDC measured in vitro is determined by the relative rates of synthesis and degradation of the protein and not by changes in the intrinsic activity of preexisting enzyme.

Because of extremely low levels of AdometDC, even in maximally induced tissues, it has generally been impossible to directly measure the rate of synthesis of the protein by isotopic incorporation experiments, particularly in intact animals. Thus, changes in synthetic rate are generally inferred from the level of the protein and its half-life in the presence of protein synthesis inhibitors (discussed in References 31, 46, and 61). In the one instance where the rate of AdometDC synthesis has been measured directly, the tenfold increase in enzyme activity within a few hours after mitogenic activation of lymphocytes could be fully accounted for by a corresponding elevation of the rate of ^{3}H-leucine incorporation into the protein.[38]

B. Protein Half-Life

It was recognized quite early that AdometDC probably had a short half-life. Measurements of the decay of enzymatic activity after inhibition of protein synthesis yielded half-lives of 20 to 120 min, depending on the biological system examined (summarized in References 31 and 61). These experiments were refined somewhat by following the disappearance of immunologically reactive material, rather than activity (e.g., see Reference 54). However, either measurement is open to the criticism that the protein synthesis inhibitor employed may alter the rate of degradation of the enzyme, either through inhibition of protein synthesis or through a secondary effect. This concern may have some validity, since in lymphocytes 24 hr after mitogenic activiation, the half-life of enzyme activity in the presence of cycloheximide is 43 min,[3] while a pulse-chase experiment gave a value of 170 min in the same cells.[38,58] Alternatively, the difference in half-life obtained by the two methods may reflect the mechanism of intracellular inactivation and degradation (see below). Unless stated otherwise, the enzyme turnover studies discussed below were carried out in the presence of cycloheximide, and this should be kept in mind in interpreting the data.

The apparent short half-life of AdometDC has two ramifications. First, since the kinetics of induction, or deinduction, of a protein are dependent on half-life,[62] the rapid turnover of AdometDC results in the rapid changes in enzyme level that have been observed. The second ramification of a short half-life is that it provides an additional mechanism for regulating enzyme level through stabilization. Several examples of this mode of regulation follow.

The first reported instance of modulation of the half-life of AdometDC resulted from an examination of the apparent paradox that treatment of cells or animals with the inhibitor of the enzyme MGBG led to substantial elevation of enzyme activity in dialyzed extracts. This was subsequently shown to be due to elevation of AdometDC protein.[63] Studies in rat liver[33] and in lymphocytes[3] demonstrated that MGBG strongly stabilized enzyme activity in the presence of cycloheximide. The conclusions from these activity measurements have been verified measuring immunoreactive protein.[63] The mechanism of the stabilization of the enzyme by MGBG is not yet understood (discussed in Reference 54).

It has been known for some time that AdometDC activity is modulated by the intracellular level of spermidine (reviewed in References 31, 46, and 61). Immunological studies in 3T3 cells[64] and in the prostate and liver of rats[54] have shown that this is due solely to changes in the level of AdometDC protein. In a rigorous study of AdometDC in the prostate of DFMO-treated rats, Shirahata and Pegg[54] found that the half-life of activity, or of titratable active sites, was extended from 35 to 108 min in the presence of drug. On the other hand, the half-life of immunoreactive material was 139 and 213 min under the two conditions. The authors suggest that this interesting result can be explained by a degradation pathway involving an early, spermidine-affected step, which results in loss of the active site, followed by a slower degradation of the protein. No intermediates were detected, suggesting that the first step involves little or no breakdown of the protein. This effect was tissue specific, since the rates of decay of activity and cross reactive material in the presence of cycloheximide were similar in the liver.[54] It should be noted that in the prostate the increased stability of the enzyme was insufficient to account for the elevation of enzyme activity, implying that polyamine deprivation also influences the rate of synthesis of AdometDC.

Because of the flexibility afforded by a cell culture system, it has been possible to monitor directly by isotope incorporation the rates of synthesis and degradation of AdometDC in mitogen-activated lymphocytes.[38,58] AdometDC activity is elevated biphasically in this system, as mentioned above, with a tenfold increase in the first 10 hr after activation and a second increase of twofold after about 20 hr, as the cells enter S-phase.[3] The early elevation of enzyme activity is accounted for entirely by a tenfold increase in rate of synthesis. The second phase of accumulation occurs with no futher increase in rate of synthesis, but is accounted for by a lengthening of the half-life from 80 to 170 min. Although the mechanism of modulation of the half-life is not understood, the tenfold increase in synthesis rate is accomplished by elevation of both the level of AdometDC mRNA and the efficiency of its translation (discussed below).

V. ADOMETDC GENE EXPRESSION

A. Mutants

Over the last few years a number of groups have begun to decipher the molecular biology of AdometDC. Recently the structural gene of *E. coli* has been cloned and amplified[65] and AdometDC mutants have been generated in *E. coli*[66] and in yeast.[67-69] The *S. cerevisiae* spe2 mutants have no detectable AdometDC activity and the mutation has been mapped very close to the arg1 gene.[67] These cells require exogenous polyamines, specifically spermidine or spermine, for growth and diploid spe2 mutants are unable to sporulate without exogenous polyamines.[67] Other *Saccharomyces* strains with mutations in ODC or the aminopropyltransferases[69] show similar phenotypes to the AdometDC-spe2 mutants and demonstrates clearly the absolute requirement of polyamines for the growth of these cells.

B. Cloning of the AdometDC cDNA

Several laboratories have recently cloned the cDNAs for mammalian AdometDC and ODC. Although the work on ODC will not be discussed here except in comparison to AdometDC, the majority of the current studies have dealt with this gene. This has principally been due to the use of the drug difluoromethylornithine (DFMO) to produce variants with amplified ODC genes and the subsequent cloning of the ODC cDNAs from these resistant cell lines.[70-73] DFMO is an irreversible inhibitor of ODC and suited for this purpose since it is very specific and not cytotoxic. Unfortunately, there is no inhibitor of AdometDC with similar properties and the AdometDC cDNA could not be cloned through this type of strategy. Calf liver AdometDC was purified to homogeneity[74] in sufficient quantities to allow partial amino acid sequence analysis. Based on[44] this analysis, a 17-base synthetic oligonucleotide pool was prepared and used to screen a λgt11 bovine cDNA library. One of the positive clones isolated from this screening contained a 1350 bp insert whose nucleic acid sequence matched precisely the amino acid sequence of two peptides. Based on this criterion the clone was designated, pSD-1.35, and was established as a cDNA clone encoding bovine AdometDC.

C. Changes in Message Levels

Cloning of the bovine AdometDC cDNA has made possible the examination of the changes in AdometDC message levels in mitogen-activated lymphocytes.[44] Two AdometDC mRNAs, 3.6 and 2.4 kilobase (kb), are observed on northern blots of total RNA or cytoplasmic RNA prepared from bovine lymphocytes.[44] Both forms increase coordinately following ConA activation and together are maximally elevated fourfold with respect to total RNA by 10 hrs after ConA addition. During this period of ConA stimulation the rate of AdometDC biosynthesis, as well as enzyme activity, are stimulated to a maximum of tenfold.[3] Thus, a discrepancy exists between the change in AdometDC mRNA level and the change in the rate of AdometDC synthesis. These results suggest that increases in the rate of AdometDC synthesis are not regulated by a straight forward change in mRNA level, but that an additional component of translational control exists.

D. Message Utilization

There are few clear examples of translational control in mammalian cells other than such specialized viral systems such as poliovirus[75] and adenovirus.[76] AdometDC represents one of the first clear examples of translational control of a "housekeeping" type gene. Several models can be developed to explain this type of regulation:

1. One potential model involves the recruitment or the redirection of untranslated mRNA into polysomes or translational particles. This model predicts that the amount of mRNA in polysomes, not the size of polysomes containing the mRNA, would increase after mitogen activation.

2. Another model would involve an increase in the rate of protein synthesis initiation relative to the rate of elongation which would lead to an increase in the average number of ribosomes per mRNA and therefore a larger polysome size. Larger polysome size could also be achieved by a decrease in the rate of peptide elongation relative to initiation; however, this would not lead to an increase in the overall rate of synthesis.

A comparison of the distribution of AdometDC mRNA across sucrose gradients of polysomes from resting vs. activated cells would be diagnostic of which mechanism is operative. We have tested these models in polysomes prepared from resting and ConA-activated lymphocytes.[44] The analysis of RNA purified from the sucrose gradient fractions revealed two major observations.

1. The largest portion of the AdometDC mRNA from both resting (75% of total) and activated lymphocytes (58% of total) was found associated with monosomes. The localization of AdometDC mRNA on monosomes is much smaller than would be expected based on β-globin (12,000 MW) which is found on pentameric polysomes[77] and actin (45,000 MW) which is found on polysomes of roughly 15 ribosomes.[78] We would predict from these two proteins that mRNA coding for AdometDC (subunit, 32,000 MW) should be contained on polysomes of 10 to 11 ribosomes. It should be noted that mature AdometDC may be a processed protein[58,59] with a larger precursor. It is clear from these data that, in general, AdometDC mRNA is very inefficiently translated in these cells and may account for the difficulty in translating this message in vitro (unpublished results).
2. The level of AdometDC mRNA in the region of the sucrose gradients corresponding to polysomes containing an average of eight to ten ribosomes increases tenfold with respect to resting cells by 8 hr after conA addition. As a consequence of this increase, the average size polysome containing AdometDC mRNA in 8-hr-activated lymphocytes, which was calculated to be 2.7 ribosomes per mRNA, was twofold larger than the average size polysome (1.4 ribosomes per mRNA) in resting cells.

Thus the shift in the distribution of AdometDC mRNA and the subsequent increase in polysome size easily accounts for the twofold discrepancy between the change in AdometDC mRNA level and the rate of AdometDC synthesis following conA stimulation. In addition, the mechanism operating in AdometDC translational control appears to be equivalent to the second model discussed above; following lectin-stimulation there is an elevation in the rate of translational initiation which leads to an increase in the number of ribosomes per AdometDC mRNA.

Thus, the synthesis of AdometDC in mitogen-activated lymphocytes is controlled at two levels, i.e., changes in the amount of AdometDC message and the efficiency of its translation. Recently we have observed in the same system that synthesis of ODC is also controlled at both the mRNA level and at the level of mRNA utilization (unpublished results). Unlike AdometDC, however, the mechanism through which ODC mRNA is translationally controlled appears to be similar to the first model described above. It seems likely that the common regulatory pattern which is emerging for these key polyamine enzymes, rather than being a simple coincidence, has some functional significance.

VI. FUTURE PROSPECTS

Although the tools to study the molecular biology of the enzymes of the polyamine pathway have just begun to be acquired, it is clear the story concerning regulation of message expression is rapidly unfolding. Yet the acquisition of the AdometDC cDNA clone has raised a number of new questions. What is the source of the two distinct AdometDC mRNA species? Two forms of AdometDC mRNA, similar in size to the bovine mRNA species,[43,44] have also been found in mouse 3T3 fibroblasts, mouse myoblasts, and human B-lymphocytes (unpublished results). The two mRNAs are coordinately regulated in serum-stimulated 3T3 fibroblasts (unpublished results) as well as in mitogen-induced bovine T-lymphocytes. Several mechanisms could be proposed to explain the generation of the two mRNA species including alternate splice sites, alternate polyadenylation sites, or multiple genes. Based on our current knowledge of the AdometDC gene in bovine, discussed below, this last alternative appears unlikely.

What structure in the AdometDC mRNAs identifies them for translational regulation? In the case of mouse ODC it has been suggested that structures in the unusually long 5′-untranslated region of ODC mRNA might be involved in translational control. Whether ODC and AdometDC share common translational control elements awaits the complete elucidation of the AdometDC mRNA sequence. Based on the few clear examples of translational regulation in eukaryotes, such as the heat shock proteins[79] and the yeast GCN4 protein,[80] the 5′-untranslated region would seem the most likely candidate for translational control elements.

Our current knowledge of the structures of either ODC or AdometDC genomic DNA is relatively sparse. Preliminary southern blots indicate that the AdometDC has few and perhaps a single copy gene in bovine (unpublished results). In contrast, bovine ODC gives numerous bands on a genomic southern blot (unpublished results) consistent with the idea that ODC is a multigene family.[81] Currently several AdometDC genomic clones have been isolated and their characterization is underway.

REFERENCES

1. **Tabor, C. W. and Tabor, H.,** 1,4-Diaminobutane (putrescine), spermidine, and spermine, *Annu. Rev. Biochem.*, 45, 285, 1976.
2. **Korpela, H., Hölttä, E., Hovi, T., and Jänne, J.,** Response of enzymes involved in the metabolism of polyamines to phytohaemagglutinin-induced activation of human lymphocytes, *Biochem. J.*, 196, 733, 1981.
3. **Fillingame, R. H. and Morris, D. R.,** *S*-Adenosylmethionine decarboxylase during lymphocyte transformation: decreased degradation in the presence of a specific inhibitor, *Biochem. Biophys. Res. Commun.*, 52, 1020, 1973.
4. **Marton, L. J. and Morris, D. R.,** Molecular and cellular functions of the polyamines, in *Inhibition of Polyamine Biosynthesis: Biological Significance and Basis for New Therapies,* McCann, P. P., Pegg, A. E., and Sjoerdsma, A., Eds., Academic Press, New York, in press.
5. **Richards, J. F., Beer, C. T., Bourgeault, C., Chen, K., and Gout, P. W.,** Biochemical response of lymphoma cells to mitogenic stimulation by prolactin, *Mol. Cell. Endocrinol.*, 26, 41, 1982.
6. **Sakai, T., Lundgren, D. W., and Oka, T.,** Polyamine biosynthesis and DNA synthesis in cultured mammary gland explants from virgin mice, *J. Cell Physiol.*, 95, 259, 1978.
7. **Paretsky, D., Gonzales, F. R., and Berquist, K.,** Temporal studies of factors associated with changes in transcription during Q fever, *J. Gen. Microbiol.*, 122, 227, 1981.
8. **Bachrach, U. and Weiner, H.,** Increase in *S*-adenosyl-L-methionine decarboxylase activity during the transformation of chick embryo fibroblasts by Rous sarcoma virus, *Int. J. Cancer,* 26, 75, 1980.
9. **Goldstein, D. A., Heby, O., and Marton, L. J.,** Biphasic stimulation of polyamine biosynthesis in primary mouse kidney cells by infection with polyoma virus: uncoupling from DNA and rRNA synthesis, *Proc. Natl. Acad. Sci. U.S.A.*, 73, 4022, 1976.
10. **Heby, O., Marton, L. J., Wilson, C. B., and Martinez, H. M.,** Polyamine metabolism in a rat brain tumor cell line: its relationship to the growth rate, *J. Cell. Physiol.*, 86, 511, 1975.
11. **Bethell, D. R., Hibasami, H., and Pegg, A. E.,** Regulation of polyamine content in cultured fibroblasts, *Am. J. Physiol.*, 243, C262, 1982.

12. **O'Brien, T. G., Simsiman, R. C., and Boutwell, R. K.,** Induction of the polyamine-biosynthetic enzymes in mouse epidermis by tumor-promoting agents, *Cancer Res.*, 35, 1662, 1975.
13. **O'Brien, T. G., Simsiman, R. C., and Boutwell, R. K.,** Induction of the polyamine-biosynthetic enzymes in mouse epidermis and their specificity for tumor promotion, *Cancer Res.*, 35, 2426, 1975.
14. **Scalabrino, G. and Ferioli, M. E.,** Degree of enhancement of polyamine biosynthetic decarboxylase activities in human tumors: a useful new index of degree of malignancy, *Cancer Detct. Prev.*, 8, 11, 1985.
15. **Matsuzaki, S.,** A potential role of polyamines in goiter formation (transl.), *Nippon Naibumpi Gakkai Zasshi*, 57, 1554, 1981.
16. **Pegg, A. E. and Hibasami, H.,** Polyamine metabolism during cardiac hypertrophy, *Am. J. Physiol.*, 239, E372, 1980.
17. **Lavia, L. A. Stohs, S. J., and Lemon, H. M.,** Polyamine biosynthetic decarboxylase activities following estradiol-17 beta or estriol stimulation of the immature rat uterus, *Steriods*, 42, 609, 1983.
18. **Piik, K., Rajamaki, P., Guha, S. K., and Jänne, J.,** Regulation of L-ornithine decarboxylase and *S*-adenosyl-L-methionine decarboxylase in rat ventral prostate and seminal vesicle, *Biochem. J.*, 168, 379, 1977.
19. **Pajunen, A. E., Isomaa, V. V., Jänne, O. A., and Bardin, C. W.,** Androgenic regulation of ornithine decarboxylase activity in mouse kidney and its relationship to changes in cytosol and nuclear androgen receptor concentrations, *J. Biol. Chem.*, 257, 8190, 1982.
20. **Williams-Ashman, H. G. and Canellakis, Z. N.,** Polyamines in mammalian biology and medicine, *Perspect. Biol. Med.*, 22, 421, 1979.
21. **Lowkvist, B., Emanuelsson, H., and Heby, O.,** Changes in polyamine synthesis and concentrations during chick embryo development, *J. Exp. Zool.*, 234, 375, 1985.
22. **Oliva, R., Vidal, S., and Mezquita, C.,** Cellular content and biosynthesis of polyamines during rooster spermatogenesis, *Biochem. J.*, 208, 269, 1982.
23. **Russell, D. H. and Lombardini, J. B.,** Polyamines. I. Enhanced AdometDC in rapid growth systems and polyamines. II. The relationship between polyamine concentration and RNA accumulation, *Biochim. Biophys. Acta*, 240, 273, 1971.
24. **Luk, G. D. and Baylin, S. B.,** Polyamines and intestinal growth; increased polyamine biosynthesis after jejunectomy, *Am. J. Physiol.*, 245, G656, 1983.
25. **Hopkins, D. and Manchester, K. L.,** The influence of nerve section on the metabolism of polyamines in rat diaphram muscle, *Biochem. J.*, 196, 603, 1981.
26. **Moore, P. and Swendseid, M. E.,** Dietary regulation of the activities of ornithine decarboxylase and *S*-adenosylmethionine decarboxylase in rats, *J. Nutr.*, 113, 1927, 1983.
27. **Steeves, R. M. and Lawson, D. E.,** Effect of 1,25-dihydroxyvitamin D on *S*-adenosylmethionine decarboxylase in chick intestine, *Biochim. Biophys. Acta*, 841, 292, 1985.
28. **Sturman, J. A. and Gaull, G. E.,** Polyamine metabolism in the brain and liver of the developing monkey, *J. Neurochem.*, 25, 267, 1975.
29. **Laitinen, S. I., Laitinen, P. H., Hietala, O. A., Pajunen, A. E., and Piha, R. S.,** Developmental changes in mouse brain polyamine metabolism, *Neurochem. Res.*, 7, 1477, 1982.
30. **Russell, D. H., Gfeller, E., Marton, L. J., and Legendre, S. M.,** Distribution of putrescine, spermidine, and spermine in Rhesus monkey brain: decrease in spermidine and spermine concentration in motor cortex after electrical stimulation, *J. Neurobiol.*, 5, 349, 1974.
31. **Pegg, A. E.,** *S*-Adenosylmethionine decarboxylase: a brief review, *Cell Biochem. Funct.*, 2, 11, 1984.
32. **Mamont, P. S., Joder-Olenbusch, A. M., Nussli, M., and Grove, J.,** Indirect evidence for a strict negative control *S*-adenosyl-L-methionine decarboxylase by spermidine in rat hepatoma cells, *Biochem. J.*, 196, 411, 1981.
33. **Pegg, A. E., Corti, A., and Williams-Ashman, H. G.,** Paradoxical enhancement of *S*-adenosylmethionine decarboxylase in rat tissues following administration of the specific inhibitor methylglyoxal bis(guanylhydrazone), *Biochem. Biophys. Res. Commun.*, 52, 696, 1983.
34. **Pegg, A. E.,** Purification of rat liver *S*-adenosyl-L-methionine decarboxylase, *Biochem. J.*, 141, 581, 1974.
35. **Pegg, A. E.,** Evidence for the presence of pyruvate in rat liver *S*-adenosylmethionine decarboxylase, *FEBS Lett.*, 84, 33, 1977.
36. **Demetriou, A. A., Cohn, M. S., Tabor, C. W., and Tabor, H.,** Identification of pyruvate in *S*-adenosylmethionine decarboxylase from rat liver, *J. Biol. Chem.*, 253, 1684, 1978.
37. **Sakai, T., Hori, C., Kano, K., and Oka, T.,** Purification and characterization of *S*-adenoxyl-L-methionine decarboxylase from mouse mammary gland and liver, *Biochemistry*, 18, 5541, 1979.
38. **Seyfried, C. E., Oleinik, O. E., Degen, J. L., Resing, K., and Morris, D. R.,** Purification, properties, and regulation of the level of bovine *S*-adenosylmethionine decarboxylase during lymphocyte mitogenesis, *Biochim. Biophys. Acta*, 716, 169, 1982.
39. **Poso, H. and Pegg, A. E.,** Comparison of *S*-adenosylmethionine decarboxylase from rat liver and muscle, *Biochemistry*, 21, 3116, 1982.

40. **Cohn, M. S., Tabor, C. W., and Tabor, H.,** Identification of a pyruvoyl residue in *S*-adenosylmethionine decarboxylase from *Saccharomyces cerevisiae, J. Biol. Chem.*, 252, 8212, 1977.
41. **Pegg, A. E. and Jacobs, G.,** Comparison of inhibitors of *S*-adenosyl-L-methionine decarboxylase from different species, *Biochem. J.*, 213, 459, 1983.
42. **Markham, G. D., Tabor, C. W., and Tabor, H.,** *S*-Adenosylmethionine decarboxylase of *Escherichia coli, J. Biol. Chem.*, 257, 12063, 1982.
43. **Mach, M., Neubauer, M. G., Degen, J. L., Seyfried, C. E., and Morris, D. R.,** Isolation of a cDNA clone encoding *S*-adenosylmethionine decarboxylase from bovine lymphocytes, in *Polyamines: Basic and Clinical Aspects,* Imahori, K., Suzuki, O., and Bachrach, U., Eds., VNU Science Press, Utrecht, The Netherlands, 1985.
44. **Mach, M., White, M. W., Neubauer, M., Degen, J. L., and Morris, D. R.,** Isolation of a cDNA clone encoding *S*-adenosylmethionine decarboxylase; expression of the gene in mitogen-activated lymphocytes, *J. Biol. Chem.*, in press.
45. **Allen, R. R. and Klinman, J. P.,** Stereochemistry and kinetic isotope effects in the decarboxylation of *S*-adenosylmethionine catalyzed by the pyruvyl enzyme, *S*-adenosylmethionine decarboxylase, *J. Biol. Chem.*, 256, 3233, 1981.
46. **Pegg, A. E.,** Recent advances in the biochemistry of polyamines in eukaryotes, *Biochem. J.*, 234, 249, 1986.
47. **Poso, H., Hannonen, P., Himberg, J. J., and Jänne, J.,** Adenosylmethionine decarboxylase from various organisms: relation of the putrescine activation of the enzyme to the ability of the organism to synthesize spermine, *Biochem. Biophys. Res. Commun.*, 68, 227, 1976.
48. **Pegg, A. E. and Williams-Ashman, H. G.,** On the role of *S*-Adenosyl-L-methionine in the biosynthesis of spermidine by rat prostate, *J. Biol. Chem.*, 244, 682, 1969.
49. **Wickner, R. B., Tabor, C. W., and Tabor, H.,** Purification of adenosylmethionine decarboxylase from *Escherichia coli* W: evidence for covalently bound pyruvate, *J. Biol. Chem.*, 245, 2132, 1970.
50. **Wu, W. H. and Morris, D. R.,** Biosynthetic arginine decarboxylase from *Escherichia coli, J. Biol. Chem.*, 248, 1696, 1973.
51. **Pegg, A. E.,** Role of pyridoxal phosphate in mammalian polyamine biosynthesis. Lack of requirement for mammalian *S*-adenosylmethionine decarboxylase activity, *Biochem. J.*, 166, 81, 1977.
52. **Recsei, P. A. and Snell, E. E.,** Pyruvoyl enzymes, *Annu. Rev. Biochem.*, 53, 357, 1984.
53. **Pankaskie, M. and Abdel-Monem, M. M.,** Inhibitors of polyamine biosynthesis. VII. Evaluation of pyruvate derivatives as inhibitors of *S*-adenosyl-L-methionine decarboxylase, *J. Pharm. Sci.*, 69, 1000, 1980.
54. **Shirahata, A. and Pegg, A. E.,** Regulation of *S*-adenosylmethionine decarboxylase activity in rat liver and prostate, *J. Biol. Chem.*, 260, 9583, 1985.
55. **Pegg, A. E.,** Inhibition of mammalian *S*-adenosylmethionine decarboxylase activity by 1,1′-[(methylethenediylidene)dinitrilo]bis(3-aminoguanidine), *J. Biol. Chem.*, 253, 539, 1978.,
56. **Aures, D. and Hakanson, R.,** Histidine decarboxylase (mammalian), *Methods Enzymol.*, XVIIB, 667, 1971.
57. **Recsei, P. A., Huynh, Q. K., and Snell, E. E.,** Conversion of prohistidine decarboxylase to histidine decarboxylase: peptide chain cleavage by nonhydrolytic serinolysis, *Proc. Natl. Acad. Sci. U.S.A.*, 80, 973, 1983.
58. **Morris, D. R., Degen, J. L., Oleinik, O. E., and Seyfried, C. E.,** Regulation of the level of *S*-adenosylmethionine decarboxylase during lymphocyte mitogenesis, *Med. Biol.*, 59, 314, 1981.
59. **Pegg, A. E. and Shirahata, A.,** Synthesis of *S*-adenosylmethionine decarboxylase *in vitro, Fed. Proc. Fed. Am. Soc. Exp. Biol.*, 45, 1731, 1986.
60. **Poso, H. and Pegg, A. E.,** Differences between tissues in the response of *S*-adenosylmethionine decarboxylase to administration of polyamines, *Biochem. J.*, 200, 629, 1981.
61. **Tabor, C. W. and Tabor, H.,** Methionine adenosyltransferase (*S*-adenosylmethionine synthetase) and *S*-adenosylmethionine decarboxylase, *Adv. Enzymol.*, 56, 251, 1984.
62. **Schimke, R. T.,** On the roles of synthesis and degradation in regulation of enzyme levels in mammalian tissues, *Curr. Topics Cell. Regul.*, 1, 77, 1969.
63. **Pegg, A. E.,** Investigation of the turnover of rat liver *S*-adenosylmethionine decarboxylase using a specific antibody, *J. Biol. Chem.*, 254, 3249, 1979.
64. **Shirahata, A., Christman, K. L., and Pegg, A. E.,** Quantitation of *S*-adenosylmethionine decarboxylase protein, *Biochemistry,* 24, 4417, 1985.
65. **Markham, G. D., Tabor, C. W., and Tabor, H.,** *S*-Adenosylmethionine decarboxylase of *Escherichia coli, J. Biol. Chem.*, 257, 12063, 1982.
66. **Tabor, C. W., Tabor, H., and Hafner, E. W.,** *Escherichia coli* mutants completely deficient in adenosylmethionine decarboxylase and in spermidine biosynthesis, *J. Biol. Chem.*, 253, 3671, 1978.

67. **Cohn, M. S., Tabor, C. W., and Tabor, H.,** Isolation and characterization of *Saccharomyces cerevisiae* mutants deficient in *S*-adenosylmethionine decarboxylase, spermidine, and spermine, *J. Bacteriol.*, 134, 208, 1978.
68. **Whitney, P. A. and Morris, D. R.,** Polyamine auxotrophs of *Saccharomyces cerevisiae, J. Bacteriol.*, 134, 214, 1978.
69. **Tabor, C. W.,** Mutants of *Saccharomyces cerevisiae* deficient in polyamine biosynthesis: studies on the regulation of ornithine decarboxylase, *Med. Biol.*, 59, 272, 1981.
70. **Kahana, C. and Nathans, D.,** Isolation of a cloned cDNA encoding mammalian ornithine decarboxylase, *Proc. Natl. Acad. Sci. U.S.A.*, 81, 3645, 1984.
71. **McConlogue, L. and Coffino, P.,** Ornithine decarboxylase in difluoromethylornithine-resistant mouse lymphoma cells, *J. Biol. Chem.*, 258, 8384, 1983.
72. **McConlogue, L. and Coffino, P.,** A mouse lymphoma cell mutant whose major protein product is ornithine decarboxylase, *J. Biol. Chem.*, 258, 12083, 1983.
73. **Choi, J. H. and Scheffler, I.,** Chinese hamster ovary cells resistant to α-difluoromethylornithine are overproducers of ornithine decarboxylase, *J. Biol. Chem.*, 258, 12601, 1983.
74. **Seyfried, C. E., Oleinick, O. E., Degen, J. L., Resing, K., and Morris, D. R.,** Purification, properties and regulation of the level of bovine *S*-adenosylmethionine decarboxylase during lymphocyte mitogenesis, *Biochim. Biophys. Acta,* 716, 169, 1982.
75. **Ehrenfeld, E.,** Picornavirus inhibition of host cell protein synthesis, in *Comprehensive Virology,* Vol. 19, Fraenkel-Conrat, H. and Wagner, R. E., Eds., Plenum Press, New York, 1984, 177.
76. **Siekierka, J., Mariano, T. M., O'Malley, R. P., and Mathews, M. B.,** Translational control by adenovirus: VA RNA_I prevents activation of host double-stranded RNA-activated protein kinase during viral infection, in *Current Communications in Molecular Biology; Translational Control,* Mathews, M. B., Ed., Cold Spring Harbor Laboratory, Cold Spring Harbor, New York, 1986, 98.
77. **Warner, J. R., Rich, A., and Hall, C. E.,** Electron microscope studies of ribosomal clusters synthesizing hemoglobin, *Science,* 138, 1399, 1962.
78. **Degen, J. L., Neubauer, M. G., Friezner-Degen, S. J., Seyfried, C. E., and Morris, D. R.,** Regulation of protein synthesis in mitogen-activated bovine lymphocytes, *J. Biol. Chem.*, 258, 12153, 1983.
79. **McGarry, T. J. and Lindquist, S.,** The preferential translation of *Drosophila* hsp70 mRNA requires sequences in the untranslated leader, *Cell,* 42, 903, 1985.
80. **Mueller, P. P. and Hinnebusch, A. G.,** Multiple upstream AUG codons mediate translational control of GCN4, *Cell,* 45, 201, 1986.
81. **McConlogue, L., Gupta, M., Wu, L., and Coffino, P.,** Molecular cloning and expression of the mouse ornithine decarboxylase gene, *Proc. Natl. Acad. Sci. U.S.A.*, 81, 540, 1984.

Index

INDEX

A

B

C

D

E

F

G

H

I

N

O

For Product Safety Concerns and Information please contact our EU
representative GPSR@taylorandfrancis.com
Taylor & Francis Verlag GmbH, Kaufingerstraße 24, 80331 München, Germany

www.ingramcontent.com/pod-product-compliance
Ingram Content Group UK Ltd.
Pitfield, Milton Keynes, MK11 3LW, UK
UKHW052218190926
475702UK00006B/1086

* 9 7 8 0 8 4 9 3 6 8 0 8 0 *